Power Energy and Secure Smart Technologies

Power Energy and Secure Smart Technologies

Edited by

Thanikanti Sudhakar Babu
Mallala Balasubbareddy
Subramaniam M
Nnamdi Nwulu
Vigna Kumaran Ramachandaramurthy
Renu Sharma

CRC Press
Taylor & Francis Group

First edition published 2025
by CRC Press
4 Park Square, Milton Park, Abingdon, Oxon, OX14 4RN

and by CRC Press
2385 NW Executive Center Drive, Suite 320, Boca Raton FL 33431

CRC Press is an imprint of Informa UK Limited

British Library Cataloguing-in-Publication Data
A catalogue record for this book is available from the British Library

ISBN: 9781041118510 (hbk)
ISBN: 9781041118527 (pbk)
ISBN: 9781003661917 (ebk)

DOI: 10.1201/9781003661917

Font in Sabon LT Std
Typeset by Ozone Publishing Services

Contents

List of Figures

List of Tables

Author/Editor Biography(ies)

Thanikanti Sudhakar Babu (Senior Member, IEEE) received the M.Tech. degree in power electronics and industrial drives from Anna University, Chennai, India, in 2011, and the Ph.D. degree from VIT University, Vellore, India, in 2017. He has completed his Postdoctoral Research fellowship from the Department of Electrical Power Engineering, Institute of Power Engineering, University Tenaga Nasional (UNITEN), Malaysia. He was also associated with University of Johannesburg, South Africa as Senior Research Associate.

Currently he is working as Associate Director R & D, and Associate professor in the Department of EEE, Chaitanya Bharati Institute of Technology, Hyderabad, India. He has published more than 200 research articles in various renowned international journals. He has receive various awards like IEI Young Engineer award, Vice-Chancellors Research Excellence Award, and Listed among the top 2% of researchers in the world - electrical & electronics engineering, power, and energy areas, Information provided by Elsevier BV, Stanford University. He has been acting as an Editorial Board Member and a Reviewer for various reputed journals, such as the IEEE and IEEE Access, IET, Elsevier, EPCS and Taylor and Francis.

Balasubbareddy Mallala (Senior Member, IEEE) received the Bachelor of Engineering degree in electrical and electronics engineering from Madras University, in 2000, the master's degree from NIT Trichy, in 2004, and the Ph.D. degree from JNTUK, Kakinada, in 2015. Currently, he is a Professor in EEE with the Chaitanya Bharathi Institute of Technology, Hyderabad. His research interests include power quality, machine learning techniques, soft computing techniques, and FACTS controllers. He received "Best Researcher of the Year 2009" Award from JNT University, Kakinada, and National Award for Teaching Excellence—2013 from Indus Foundation

Subramaniam M (1974) is a Professor & Head, Dept. of Computer Engineering and Technologyg, Chaitanya Bharathi Institute of Technology, Hyderabad - 500075, (INDIA). He obtained his Bachelor's degree (B.E) in Computer Science and Engineering from the University of Madras (1998), Master's degree (M.E) in Software Engineering, and Ph.D from College of Engineering Guindy (CEG), Anna University Main Campus, Chennai -25 in the year 2003 and 2013 respectively. His research focuses are Computer Networks, Cloud, Big-data and Software Engineering, AI & ML. He is an active life member of the Computer Society of India (CSI) and the Indian Society for Technical Education (ISTE). He has produced four Ph.D's and currently two Research scholars are pursuing Ph.D. under his guidance. He published many research papers in reputed journals. He is also a reviewer in IEEE- International Journal of Communication.

Nnamdi Nwulu (Senior Member, IEEE) is currently a Full Professor with the Department of Electrical and Electronic Engineering Science, University of Johannesburg, and the Director of the Centre for Cyber-Physical Food, Energy and Water Systems (CCP-FEWS). His research interests include the application of digital technologies, mathematical optimization techniques, and machine learning algorithms in food, energy, and water systems. He is a Professional Engineer registered with the Engineering Council of South Africa (ECSA), a Senior Member of South African Institute of Electrical Engineers (SMSAIEE), and a Y-Rated Researcher by the National Research Foundation of South Africa. He is the

Editor-in-Chief of the Journal of Digital Food, Energy and Water Systems (JDFEWS) and an Associate Editor of IET Renewable Power Generation (IET-RPG) and African Journal of Science, Technology, Innovation and Development (AJSTID).

Vigna Kumaran Ramachandaramurthy (Senior Member, IEEE) received the bachelor's degree (Hons.) and the Ph.D. degree in electrical engineering from The University of Manchester Institute of Science and Technology (UMIST), U.K., in 1998 and 2001, respectively. In 2002, he joined as an Electrical Engineer with Malaysian Electrical Utility. In 2005, he moved to Universiti Tenaga Nasional (UNITEN), where he is currently a Professor with the Institute of Power Engineering. He is also a Principal Consultant with RE Industry, Malaysia, and has done the due diligence study for more than 250 projects in renewable energy. He led the development of technical guidelines for the interconnection of distributed generation, solar PV, electric vehicles, and energy storage, in Malaysia. His research interests include power system-related studies, renewable energy, energy storage, power quality, electric vehicles, and smart grids. He is a fellow of the Institution of Engineers, Malaysia, and the Institution of Engineers, India, a Chartered Engineer, registered with the Engineering Council, U.K., and a Professional Engineer, registered with the Board of Engineers Malaysia. He is also the Editorial Board Member/ Associate Editor of IET Smart Grid, IET Renewable Power Generation (RPG), IEEE TRANSACTIONS ON SMART GRID, and IEEE ACCESS.

Renu Sharma (Senior Member, IEEE) is currently with the Department of Electrical Engineering, Siksha 'O' Anusandhan Deemed to be University, Bhubaneswar, India. She has published around 100 journals and conference articles of international repute. Her research interests include smart grids, soft computing, solar photovoltaic systems, power system scheduling, evolutionary algorithms, and wireless sensor networks. She has organized several national and international conferences. She is a Life Member IE (India), ISTE, and ISSE; a member of IET; and the Past Chair of WIE IEEE Bhubaneswar Sub Section. She is a guest editor of several reputed journals. She has coordinated AICTE sponsored FDP programs. She was one of the Plenary Chair PEDES 2020. She has around 21 years of leading impactful technical, professional, and educational experience.

1 A Classical H6 Topology for Modern PV Inverter Design

Namburi Nireekshana[1], A. Archana[2], K. Pullareddy[3] and Rajini kanth P.[4]

[1,2,3,4]Assistant Professor, Department of Electrical and Electronics Engineering, Methodist College of Engineering and Technology, Hyderabad

Abstract

Transformer-based inverters offer galvanic isolation, which improves safety by physically disconnecting the PV array from the grid. Simultaneously, transformers contribute to extra energy losses as a result of their intrinsic resistance and inductance, resulting in reduced overall system efficiency. Integrating a transformer within the inverter increases its size and weight, which adds complexity to the installation process and necessitates stronger mounting options. Transformers produce excess heat, which requires more advanced cooling methods and may decrease the lifespan of the inverter components owing to thermal strain. Transformers can restrict the inverter's frequency response, which can impact its capacity to rapidly adjust to variations in load or grid circumstances. Hence, in order to circumvent these intricacies, this article introduces a transformer less full-bridge PV grid structure. The proposed technique is an advanced and efficient technology that converts DC voltage generated by solar panels into AC voltage that can be safely fed into the electrical grid. They are very dependable and are ideal for current PV systems that focus on maximum energy production while reducing installation and operational expenses. This article presents the development of a H6 transformer-less photovoltaic (PV) grid-tied inverter using insulated-gate bipolar transistor (IGBT) switches in MATLAB Simulink.

Keywords: Electric grid, inverter, H6 topology, PV system, pulse width modulation technique

1. Introduction

1.1 Background

Conventional energy sources and natural gas are not viable in the long term since they will eventually run out, are expensive to produce, and have negative environmental impacts. Renewable natural energy resources, are progressively replacing conventional energy sources. However, these supplies are sporadic and require efficient power conversion. High-capacity multilevel inverters with real-time execution of control algorithms are necessary to provide uninterrupted and high-quality electricity from non-traditional sources [10].

Power electronic converters are a type of electrical circuit that uses semiconductor-based electronic switches to transform

Email: nireekshan222@gmail.com[1], archanabhupathi.mcet@gmail.com[2], pullareddy.kasireddy@gmail.com[3], rajinikanth.nitc@gmail.com[4]

DOI: 10.1201/9781003661917-1

electrical energy from one voltage, current, or frequency level to another [4]. A multilayer inverter produces a voltage waveform that closely resembles a sine wave by effectively controls the switching of semiconductor devices using isolated or non-isolated DC voltage sources. As the level of the inverter increases, so does the number of steps in the staircase waveform, resulting in an output that resembles a sinusoidal signal. Several factors, including the inverter structure, control algorithm, filter size, and switching frequency, influence the output waveforms of an inverter [1]. The popularity of multilevel inverters is increasing due to their numerous advantages, including enhanced power quality, superior harmonic performance, decreased zero-sequence voltage, and reduced stress on switching devices caused by dv/dt. Multilevel inverters are widely used in medium-voltage, high-power industrial drive applications, as well as renewable energy applications [2].

1.2 Distributed Photovoltaic (PV) Generation

Nowadays, photovoltaic (PV) methods are a significant advancement in the use of renewable energy sources. The increasing global demand for energy, outpacing the supply of fossil fuels, has led to a growing emphasis on the use of renewable and clean energy sources. Grid-connected PV systems are the preferred method for extracting energy in industrialised countries because of their numerous advantages. The power converter is a crucial component for the operation of a solar photovoltaic system. The power converter regulates the voltage source to match the inverter's operating voltage and acts as a maximum power point [6].

Distributed photovoltaic (PV) generation systems refer to decentralized solar power installations that are typically integrated into both commercial and residential structures. These systems harness solar energy locally, converting it into electrical power that can be used on-site or fed into the electrical grid [9].

Figure 1.1 illustrates a typical grid-tied PV system where solar panels generate DC electricity. This electricity is then optimized by a DC/DC converter with an MPPT controller, converted to AC finally fed into the electrical grid. This system maximizes the efficiency of solar power generation and ensures compatibility with the grid for reliable power delivery [5].

The DC-DC converter converts varying voltages from photovoltaic modules into a regulated DC terminal voltage. A more efficient DC-DC converter is specifically built for this activity. These devices have the capability to temporarily store electrical energy in order to convert direct current from one voltage level to another [7]. It acts as a bridge between demand and solar cells. By adjusting the duty ratio, the load resistances may be altered to get high power. Figure 1.2 presents a comprehensive representation of the power converter [8].

MPPT (Maximum Power Point Tracking) Controller: Ensures the solar panels operate at their maximum power point, adjusting the electrical characteristics to maximize energy extraction under

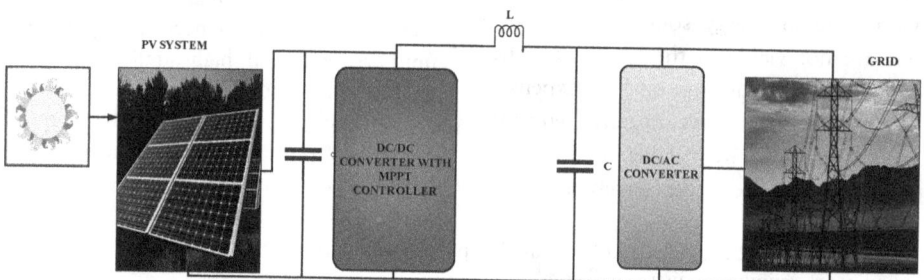

Figure 1.1 Representation of Grid-Tied PV System

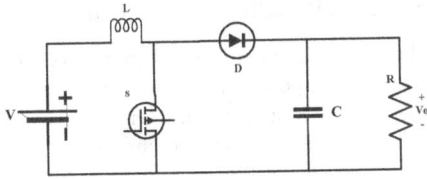

Figure 1.2 DC-DC converter equivalent circuit diagram

varying conditions (e.g., changes in sunlight intensity, temperature) [11]. DC/AC Converter (Inverter) Converts the DC electricity from the solar panels (via the DC/DC converter) into alternating current (AC) electricity that can be fed into the grid. Grid transmits the generated AC power to the electrical grid. It involves protective devices and metering to ensure safe integration with the grid, complying with regulatory standards [3].

2. POD Modulation Method

The Novel H6 Transformerless Topology with Phase Opposition Disposition (POD) Modulation Technique is a sophisticated inverter design that combines the benefits of the H6 topology with an advanced modulation strategy to further enhance performance. Here's an overview of this combination.

Phase Opposition Disposition (POD) modulation as shown in Figure 1.3 is a technique used in multi-level inverters to improve the quality of the output waveform. In POD, the carrier waves for the PWM (Pulse Width Modulation) signals are phase-shifted by 180 degrees for the positive and negative half-cycles. This reduces

the common-mode voltage and improves the overall harmonic performance.

2.1 Working Principle

DC to AC Conversion: The six switches convert DC from the PV panel to AC.

Freewheeling Path: During zero-crossing, specific switches create a freewheeling path, reducing leakage currents.

POD Modulation: The control system uses POD modulation to generate the PWM signals, ensuring the carrier waves for the positive and negative half-cycles are phase-shifted by 180 degrees.

2.2 Operation and Working

Figure 1.6 represents, the circuit structure of the proposed innovative H6 inverter topology. Photovoltaic grid-tied systems typically function with a power factor of 1. Each period of the utility grid consists of four operating modes.

From Figure 1.4 Van, Vbn are output Voltages. Where Van is voltage between A and N points. Vbn is voltage between B and N points.

Figure 1.4 represents active mode operation in positive peak. In which current circulation path is PV panel-switches S5and S1-L1-AC Source-Switch S3 and PV Panel.

Figure 1.5 depicts freewheeling operation in positive peak. In which current circulating path is L1-AC Source-L2-S2 and Switch S1.

Figure 1.6 illustrates the active mode in negative peak. In which current

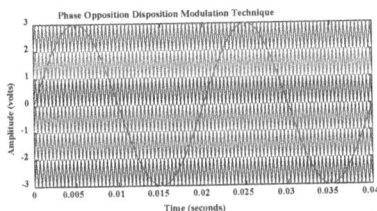

Figure 1.3 Signal Representation of Phase Opposition Disposition

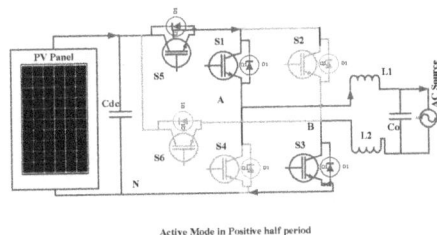

Figure 1.4 Representation of Active Mode Operation in Positive Half Cycle

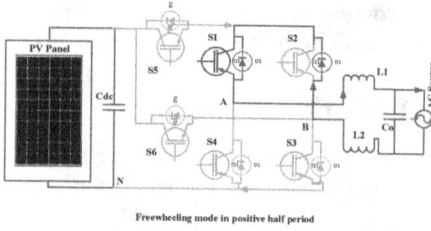

Freewheeling mode in positive half period

Figure 1.5 Representation of Freewheeling Mode Operation in Positive Half Cycle

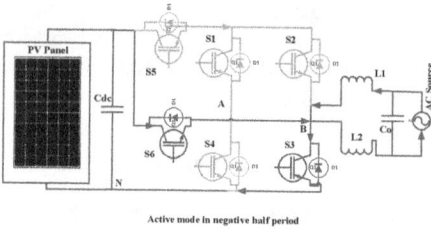

Active mode in negative half period

Figure 1.6 Representation of Active Mode Operation in Negative Half Cycle

circulating path is PV panel-Switch S6-L2-AC Source-L1-SwitchS3-PVpanel.

Figure 1.7 depicts Freewheeling mode operation in negative peak. In which current path is L2-AC source-L1-S1-and then S2.

This examination indicates that the utility grid can remove the PV array when the H6 inverter's output voltage drops to zero and interrupts the leakage current channel. Combining the H6 architecture with the POD modulation method not only makes it possible to reach a power factor of 1, but it also lets you fine-tune the phase shift between the voltage and current waveforms.

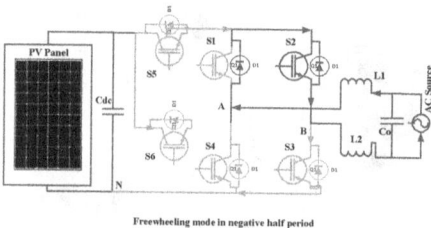

Freewheeling mode in negative half period

Figure 1.7 Representation of Freewheeling Mode Operation in Negative Half Cycle

3. Results and Discussion

The Requirement of Parameters for Proposed Topology are given in Table 1.1. and The PV system parameters are given in Table 1.2. Figure 1.8 represents I-V and P-V characteristics of PV array. Observations of Figure 1.8 is at 25°C, the current remains nearly constant until a certain voltage level (around 350V), after which it drops sharply. At 45°C, the current also remains nearly constant but at a slightly lower value compared to 25°C. The voltage at which the current drops also decrease. Higher temperatures reduce the open-circuit voltage (Voc), resulting in a decrease in the maximum voltage.

At 25°C, the array produces higher maximum power compared to 45°C.

Figure 1.9 depicts I-V and P-V characteristics of solar module with different irradiations. As the irradiance decreases to 0.5 kW/m² and 0.1 kW/m², the current also decreases proportionally, with the voltage at which the current drops remaining nearly the same. Lower irradiance levels reduce the short-circuit current

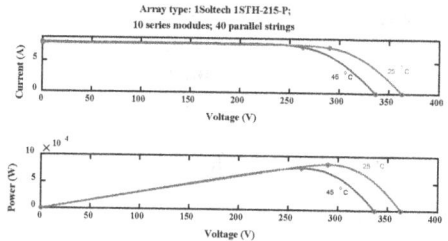

Figure 1.8 Representation of I-V and P-V Characteristics of Solar Module with Different Temperature

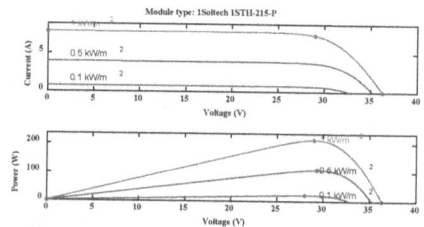

Figure 1.9 Representation of I-V and P-V Characteristics of Solar Module with Different Irradiations

Table 1.1 Requirement of Parameters for Proposed Topology

Parameter	Specification
Input Voltage	378.4VDC
Input Current	6A
Grid Voltage	227.4V/50Hz one phase
Switches	IGBT with antiparallel diodes
Cdc	940 e^{-6}F
Co	100 e^{-3}F
L1	100 e^{-3}H
L2	100 e^{-3}H

Table 1.2 PV System parameters

Sl. No.	Parameter
1	1 Soltech 1STH 215P Model
2	10-Mod-string
3	40 Strings
4	45°C Temperature
5	1000 Irrad

(Isc), but the open-circuit voltage (Voc) does not significantly change with irradiance. The P-V curve shows the maximum power point (MPP) occurs where the product of current and voltage is maximized. At the highest irradiance level (1 kW/m²), the module produces the highest maximum power. As the irradiance decreases, the maximum power point shifts lower, and the overall power output decreases proportionally.

Figure 1.10 depicts the voltage is nearly constant at around 378.4045 volts

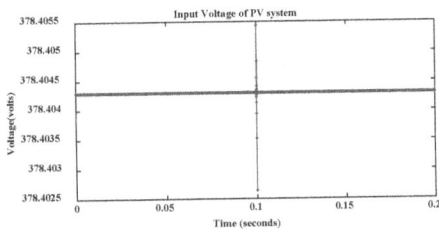

Figure 1.10 Representation of PV Array Voltage

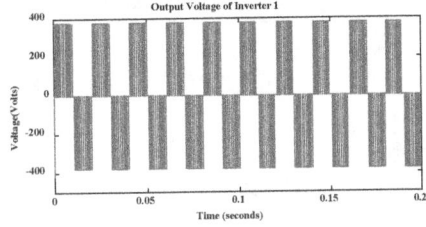

Figure 1.11 Representation of Inverter Voltage

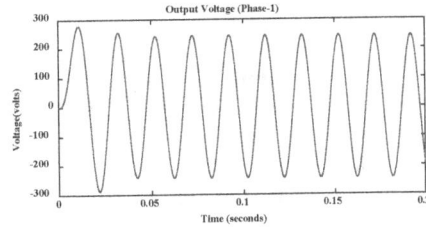

Figure 1.12 Representation of Grid Voltage

throughout the time interval, indicating a highly stable input voltage to the PV system. There is a very slight variation at around 0.1 seconds, but the change is minimal, and the voltage quickly returns to its stable value.

Figure 1.11 depicts two level inverter voltages at output terminals of proposed topology is about 375.09 V.

Figure 1.12 depicts grid voltage of three phases separately, the voltage about 227.58V per phase at 50 Hz. This waveforms are getting from proposed topology after filtering. The waveform appears smooth and regular, indicating a well-regulated and filtered AC output. Therefore in this topology total harmonic distortion (THD) is reduced.

Figure 1.13 demonstrates that the proposed method quickly reaches and maintains 98% efficiency within 1 millisecond and remains stable thereafter. This suggests that the method is highly effective and reliable for its intended application. The rapid rise to maximum efficiency and the steady maintenance of that efficiency indicate that this method is well-optimized and can deliver high performance in real-time applications.

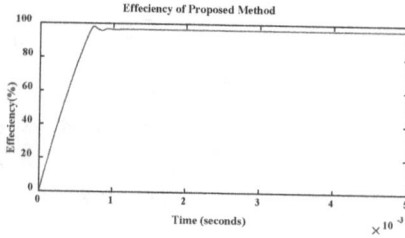

Figure 1.13 Efficiency of Proposed Topology

4. Conclusion

Based on the results and analysis, it is evident that the output voltages of the three-phase H6 topologies exhibit sinusoidal waveforms with a 120-degree phase shift and a low THD of 0.09%. The LC filter positioned at the converter's output effectively reduces the THD in the H6 topology, thereby enhancing the output voltage. The H6 transformer-less full bridge PV grid-tied inverter presents a significant advancement in photovoltaic (PV) technology. The main benefits of this inverter include increased efficiency, reduced size and weight, and lower costs compared to traditional transformer-based systems. By eliminating the transformer, these inverters not only reduce material and manufacturing costs but also improve the overall energy conversion efficiency. The design of the H6 topology ensures high reliability and performance in converting DC power generated by solar panels into AC power that can be fed into the grid. This is achieved through this proposed technique that minimize power losses and enhance the inverter's lifespan.

5. Future Scope

The future scope of H6 transformer-less full bridge PV grid-tied inverters is promising, given the ongoing advancements in renewable energy technology and the increasing adoption of solar power systems.

Integration with Smart Grids: As smart grid technology evolves, H6 inverters can be designed to seamlessly integrate with advanced grid management systems.

Energy Storage Integration: The incorporation of energy storage systems, such as batteries, with H6 inverters can provide a more stable and reliable power supply. Future research can focus on developing hybrid inverters that efficiently manage both PV generation and energy storage, optimizing the use of renewable energy and reducing dependency on the grid.

References

[1] Anand, V., Singh, V., & Mekhlief, S. (2022). Power electronics for renewable energy systems. In *Renewable Energy for Sustainable Growth Assessment*, edited by Nayan Kumar and Prabhansu, 1st ed., 81–117. Wiley. https://doi.org/10.1002/9781119785460.ch4.

[2] Freire, R., Delgado, J., Santos, J. M., & De Almeida, A. T. (2010). Integration of renewable energy generation with EV charging strategies to optimize grid load balancing. In *13th International IEEE Conference on Intelligent Transportation Systems*, 392–396. IEEE. https://ieeexplore.ieee.org/abstract/document/5625071/.

[3] Kolantla, D., Mikkili, S., Pendem, S. R., & Desai, A. A. (2020). Critical review on various inverter topologies for PV system architectures. *IET Renewable Power Generation*, 14(17), 3418–3438. https://doi.org/10.1049/iet-rpg.2020.0317.

[4] Mouli, G. R. C., Schijffelen, J., van den Heuvel, M., Kardolus, M., & Bauer, P. (2018). A 10 kW solar-powered bidirectional EV charger compatible with chademo and COMBO. *IEEE Transactions on Power Electronics*, 34(2), 1082–1098.

[5] Namburi Nireekshana, N., & Rajesh Kumar, K. (n.d.). A modern distribution power flow controller with a PID-fuzzy approach: Improves the power quality. Accessed April 25, 2024. https://ijeer.forex-journal.co.in/papers-pdf/ijeer-120124.pdf.

[6] Namburi, N., Nerlekar, T. H., Kumar, P. N., & Bajaber, M. M. (2023). An innovative solar based robotic floor cleaner. May. https://doi.org/10.5281/ZENODO.7918621.

[7] Nireekshana, N., Ramachandran, R., & Narayana, G. V. (2023a). A new soft computing fuzzy logic frequency regulation scheme for two area hybrid power systems. *International Journal of Electrical and Electronics Research*, 11(3), 705–710. https://doi.org/10.37391/IJEER.110310. 2023b.

[8] Nireekshana, N., Ramachandran, R., & Narayana, G. (2023). A novel swarm approach for regulating load frequency in two-area energy systems. *International Journal of Electrical and Electronics Research*, 11(2), 371–377. https://doi.org/10.37391/ijeer.110218.

[9] Nireekshana, N., Ramachandran, R., & Narayana, G. V. (2023b). An intelligent technique for load frequency control in hybrid power system. https://www.researchsquare.com/article/rs-2452347/latest.

[10] Seddig, K., Jochem, P., & Fichtner, W. (2017). Integrating renewable energy sources by electric vehicle fleets under uncertainty. *Energy*, 141, 2145–2153.

[11] Zeb, K., Uddin, W., Khan, M. A., Ali, Z., Ali, M. U., Christofides, N., & Kim, H. J. (2018). A comprehensive review on inverter topologies and control strategies for grid connected photovoltaic system. *Renewable and Sustainable Energy Reviews*, 94, 1120–1141.

2 Enhancing Solar Power Generation Efficiency and Grid Interfacing through Improved MPPT and OFOPID PWM Controllers

Tippa Venkateswara Rao[1,], Ch Venkateswara Rao[2] and Kappagantula BVSR Subrahmanyam[3]*

[1*]Research Scholar, EEE Department, GIET University, Gunupur, Odisha
[2]Professor, EEE Department, GIET University, Gunupur, Odisha
[3]Professor and Head, EEE Department, DNR College of Engineering and Technology, Bhimavaram, Andhra Pradesh, India

Abstract

This paper presents an novel approach to enhance performance of Photovoltaic (PV) systems within Renewable Energy Systems (RES). To optimize power extraction under varying solar irradiance and temperature conditions, an Improved Perturb and Observe (IP&O) based Maximum Power Point Tracking (MPPT) technique is implemented. To mitigate power quality issues associated with power electronic converters, an Osprey Optimized Fractional Order Proportional Integral Derivative (OFOPID) Pulse Width Modulation (PWM) controller is proposed. Osprey Optimization Algorithm (OOA) is included to tune gain parameters of FOPID controller, this combination ensures precise control and improved system dynamics. This OFOPID method is implemented in MATLAB/Simulink tool. Efficacy of proposed approach is validated through simulations under diverse solar irradiance scenarios, demonstrating significant improvements in power extraction.

Keywords: RES, MPPT, PV, DC-DC converter, OOA, Modified P&O.

1. Introduction

All energy resources that can be continuously replenished and produce energy indefinitely are included in renewable energy sources (RES), also referred to as alternative energy sources [11]. Usually, RES consists of load, storage, and energy generating units. RESs may include biogas, solar PV, biomass, geothermal, solar thermal, small hydro, and wind [12]. By implementing renewable energy-powered micro grids remote areas can achieve energy self-sufficiency. Such micro grids have been deployed worldwide utilizing either single or hybrid configurations of RES in attempt to reach this purpose [27]. Solar and wind power's grid integration raises concerns about power quality. To ensure that sensitive equipment operates at its best, it is important to have access to high-quality power and grounding options [26].

Email: venkateswararao.tippa@giet.edu[1], venkateswararao@giet.edu[2], kappagantulasubbu@gmail.com[3]

DOI: 10.1201/9781003661917-2

Renewable power through its distribution-level penetration and essential characteristics offers various benefits including the reduction of power losses, mitigation of environmental pollution, elimination of system upgrades, lowered operating costs, and enhancement of voltage profiles [6]. Power quality refers to provide the power and grounding solutions that are conducting to the optimal functioning of sensitive equipment [15]. A number of variables and limitations affects microgrids' (MGs) ability to generate power. Passive filters (PF), hybrid filters (HF), custom power devices (CPDs), and active filters (AF) are among the options that can be used to improve power quality [24]. Power quality (PQ) issues associated with voltage, a dynamic voltage restorer (DVR) is employed [22]. DVR injects a voltage with specific phase angle and magnitude in series with distribution line to produce a rated sinusoidal load voltage [4]. By mitigating voltage sag and swell issues, the DVR enhances energy consumption. Static synchronous compensator (STATCOM) to reduce sag events and perform low voltage ride-through (LVRT) requirements [2].

Additionally, the Static var compensator (SVC) and energy storage system are introduced to assemble LVRT measure introduce reactive current to support the grid and mitigate sag incidents [21]. Within power networks, voltage at point of common coupling (PCC) is regulated by reactive power that is drawn from the distribution network [20]. Furthermore, the D-STATCOM was used to decrease harmonics at PCC [23]. When dealing with power quality difficulties, it is common practice to use filters and other FACTS devices, which stand for Flexible Alternating Current Transmission Systems [17]. Nevertheless, these techniques demonstrate limitations such as inadequate convergence and elevated total harmonic distortion [18]. The major contribution of this work is given below.

- Develop and implement an IP&O based MPPT controller to optimize power output of PV cells.

- To maximize power output of PV cells, develop and implement an IP&O MPPT controller.
- Using OOA, adjust the OFOPID PWM controller's gain parameters.

This work is separated into several sections. Section 2 discusses the survey of power quality enhancement, and section 3 illustrates the proposed methodology of MPPT techniques used in PV systems. Results and discussion derived from the simulation are presented in section 4. Paper's conclusion is provided in Section 5.

2. Literature Survey

Some of the existing works related to discussing the varying ranges of solar irradiances in power generation are shown in the below section.

Adware et al. [3] introduced fuzzy logic control (FLC) based STATCOM to mitigate Power Quality issues in a Wind Farm. A voltage controller utilizing fuzzy logic was developed to enhance power quality. This research focuses on mitigating voltage sags/swells, enhancing active and reactive power, and controlling voltage flickers in a multi-terminal load system integrated with wind generators. Proposed solution employs two controllers to achieve these objectives. Prashant et al. [19] introduced the utilization of Fuzzy Logic Controlled based D-STATCOM as a means to enhance the positioning and sizing of distributed Generators. Switching and control of the D-STATCOM are evaluated using a FLC to enhance voltage profile and total harmonic distortion at specified bus. Adware et al. [3] suggested optimized ANFIS based on PI controllers to improve grid power quality. By contrasting the outcomes with those achieved using PI, PI-RC, and fuzzy current controller to confirm the efficacy of the suggested controller.

Das et al. [7] suggested a fuzzy logic method in solar photovoltaic (PV) systems that incorporate shunt hybrid active filters (SHAFs) and adaptive controllers. In order to maximize power generation, the solar tracker is equipped with a MPPT technology that is based on fuzzy logic. Wang et al. [25] recommended implementation of a unified power quality conditioner that incorporates a control technique aimed at enhancing power quality. The efficacy of control technique employed by UPQC utilizing dq0, this method enhanced performance of system.

3. Proposed Methodology

The solar power generation is considered because there are plenty of available sources. Photovoltaic (PV) cells are used in solar-powered PS systems to convert energy. An Improved P&O (IP&O) MPPT controller is suggested in order to maximize the power generated by solar PV. The OFOPID controller is suggested as a solution to these problems. OOA is used to fine-tune proposed OFOPID. Figure 2.1 shows the photovoltaic system that has been suggested.

3.1 PV Modelling

Semiconductor materials make up the solar photovoltaic cell, with the negative side facing the sun and the positive side facing the back [14]. The PV materials produce electrons when sunlight strikes radiation. These electrons flow through the external circuit and are referred to as photocurrent or short circuit current resistances, which improve the PV module's performance. The solar photovoltaic cell's single diode model is

shown in Figure 2.2. When output terminal is open-circuited, a voltage appears, known as the open circuit voltage (Voc). This voltage induces a current flow across P-N junction, mimicking behavior of a diode. In model, a parallel combination of this diode and a current source (Iph) is depicted in Figure 2.2. As photocurrent generation commences, some electron-hole recombination occurs, reducing initial electron count. This current loss is represented by a shunt resistance (Rsh). Additionally, a series resistance (Rs) accounts for resistance encountered by current as it traverses bulk material, external metal contacts, and load. Minimizing impact of both these resistances is crucial for optimizing PV module's performance.

In equation (1) relationship between voltage and current of solar cell is described as,

$$I = I_R \left[\exp\left(\frac{vq}{KT} \right) - 1 \right] - I_{HP} \qquad (1)$$

Where, I_R signifies saturation current, q denotes charge of electron, V signifies voltage across diode, I_{HP} represents light-generated current, k represents Boltzmann constant and T denotes absolute temperature (in Kelvin).

Output current of a photovoltaic (I_{PV}) cell is determined using Kirchhoff's law, as demonstrated in equation (2).

$$I_{PV} = I_{HP} - I_D - I_{HS}$$
$$= I_{HP} - I_D - \frac{V_{PV} + R \cdot I_{PV}}{R_{HS}} \qquad (2)$$

Current, denoted as I_D, can be characterized using following Shockley equation, as determined in (3).

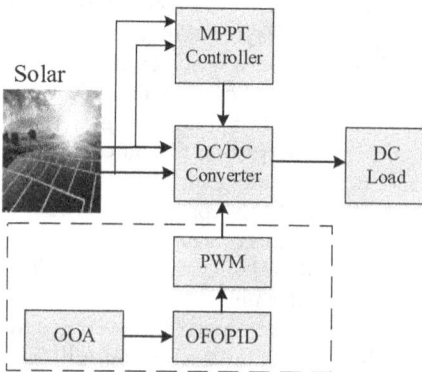

Figure 2.1 Basic Workflow of the Work

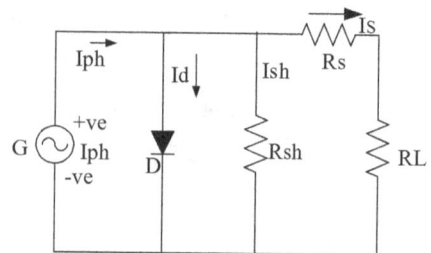

Figure 2.2 A PV Cell's Electrical Model

$$I_D = I_O \cdot \left(e^{\frac{qV_k}{nKT-1}} \right)$$

$$= I_O \cdot \left(e^{\frac{q(V_{PV} + R_s \cdot I_{PV})}{nKT-1}} - 1 \right) \quad (3)$$

Here, I_O indicates reverse saturation current, T signifies temperature and n denotes ideality factor.

3.2 DC-DC Converter

An uncontrolled DC input's voltage level is to be regulated by the DC-DC converter, which transforms it into a regulated DC output voltage at a predetermined level [9]. It comprises components such as diodes, transistors, inductors, and capacitors. Boost converter topology is employed among the three main converter topologies, namely buck, buck-boost, and boost converter. By modifying the duty cycle, the PV module can achieve its ideal load impedance. Figure 2.3 presents schematic of a DC-DC converter. When switch (S) is closed, inductor charged by input source, while diode remains reverse-biased, isolating input and output sections. Upon opening switch, energy stored in inductor, coupled with energy from PV is transmitted to load. Resulting modeling equations are derived from Kirchhoff's laws, as detailed in [10].

$$\frac{dI_K}{dt} = \frac{1}{K} \left[V_{in} - (1-d)V_{out} \right] \quad (4)$$

$$\frac{dV_{out}}{dt} = \frac{1}{C_{DC}} \left[(1-d)I_L - I_{out} \right] \quad (5)$$

Where, V_{in} represents input voltage, V_{out} denotes average output voltage, I_{out} represents input current, d denotes duty cycle, which is percentage of overall switching period that is devoted to the on-state operation.

The correlation between input and output voltages is expressed as:

$$V_{in} R_{on} + (V_{in} - V_{out}) T_{off} = 0 \quad (6)$$

$$\frac{V_{out}}{V_{in}} = \frac{R_{on} + R_{off}}{R_{off}} = \frac{1}{1-d} \quad (7)$$

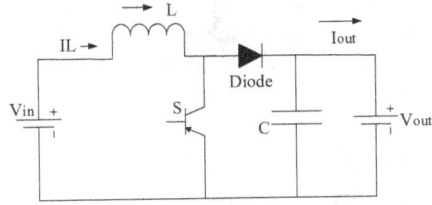

Figure 2.3 Diagram of DC to DC Boost Converter Circuit

Where, R_{on} represents on resistance and R_{off} denote as off resistance, T_{off} represents turn off input and output voltage.

To maximize power generated by PV generator, a converter is inserted between PV generator and load. A boost-type converter is selected for this purpose, which elevates input voltage to required level at its output. A boost converter, as depicted in Figure 2.3. Transistor (T) is rapidly switched at a high frequency by applying a PWM signal to its gate. Duty cycle value is adjusted to control output voltage level, as shown in equation (8).

$$V_{out} = \frac{1}{1-\eta} V_{in} \quad (8)$$

Where, η represents the Control output voltage using selected duty cycle.

3.3 MPPT

MPPT method is utilized to enhance power output from turbines and PV solar modules across various environmental conditions. By continuously monitoring generated voltage of PV in real-time, an MPPT controller can precisely track maximum voltage and current values consequently optimizing charging process of batteries to achieve maximum power output. This technology finds application in photovoltaic inverters to effectively synchronize with solar panels, batteries, and various loads.

MPPT is employed to maximize output of wind turbines and solar panels. There is a complex link between operating voltage of MPPT controller and output power of a PV cell. Position of this point is influenced by meteorological circumstances, which in turn affect the power extension. Increased irradiance leads to higher power extension, while lower temperatures result in more efficient performance of a PV panel compared to higher temperatures [13].

3.3.1 P&O

This method is generally employed in commercial MPPT controllers, involves constantly monitoring power variation (dp) of PV module. At the same time, direction of voltage (dv) in PV module is analyzed in order to modify duty cycle (d) for future updates and corrections. Voltage perturbation in PV is evaluated using equations (9)–(11). Variations arise primarily in step size for duty cycle control (adaptive control). Following measurement of PV power $p(t)$, it is adjacent with preceding MPP $p(t-1)$. Duty ratio regulating converter is created using resulting difference (Δ), which causes voltage to either rise or drop by $V(t)$ by ΔV.

$$P(t) = V(t) \times I(t) \tag{9}$$

$$\text{Delta} = p(t) + p(t-1) \tag{10}$$

$$I(t) = I(t-1) \pm \Delta V \tag{11}$$

3.3.2 Improved (P&O) Technique

Using a constant step size to track MPP, value that determines duty cycle required to switch the DC-DC converter improves P&O method. A key component of enhanced P&O technique is intentional control of PV array's output voltage. Monitoring the power change that results is the next stage in reaching MPPT.

$$\begin{cases} \dfrac{dP_{PV}}{dV_{PV}} = 0 \text{ at } MPP \\[2mm] \dfrac{dP_{PV}}{dV_{PV}} > 0 \text{ at left side of } MPP \\[2mm] \dfrac{dP_{PV}}{dV_{PV}} < 0 \text{ at right side of } MPP \end{cases} \tag{12}$$

It is impossible to achieve precise MPPT $\left(\dfrac{dP_{PV}}{dV_{PV}} = 0 \right)$ frequently. Therefore, a more practical approach is to aim for a target

range, denoted as $\dfrac{dP_{PV}}{dV_{PV}} < \delta$ where a small error (δ) is acceptable. Equations (11) and (12) show how the PV system's sensitivity is determined by the size of this allowable error. Variable step size (VSS) is used to improve system's performance.

$$VSS = \left| \dfrac{dP_{PV}}{dV_{PV}} \right| \times SF \tag{13}$$

Where, SF represents calling factor for regulating step. Equation for determining duty cycle of converter at immediate $d(L)$ is set as follows:

$$d(L) = d(L-1) \pm \left| \dfrac{dP_{PV}}{dV_{PV}} \right| \times SF \tag{14}$$

This method employs a VSS dependent simply on the change in PV power (dP_{PV}) rather than the derivative $\dfrac{dP_{PV}}{dV_{PV}}$. In this issue, the VSS can be computed as follows:

$$VSS = |dP_{PV}| \times SF \tag{15}$$

3.4 Osprey Optimization Algorithm

OOA was developed based on the behavioral patterns seen in the osprey, which is also referred to as the fish hawk [16]. This study focuses on addressing the fluctuations in power networks caused by abrupt disruptions by using the intelligent foraging approach utilized by Ospreys.

3.4.1 Initialization of Population

The positions of ospreys are randomly initialized in the search space using equation (13)

$$P_{ji} = lb_i + r \text{ and } {}^* (ub_i - lb_i) \tag{16}$$

Where i represents amount of ospreys and j symbolizes quantity of problem variables.

3.4.2 Phase 1: Exploration

As part of the OOA, each osprey was compared against underwater fish and other ospreys whose positions in the search space had higher objective function values. The amount of fish caught by each osprey in connection to the MPPT controller was calculated using equation (17).

$$FP_j = \{P_K \mid K \in \{1, 2, 3, ...n\} \wedge F_K < F_j\} \cup \{P_{bst}\}$$

(17)

Where FP_j represents the set of fish positions for the jth osprey and P_{bst} denotes the best osprey or solution. The updated position of the osprey is calculated using equation (18)

$$P_{ji}^{HP2} = P_{ji} + r \text{ and } * (SF_{ji} - I_{ji} * P_{ji}) \quad (18)$$

3.4.3 Phase 2: Exploitation

Similar to ospreys, MPPT algorithms in the OOA control solar panel to operate at its extreme power point while figuring out best place for them to feed. Equation (19) states that the MPPT algorithm replaces the prior operating point with the updated one if it notices an improvement in value of objective function at the new operating point.

$$P_{ji}^{HP2} = P_{ji} + \frac{lb_i + rdm * (ub_i - lb_i)}{t} \quad (19)$$

The updated position of the osprey can be determined by the equation

$$P_j = \begin{cases} P_j^{HP2}, & F_j^{HP2} < F_j \\ P_j, & else \end{cases} \quad (20)$$

Phases 1 and 2 are continued in this process of updating the osprey's position until the maximum iteration is reached.

3.5 FOPID Controller

FOPID controller extends traditional PID controller by utilizing fractional calculation. Its effectiveness lies in its ability to handle time-varying dynamics in industrial processes due to the inclusion of time accumulation in the transfer function, a feature lacking in traditional PID controllers [5]. In t controlled system employing the FOPID controller, as depicted in Figure 2.4, closed-loop control system consists of FOPID controller and feedback control system of controlled object model. FOPID controller's non-integer order gives it more ability to change controller

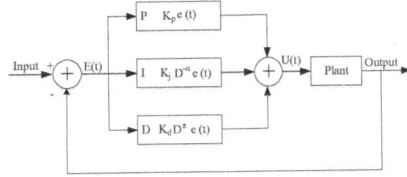

Figure 2.4 Structure of FOPID Controller

gain levels than integer-order PID controller. Integral, fractional-order differential, and proportional operations are performed on the error signal in order to compute output U(t) of FOPID controller [1].

$$U(t) = K_p e(t) + K_i D^\alpha e(t) + K_d D^\delta e(t) \quad (21)$$

Where $U(t)$ represents controller output, $e(t)$ denotes error utilized as controller input K_p, Ki and K_d stands for proportional, integral, and differential gains, respectively.

Output of FOPID controller is composed of three terms: error signal, integral, and derivative of error. Equation (22) calculates error signal.

$$\frac{E(s)}{U(t)} = K_p + \frac{K_i}{s} + K_d S \quad (22)$$

Where, $U(t)$ denotes the output control referring to the torque in the control drive.

3.6 PWM Controller

PWM is an "effective" method widely employed in current control systems. This modulation method finds applications across various fields, such as speed control, power regulation, measurement, and communication [8]. Essentially, PWM involves rapidly alternating between turning on (ON) and off (OFF) a DC motor, with accurate control over the duration of these ON and OFF states. Duty cycle value, indicating ratio of on period (Ton) to fixed value, is determined by comparing the time spent in the logic High (Ton) and logic Low (Toff) states.

4. Result and Discussion

In order to conserve a constant output voltage regardless of changes in irradiation and temperature-induced load and input voltage variations, a complete PV system's DC converter employs a PID controller based on MPPT. Both the OPID and MPPT-based PID controllers incorporate the objective function and the OOA approach for designing their parameters. A comparison of the P&O method with modified P&O MPPT approach is also shown in Table 2.1.

Figure 2.5 shows the comparative analysis of PV output with P&O and IP&O methods power, current, voltage and efficiency. The dynamic relationship between solar irradiation and power output in a photovoltaic

(PV) panel system is depicted in the graph. Both systems exhibit a rapid initial power increase within the first 0.05 seconds, followed by a stable power output. Notably, the IP&O system consistently maintains a higher power output compared to the P&O system. This suggests that the IP&O system may be more efficient and faster response time, leading to enhanced power delivery.

Figure 2.6 shows the convergence characteristic curves for the three algorithms. It was found that, among the algorithms, TLBO similarity stabilizes around the slowest convergence speed, GSA similarity

Table 2.1 Comparison of P&O and IP&O Mppt Techniques

S. NO	MPPT Techniques	Voltage	Current	Power	Efficiency (%)
1	P&O	138.8	7.588	1053.21	91
2	Improved P&O	136.2	8.105	1103.9	95.39

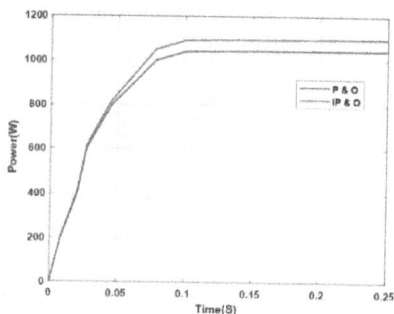

(a)

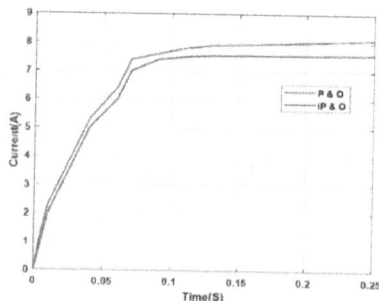

(b)

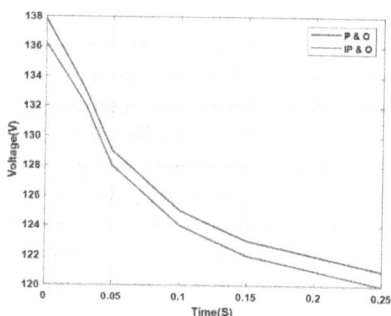

(c)

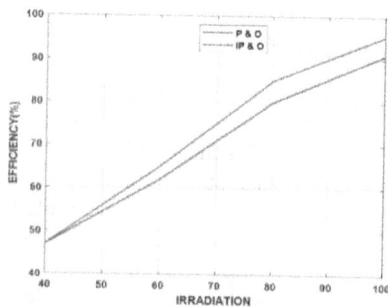

(d)

Figure 2.5 Comparison of P&O and IP&O PV Output (a) Power, (b) Current, (c) Voltage and (d) Efficiency

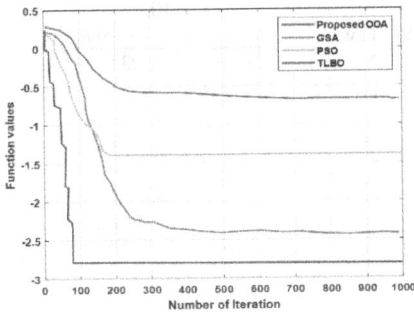

Figure 2.6 Convergence Curve of Proposed OOA

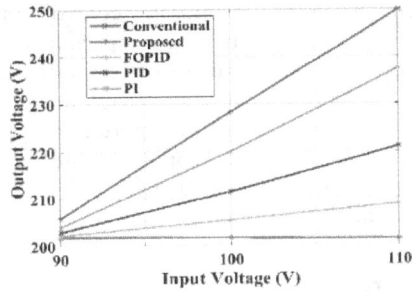

Figure 2.7 Comparison of Percentage Overshoot Between Conventional and Proposed DC-DC Converters

Table 2.2 Comparison of Conventional and Proposed Dc-Dc Converter

Controller Input voltage	Conventional	PI	PID Output voltage	FOPID	Proposed
90	205.8	203.94	202.86	202.15	201.8
100	228.3	218.5	210.5	205.25	201.7
110	249.9	237.47	221	208.83	201.4

stabilizes around the lowest value, PSO similarity stabilizes around the lowest value, and OOA probabilities stabilize around the highest value. This indicates average, minimum, maximum, and stable simulation time. The optimal inverse learning curves presented in this paper for the OOA perform better than the other techniques. Table 2.2 also includes a comparison of the proposed controller for dc-dc converters with the existing methods.

MATLAB was used to show the simulation results, which are listed in Table 2.2. A research was conducted to compare the proposed and standard DC-to-DC converters in terms of output voltage; the findings are displayed in Figure 2.7. The figure illustrates a comparison of the output voltage response for five different control strategies: Conventional, Proposed, FOPID, PID, and PI. The comparison is made across a range of input voltages, from 90V to 110V. The Proposed method consistently maintains the lowest output voltage, suggesting improved efficiency and reduced power losses compared to the other methods.

While FOPID method also exhibits lower output voltages than the Conventional and PID/PI methods, the proposed method demonstrates superior performance in terms of voltage regulation.

Results from experiments demonstrate that DC-DC converter's performance is enhanced when suggested OFOPID controller is used, leading to better regulation of output voltage and less overshoot.

5. Conclusion

This paper proposed an OFOPID controller for boost converter that uses MPPT to keep output voltage of a standalone photovoltaic system constant throughout a wide range of disturbances. With respect to voltage, current, and power dynamics, MPPT and OOA provide superior system responses to the proposed controller when subjected to different disturbances. The proposed controller exhibits faster settling time, less steady-state error, and reduced overshoot when contrasted with an osprey optimized FOPID controller that employs the identical optimization technique. The proposed controller's efficacy is proven through experimentation with numerous disturbances such as temperature fluctuations, variable irradiation, and load changes. Future research will ensure that the proposed system is improved with additional RES and other controllers.

References

[1] Abdullah, A. N., & Ali, M. H. (2020). Direct torque control of IM using PID controller. *International Journal of Electrical and Computer Engineering (IJECE)*, 10(1), 617–625.

[2] Adware, R., & Chandrakar, V. (2023). A hybrid STATCOM approach to enhance the grid power quality associated with a wind farm. *Engineering, Technology & Applied Science Research*, 13(4), 11426–11431.

[3] Adware, R., & Chandrakar, V. (2023). Power quality enhancement in a wind farm connected grid with a fuzzy-based STATCOM. *Engineering, Technology & Applied Science Research*, 13(1), 10021–10026.

[4] Alhattab, A. S., Alsammak, A. N. B., & Mohammed, H. A. (2023). A review on D-STATCOM for power quality enhancement. *Al-Rafidain Engineering Journal (AREJ)*, 28(1), 207–218.

[5] Benbouhenni, H., Bizon, N., Colak, I., Thounthong, P., & Takorabet, N. (2022). Application of fractional-order PI controllers and neuro-fuzzy PWM technique to multi-rotor wind turbine systems. *Electronics*, 11(9), 1340.

[6] Chahine, K., Tarnini, M., Moubayed, N., & El Ghaly, A. (2023). Power quality enhancement of grid-connected renewable systems using a matrix-pencil-based active power filter. *Sustainability*, 15(1), 887.

[7] Das, S. R., Hota, A. P., Pandey, H. M., & Sahoo, B. M. (2022). Industrial power quality enhancement using fuzzy logic based photovoltaic integrated with three phase shunt hybrid active filter and adaptive controller. *Applied Soft Computing*, 121, 108762.

[8] Faizal, E., Winoko, Y. A., Mustapa, M. S., & Kozin, M. (2022). Solar charger controller efficiency analysis of type pulse width modulation (PWM) and maximum power point tracking (MPPT). *Asian Journal Science and Engineering*, 1(2), 90–102.

[9] Fawzy, I. Y., Mohamad, Y. S., Shehata, E. G., & Abd El Sattar, M. (2021). A modified perturb and observe technique for MPPT of intgrated PV system using DC-DC boost converter. *Journal of Advanced Engineering Trends*, 40(1), 63–77.

[10] Gaikwad, D. D., Chavan, M. S., & Gaikwad, M. S. (2014, December). Hardware implementation of dc-dc converter for mppt in pv applications. In *2014 IEEE Global Conference on Wireless Computing & Networking (GCWCN)* (pp. 16–20). IEEE.

[11] Halkos, G. E., & Gkampoura, E. C. (2020). Reviewing usage, potentials, and limitations of renewable energy sources. *Energies*, 13(11), 2906.

[12] Husin, H., & Zaki, M. (2021). A critical review of the integration of renewable energy sources with various technologies. *Protection and Control of Modern Power Systems*, 6(1), 1–18.

[13] Idrissi, Y. E. A., Assalaou, K., Elmahni, L., & Aitiaz, E. (2022). New improved MPPT based on artificial neural network and PI controller for photovoltaic applications. *International Journal of Power Electronics and Drive Systems*, 13(3), 1791–1801.

[14] Jin, T., Chen, Y., Guo, J., Wang, M., & Mohamed, M. A. (2020). An effective compensation control strategy for power quality enhancement of unified power quality conditioner. *Energy Reports*, 6, 2167–2179.

[15] Jumani, T. A., Mustafa, M. W., Alghamdi, A. S., Rasid, M. M., Alamgir, A., & Awan, A. B. (2020). Swarm intelligence-based optimization techniques for dynamic response and power quality enhancement of AC microgrids: A comprehensive review. *IEEE Access*, 8, 75986–76001.

[16] Kumar, P., Kumar, K., Bohre, A. K., & Adhikary, N. (2023). An osprey inspired optimization-based planning of controllers for wind energy enriched hybrid power system. *International Journal of Advance Research and Innovative Ideas in Education*, 9, 2395–4396..

[17] Naik, P. L., Jafari, H., Babu, T. S., Anil, A., Padmavathi, S. V., & Nazarpour, D. (2023). Enhancement of power quality in grid integrated system using DC-link voltage PI controlled VSC based

STATCOM. *Iranian Journal of Electrical & Electronic Engineering*, 19(2).

[18] Patro, R., Varaprasad, K. S. B., & Santhosh, T. STATCOM based Power Quality Improvement in Hybrid Power System.

[19] Prashant, Siddiqui, A. S., Sarwar, M., Althobaiti, A., & Ghoneim, S. S. (2022). Optimal location and sizing of distributed generators in power system network with power quality enhancement using fuzzy logic controlled D-STATCOM. *Sustainability*, 14(6),3305.

[20] Reddy, N. M., Sri, A. T., Pavithra, R., Kumar, B. A., Chandra, B. R., & Santhosh, D. Power quality improvement of hybrid power system by using D-STATCOM.

[21] Sai Sandeep, M., Balaji, N., Ali, M., & Srinivasan, S. (2023). Power quality improvement and performance enhancement of distribution system using D-STATCOM. *Smart Grids for Smart Cities*, 1, 357–375.

[22] Satapathy, A., Nayak, N., & Parida, T. (2022). Real-time power quality enhancement in a hybrid micro-grid using nonlinear autoregressive neural network. *Energies*, 15(23), 9081.

[23] Sirohi, V., Saggu, T. S., & Singh, M. (2023). Performance analysis of SRFT based D-STATCOM for power quality improvement in distribution system under different loading conditions. *International Journal of Emerging Electric Power Systems*, (0).

[24] Solanki, A. Y., & Vyas, S. R. (2021). A review on power quality enhancement using custom power devices. *International Research Journal of Engineering and Technology (IRJET)*, 8(2), 287–290.

[25] Wang, G., Jiang, Y., Liu, Z., Ma, Y., & Wang, J. (2024). Predictive direct control of nine-switch converter unified power quality conditioner based on timesharing cooperative control. *IEEE Transactions on Power Electronics*.

[26] Zastempowski, M. (2023). Analysis and modeling of innovation factors to replace fossil fuels with renewable energy sources-Evidence from European Union enterprises. *Renewable and Sustainable Energy Reviews*, 178, 113262.

[27] Zheng, J., Du, J., Wang, B., Klemeš, J. J., Liao, Q., & Liang, Y. (2023). A hybrid framework for forecasting power generation of multiple renewable energy sources. *Renewable and* Chapter 1

3 Implementation and Analysis of Fuzzy Logic Control and Direct Torque Control in Brushless DC Motors

Ramya L. N.[1],*, Abhishek Raj[2], Mordhwaj Thakur[2] and MD Zibran Alam[2]

[1]Assistant Professor, Department of EEE, Aarupadai Veedu Institute of Technology, Vinayaka Mission's Research Foundation (Deemed to be University), Paiyanur, Tamil Nadu

[2]Student, Department of EEE, Aarupadai Veedu Institute of Technology, Vinayaka Mission's Research Foundation (Deemed to be University), Paiyanur, Tamil Nadu

Abstract

Direct Torque Control (DTC), known for its simplicity and robustness, offers direct control over torque and flux in motor drives by selecting appropriate inverter switching states. However, conventional DTC methods may suffer from high torque ripple and flux ripple, especially at low speeds and under load variations. The proposed work proposes a novel approach to improve the control efficiency of Brushless Direct Current (BLDC) motors through the integration of Fuzzy Logic Control (FLC) into the DTC scheme. By incorporating Fuzzy Logic principles, the proposed strategy aims to mitigate these issues and improve overall motor drive performance. The FLC is designed to dynamically adjust the inverter switching states based on fuzzy rules derived from expert knowledge and system characteristics. This allows for controlled responses in motor operating conditions, resulting in smoother torque and flux outputs. Simulation results showcase the efficiency of the proposed approach in reducing torque and flux ripple, improving dynamic response, and enhancing the robustness of BLDC motor drives. Experimental validation on a real-time hardware platform verifies the feasibility and practicality of the proposed Fuzzy Logic-based DTC strategy for various industrial applications.

Keywords: Brushless Direct Current motor, direct torque control, field oriented control, fuzzy logic

1. Introduction

Brushless Direct Current (BLDC) motors have garnered considerable interest in many industrial and consumer sectors owing to their notable attributes of superior efficiency, dependability, and manageability. The Direct Torque Control (DTC) methodology has become widely used for managing the torque and flux of motor drives due to its benefits, including rapid dynamic response, ease of implementation, and less sensor needs [8, 9, 15]. Nevertheless, traditional DTC techniques may encounter limitations

Email: *ramyalnraja@gmail.com

DOI: 10.1201/9781003661917-3

such as the presence of fluctuations in torque and flux, especially when operating at low speeds and in the introduction of load changes. These shortcomings can negatively affect the overall performance and efficiency of the system.

This work suggests incorporating Fuzzy Logic Control (FLC) into the DTC scheme [7], as a means to enhance the operating efficiency of BLDC motors for its control technique and tackle the existing problems. Initially the essential aspects of BLDC motor functioning [2], DTC principles, and the core concepts of Fuzzy Logic Control are examined. Subsequently, the procedure for incorporating FLC into DTC scheme, encompassing the formulation of input variables, membership functions, rule base, and defuzzification procedures are elucidated. Table 3.1 showcases the necessity of DTC for BLDC.

Table 3.1 The Necessity of DTC for BLDC

Aspect	Need for DTC for BLDC Motor
Torque control	DTC enables accurate and quick torque control, that is important for the enhanced performance
Dynamic response	Provides fast response to changes in load, supporting dynamic performance
Torque ripple	Minimizes torque ripple, thus reducing the mechanical vibrations in the machine
Complexity in the system	Simplifies the system, thus eliminating the need for complex controllers

2. Existing System

Various control mechanisms are employed in current BLDC motor drive systems to manage torque and flux [1, 4, 13]. Among these schemes, DTC is particularly notable. The control algorithm in DTC-based BLDC drives directly modifies the inverter switching states by comparing the

actual and desired flux and torque values. In order to achieve this, academicians have investigated various approaches. Researchers have explored advanced Pulse Width Modulation (PWM) approaches in order to minimize ripple. Parameter optimization algorithms, such as Genetic Algorithms (GA) and Particle Swarm Optimization (PSO), are used to optimize control parameters [10, 11]. In addition, sensorless control approaches, such as back electromotive force (EMF) estimation methods have been created to eliminate the requirement for speed and position sensors. This helps to decrease system complexity and expense. However, there is still a need for strong and efficient control systems that can successfully reduce torque and flux fluctuations in BLDC motor drives [12]. The drawbacks of the existing system are notably the following:

- Conventional control methods lead to fluctuations in torque and flux output, causing vibrations and noise in the motor.
- The presence of ripple increases excessive mechanical loading on motor parts causing accelerated degradation and reduced system lifespan.
- Ripple in torque and flux results in energy losses and decreased overall system efficiency, impacting the motor's performance.
- High levels of ripple can cause instability and degrade motor control efficacy, affecting its accuracy and responsiveness.

3. Proposed System

The objective of the proposed system is to enhance the efficiency and reliability of BLDC motor drive systems by incorporating FLC into the DTC scheme. This integration offers an innovative solution to overcome the drawbacks of traditional BLDC motor drives. The suggested system aims to design a Fuzzy Logic Controller specifically for controlling BLDC motors. This controller would utilize input variables such as torque error, flux error, and their rates of change. The language descriptions of system parameters and control actions will be captured using membership functions and a rule base. This will allow for adaptive and resilient control responses. The output of the FLC will determine the specific output voltage states of the inverter in the DTC system, enabling accurate control of torque and flux. The suggested system will undergo thorough modelling studies and experimental validation to assess its performance, efficiency, and applicability across different operating situations. Advanced methods of fine-tuning and optimization will be utilized to further improve the effectiveness of the control algorithm [3, 6]. Figure 3.1 shows the Block diagram of the proposed system.

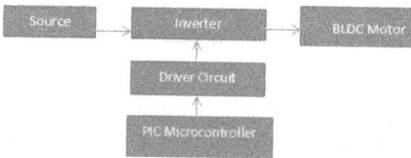

Figure 3.1 Block Diagram of the Proposed System

The working principle of the proposed Fuzzy Logic-based DTC strategy for BLDC motor drives involves the following steps:

- Identify the relevant input variables for the FLC. These variables typically include torque error, flux error, and possibly their rates of change, which are crucial for accurate control.
- Define membership functions for each input and output variable. These functions partition the input/output space into linguistic terms (e.g., low, medium, high) to facilitate fuzzy reasoning.
- Construct a rule base that maps the fuzzy input variables to appropriate control actions. These rules encode expert knowledge or system behavior, specifying the control actions corresponding to different combinations of input variables.
- Apply fuzzy inference to determine the exact control actions based on the present state of the system. Fuzzy logic reasoning evaluates the fuzzy input variables and

the rule base to initiate fuzzy output control actions.
- Convert the fuzzy output of the inference process into a crisp control action. Defuzzification methods such as centroid or weighted average are used to derive a single numerical value representing the control action.
- Integrate the Fuzzy Logic Controller into the DTC scheme to regulate the flux and torque of the motor. The crisp control action obtained from defuzzification is used to select the appropriate inverter switching states, enabling precise control over motor performance.

4. Simulation Results

The simulation of the proposed work was carried out in MATLAB/Simulink environment. The outputs are shown in Figure 3.3. The inverter output voltage and stator current waveforms are shown in Figures 3.3a and 3.3b respectively. The speed waveform is shown in Figure 3.3c and the torque in Figure 3.3d.

It can be seen from Figure 3.3c, there is a speed change in the motor from 700 rpm to 1200 rpm. Figure 3.3d showcases the quick retrieval of the torque, with a quick settling to the static position, from the transition state. This shows the effectiveness of the proposed work that when there is an abrupt variation in the speed, the system settles from the dynamic state quickly as expected. The hardware kit is shown in Figure 3.2. The hardware outputs also confirm the simulation outputs

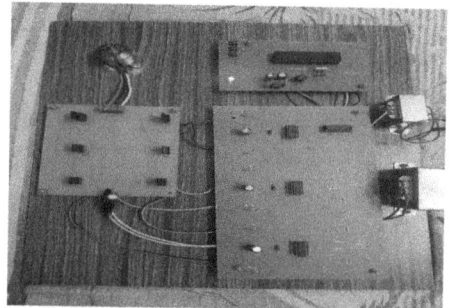

Figure 3.2 Hardware Circuit Diagram Kit

Figure 3.3 MATLAB Simulation Output of Voltage, Current, Speed and Torque

that the system instantly retrieves from the disturbances. The advantages of the proposed method are,

- Minimizes fluctuations in torque and flux for smoother motor operation.
- Enhances the motor's ability to quickly respond to speed and torque changes.
- Adjusts control actions in real-time for optimal performance under varying conditions.
- Increases overall efficiency, extends motor lifespan, and reduces maintenance needs.

5. Conclusion

The integration of FLC into the DTC scheme provides a breakthrough approach to overcome the shortcomings of conventional BLDC motor drive systems. By leveraging FLC's adaptive and robust control capabilities, the proposed system effectively reduces torque and flux ripple, improves dynamic response, and enhances the overall efficiency and reliability of BLDC motor drives. Moreover, the system's reduced dependency on sensors and ability to handle parameter variations make it a cost-effective and practical solution for a broad applicability across industrial and consumer domains. Through comprehensive simulation studies and experimental validation, the proposed system demonstrates its superiority over conventional DTC methods, paving the way for advancements in BLDC motor drive technology. With its numerous advantages, the proposed system holds great potential for revolutionizing the future of BLDC motors including enhanced performance capabilities and energy efficiency with increased operational reliability.

References

[1] Kaf, A. A., Cheng, X., Zhang, C., Almadwami, A., Abdullah, A., & Almadwami, H. (2024). Sensorless direct torque control in brushless DC motor using sliding mode observer, 4th International Conference on Emerging Smart Technologies and Applications (eSmarTA). Sana'a, Yemen (pp. 1–8).

[2] Nicola, C. -I., Nicola, M., Iacob, A., & Pîrvu, C. (2024). PMSM sensorless control system based on DTC using FLC and Luenberger-PLL observer, 6th Global Power, Energy and Communication Conference (GPECOM). (pp. 138–143).

[3] Mohanraj, D., et al. (2022). A review of BLDC motor: State of art, advanced control techniques, and applications. *IEEE Access*, 10, 54833–54869.

[4] G. R, S. V. D, C. E, P. A, S. E M., & A. M. (2024). Electric scooter ride ability: Initial torque optimization, 5th International Conference on Mobile Computing and Sustainable Informatics (ICMCSI). (pp. 829–832).

[5] Sajitha, G., Mayadevi, N., Mini, V. P., & Harikumar, R. (2019). Fault-tolerant control of BLDC motor drive for electric vehicle applications. International Conference on Power Electronics Applications and Technology in Present Energy Scenario (PETPES). (pp. 1–6).

[6] H. K. U., & R. P. P. (2022). A direct torque control scheme for BLDC motor drives with open-end windings. IEEE 1st Industrial Electronics Society Annual On-Line Conference (ONCON). (pp. 1–6).

[7] Abirami, M., & Balakrishnan, P. (2023). DTC-FOC hybrid controller to control the speed and torque of BLDC motor. Innovations in Power and Advanced Computing Technologies (i-PACT). (pp. 1–6).

[8] Depenbrock, M. (1988). Direct self control of inverter-fed induction machines. *IEEE Transactions in Power Electronics*, PE-3(4), 420–42).

[9] Patil, M. S., Medhane, R., & Dhamal, S. S. (2020). Comparative analysis of various DTC control techniques on BLDC motor for electric vehicle, 7th International Conference on Smart Structures and Systems (ICSSS). (pp. 1–6).

[10] Sangeetha, M., Jasmine, M., & Philomina, S. (2019). Fuzzy-logic method and sliding-mode control for induction motor. *International Journal of Engineering and Advanced Technology (IJEAT)*, 8 (6S2).

[11] Prabhu, N., Thirumalaivasan, R., & Ashok, B. (2023). Critical review on torque ripple sources and mitigation control strategies of BLDC motors in electric vehicle applications. *IEEE Access*, 11, 115699–115739.[12] Agrawal, P., Dubey, S. P., & Bharti, S.(2014). Comparative study of fuzzy logic based speed control of multilevel inverter fed brushless DC motor drive. *International Journal of Power Electronics and Drive System*, 4(1), 70–80.

[13] Muthamizhan, T., Saravanan, P., & Maharana, R. (2021). Sensorless control of Z source inverter fed BLDC motor drive by FOC - DTC hybrid control strategy using fuzzy logic controller, 7th International Conference on Electrical Energy Systems (ICEES). (pp. 358–363).

[14] Sharma, T. R., & Pal, Y. (2019). Direct torque control of BLDC drives with reduced torque pulsations, 3rd International conference on Electronics, Communication and Aerospace Technology (ICECA). (pp. 1242–1247).

[15] Takahashi, I., & Noguchi, T. (1986). A new quick-response and high efficiency control strategy of an induction motor. *IEEE Transactions on Industrial Applications*, 22(5), 820–827.

4 A Multi Objective Cost Effective Approach of Energy Management of Multi-Microgrid Environment using Hybrid Jaya-PSO Optimization Technique

M. Sri Suresh[1], Mandla Chandana[2], Essam Bhargavi[2], Solleti Rahul[2], Vaddi Mithilesh[2] and Thamatapu Eswara Rao[3]

[1]Assistant Professor, Chaitanya Bharathi Institute of Technology, Hyderabad, India
[2]Student Scholar, Chaitanya Bharathi Institute of Technology, Hyderabad, India
[3]Associate Professor, Vignan Institute of Information Technology, Visakhapatnam, India

Abstract

Integrating wind energy into power systems is key to reducing carbon emissions but requires addressing its impact on reliability due to wind's unpredictability. This integration must also consider generation costs, system dependability, and sector-specific energy demands, which increase pressure on reliability. Demand response can help manage load fluctuations and reduce peak demand, but rescheduling energy use incurs utility costs that must be factored in. This study proposes a comprehensive approach to optimize wind power integration, considering these challenges. As power demand rises and distributed generation grows, microgrids have emerged to meet these needs. However, renewable energy's volatility complicates scheduling. A chance-constrained model for day-ahead scheduling in multi-microgrid systems is developed, accounting for emissions from conventional and renewable sources. Using a hybrid Jaya-Particle Swarm Optimization (JPSO) approach, power loss costs are minimized. Simulations on a modified 33-bus system show this strategy effectively balances economic and environmental goals, reducing emissions while ensuring cost-efficient energy planning.

Keywords: Multi-microgrids, modified 33bus system, carbon emission, day ahead scheduling, Jaya-PSO algorithm (JPSO).

1. Introduction

Renewable energy-based distributed generation (DG) systems have shown rapid growth during the last decade. In addition to cheaper costs and better economics, the growth has been driven by benefits for the environment from less greenhouse gas emissions [1]. The establishment of microgrids (MGs) [2–3] is a rising answer to the operational problems presented by these changes in the electrical system. Integration

Email: [1]mavurisrisuresh_eee@cbit.ac.in, [2]reddychandana402@gmail.com, [2]essambhargavi74@gmail.com, [2]srahul2003r@gmail.com, vaddimithilesh@gmail.com, [3]eswareee@vignaniit.edu.in

DOI: 10.1201/9781003661917-4

adds to the complexity of MGs, but it also carries with it issues with major investment expenditures, market dynamics, and regulatory considerations. A strong control system is consequently important for both short-term power management [4] and long-term energy efficiency. The authors in [5] able to address about EMS's capacity to generate conclusions that enable the system to be run with more cheap and efficient parameters is vital in an MG. There are several ways that have been used to tackle the MG energy management issue, ranging from metaheuristic to mathematical [6–7]. Nevertheless, most research admits uncertainty in the creation of renewable energy. Because renewable energy generation is unpredictable, this technique has a substantial influence on power systems and converts energy management into a stochastic programming issue. In order to solve such issues, this work offers a stochastic chance-constrained model in this paper for the best feasible scheduling of MMG.

Our research from [8–10] attempts to address these issues by putting forth a stochastic chance-constrained model for MMG system optimal scheduling. In this context, reducing emissions is also a top priority, as per international environmental standards, along with managing uncertainties. As was previously said, the recommended strategy makes use of chance-constrained programming to decrease expenditures and emissions. However, this development has generated significant technical issues for electricity networks, which microgrids assist reduce. As microgrid (MG) technology progresses, controlling these systems becomes more challenging. Typically, microgrids are arranged a day in advance, with financial goals being the major emphasis [11]. As MGs advance, they encounter hurdles such as high investment costs, commercial restraints, and regulatory limitations [12]. For instance, [13–14] explored MINLP-based techniques for energy management in islanded MGs, albeit these studies did not account for uncertainties or renewable-based DGs [15]. Authors also developed a solver-friendly MINLP model for optimum

power flow but similarly overlooked uncertainties. The insufficient treatment of uncertainty in these investigations shows a substantial research need [13]. Authors are employed Monte Carlo simulations and robust chance-constrained programming (CCP) to address emission reduction under uncertainty, however their attention was confined to single MG systems and did not account for network restrictions. There is a paucity of thorough study on MMG day-ahead scheduling [7] that incorporates security limitations for minimizing emissions and operating expenses.

This work presents a hierarchical strategy for controlling energy in MMG systems under uncertainty using hybrid optimization JPSO. The proposed framework contains convex power flow limitations to maintain network security and analyzes emission costs to encourage renewable-based distributed production. Below are the novelty and key contributions of this research:

1. Development of an energy management framework for MMG systems utilizing a hybrid optimization Jaya-PSO (JPSO).
2. Incorporation of uncertainties due to variable loads, time-varying wind, and photovoltaic.
3. Consideration of emissions for day-ahead scheduling to improve unit scheduling.
4. Consider reliability index with 95% to validate the proposed approach under different case studies.

2. Multi-Microgrid Structure and Problem Formulation

Multi-microgrid (MG) distribution networks are made up of several MGs and other independent entities, including batteries, energy storage devices, renewable energy sources (like PV and wind turbines), dispatchable generation units (like internal combustion generators), and various kinds of loads. Figure 4.1 shows an proposed of a typical MMG system [14], which consists

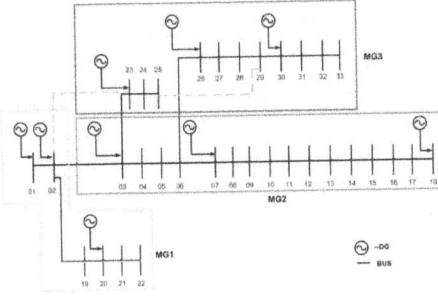

Figure 4.1 A modified 33 Bus System with Three Microgrids and Locations of Distributed Generation (DGs)

of three MGs and separate independent units. Wind turbines (WTs), photovoltaic (PV) systems, energy storage devices, internal combustion generators, and loads are all included in each MG. The mathematical modelling of these different components is shown in this section.

2.1 IC Engine Systems

Because of their controlled output power, generators powered by internal combustion (IC) engines are regarded as dispatchable generation systems. The following equations illustrate the pertinent constraints and interactions for these systems:

$$\text{Cos}\, t_{ICS,tls} = \text{Cos}\, t_{ICS,tls}^{operation}(P_{ICS}) + \text{Cos}\, t_{ICS,tls}^{stup} \quad (1)$$

2.2 Wind Turbines

One of the most promising and quickly developing methods for generating electricity from renewable sources is wind power generation. These devices are made up of wind turbines (WTs), which transform wind energy from kinetic energy to electrical power. Because wind speed is a naturally stochastic variable, their output power depends on it, which makes the electrical energy they create intrinsically unreliable.

As shown below (Aghdam et al., 2018a), the form factor and the scale factor—two characteristics unique to the wind turbine site—define the Weibull probability density function (PDF), which describes wind speed [16].

$$P_{wt} = \begin{cases} 0 & v \leq v_{in}, v \geq v_0 \\ \dfrac{v - v_{in}}{v^{rated} - v_{in}} P_{wt}^{rated} & v_{in} \leq v \leq v^{rated} \\ P_{wt}^{rated} & v^{rated} \leq v \leq v_0 \end{cases} \quad (2)$$

2.3 PV Systems

Photovoltaic (PV) systems are widely regarded as one of the most popular and straightforward means of generating renewable energy. This is explained by the fact that they are less expensive and smaller than wind turbines (WTs), which has resulted in a notable increase in the installed capacity of PV panels

$$P_{PVM} = \begin{cases} P_{PV}^{rated} \times \left(\dfrac{Lr^2}{Lr_{std} \cdot Lr_{cer}} \right) & Lr \leq Lr_{cer} \\ P_{PVM}^{rated} \times \left(\dfrac{Lr^2}{Lr_{std}} \right) & Lr_{cer} \leq Lr \end{cases} \quad (3)$$

2.4 Objective Function

The optimization objective is to simultaneously minimize:

1. Economic Energy Curtailment Cost (ECC)
2. Economic Generation System Cost (EGC)

This multi-objective approach enables the identification of Pareto-optimal solutions, providing a range of optimal trade-offs among the conflicting objectives.

$$\min[f_1(t), f_2(t)] = \min[ECC, EGC] \quad (4)$$

Objective 1: To minimize the Economic Energy Curtailment Cost (ECC)

$$\min f_1(t) = \min \left[\sum_{Lc=1}^{n} P_{Lc}^{loss} LL \right] \quad (5)$$

$$\min f_2(t) = \min[EGC] \quad (6)$$

Objective 2: To minimize the Economic Generation System Cost (EGC)

$$F_l(x) = \begin{cases} 0 & \text{if } F_l \leq F_{l\,min} \\ \dfrac{F_l - F_{l\,min}}{F_{l\,max} - F_{l\,min}} & \text{if } F_{l\,min} < F_l < F_{l\,max} \\ 1 & \text{if } F_l > F_{l\,max} \end{cases}$$

$$(7)$$

2.5 Algorithm Hybrid Jaya-Pso Optimization

Step 1: Initialization

1. Initialize a population of N particles with random positions x[i] and velocities v[i].
2. Compute the fitness f(x[i])f(x[i])f(x[i]) for each particle.
3. Set personal best positions p[i] = x[i].
4. Identify the global best position g = min(f(p[i])).

Step 2: Iterative Optimization

$$\text{For } t = 1 \text{ to } T_{max}$$

A. PSO Phase

For each particle i=1 to N:

Update velocity: v[i]=w·v[i]+c1·random()· (p[i] − x[i]) + c2 · random() · (g−x[i])

Update position: x[i]=x[i]+v[i]x[i]=x[i]+ v[i] x [i]=x[i]+v[i]

Enforce constraints on x[i] (repair if out of bounds).

Compute fitness f(x[i])

Update personal best: If f(x[i])<f(p[i]), then p[i]=x[i]

Update global best: g = min f(p[i])

B. Jaya Phase

1. Identify the best solution xbestx_ {best}xbest and worst solution xworstx_{worst}xworst in the current population.
2. For each particle i=1 to N:
3. Update position: x[i]=x[i]+random()·(xbest−|x[i]|)−random()· (xworst − |x[i]|)
4. Enforce constraints on x[i].
5. Compute fitness f(x[i]).
6. Update global best:
7. g = min(f(x[i]))

C. Hybridization

If tmod k = 0 (every k-th iteration):
   Combine solutions from PSO and Jaya phases.
   Retain the top-performing solutions based on fitness for the next iteration.

Step 3: Termination

1. Stop when t = Tmax
2. Return g as the optimized solution.

3. Hierarchical Day-Ahead Scheduling of MMG System

According to the proposed procedure, each microgrid (MG) [8] determines how much electricity it can supply to the upstream network by first carrying out optimal programming for its day-ahead horizon. In Figure 4.3 shows the upstream network then starts its optimisation process based on the data gathered and relays the different programs that different entities need to execute the next day. Based on the amount of distributed generators (DGs) with renewable energy sources included into its network, the MG operator has to calculate its reliability level £. This value is used to compute £ for each bus using the method shown in Figure 4.2. It is significant to remember that every node may have a distinct source of uncertainty. As an illustration, one node might contain a wind generator, and another might serve as a load bus. In the end, the following optimisation issue must be resolved for MG energy management: After gathering data from the microgrids' (MGs) energy management, the DN operator starts global day-ahead scheduling, taking into account a predefined reliability level £. At this point, the optimisation issue is: The quantity of power that can be purchased or sold from that microgrid (MG) is represented by MG. Every MG rearranges its entities in accordance with this data. Figure 4.3 displays hierarchical energy management that explains the procedure of optimizing the cost and power balance.

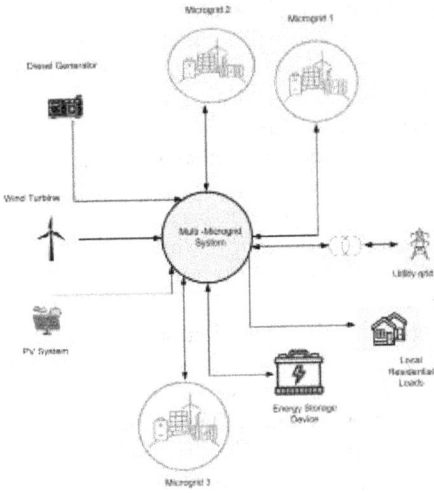

Figure 4.2 Structure of Multi-Microgrid System

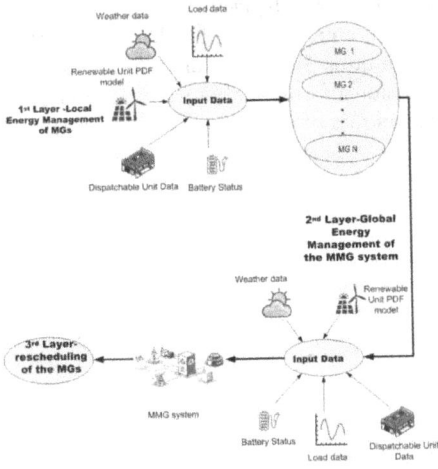

Figure 4.3 Hierarchical energy management structure

4. Simulation Results

The simulations are run on a PC with a Core i7 CPU and 8 GB of RAM within the MATLAB environment. Furthermore, £ has been calculated using MATLAB program. Three microgrids (MGs), multiple independent generators, an energy storage system, and a range of loads make up the test system. The designated units are also included in every MG. Table 4.1 provides a summary of the unit characteristics, whereas Figure 4.4 shows the prediction of PV and wind of MMG through JPSO. In addition, Figure 4.5 shows the prices of electrical energy in the multi-microgrid. Table 4.1 displays comparison of various costs optimized by JPSO. In the optimization process the following work is presented.

Two case studies are considered in this work as follows:

Case-I: Including emission costs in the cost function with reliability index £ = 0.95.

Case-II: Excluding emission costs in the cost function with reliability index £ = 0.95.

Case-I: In this instance, the simulation process accounts for the cost of emissions, and the CCP model's reliability level is set at 95%. Local day-ahead scheduling is done by each microgrid (MG) during the initial step of optimisation. When energy prices are higher, the MGs want to provide the distribution network (DN) with as much power as possible; during times when prices are lower, they try to buy power. The judgements made for battery charging and discharging, as well as the curves for surplus and shortage power, all exhibit this behaviour. For instance, the IC generators in every MG run at 90% of their maximum capacity around noon, when electricity rates are at their highest, and they give the excess power to the DN to increase their profits. Their rated output is restricted due to efforts to minimise emissions, which is the reason for their inability to operate at full capacity. Table 4.1 displays the emissions and the objective function value for every microgrid (MG). MG2, with its higher emissions than MG1, has the greatest objective function value. The third MG, on the other hand, has a negative value, which indicates that clean production is predominant in its units. The total amount of pollutants released into the atmosphere by the entire system during a 24-hour period is 26.8 tCO2. In Figure 4.6 represents active power loss of the multi-microgrid, the proposed system is tested with hybrid optimization Jaya-PSO (JPSO). The proposed approach gives the best optimal solution in comparing with other optimization techniques.

Case-II: In order to investigate how the cost of emissions affects the scheduling of the system, it has been removed from the objective function in this instance. A 95% reliability threshold has been established. For instance, in contrast to Case- I, the IC generators of MGs #1 and #2 run continuously from 6 p.m. till dusk, but in Case I, they only reached 90% capacity for a single hour. It is evident that this raises emissions across the board for the system. The elimination of emission costs has resulted in a significant decrease in the objective function values as compared to Case-I. But overall emissions have gone up, which is bad for the environment. 50.1 tCO2 was a noteworthy sum of emissions during the schedule period. MG-3's goal function values remain unchanged, despite MG-1 and MG-2 objective function values declining. This is explained by the increased

proportion of distributed generation (DG) based on renewable energy sources in its structure, which reduces the output's reliance on emission costs. This is in line with laws intended to lessen greenhouse gas emissions and stop global warming.

In Figure 4.7 represents voltage profile of the multi-microgrid, the proposed system is tested with hybrid optimization Jaya-PSO (JPSO). The proposed approach gives the best optimal solution in comparing with other optimization techniques. A multi-objective pareto based optimization is utilized for the MMG system is to get best optimal solution. In Figure 4.8a displays the comparison of the EEC and EGC with jaya algorithm, which gives better pareto optimal solution. In Figure 4.8b shows that multi-objective cost comparison with hybrid Jaya-PSO, which gives the best optimal solution in comparing with other optimization techniques. Table 4.1 summarizes the comparision of cost with various optimization techniques of the multi-microgrid system.

Table 4.1 Comparison of total annual savings of MMG system with different optimization techniques

Case	Optimization technique	Voltage deviations	Economic Energy curtailment cost ($/kW)	Economic generation system cost ($/kW)	Annual cost of power loss ($/kW)	Annual cost of load demand ($/kW)	Total Annual cost savings ($/kW)
Case-I with £ = 0.95	GA	2.05*E-03	0.95	9524	82.5	8594	847.5
	PSO	1.55*E-03	0.9	10258	75.2	9246	936.8
	WOA	1.36*E-03	0.87	11025	71.5	10011	942.5
	Jaya	1.15*E-03	0.83	11325	65.3	10246	1015.5
	Proposed hybrid Jaya-PSO	1.05*E-03	0.6	11523	45.6	10145	1332.4
Case-II with £ = 0.95	GA	2.55*E-03	0.8	9324	75.6	8267	981.4
	PSO	1.45*E-03	0.77	10158	72.3	9014	1071.7
	WOA	1.25*E-03	0.72	13156	65.8	11964	1126.2
	Jaya	1.10*E-03	0.65	13256	54.2	11968	1233.8
	Proposed hybrid Jaya-PSO	1.01*E-03	0.51	14536	45.2	12964	1526.8

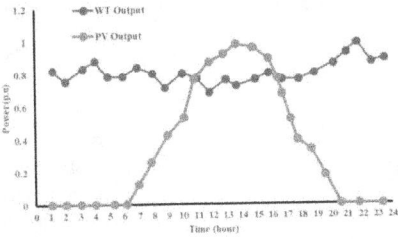

Figure 4.4 Prediction of Outputs of Wind and PV of MMG

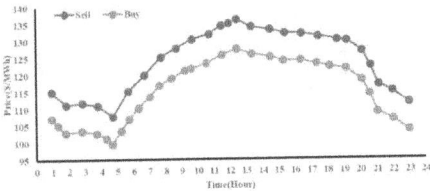

Figure 4.5 Energy Trading Prices During Energy Management

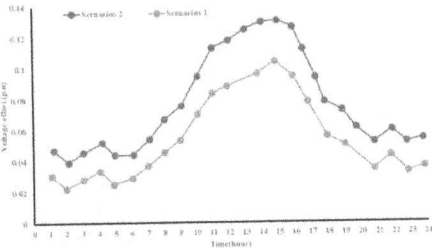

Figure 4.6. Active power loss under different scenarios

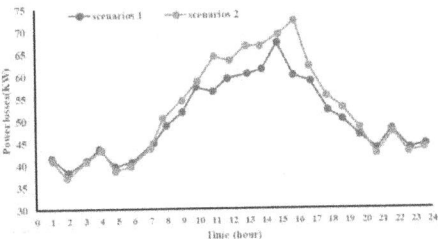

Figure 4.7. Voltage profile of MMG under different scenarios

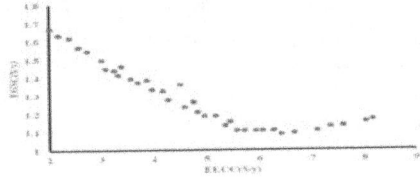

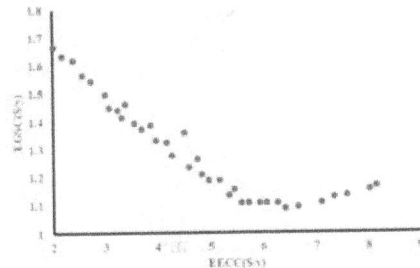

Figure 4.8 (a) Pareto Based Comparison of ECC and EGC with Jaya. (b) Pareto Based Comparison of ECC and EGC with JPSO

5. Conclusion

A novel idea known as MMG systems has emerged as a result of the rising use of MGs in distribution networks. These systems present substantial issues for energy management, in part because of the inherent uncertainties. This research uses a hybrid jaya–pso based method for day-ahead scheduling of MMG systems. The Jaya-PSO approach reduces computing overhead and improves system reliability, among other benefits. Each MG in this method starts by doing a local optimisation in the first layer. The DN operator then compiles this stage's data for global scheduling and sends out schedule notifications to the MGs and other organisations. According to the simulation results, there was a noticeable decrease in greenhouse gas emissions when the emission cost was taken into account while creating the objective function benefits. Each MG in this method starts by doing a local optimisation in the first layer. The DN operator then compiles this stage's data for global scheduling and sends out schedule notifications to the MGs and other organisations. According to the simulation results, there was a noticeable decrease in greenhouse gas emissions when the emission cost was taken into account while creating the objective function. The simulations also showed that clean generation units were being used more frequently. Moreover, the objective function value decreased with decreasing the reliability level and increasing risk tolerance. It is advised that in the future, the problem be modelled as a linear optimisation problem using linearization techniques, which may result in a global optimum. Furthermore, it could be worthwhile to look into creating an MMG system bidding strategy that uses hybrid JPSO optimization approach and takes emissions into account. While maintaining the MGs' anonymity, taking into account a decentralised method of problem solving within the hybrid JPSO framework gives best optimal values in comparing with other optimization techniques.

References

[1] Zhou, B., Zou, J., Chung, C. Y., Wang, H., Liu, N., Voropai, N., & Xu, D. (2021). Multi-microgrid energy management systems: Architecture, communication, and scheduling strategies. *Journal of Modern Power Systems and Clean Energy*, 9(3), 463–476.

[2] Mavuri, S. S., & Nakka, J. (2024). Economic scheduling and dispatching of distributed generators considering uncertainties in modified 33-bus and modified 69-bus system under different microgrid regions. *Transactions on Energy Systems and Engineering Applications*, 5(2), 1–22. https://doi.org/10.32397/tesea.vol5.n2.570.

[3] Mavuri, S. S., Nakka, J., & Kotla, A. (2023). Deep neural network based intelligent multi-microgrid energy management. *IEEE 3rd International Conference on Sustainable Energy and Future Electric Transportation (SEFET)*. Bhubaneswar, India, pp. 1–6, doi: 10.1109/SeFeT57834.2023.10245648.

[4] Chen, J. J., Qi, B. X., Rong, Z. K., Peng, K., Zhao, Y. L., & Zhang, X. H. (2021). Multi-energy coordinated microgrid scheduling with integrated demand response for flexibility improvement. *Energy*, 217, 119387.

[5] Rao, T. E., Naidu, B., Ankamma Rao, J., Zelie, G. T., Avinash, K., & Srisuresh, M. (2024). Soft computing based hybrid MPPT for grid connected photovoltaic system. *International Conference on Electronics, Computing, Communication and Control Technology (ICECCC)*. Bengaluru, India, pp. 1–6, doi: 10.1109/ICECCC61767.2024.10593845.

[6] Mavuri, S. S., Nakka, J., & Kotla, A. (2022). Interconnected microgrids: A review and future perspectives. *IEEE 2nd International Conference on Sustainable Energy and Future Electric Transportation (SeFeT)*. Hyderabad, India, pp. 1–7, doi: 10.1109/SeFeT55524.2022.9908988.

[7] Mavuri Sri Suresh, S. V. D. Anil Kumar. (2016). Advanced intelligent control for power quality with reduced rating DVR. *International Journal of Electrical and Electronics Engineers*, 8(2), ISSN: 2321–2055.

[8] Cao, Y., Taslimi, M. S., Dastjerdi, S. M., Ahmadi, P., & Ashjaee, M. (2022). Design, dynamic simulation, and optimal size selection of a hybrid solar/wind and battery-based system for off-grid energy supply. *Renewable Energy*, 187, 1082–1099. doi: 10.1016/j.renene.2022.01.112.

[9] Kumar, B. A., et al. (2023). A novel framework for enhancing the power quality of electrical vehicle battery charging based on a modifed Ferdowsi converter. *Energy Reports*, 10, 2394–2416. https://doi.org/10.1016/j.egyr.2023.09.070.

[10] Misra, S., Panigrahi, P. K., Ghosh, S., & Dey, B. (2023). Economic operation of a microgrid system with renewables considering load shifing policy. *International Journal of Environmental Science and Technology*, 21(3), 2695–2708.

[11] Alam, M. S., Almehizia, A. A., Al-Ismail, F. S., Hossain, M. A., Islam, M. A., Shafiullah, M., & Ullah, A. (2022). Frequency stabilization of AC microgrid clusters: An efficient fractional order supercapacitor controller approach. *Energies*, 15(14), 5179.

[12] Dashtaki, A. A., Hakimi, S. M., Hasankhani, A., Derakhshani, G., & Abdi, B. (2023). Optimal management algorithm of

microgrid connected to the distribution network considering renewable energy system uncertainties. *International Journal of Electrical Power & Energy Systems*, 145, Art. no. 108633.

[13] Zulu, M. L. T., Carpanen, R. P., & Tiako, R. (2023). A comprehensive review: Study of artificial intelligence optimization technique applications in a hybrid microgrid at times of fault outbreaks. *Energies*, 16(4), 1786.

[14] Battistelli, C., Agalgaonkar, Y. P., & Pal, B. C. (2017). Probabilistic dispatch of remote hybrid microgrids including battery storage and load management. *IEEE Transactions on Smart Grid*, 8(3), 1305–1317.

[15] Li, Z., Shahidehpour, M., Aminifar, F., Alabdulwahab, A., & Al Turki, Y. (2017). Networked microgrids for enhancing the power system resilience. *Proceedings of the IEEE*, 105(7), 1289–1310.

[16] Du, Y., & Li, F. (2018). A hierarchical real-time balancing market considering multi-microgrids with distributed sustainable resources. *IEEE Transactions Sustainable Energy*.

5 Implementation of Automatic Speed Control System for Electric Vehicles

Pydikalva Padmavathi[1], Yaganti Krishna priya[2], Talari Manohar[3], Gautham L.[4], Navya Sree A.[4] and Pasala Pratheep[5]

[1]Assistant Professor, Srinivasa Ramanujan Institute of Technology, Anantapur, India
[2]Associate Professor, Anantha Lakshmi Institute of Technology and Sciences, Anantapur, India
[3]Assistant Professor, Anantha Lakshmi Institute of Technology and Sciences, Anantapur, India
[4]UG Student, Srinivasa Ramanujan Institute of Technology, Anantapur, India
[5]UG Student, Anantha Lakshmi Institute of Technology and Sciences, Anantapur, India

Abstract

These days most of the people are moving towards the electric vehicles leaving behind the fuel-based mechanisms. Due to driver's unawareness or over speeding and several reasons, the number of accidents has grown in a faster rate in certain zones such as Schools, U-Turns, Steep curves, and hospitals too. There is no such kind of feature in E-Vehicles to slow down the speed in restricted zones automatically. So, our first and foremost work is to prepare a system for the sake of speed controlling for an E-Vehicle which will do certain operations such as getting throttle inputs, controlling motor speed as per throttle, getting speed values and displaying it. This system will be inserted in our electric vehicles to limit the speed of our vehicles automatically. This will reduce the accidents growth rate up to a certain extent. This can be done in fuel-based vehicles also but the reason to perform this on E-Vehicles is the rapid growth of them these days and in future probably. Whenever the electric vehicle (it may be either bike or car) is going in a full speed, at that time if the speed limit area is there then the vehicle's speed is limited according to the speed limit in that zone. The proposed work is done with the help of an E-Vehicle motor, Radio Frequency (RF) using ESP32 Microcontroller. The microcontroller has been coded in a specific way where the required speed limits will be inserted in the transmitter circuit by using the switches and their job is to send the signal through the antenna which is accepted by receiver. The microcontroller will monitor the speed of vehicle by pulse width modulation (PWM) technique and then the subsequent process is executed by limiting the speed of the vehicle according to the signal sent by the receiver. At this time, the control will be in the hands of microcontroller and the transmitter circuit.

Keywords: Automatic speed control, ESP32 micro-controller, EV motor, RF receiver, RF transmitter.

Email: [1]Padmavathi.eee@srit.ac.in, [2]204g1a0218@srit.ac.in, [2]204g1a0259@srit.ac.in, [2]yagantikrishnapriya5@gmail.com, [3]talarimanohar1207@gmail.com, [4]222g5a0241@gmail.com.

DOI: 10.1201/9781003661917-5

1. Introduction

In recent years, the global push for sustainable transportation solutions has spurred the rapid development of electric vehicles. As the adoption of e-vehicles continues to grow, ensuring safe and efficient operation within speed-limited areas becomes paramount [1, 2, 3]. Since the population has increased, the usage of vehicles has also increased and the number of accidents has also increased because of violating the rules set by the government [4]. If we dig deep, majorly these accidents have occurred in some specific areas such as sharp turns, steep curves, or any high alert areas due to the driver's unawareness, over speeding. Addressing these challenges requires innovative solutions that leverage advancements in technology to enhance the safety features of electric vehicles. Many Electric Vehicles comes with a feature of cruise control where the driver can set constant speed for limiting the vehicle by manual method. So, still there is no automatic speed limitation feature without the involvement of the driver. In response to this imperative, this paper is proposing a system which is made for the control of speed of electric vehicles which has been tailored especially in navigating speed-restricted zones. The proposed system aims to mitigate the risk of accidents by automatically adjusting the speed of electric vehicles to adhere to prescribed speed limits in designated zones [5]. Leveraging Radio Frequency (RF) communication and microcontroller programming, the system facilitates real-time interaction between a transmitter unit, which holds the prescribed speed limits, and a receiver unit installed within the EV. This paper proposes a comprehensive explanation and study of the design and working of our system, encompassing key components such as throttle input processing, motor speed control [6], speed detection using pulse width modulation (PWM) techniques, and generation of control signals for activating the vehicle's speed reduction mechanism. The integration of these components facilitates precise and responsive speed regulation, allowing e-vehicles to adhere to speed limits while ensuring smooth acceleration and deceleration. The ESP32 microcontroller serves as the central processing unit, orchestrating the communication and coordination between various subsystems [7]. Its versatility and computational power make it an ideal choice for managing complex control algorithms and interfacing with peripheral devices. Through the deployment of this Electric Vehicle Speed Control System, it is anticipated that significant strides can be made towards enhancing road safety, reducing the incidence of accidents, and fostering the widespread acceptance of electric vehicles as a safe and sustainable mode of transportation in urban environments.

2. Main Objective

From the problem identification, we have come to know that the need of speed control in speed limit areas is essential to create safer, more livable, and more sustainable communities by reducing speeding-related risks and promoting responsible driving behaviour among motorists. This requires a comprehensive approach that combines engineering solutions, enforcement measures, educational initiatives, and community engagement strategies. Speed control measures aim to create safer road environments by reducing the likelihood of accidents, minimizing the impact of collisions, and providing adequate protection for all road users. This includes implementing traffic calming measures, enhancing visibility, and improving infrastructure design.

3. System Architecture

The Electric Vehicle Speed Controlling model duty is to limit the speed of the vehicle whenever it goes in excessive speed. This model will be more helpful to the electric vehicles because of their rapid growth in market. Whenever the electric vehicle (it may be either bike or car) is going in a full speed, at that time if the speed limit area is there then the vehicle's speed is limited according to the speed limit in that zone. The proposed work can be achieved with the help of Embedded System which is a microcontroller, Radio Frequency (RF) Transmitter and Receiver. The microcontroller has been coded in a specific way where the required speed limits will be inserted in the transmitter circuit by using the switches and their job is to send the signal through the antenna which is accepted by receiver. Then micro-controller will decode the signal and take immediate action by generating a control signal which is analyzed by the receiver, then it will initialize the process of the controlling of the speed in the e-vehicle and after doing that, the e-vehicle's speed will be limited as per the instruction from the transmitter.

The Figure 5.1 represents the functioning of the working model with the help of diagram. The speed control process begins with the driver input, which is typically received via a throttle pedal. The microcontroller interprets these inputs and generates corresponding signals to adjust the motor's speed through the motor driver. During operation, the microcontroller continuously monitors the vehicle's speed and this feedback loop allows the system to make real-time adjustments to maintain the desired speed and ensure smooth acceleration and deceleration The RF communication capability of the ESP32 enables remote control functionalities, allowing users to adjust the vehicle's speed wirelessly from a distance. Moreover, it facilitates data exchange with external devices for monitoring performance metrics and diagnosing issues remotely.

4. Hardware Components

4.1 DC Motor

In the prototype model, the EV's DC motor is taken instead of an Electric Vehicle. For better usage, we take a 24V DC Brushed motor because their construction is simple to understand when they are compared with other motor types and they are low in price to manufacture and in maintenance. They consist of fewer components, such as brushes, a commutator, and a rotor, which contributes to their ease of use and reliability. At the core of a DC motor lies the principle of electromagnetic induction, where the interaction between magnetic fields generates rotational motion. At the tip of the motor, we attach a wheel to it to understand the rotational speed of the motor. DC motors provide high starting torque, making them well-suited for applications requiring rapid acceleration or the ability to overcome initial resistance.

4.2 Motor Driver

A motor driver is a specialized circuit or module that interfaces between a microcontroller or other control circuitry and the motor itself. And the driver runs the motor smoothly and efficiently by providing required power. We are taking an L298N Driver because of its suitability and we are using pulse width modulation technique in which the input voltage value is maintained using ON and OFF pulse. Another mostly used method is H-Bridge.

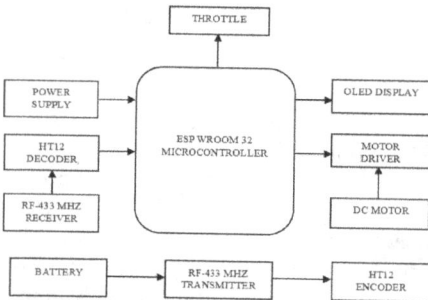

Figure 5.1 Prototype Model Block Diagram

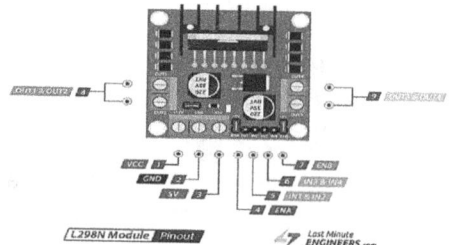

Figure 5.2 L298N Driver IC Pinout

Fig. 5.2 depicts the L298N Driver is one of the appropriate motor drivers because it has dual channeling facility which is suitable for controlling many of the DC motors or any stepper motor. Its design facilitates the independent operation of one or more motors, which makes it an ultimate choice for the construction of motor driving applications. Beyond motor control, the L298N Driver can also effectively drive different types of loads such as inductive and power transistors. If the duty cycle is higher, the average voltage will also be higher which is applied to the Dc motor and if we decrease the duty cycle, then the voltage will also be reduced which means resulting in lower speed.

4.3 ESP32 Microcontroller

We are using an ESP32 development board for the programming purpose. The model we have taken is ESP WROOM 32. This type of microcontroller has a micro-USB feature which is connected to our computer to do the programming. To perform the coding part, we are using Arduino IDE which supports the ESP32 packages and we are using C/C++ language for installing the code into the microcontroller. We generally think that ESP32 chips are easy to use, but it is not like that really, more importantly at the time of testing it and doing the programming. The ESP32 development boards which we are using is already designed with all the connections in it, after that we connect them to the computer, and we boot load them by using the software. The pins are used to connect peripherals, an antenna is provided in the board for the signaling purpose, and they have LEDs for the power and other useful things are provided in these boards. Other than this, we have different types of development boards which are coming with extra specifications added to it like the display part, sensors, and camera too.

The board we are using is also having a feature which is a Reset button installed in it and other button is used for booting the board which is called as BOOT button. We can see that many other boards do not have that button in common. The ESP32 takes the inputs from all the components and decode each problem statement and helps in running the prototype model smoothly. Fig. 5.3 indicates the ESP32 Development Board.

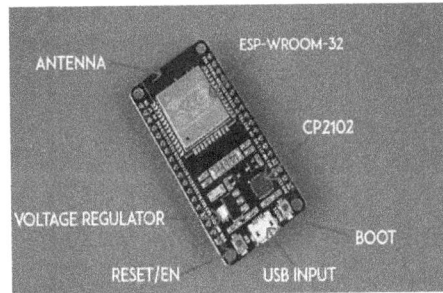

Figure 5.3 ESP32 Development Board

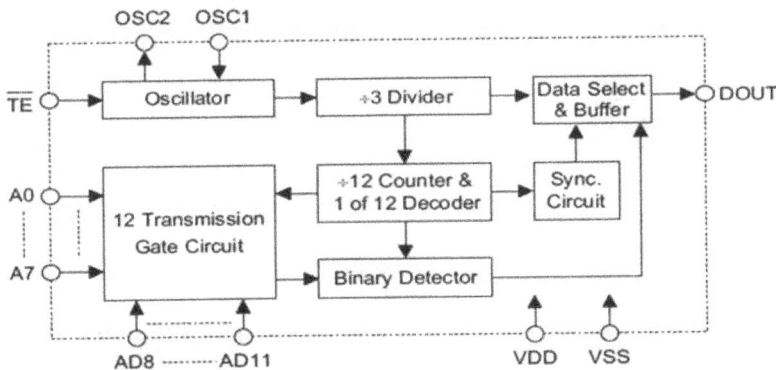

Figure 5.4 Block Diagram of Internal Structure of HT12 Encoder

4.4 Rf Transmitter Kit

Fig. 5.4 depicts the Block Diagram of Internal structure of HT12 Encoder. The RF Transmitter Kit consists of a Radio Frequency (RF) Transmitter with HT12 Encoder. We will give some speed limit switches and the kit has been given a 9V supply with the help of HW Battery. The RF 433MHz transmitter module is a compact electronic device that converts digital signals into radio waves. The transmitter module accepts digital data input from a microcontroller and then the modulated signal is then amplified and transmitted through the attached antenna.

The HT12 Encoder will use the logic states as data and address inputs and sends the signal. We give speed limits by using the switches and this kit is kept in speed limit zones. Usually, the transmitter has a range from 50m to 100m. The ranges could be increased with the help of an antenna.

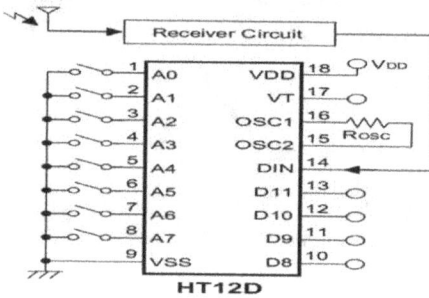

Figure 5.5 HT12 Decoder Pinout Diagram

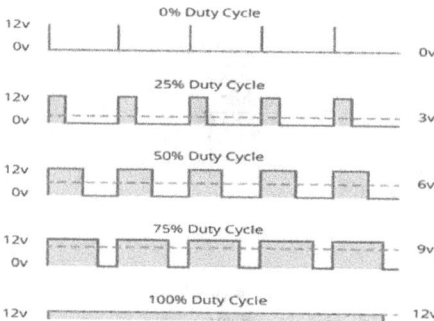

Figure 5.6 PWM Chart for Different Voltages and Duty Cycles

4.5 RF Receiver Kit

The RF Receiver Kit consists of a Radio Frequency (RF) Receiver along with HT12 Decoder. Its function is to receive the signal sent by the transmitter circuit. And it is directly interfaced with microcontroller which will check the speed. Here also, we use an RF Receiver which ranges between 50m to 200m. And the HT12 decoder IC decodes the received RF signal based on the encoding scheme used by the HT12E encoder.

It compares the received signal with its internal address and data settings to determine if the transmitted data is intended for it. Once the HT12D decoder successfully decodes the received RF signal, it outputs the decoded data in parallel format on its data output pins. This data can then be interfaced with the ESP32 for further processing. Fig. 5.5 represents the HT12 Decoder Pinout Diagram.

4.6 OLED Display

Fig. 5.6 indiactes the PWM Chart for different voltages and duty cycles. Authors are using a display for the sake of observing speed values. For that, we took a colour display which comes with a screen resolution of 128×160 pixels and has a size of 1.8 inches which is measured diagonally. The LED display communicates with its master by using SPI protocol. And the receiver module which we are using needs 3.3V supply for working, by connecting directly to a 5V system, it will not work and there is a chance of controller circuit damage. The receiver module also provides a single RGB controlling option which gives a lot better result than the Black and White displays give us usually. The touch sensing for the screen is resisted and already installed as an add-on for that. So, it has the capability to find out if anyone press the screen.

5. System Implementation

The system implementation is done by preparing a hardware prototype model. This can be done in fuel-based vehicles also but the reason to perform this on E-Vehicles is the rapid growth of them these days and in future probably. The prototype model has a Transmitter Kit and Receiver Kit. The whole speed control mechanism in this system is done with the help of Pulse Width Modulation (PWM) technique. If the duty cycle is adjusted in PWM signal, then the voltage which is given to the motor can be decreased so that the speed could also be controlled. The pulse signals generated by the ESP32 are sent to the motor driver, which interprets these signals and adjusts the power supplied to the electric motor accordingly.

Fig. 5.7 shows the Algorithmic Flow Chart of the Prototype. The RF communication module in the ESP32 allows the user to wirelessly send commands to the microcontroller, such as speed adjustments or directional changes. The microcontroller receives these commands and adjusts the PWM signals sent to the motor driver accordingly, thereby controlling the speed of the e-vehicle. There is a flowchart below explaining the whole mechanism of the prototype model.

5.1 Working Model

The working of the prototype starts with the increase of the speed by using the

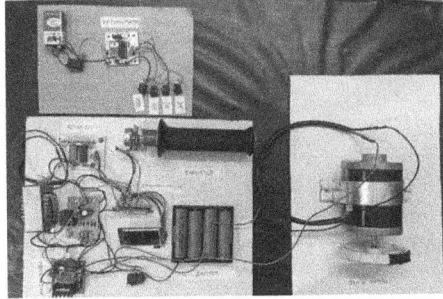

Figure 5.8 Prototype Model for Speed Control of Electric Vehicle

Figure 5.9 Speed Monitoring at the Time of Varying the throttle

Figure 5.10 Rotation of the Wheel When the Throttle is Varied

throttle which acts as a potentiometer. So, by increasing that, the motor starts to rotate and the speed will be displayed on LED Screen correspondingly. When the transmitter kit is switched on and by setting a particular speed limit, the transmitter will send a signal with the help of an encoder and the receiver gets the signal and understands it by using the decoder.

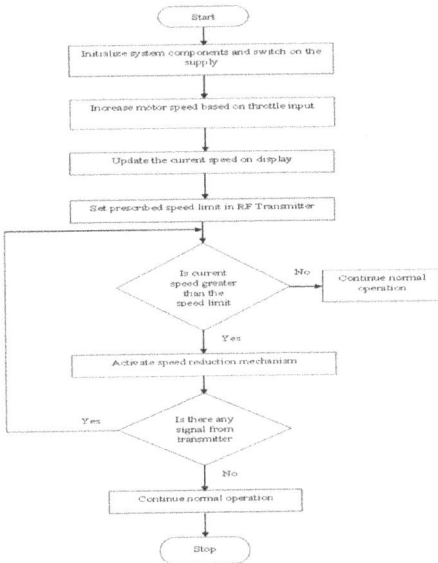

Figure 5.7 Algorithmic Flow Chart of the Prototype

Fig. 5.8 depicts the Prototype Model for Speed Control of Electric Vehicle. So now, the ESP32 understands the signal given from the transmitter and automatically reduces the vehicle's speed with the help of PWM technique. At this time, even if we disturb the throttle by rotating it there will be no use of it. Because the whole control will be in the hands of the microcontroller.

6. Result

The hardware of our working model was built by using a wooden board which consists of Power Supply, ESP32 Development Board, Receiver with its decoder, Motor Driver, LED Display, Throttle and Brushed Dc Motor on other board. And the Transmitter Kit is placed somewhere else within its permissible range. All the hardware components are connected to ESP32 Development board.

Throttle inputs are interpreted by the microcontroller, which then generates corresponding signals to adjust the motor's speed via Pulse Width Modulation (PWM). Crucially, the system enables wireless transmission of speed limit signals from the RF transmitter unit to the receiver unit installed within the electric vehicle. Fig. 5.9 shows the Speed monitoring at the time of varying the throttle.

The successful implementation of the prototype model underscores the feasibility and effectiveness of the proposed approach. Fig. 5.10 depicts the Rotation of the wheel when the throttle is varied. The RF Communication is quite suitable because of its accurate signal passing without any delay. Overall, the Electric Vehicle Speed Control System represents a significant step forward in enhancing road safety and promoting responsible driving behaviour. Since the microcontroller is already preprogrammed with required code, there is no need of running the code every time when the prototype working needs to be performed.

7. Conclusion

The implementation of the prototype model demonstrates the feasibility and it offers a vital solution for prevention of the accidents in speed limit areas. Its integration of RF communication and microcontroller technology presents a promising approach to regulating vehicle speed in speed-limited zones. With further development, this system holds significant potential to create safer roads and encourage wider adoption of electric vehicles, fostering a more sustainable future for transportation. Then the accidents growth rate will be reduced to some extent with the help of this prototype model. Along with this, ESP32 offers a better feature when compared to its equivalent devices. So, it is an appropriate choice for designing this system with expected outputs in short time.

Acknowledgment

We extend our heartfelt gratitude to our mentor Dr. P. Padmavathi, Assistant Professor in the Electrical and Electronics Department at Srinivasa Ramanujan Institute of Technology, for her unwavering encouragement and invaluable assistance in doing this work. Her guidance has complemented our work very well and we completed our task successfully in less time. We also thank Electrical and Electronics Engineering department for providing the platform and resources necessary for conducting this research work. Their support has been essential in facilitating hands-on experimentation and fostering innovative ideas. Furthermore, we express our appreciation to all individuals who have contributed to this project directly or indirectly, for their assistance, guidance, and cooperation.

References

[1] Magdici, S., & Althoff, M. (2017). Adaptive cruise control with safety guarantees for autonomous vehicles. Science Direct.

[2] Sugeno, M., & Nishida, M. (1985). Fuzzy-control of model car. *Journal of Fuzzy Sets and Systems*, 16, 103–113.

[3] Prakash, S. A, Aravind Mohan, R., Warrier, R. M., Krishna, R. A., Sooraj Bhaskar, A., & Nair, A. K. (2018). Real Time

Automatic Speed Control Unit For Vehicles. IEEE Xplore CFP18OZV-ART and ISBN 978-1-5386-1442-6.

[4] Senthilkumar, R., Dhanu Prakash, T., Kannan, A., Selvaprasath, S., & Karthikrajan, P. T. (2019). Embedded system for automatic vehicle speed control and data analysis using python software. *Journal of Control System and Control Instrumentation*, 5(2), 14–20.

[5] George, M. A., Kamat, D. V., & Kurian, C. P. (2024). Electric vehicle speed tracking control using an ANFIS-based fractional order PID controller. King Saud University *Journal—Engineering Sciences*, 36(4), 256–264.

[6] Qi, Y. (2016). Separately excited DC motor for electric vehicle controller design. 6th International Conference on Sensor Network and Computer Engineering (ICSNCE).

[7] Sharma, A. (2016). Speed controlling of electric vehicle using thyristor: A comparative analysis. *IJARSE* (ISSN: 2319–8354), 05(05).

[8] Zunjani, S. N. Shrivastava, A., & Barve, A. (2015). Speed control of DC motor using dual converter. *IJSPR*, 14(2).

6 Predicting Optimal Crops Using Environmental and Soil Data

Swarup Parua[1], Subir Gupta[2], Dhiman Debsarma[3], Soumitra Samanta[4], Sudipta Hazra[5] and Devareddy Harsha[6]

[1,2,3,4]Haldia Institute of Technology, Haldia, W.B. India
[5]Asansol Engineering College, Asansol, West Bengal, India
[6]Chaitanya Bharathi Institute of Technology (A), Hyderabad, India

Abstract

The goal of this research is to develop a predictive model using machine learning to identify the most suitable crop types based on environmental and soil variables. The research aims to address the problem of optimizing agricultural practices in the face of varying environmental conditions, which can have a significant impact on crop yields. Traditional methods of crop selection are often inefficient, relying on empirical knowledge that may not account for the full range of variables that influence agricultural success. As a result, there is a need for a more accurate, data-driven approach to predicting crop suitability. The proposed methodology involves the use of a Random Forest Classifier, a machine learning algorithm that excels in handling complex classification tasks with multiple features. This method is well-suited for agricultural data, which involves a variety of input parameters such as nitrogen, phosphorus, potassium, temperature, humidity, pH levels, and rainfall. The research demonstrated that the Random Forest Classifier effectively predicts crop types with high accuracy, achieving a performance rate of 99.58% on the test set, indicating the model's reliability in agricultural decision-making. This approach offers a promising solution to improving crop management and resource optimization in farming practices.

Keywords: Crop prediction, machine learning, Random Forest, environmental data, precision agriculture.

1. Introduction

Objective of the research is to grow a machine learning-based model capable of predicting optimal crop types based on various environmental and soil parameters [1]. This approach aims to support agricultural and farmers experts in making informed decisions about crop selection, improving agricultural productivity and resource use [2]. By leveraging machine learning techniques, the study addresses the increasing complexity of farming environments, where traditional methods often fall short in accounting for all the necessary variables, such as nitrogen, phosphorus,

Email: [1]swarupparua2001@gmail.com, [2]subir2276@gmail.com, [3]dhimandebsarma2003@gmail.com, [4]soumitrasamanta69@gmail.com, [5]sudiptahazra.nitdgp@gmail.com, [6]dharsha_eee@cbit.ac.in

DOI: 10.1201/9781003661917-6

potassium, temperature, humidity, pH, and rainfall [3].

The proposed methodology employs the Random Forest Classifier, a well-known machine learning algorithm effective in handling complex classification problems with multiple input features [4]. This method begins with collecting relevant environmental and soil data, followed by preprocessing to ensure quality and consistency. The dataset is then split into training and testing sets, with the Random Forest model trained on the former to make predictions based on the input parameters [5]. By using multiple decision trees and averaging their outcomes, the model provides robust predictions of crop types, minimizing issues such as overfitting and ensuring high accuracy, with a reported performance rate of 99.58%.. In an era where precision farming is crucial for meeting the global demand for food amidst environmental challenges, accurate crop prediction models like the one developed in this study offer a promising solution [6]. By enabling farmers to select crops suited to specific environmental conditions, the research contributes to more sustainable and efficient farming practices, which are critical for addressing the growing pressures on food security and agricultural productivity [7].

2. Literature Review

The existing body of research on crop prediction models has expanded significantly in recent years, with a focus on leveraging environmental and soil data to improve agricultural decision-making [8]. Various machine learning algorithms have been applied to predict crop yields and suitability [9]land use change and forestry (LULUCF. The traditional approaches discussed above are based only on empirical techniques and expert know-how, which, as mentioned earlier, tend to be inefficient due to the intricate and constantly changing characteristics of agricultural ecosystems. Now, the meteorological field has begun to adapt machine learning models, which have showed good performance

for assimilating multidimensional data. However, many of these studies do not present field tests and are oriented almost exclusively to one or two crops or environment sectors [10,it is vital to know what impacts the preferences for electric vehicles over conventional fuel-based cars. To address this, a discrete choice experiment is developed and integrated into a survey. An online survey was conducted in Canada with 2062 valid responses. Different labels are designed for the survey to determine the most effective GHG information framing to increase the influence of such information on decisions. In this study, the influence of lifecycle emissions is considered. Three ensemble learning techniques are applied and they are compared based on prediction accuracy, and the most accurate technique is applied to determine the relative influence of variables on the intention to buy electric vehicles. Further, the interaction of variables is investigated using xgbfir. Subsequently, Accumulated Local Effect (ALE 11]. The algorithm's capacity to integrate various features, including soil composition and climatic conditions, makes it suitable for precision agriculture. However, existing models often fall short when applied to regions with highly variable environmental conditions or less represented crops, as they typically rely on static datasets that do not capture the full spectrum of possible agricultural environments. The research gap lies in the need for a more adaptable and scalable crop prediction model that not only incorporates a wider range of environmental variables but also addresses the limitations observed in prior studies concerning underrepresented crops and varying climatic zones. This research aims to fill this gap by employing a Random Forest Classifier that integrates multiple environmental and soil features to predict optimal crop types across diverse agricultural landscapes [12]. The model's adaptability to different conditions and its high accuracy in predicting underrepresented crops make it a valuable addition to the existing literature on precision agriculture [13–15].

3. Methodology

In conducting this study, a Random Forest algorithm was employed given its ability to handle multi-classification tasks with more than one target feature. The dataset also incorporated fundamental agricultural and environmental parameters required for crop yield and management which included nitrogen, phosphorus, potassium, temperature, humidity, pH and rainfall. All these parameters have an impact on crop selection and growth. The next processes included data preprocessing where the dataset was checked against missing or inconsistent values. Because the dataset contains multiple variables with different units, normalization and scaling methods were applied in order to standardize the data. Afterward, the data was split into two subsets, with 70% reserved for building the model and the remaining 30% for testing the model Built. The decision trees were then constructed by the Random forest algorithm using different subsets of features from the dataset. During the construction of the tree at each decision node, the feature which best separated differential crop classes was employed. The combination of all decision trees' predictions determined the final classification, with the class that was predicted most often being selected. This approach reduces the likelihood of overfitting and improves the overall robustness and accuracy of the crop prediction model. The splitting criterion used was the Gini impurity measure, given by equation 1

$$Gini(D) = 1 - \sum_{i=1}^{n} q_i^2 \qquad (1)$$

Where qi represents the probability of class i in dataset D. The feature with the lowest Gini impurity is selected at each node for splitting. After constructing the decision trees, the Random Forest model combines their predictions. The final crop classification is determined by majority vote, minimizing overfitting. This ensemble method enhances accuracy, addressing issues

commonly faced by individual decision trees.

$$RF(x) = \text{argmax}_c \sum_{t=1}^{T} I(Z_t(x) = c) \qquad (2)$$

Where $RF(x)$ is the final prediction for input x, $z\_t(x)$ is the prediction from the tth decision tree, c represents the class, and I is an indicator function that equals 1 if the prediction matches the class and 0 otherwise. The final crop type is the class that receives the majority of votes across all decision trees. After training, the model's accuracy was assessed using the test dataset, calculated according to the formula in equation 3.

$$Accuracy = \frac{TP + TN}{TP + TN + FP + FN} \qquad (3)$$

Where: TP->True Positives, TN->True Negatives, FP ->False Positives, and FN -> False Negatives represents the cases incorrectly predicted as negative. Besides accuracy, precision, recall, and F1-score were used to evaluate model performance, as shown in equations 4, 5, and 6.

$$precision = \frac{TP}{TP + FP} \qquad (4)$$

$$Recall = \frac{TP}{TP + Fn} \qquad (5)$$

$$F1 = 2 \times \frac{Precision \times Recall}{Precision + Recall} \qquad (6)$$

These metrics provided a comprehensive evaluation of the model's performance, showing that it was highly effective in predicting crop types based on the given set of features. The F1-scores were close to 1.00 for most crop types, indicating near-perfect performance in balancing precision and recall. One challenge noted in the methodology was the slight deviation in performance for crops labeled as class 6 and class 7. These crops had lower precision and recall scores compared to other crop types. This issue may have been caused by a smaller sample size for these particular classes or

Table 6.1 Pseudo Code

Step	Pseudocode Description	Mathematical Expression
1	Initialize dataset D and define features	$D = \{(x_1, y_1),(x_2, y_2),...,(x_n, y_n)\}$
2	Split dataset into training and test sets	$D_{train} = 0.7D, D_{test} = 0.3D$
3	Select Random Forest parameters (number of trees T)	$T = \text{number of trees}$
4	For each tree, randomly select subset of features	$F_{sub} = \text{RandomSubset}(F)$
5	Build decision tree by minimizing Gini impurity	$\text{Gini}(D) = 1 - \sum_{i=1}^{n} p_i^2$
6	Aggregate predictions across all decision trees	$H(x) = \text{argmax}_c \sum_{t=1}^{T} I(h_t(x) = c)$
7	Calculate accuracy on test data	$\text{Accuracy} = \dfrac{TP + TN}{TP + TN + FP + FN}$
8	Return final classification results	$Output = H(x)$

a higher degree of variability in their features, which made them harder to classify accurately. In the upcoming species, in order to rectify the current issue, either more feature selection techniques would need to be devised to gather better data for the more underrepresented crops of choice or a more robust feature selection technique may need to be set in order to gather additional data. All in all, the approach taken in this study in terms of utilizing machine learning computational methods including Random Forests in the context of agriculture is evident. By effectively utilizing a broad array of soil features as well as other similar factors, the model was predicting crop types with a great rate of success. The algorithm of the Random Forests, which encompasses several distinct elements such as Gini impurity, and ensemble of multiplicity of decision trees performed amply on a problem of such kind, which was multiclass classification. Although some elements presented hurdles, the crop included in the challenge offered an effective methodology for augmenting agricultural processes that will result in a proficient model for decision making and managing resources. The pseudocode is presented in Table 6.1.

4. Results

This method employed a Random Forest Classifier to estimate the generative types of the crops based on the environmental and soil properties with the dataset comprising of other key variables like nitrogen, phosphorus, potassium, temperature, humidity, Ph levels, and precipitation data which together enhance the classification process of crops. The data was partitioned in the ratio of 70% to 30% between training and testing set respectively, which is the most ideal for machine learning applications in sustaining validation and evaluation of the model. This accuracy of the model is quite high and it is able to achieve 99.58% accuracy ideally classifying and predicting types of crops with their environmental, soil variables, and other characteristics.

The metrics of precision, recall, and the F1-score for identifying the types of crops correct were almost perfect with 1.00 score in most of the crops' set test individual scores indicating reliability and effectiveness of the model for its multi-class classification tasks. However, in this study, it has been observed that a range of factors can affect the performance of these crops and the related tasks which explains evaluation of model involving multi-smooth layers, specifically the layer denoted by crop 6 and crop 7. In general the Random Forest Classifier performed quite well, but the metrics of precision, recall and F1-scores related to these internal crops were slightly less than the rest. The underlying cause for most of the crops set could be the environmental and soil variables which in case of these crops perhaps were not distinct and therefore did not have the same feature.

There may be less reliable predictions when there is a small sample size as trained models have fewer examples to learn from. This, however, can be attributed to the shrinking of sample sizes of certain crops in a dataset. Notwithstanding a few deviations, the performance of random forest classifiers remained satisfactory since the classifier can handle intricate relationships amongst multiple inputs. This also goes on to show why random forest classifier is most suited towards agricultural classification tasks as it offers high accuracy while factoring in multiple crops. Minor refinements of the dataset may be needed to tackle the specific performance shortcomings of crop 6 and crop 7 gotten from the model due to model not fully capturing the crop's features. One could make up for this by increasing the sample of the crops further needed or additionally deepening the processes of feature selection. It can also be predicted with certainty that a random forest classifier developed using environmental and soil data with high precision, accuracy and F1 scores is quite applicable to use in agriculture. Despite the presence of some performance-related weaknesses in certain crops, my understanding leads me to believe that this can be resolved with further data

optimization or with improvement of the model. Overall, the research substantiates the claim that machine learning algorithms such as Random Forest Classifier can accurately predict, and therefore help improve the quality of agricultural decision-making on the basis of complex information.

In supposition, the Random Forest Classifier proved to be an excellent choice for predicting crop types based on environmental and soil data. Its high accuracy, coupled with strong precision, recall, and F1-scores for most crop classes, demonstrates its effectiveness in agricultural applications.

Although slight performance issues were noted for certain crops, these could likely be addressed through further data optimization or model refinement. Overall, the study confirms that machine learning algorithms like the Random Forest Classifier can play a crucial role in enhancing agricultural decision-making by providing accurate predictions based on complex environmental data. The results suggest significant potential for using such models to improve crop management and optimize resource use in farming practices. Figures 6.1, 6.2, and 6.3 respectively illustrate the Confusion Matrix Figure 6.1, Feature Importance in the Random Forest model Figure 6.2, and Prediction Accuracy by crop type Figure 6.3.

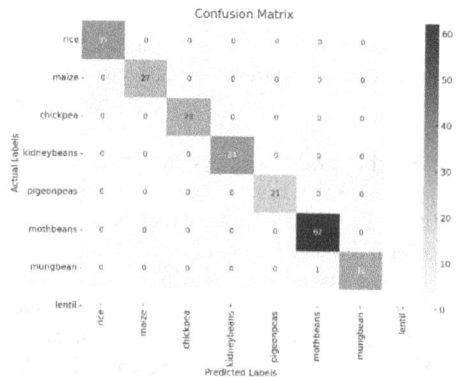

Figure 6.1 Confusion Matrix

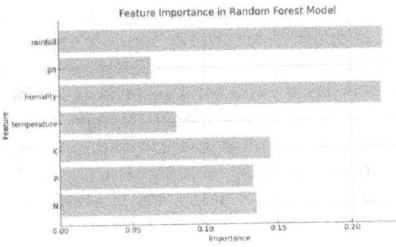

Figure 6.2 Feature Importance in RF model

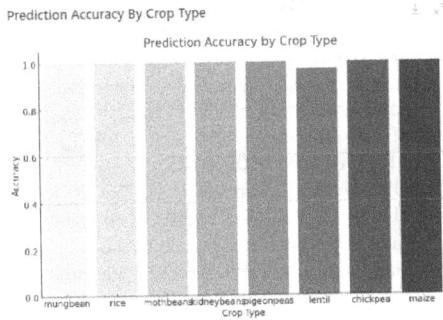

Figure 6.3 Prediction Accuracy by crop type

5. Conclusion

This study aimed to create a machine learning model that can accurately identify the specific type of crop that needs to be planted by using a variety of soil and environmental data in order to strengthen informed decisions in the agricultural sector. It has been established that in the regions characterized by a wide range of environmental settings, crop selection is at most times based on knowledge and experience, which is not optimal. This study aimed to overcome this challenge by developing a Random Forest Classifier which is explained to perform well given high dimensional data. In this instance, sufficient agricultural data such as nitrogen, phosphorus, potassium, temperature and humidity, pH and rainfall were used to select the suitable crops. The accuracy of the results achieved will also be discussed with emphasis being on the high predictive power of the model, which was able to achieve a test performance of 99.58%, thereby giving credence to this model. The investigative component of this research confirmed that machine learning models like the Random Forest Classifier are effective and practically applicable in agriculture. The model was successful in combining different factors within one equation and achieved decent precision accompanied by high recall and F1-scores for most crops. Only certain crops showed slight discrepancies in their performance which might be due to variations in features or small sizes of the dataset.

This stresses the necessity of future research aimed at enhancing the dataset and model in order to respond to these particular issues. Certainly, this study presents some initiatives that can be pursued in the future. Creating larger and more representative samples in terms of factors such as growth environments and crop types along with a better balance in the dataset could lead to enhanced model performance. Also, such techniques would improve the model's performance in learning the non-linear relationships between the environmental factors and the agricultural output. On the other hand, the use of other machine learning algorithms and hybrid models in subsequent work would also be useful in comparing performance and improving the results. Such advances would also improve the level of accuracy in prediction and enable farmers to get more informative guidance on achieving optimal yields. The social impact of this research is important. The model could help use resources better, increase food production and decrease crop failure by allowing farmers and agricultural policymakers to use a strong, data-oriented model for crop prediction. Where farmers are largely reliant on the weather, knowing how to use the right crop in such situations would encourage proper farming practices which are useful in solving food problems brought about by the changing weather or the increasing population. Global warming and food insecurity have become the most pressing issues of our time and this work will, for the sake of humanity, allow the modernization of agricultural methods, all while guaranteeing profitability and sustainability."

References

[1] Gomez Selvaraj, M., et al. (2020). Detection of banana plants and their major diseases through aerial images and machine learning methods: A case study in DR Congo and Republic of Benin. *ISPRS J. Photogramm. Remote Sens.*, 169(April), 110–124. doi: 10.1016/j.isprsjprs.2020.08.025.

[2] Ngugi, L. C., Abdelwahab, M., & Abo-Zahhad, M. (2021). A new approach to learning and recognizing leaf diseases from individual lesions using convolutional neural networks. *Inf. Process. Agric.*, xxxx. doi: 10.1016/j.inpa.2021.10.004.

[3] Cheng, X., Peng, J., Dong, J., Liu, Y., & Wang, Y. (2022). Non-linear effects of meteorological variables on cooling efficiency of African urban trees. *Environ. Int.*, 169(5), 107489. doi: 10.1016/j.envint.2022.107489.

[4] Srivastava, R., & Yadav, B. C. (2012). Ferrite materials: Introduction, synthesis techniques, and applications as sensors. *Int. J. Green Nanotechnol. Biomed.*, 4(2), 141–154. doi: 10.1080/19430892.2012.676918.

[5] San Emeterio de la Parte, M., Martínez-Ortega, J. F., Castillejo, P., & Lucas-Martínez, N. (2023). Spatio-temporal semantic data management systems for IoT in agriculture 5.0: Challenges and future directions. *Internet of Things (Netherlands)*, 25(November). doi: 10.1016/j.iot.2023.101030.

[6] True, J., Bassiouny, R., Razfar, N., Venkatesh, V., & Kashef, R. (2022). Weed detection in soybean crops using custom lightweight deep learning models. *J. Agric. Food Res.*, 8(April), 100308. doi: 10.1016/j.jafr.2022.100308.

[7] Federgruen, A., Lall, U., & Serdar Şimşek, A. (2019). Supply chain analysis of contract farming. *Manuf. Serv. Oper. Manag.*, 21(2), 361–378. doi: 10.1287/msom.2018.0735.

[8] Anim-Ayeko, A. O., Schillaci, C., & Lipani, A. (2023). Automatic blight disease detection in potato (Solanum tuberosum L.) and tomato (Solanum lycopersicum, L. 1753) plants using deep learning. *Smart Agric. Technol.*, 4(November 2022), 100178. doi: 10.1016/j.atech.2023.100178.

[9] Garcia-Quijano, J. F., Deckmyn, G., Moons, E., Proost, S., Ceulemans, R., & Muys, B. (2005). An integrated decision support framework for the prediction and evaluation of efficiency, environmental impact and total social cost of domestic and international forestry projects for greenhouse gas mitigation: Description and case studies. *For. Ecol. Manage.*, 207(1–2 SPEC. ISS.), 245–262. doi: 10.1016/j.foreco.2004.10.030.

[10] Naseri, H., Waygood, E. O. D., Patterson, Z., & Wang, B. (2024). Who is more likely to buy electric vehicles?. *Transp. Policy*, 155(May), 15–28. doi: 10.1016/j.tranpol.2024.06.013.

[11] Piryonesi, S. M., & El-Diraby, T. E. (2020). Role of data analytics in infrastructure asset management: overcoming data size and quality problems. *J. Transp. Eng. Part B Pavements*, 146(2), 04020022. doi: 10.1061/jpeodx.0000175.

[12] Handhika, T., Fahrurozi, A., Zen, R. I. M., Lestari, D. P., & Sari, I. (2019). Modified average of the base-level models in the hill-climbing bagged ensemble selection algorithm for credit scoring. *Procedia Comput. Sci.*, 157, 229–237. doi: 10.1016/j.procs.2019.08.162.

[13] Bhattacherjee, N. K., Samanta, S., Roy, B., Mukherjee, S., Lahiri, G., & Gupta, S. (2023). Analytical review of the trends of microfinance sector in India on lending assets. In 2023 3rd International Conference on Innovative Sustainable Computational Technologies (CISCT), 1–6. doi: 10.1109/CISCT57197.2023.10351406.

[14] Hazra, S., Mondal, A., Dey, P., Prabhakar, S., Jha, A. K., & Rakshit, N. (2024). Future prospects of agriculture using IoT and machine learning. 2024 IEEE International Conference on Smart Power Control and Renewable Energy (ICSPCRE), Rourkela, India, 1–6. doi: 10.1109/ICSPCRE62303.2024.10674786.

[15] Hazra, S., Ghosal, S., Mondal, A., & Dey, P. (2024). Forecasting of rainfall in subdivisions of India using machine learning. In: Bhattacharyya, S., Das, G., De, S., & Mrsic, L. (eds), Recent Trends in Intelligence Enabled Research. DoSIER 2023. Advances in Intelligent Systems and Computing, 1457. Springer, Singapore. https://doi.org/10.1007/978-981-97-2321-8_18

7 An Improved Neutral Point Clamped Heric Based Transformerless PV Inverter Topology

N. Santosh Kumar[1], Ahmad Syed[2], P. Hemeshwar Chary[3], Syed Sarfaraz Nawaz[4], Abu Tayab[5] and Zia Ur Rehman[6]

[1,3]Chaitanya Bharathi Institute of Technology (A), India
[2,5,6]Postdoctoral Researcher, Yanshan University, China
[4]Gokaraju Rangaraju Institute of Engineering & Technology, Hyderabad, India

Abstract

Neutral-point-clamped (NPC) topologies have recently gained considerable attention in transformerless photovoltaic (PV) inverter applications due to their advantages in reducing weight, minimizing size, enhancing efficiency, and improving reliability. Numerous studies in the literature have focused on the leakage current elimination with constant common mode voltage (CMV) throughout the entire grid cycle. Many of these topologies utilize bidirectional switches at the middle of the DC capacitors point. In the proposed approach, an improved clamping section replaces these high-frequency bidirectional switches with a simple diode bridge rectifier in the proposed H-NPC topology based on the Heric structure. The accuracy of the proposed circuit has been thoroughly analysed through simulation platforms. A proposed H-NPC is compared with conventional topologies was conducted, focusing on common-mode performance, leakage current, device count, and %THD. The results validate the proposed topology's viability for practical grid-connected systems.

Keywords: Transformerless NPC inverter, photovoltaic system, common mode voltage, leakage current, %THD.

1. Introduction

The rapid progress in renewable energy utilization has significantly impacted various applications, including power electronics systems and distributed power smart grid systems, due to its eco-friendly nature and the availability of unlimited power year-round. In photovoltaic (PV) systems, a power conversion component, specifically an inverter, is essential for converting solar energy into usable electricity and connecting the PV panel to the grid. These inverters are classified into two types: isolated (with transformer) and non-isolated (without transformer) systems [1]. In non-isolated grid systems, the transformer is omitted, making the system

Email: [1]santoshkumar_eee@cbit.ac.in, [2]asd@ysu.edu.cn, [3]hemeshwar_eee@cbit.ac.in
[4]sarfaraz86nawaz@gmail.com [5]abtayababu2591@gmail.com, [6]ziaurrahman4h@gmail.com

DOI: 10.1201/9781003661917-7

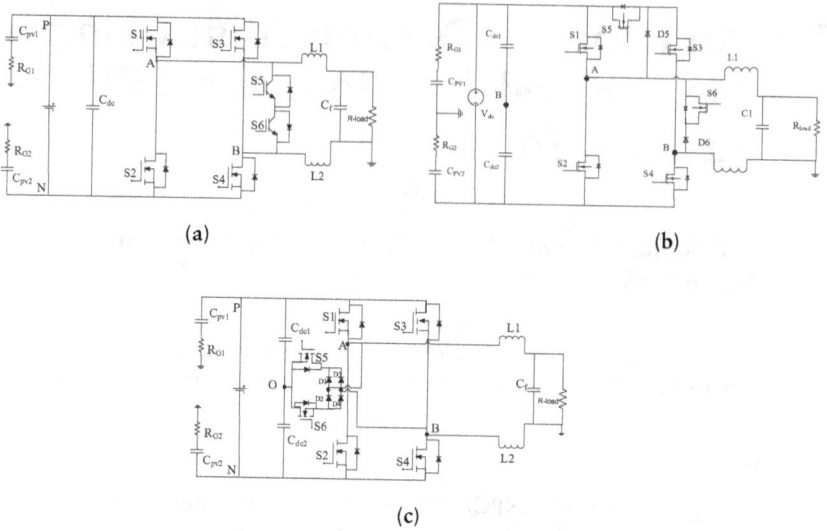

Figure 7.1 Several traditional Heric based topologies (a) Heric, (b) Heric1, (c) HBZVSCR

more cost-effective, compact, and efficient. Transformerless inverters are commonly used in distributed power applications due to their advantages, such as reduced size, weight, and improved efficiency compared to systems with transformers. However, the absence of a transformer introduces critical challenges, particularly concerning galvanic isolation. This issue arises between the PV module's surface metal and the ground stray capacitance, potentially leading to dangerous abnormal currents, also known as leakage currents. To ensure safety and prevent current distortion and electromagnetic interference [3], it is essential to control and limit leakage currents. As a result, the design of transformerless PV systems must adhere to safety standards set by regulatory bodies such as IEEE 1547.1, VDE0126-1-1, IEC61727, and AS/NZS5033 (DIN VDE, 2005) to address these issues and protect both equipment and personnel (Mathi et al., 2018).

Based on the critical need to eliminate leakage currents in transformerless PV systems, various configurations have been proposed in the literature using techniques such as DC-isolated, AC-isolated, and neutral-point-clamped (NPC) PV systems (Kumar, K.S, 2020, Li, H.; Zeng 2018, Y. Wei 2022). Basic H5 and HERIC inverter topologies were introduced for PV applications by [7] and [8]. Although several HERIC-based topologies have emerged recently, many still suffer from high switching counts and are limited in their range of applications. For instance, Yang et al. (2018) introduced HB-ZVR converter with a rectifier bridge, but it exhibited poor CMV performance, resulting in higher leakage currents.

More recently, improved HBZVR topologies have been introduced, including HBZVR-D, E-Heric, and HBZVR-S, which incorporate active and passive clamping methods, as shown in Figure 7.1. These topologies demonstrate better CM characteristics and reduced leakage currents, but they are not reliable for high-load applications. To address these limitations, this paper proposes an improved HERIC-based NPC topology with a diode rectifier bridge, termed H-NPC. This configuration uses only four diodes and one switch, replacing the two bidirectional switches typically found at the center-point of the

DC-link capacitor. The proposed topology has been validated through simulations to ensure accuracy and practicality for real-world applications. A thorough comparison between the conventional HERIC topology and the proposed H-NPC is provided, focusing on key metrics such as CMV behavior, leakage flow current, total devices, and total harmonic analysis (%THA) as well.

This article is organized into five sections. Section 2 discusses the modes of operation of the H-NPC topology. Simulation studies are given in section 3 and summary and key findings are highlighted in Section 4. Finally, the references are listed in Section 5.

2. Proposed Novel Circuit

In this article a novel improved heric based transformerless NPC inverter is proposed with very low leakage current and stable common mode voltage (CMV). Figure 7.1, shows the inverter topology, The proposed topology replacing the traditional two bidirectional switches with

a single power switch and four diodes only namely H-NPC. A diode bridge rectifier is connected at the ground N and phases 'A' terminal. The H-NPC topology merges the advantages of Heric technology with low loss and the NPC scheme of near zero leakage current circuit and

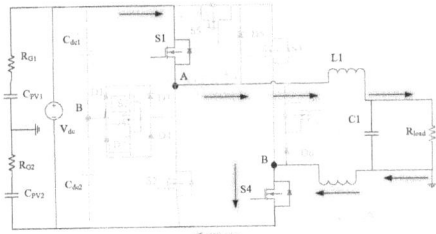

(a)

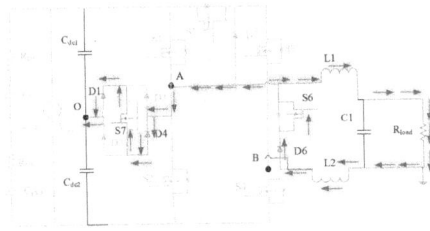

(b)

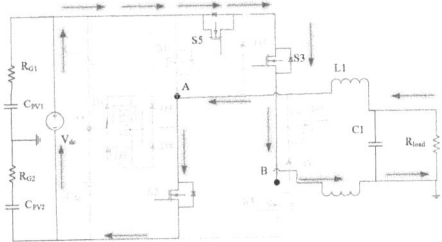

(c)

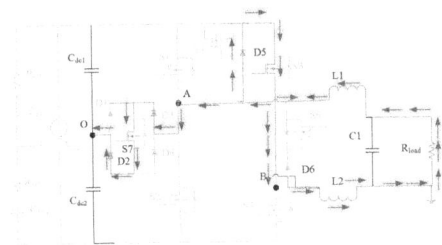

(d)

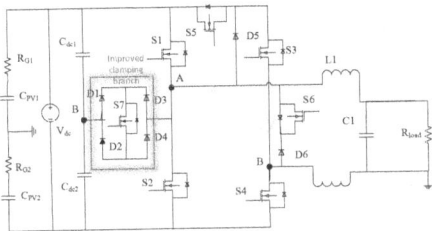

(a)

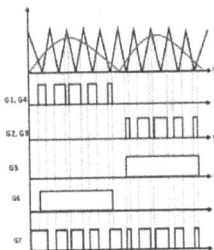

(b)

Figure 7.1 Proposed Circuit (a) H-NPC, (b) Control scheme

Figure 7.2 H-NPC Operating periods (a) Active-Stage 1, (b) Freewheel-Stage 2, (c) Active-Stage 3, (d) Freewheel-Stage 4

its corresponding modulation signals are shown in Figure 7.2(b).

Here proposed H-NPC is designed with unity power factor mode only. Switches S1, S2, S3, S4, and S7 function at elevated switching frequencies, while S5 and S6 operate in synchronization with the grid frequency power factor, which eliminates the need for body diodes during all operational intervals.

2.1 Operating Stages

Active Stage1: During the positive switch cycle, S1 and S4 are enabled, and rests are stay off. In this phase, the current flows from S1, passes through the RL and S4, as illustrated in Figure 7.2(a). The CM and differential-mode (DM) voltages are given by the following equations:

$$V_{cm} = \frac{V_{AN} + V_{BN}}{2} = \frac{1}{2}\left(V_{dc} + 0\right) = \frac{V_{dc}}{2} \quad (1)$$

$$V_{AB} = V_{AN} - V_{BN} = (V_{dc} - 0) = V_{dc} \quad (2)$$

In State 2: Positive Half Freewheeling interval (PHFI), switch S6 and diode D6 are actively ON, while the rest of device tubes are OFF. During this period, the I_L begins to decrease and freewheels through the clamping diodes D1 and D4, as well as the active switch S7. The common-mode and differential mode (DM) behaviour is computed using the following equations:

$$V_{cm} = \frac{V_{dc} + V_{dc}}{2} = \frac{1}{2}\left(\frac{V_{dc}}{2} + \frac{V_{dc}}{2}\right) = \frac{V_{dc}}{2} \quad (3)$$

$$V_{Dm} = \left(\frac{V_{dc}}{2} - \frac{V_{dc}}{2}\right) = 0 \quad (4)$$

In Stage 3: In the negative half-cycle, switches S2, S3, and S5 are enabled, while all other switches remain disabled. During this interval, the current flows through S2, passes through the R_L load, and exits via

S3, as depicted in Figure 7.2(c). The common-mode and DM can be calculated using the following equations:

$$V_{cm} = \frac{V_{AN} + V_{BN}}{2} = \frac{1}{2}\left(0 + V_{dc}\right) = \frac{V_{dc}}{2} \quad (5)$$

$$V_{DM} = 0 - V_{dc} = -V_{dc} \quad (6)$$

In Stage 4: Negative Half Freewheeling Interval (NHFI), switch S5 and diode D5 are actively ON, while the rest are disabled. During this interval, the current begins to increase and freewheels through the clamping diodes D2 and D3, along with the active switch S7. The CM and DM is computed using the following equations:

$$V_{cm} = \frac{V_{AN} + V_{BN}}{2} = \frac{V_{dc}}{2} + \frac{V_{dc}}{2} = \frac{V_{dc}}{2} \quad (7)$$

$$V_{DM} = \frac{V_{dc}}{2} - \frac{V_{dc}}{2} = 0 \quad (8)$$

From the Equations (1)-(8) it is clearly understood that in whole operating stages CMV is equal to $0.5 V_{dc}$ and results leakage flow suppression is thoroughly eliminated.

3. Simulation Studies

This section evaluates various Heric-based conventional topologies, including the proposed H-NPC, through simulation studies. The simulation parameters are considered as Input Dc Voltage is 400V, parasitic capacitors (C_{PV1}, C_{PV2}) are 100Nf, Ground resistance (R_{G1}, R_{G2}), 11ohms, Filter inductor is 3mH, switch frequency is 10kHz and load resistance is 50Ω listed in Table 7.1 [3]. From the observations in Figure 7.3, it is evident that the unipolar output voltage varies between -400V, 0, and 400V.

Similarly, in Figure 7.3(a)-(b) shows that all output load current waveforms are sinusoidal. The CMV behavior for both the

Table 7.1 Key summary and comparison of discussed topologies

Factors	Voltage Form	Over-all switches	Over-all diodes	CMV	LC	%Harmonics
Heric1	Unipolar	Six	2	~280V	>30mA	23.62
H-NPC	Unipolar	seven	7	200V	<30mA	0.29

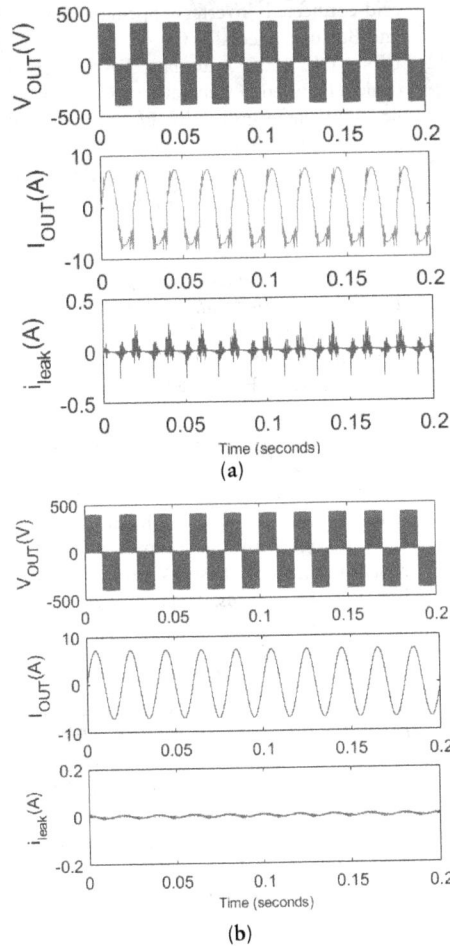

Figure 7.3 The Simulation waveforms of V_{out}, i_{out} and i_{leak} for (a) Heric1 (b) H-NPC

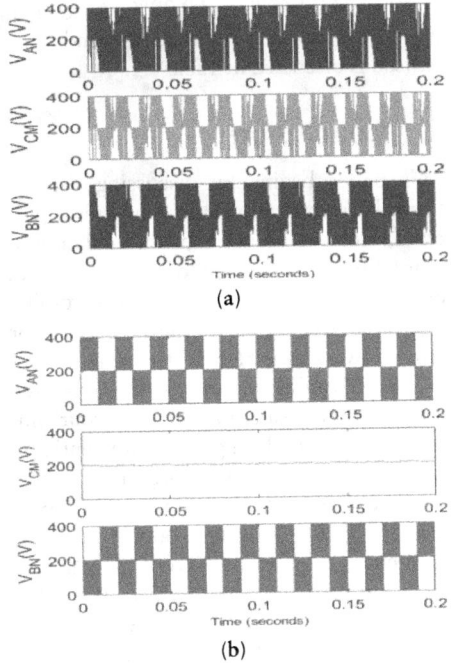

Figure 7.4 The simulation waveforms of V_{AN}, V_{CM} and V_{BN} for (a) Heric1 (b) H-NPC

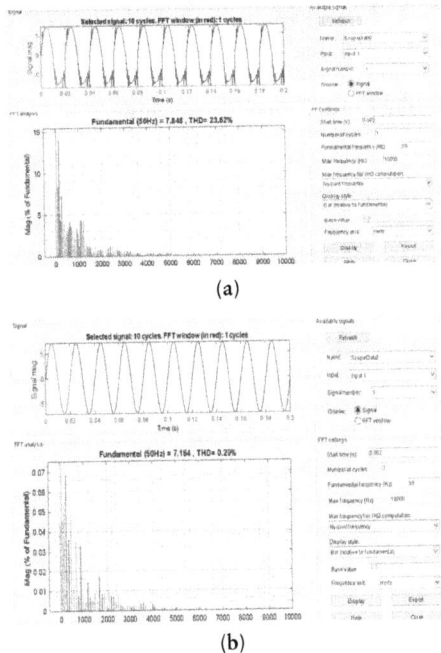

Figure 7.5 Harmonics %THD, (a) Heric 1, (b) H-NPC.

conventional and proposed H-NPC topologies is illustrated in Figure 7.4. The proposed H-NPC shows better performance compared to the conventional Heric topology.

In the conventional topology, slight oscillations are observed in V_{AN} and V_{BN}, with a magnitude of 250V, which results in floating CMV and higher leakage current, as seen in Figure 7.4(a). In contrast, the proposed H-NPC topology eliminates such oscillations due to the improved clamping branch, resulting in a constant CMV of 200V in the whole grid intervals. Hence, leakage current is significantly reduced and meets the VDE0126-1-1 standards. This significantly impacts the total harmonic distortion (%THD), resulting in values of 23.62% and 0.29% as shown in Fig. 7.5. A fair summary analysis and comparison is discussed in Table 7.1.

4. Conclusion

This paper begins by discussing various Heric-based transformerless inverter topologies. Building on the conventional circuit, a modified Heric topology is proposed, incorporating a diode rectifier clamping section at the center of the DC capacitor link known as the H-NPC topology.

The clamping section of the proposed topology uses one switch and four linear diodes, replacing the conventional two high-frequency switches. This design enhances the CMV behavior. The proposed H-NPC topology, along with its theoretical analysis, is verified through simulation studies. The results confirm that the H-NPC topology offers superior performance compared to conventional Heric1. A comprehensive comparison is presented and highlighting key aspects such as CMV behavior, leakage flow current, semiconductor power device count, and total harmonics in current ripples distortion (TCRHD). These findings indicate that new H-NPC circuit is extremely recommended for practical PV grid connected systems.

References

[1] Huafeng Xiao, G., & Wang, X. (2021). Transformerless Photovoltaic grid-connected inverters. Springer Science and Business Media LLC.

[2] DIN VDE, 2005, Automatic Disconnection device between a generator and the public low-voltage grid. DIN Electro technical Standard DIN VDE 0126–1–1.

[3] Khan, M. N. H., Forouzesh, M., Siwakoti, Y. P., Li, L., Kerekes, T., & Blaabjerg, F. (2020). Transformerless inverter topologies for single-phase photovoltaic systems: A comparative review. *IEEE J. Emerg. Sel. Topics Power Electron.*, 8(1), 805–835.

[4] Kumar, K. S., Kirubakaran, A., & Subrahmanyam, N. (2020). Bidirectional clamping-based H5, HERIC, and H6 transformerless inverter topologies with reactive power capability. *IEEE Trans. Ind. Appl.*, 56, 5119–5128.

[5] Li, H., Zeng, Y., Zhang, B., Zheng, Q., Hao, R., & Yang, Z. (2018). An improved H5 topology with low common mode current for transformerless PV grid connected inverter. *IEEE Trans. Power Electron.*, 34, 1254–1256.

[6] Wei, Y., Guo, X., Zhang, Z., Wang, L., & Guerrero, J. M. (2022). A high-voltage gain transformerless grid-connected inverter. In *Proc. 4th Int. Conf. Smart Power Internet Energy Syst.*, pp. 533–536.

[7] Schmidt, D., Siedle, D., & Ketterer, J. (2003). Inverter for transforming a DC voltage into an AC current or an AC voltage. *EP Patent*, 1, 369–985.

[8] Victor, M., Greizer, F., Bremicker, S., & Hubler, U. (2008). Method of converting a direct current voltage from a source of direct current voltage, more specifically from a photovoltaic source of direct current voltage, into an alternating current voltage. *U.S. Patent*, 7, 411–802.

8 Optimizing Energy Efficiency in Autonomous Vehicle Platoons Using Deep Reinforcement Learning

Abu Tayab[1], Yanwen Li[1], Ahmad Syed[2] and Zia Ur Rehman[3]

[1,2,3]Yanshan University, China

Abstract

This article suggested a Deep Reinforcement Learning (DRL)-based control strategy aimed at optimizing energy consumption in autonomous vehicle platoons while ensuring safety and stability. Utilizing a multi-agent approach, each vehicle in the platoon dynamically adjusts its throttle and braking based on real-time interactions and environmental changes. The suggested method leverages a Markov Decision Process (MDP) framework, allowing the DRL agent to minimize energy consumption by optimizing control inputs in both urban traffic and highway driving scenarios. Simulation results demonstrate significant reductions in energy consumption up to 20% compared to traditional control methods while maintaining safe inter-vehicle distances and smooth platoon dynamics. These outcomes highlight the possible of DRL as a scalable and efficient solution for energy optimization in autonomous driving, with broad implications for future intelligent transportation systems.

Keywords: Energy consumption, DRL, autonomous connected vehicles

1. Introduction

The development of autonomous vehicles (AVs) and vehicle platooning technology has introduced new opportunities to improve traffic efficiency, safety, and environmental sustainability. In particular, vehicle platooning, where multiple vehicles drive closely together with synchronized control, has shown great potential to reduce aerodynamic drag and improve fuel efficiency. However, traditional platoon control strategies focus more on safety and velocity tracking, often neglecting energy optimization. Transportation is a significant contributor to energy use and greenhouse gas radiations. In Europe, it accounts for 33% of energy consumption and 23% of overall emissions [1]. Highway transportation accounts for 72.8% of total greenhouse gas radiations and 73.4% of the energy consumption in the transportation sector. Consequently, decreasing fuel consumption related to transportation is an essential concern. As vehicle-to-vehicle (V2V) communication technology advances, vehicle platooning is set to become the primary mode of driving on both urban and highway roads [2]. A vehicle platoon consists of a group of vehicles that move in unison, guided by a

Email: ywl@ysu.edu.cn[1], asd@ysu.edu.cn[2], ziaurrahman4h@gmail.com[3]

DOI: 10.1201/9781003661917-8

particular set of information. Many studies to date have utilized optimal control theory, such as model predictive control (MPC), to enhance the energy efficacy of these platoons. This approach was based on a cost function aimed at minimizing fuel consumption, and it employed heuristic algorithms to determine the best actions for each vehicle [3]. Nevertheless, autonomous connected vehicles size results in a high-dimensional key space, which substantially raises computational complication and limits the practicality of real-life claims. In recent years, deep reinforcement learning (DRL) methods have been utilized in various microscopic traffic situations, such as with autonomous vehicles, and have gained considerable interest due to their impressive performance [4]. As far as we are aware, only a limited number of DRL algorithms have been utilized for controlling vehicle platoons. Effective platoon control requires determining the accelerations (or velocities) of all vehicles involved. Using a single-agent RL approach to derive the collective actions for all vehicles leads to a significant increase in the numeral of movements and situations as the size of the platoon grows, rendering the difficult unmanageable. Therefore, centralized RL methods are not viable for controlling vehicles platoons due to the excessively high dimensionality of the joint action space [5].

In the past, many multiagent RL algorithms have been developed to report the vehicles action dilemma. Some early studies employed an independent learner approach to maximize the team reward [6]. However, from the standpoint of an individual agent, the situation is only moderately visible, meaning causes might obtain misleading incentive signals that stem from the behaviors of their associates, which they cannot observe. From the viewpoint of a particular agent, the situation is only moderately visible, which means agents may obtain misleading reward signals stemming from the behavior of their teammates that they cannot observe. Independent RL often struggles due to the so-called nonstationary problem [7]. Subsequently, some scholars utilized N parallel actors guided by a centralized critic to manage N agents. This integrated preparation with decentralized execution framework allows for reinforcement learning in a multiagent setting [8]. Nevertheless, according to research, training a fully centralized critic becomes unmanageable when the number of agents exceeds a few [9]. To enhance the effectiveness of multiagent RL, several communication protocols have been suggested to boost cooperation among agents [10]. Through communication, collaborating agents become aware of each other's actions. However, in scenarios involving a large amount of agents, disseminating information from all agents may not be beneficial, as it can be challenging for an agent to distinguish valuable information from the rest of the shared data [5]. Moreover, this approach often requires substantial computational resources, particularly when dealing with complex neural networks and various types of information.

This paper addresses this gap by introducing a DRL-based control strategy that dynamically minimizes energy consumption while maintaining safe inter-vehicle distances and platoon stability. Unlike traditional control strategies, DRL can learn from dynamic traffic environments, adapting to changes in traffic flow, road conditions, and vehicle interactions in real time. The basic objectives of this research are:

1. To improve a DRL-based control strategy for autonomous vehicle platoons.
2. To optimize energy consumption by minimizing throttle and braking usage while ensuring safety and stability.
3. To evaluate the proposed method in both urban and highway driving conditions.

2. Problem Formulation

The problem addressed in this research is the optimization of energy consumption in a platoon of N autonomous vehicles traveling in a single lane. Each vehicle in the platoon must dynamically adjust its throttle and braking inputs to minimize energy usage while maintaining safety and platoon stability. The vehicles are connected through a V2V communication network, allowing them to exchange critical information such as position, speed, and control actions.

Each vehicle i in the platoon has a position $x_i(t)$ velocity $v_i(t)$, and control inputs $u_i(t)$ (throttle and braking) at time t. The total energy consumption E_{total} for the platoon is given by the sum of the energy used by all vehicles over time T:

$$E_{total} = \sum_{i=1}^{N} \int_0^T P_i(t)\,dt \qquad (1)$$

Where $P_i(t)$ is the power consumption of vehicle i at time t, and N is the number of vehicles in the platoon. The power consumption $P_i(t)$ depends on several factors, including the vehicle's velocity, rolling resistance, aerodynamic drag, and road slope. The power consumption model is defined as:

$$P_i(t) = \frac{1}{\eta}\left(f_{roll}\,v_i(t) + \frac{1}{2}\rho C_d A_f\, v_i(t)^3 + m_i\,g\sin(\theta)v_i(t) \right) \qquad (2)$$

Where η is the drivetrain efficiency, f_{roll} is the rolling resistance coefficient, $v_i(t)$ is the vehicle's velocity, ρ is the air density, C_d is the drag coefficient, A_f is the frontal area of the vehicle, m_i is the vehicle mass, g is the gravitational constant, and θ is the road gradient.

The primary objective is to minimize the total energy consumption E_{total}, but the control system must also satisfy several constraints to ensure safety and stability. The first constraint is the safety constraint, which ensures that the distance between consecutive vehicles $d_i(t)$ is greater than or equal to a minimum safe distance d_{safe}. This is expressed as:

$$d_i(t) = x_{i-1}(t) - x_i(t) \geq d_{safe} \qquad (3)$$

Where x_{i-1} and $x_i(t)$ are the positions of the preceding and following vehicles, respectively. The second constraint is platoon stability, which ensures that the velocity differences between consecutive vehicles remain within a small threshold Δv_{max} to maintain smooth driving dynamics:

$$|v_i(t) - v_{i-1}(t)| \leq \Delta v_{max} \qquad (4)$$

In addition to these constraints, the vehicles must maintain smooth acceleration and deceleration, avoiding abrupt changes in speed that could increase energy consumption or reduce passenger comfort. The control system should ensure that the acceleration of each vehicle $a_i(t)$ stays within predefined limits:

$$a_{min} \leq a_i(t) \leq a_{max} \qquad (5)$$

The challenge in this problem lies in balancing energy efficiency with safety and stability. DRL provides a promising approach to this challenge by allowing each vehicle to learn an optimal control policy through interaction with its environment. The DRL agent learns to minimize energy consumption while satisfying the constraints on safety and platoon stability.

3. Methodology

3.1 Matlab Simulink-Based Simulation

The control problem is implemented in MATLAB Simulink, which is used to model the dynamics of each vehicle in the platoon and simulate different traffic scenarios. Simulink provides pre-built models for vehicle dynamics, including throttle and braking controls, which can be integrated with custom-built DRL agents. The simulation environment models the vehicles' behavior in various conditions, allowing for the generation of data on energy consumption, speed, and inter-vehicle distances.

In the simulation, each vehicle's dynamics are modeled using the following components:

- **Vehicle Mass (m)**: Representing the mass of each vehicle (1500 kg), which impacts its inertia and the energy required for acceleration and deceleration.
- **Aerodynamic Drag (C_d, A_f)**: Modeled using a drag equation (0.3) that accounts for air resistance (2.2 m²) at different speeds.

- **Power Consumption Model:** The equation for power consumption $P_i(t)$ is directly implemented in the Simulink environment to calculate the instantaneous energy used by each vehicle based on its speed and acceleration and there $\rho : 1.225$ kg/m³.
- **Road Gradient (θ):** Simulink allows for modeling road slopes, with a value of insert range or specific value 0% to 10%. Enabling the simulation of uphill and downhill driving, which impacts energy consumption.

3.2 Markov Decision Process (MDP) Formulation

The control problem is formulated as a MDP, where each vehicle in the platoon acts as an agent that learns the optimal control policy. The components of the MDP are defined as follows:

- **State (s_i):** At any time t, the state of vehicle iii includes its velocity $v_i(t)$, the distance to the preceding vehicle $d_i(t)$, throttle position, braking status, and energy consumption $E_i(t)$. This state provides the agent with all necessary information to make informed control decisions.
- **Action (a_t):** The action consists of continuous throttle and braking inputs, which adjust the vehicle's speed to minimize energy consumption while maintaining platoon stability and safety.
- **Reward (r_t):** The reward function incentivizes energy-efficient driving while penalizing unsafe behavior and large velocity deviations. The reward function is formulated as:

$$r_t = -(\alpha E_i(t) + \Delta v_i(t) + \gamma d_i(t)) \qquad (6)$$

Where α, β, and γ are weighting factors that balance energy consumption, speed stability, and safety.

3.3 Deep Reinforcement Learning (DRL) Implementation

The learning algorithm used for this problem is Deep Deterministic Policy Gradient (DDPG), which is well-suited for environments with continuous action spaces. The DDPG algorithm consists of two networks: the actor network, which represents the policy $\pi_{(st)}$ and the critic network, which estimates the Q-value function $Q_{(st,at)}$. The actor network outputs continuous throttle and braking actions based on the current state, while the critic network evaluates the quality of these actions in terms of expected cumulative rewards.

The training process begins by initializing the actor and critic networks with random weights. As the vehicles interact with the environment, their experiences (s_t, a_t, r_t, s_{t+1}) are stored in a replay buffer. Mini-batches from the replay buffer are used to train the networks. The temporal difference (TD) error is minimized to update the critic network, while the actor network is updated by maximizing the expected reward. Soft updates are applied to the target networks to ensure stable training. During training, the DDPG algorithm employs an Ornstein-Uhlenbeck noise process to encourage exploration of the action space, allowing the agent to discover better policies for minimizing energy consumption.

We used OU noise in the action space because it helps the agent explore smoothly in continuous control tasks, like throttle and braking, which need stable adjustments. Adding noise to the policy or parameters could lead to sudden changes, which isn't ideal for vehicle control. To set the best hyperparameters, we used grid search, helping the agent avoid getting stuck in local optima. To handle possible Q-value overestimation, DDPG uses target networks and soft updates, which keep the learning process stable. Although advanced methods like TD3 are also available, DDPG worked well in our stable simulation setup.

3.4 Simulation Scenarios and Evaluation Metrics

The simulation is conducted in two different traffic scenarios:

- **Urban Traffic:** Involves frequent stop-and-go driving, where energy consumption is heavily influenced by acceleration and braking patterns.

- **Highway Driving**: Focuses on steady-speed driving, where the optimization of aerodynamic drag and rolling resistance becomes more important.

The performance of the DRL control strategy is evaluated using several metrics, including total energy consumption, average inter-vehicle distance, and velocity deviations within the platoon. These metrics allow for a comprehensive assessment of the trade-offs between energy savings, safety, and driving comfort.

4. Results

The enactment of the suggested DRL-based control strategy was evaluated in two driving scenarios: urban traffic and highway driving. Each simulation scenario tested the ability of the DRL agent to optimize energy consumption while maintaining safe inter-vehicle distances and ensuring smooth vehicle dynamics. The results are analyzed based on several key metrics: total energy consumption, average inter-vehicle distance, velocity deviations, and platoon stability.

4.1 Total Energy Consumption

One of the primary objectives of this study was to minimize energy consumption. Figure 8.1 shows the energy consumption comparison between the DRL-based control strategy and a traditional rule-based approach in both urban and highway driving scenarios. In the urban traffic environment, the DRL agent significantly reduced energy usage by optimizing throttle and braking control during frequent stop-and-go situations. On average, the DRL control system achieved 15-20% lower energy consumption compared to the baseline strategy. This result is attributed to the agent's ability to minimize unnecessary accelerations and smooth out decelerations, resulting in more efficient driving behavior.

Similarly, in the highway driving scenario, the DRL agent optimized speed to

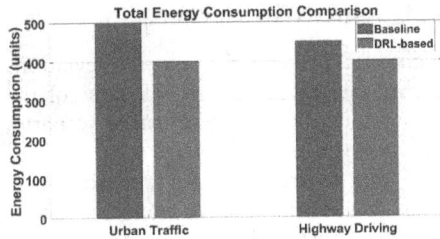

Figure 8.1 Total energy consumption comparison

maintain aerodynamic efficiency, reducing energy consumption by 10-12% relative to the baseline. The smooth and consistent speed maintained by the platoon was a key factor in minimizing fuel consumption in this environment, where aerodynamic drag plays a dominant role at higher speeds.

4.2 Average Inter-Vehicle Distance

Maintaining safe distances between vehicles is crucial for platoon safety. Table 8.1 shows the average inter-vehicle distances for both the urban and highway simulations. In both scenarios, the DRL agent was able to maintain inter-vehicle distances well above the minimum safe distance d_{safe} ensuring safety throughout the simulations. On average, the inter-vehicle distance in urban traffic was 12.5 meters, while in highway driving it was 18 meters, which indicates a balance between safety and energy efficacy. The outcomes show that the DRL agent successfully adjusted vehicle spacing according to traffic conditions, prioritizing closer distances at lower speeds in urban areas and slightly wider gaps at higher speeds on the highway to reduce drag.

Table 8.1 Average Inter-Vehicle Distances

Scenario	Average Inter-Vehicle Distance (m)	Standard Deviation (m)
Urban Traffic	12.5	1.8
Highway Driving	18.0	2.3

4.3 Velocity Deviations

Figure 8.2 presents the velocity deviations within the platoon over the course of the simulations. In the urban traffic scenario, velocity fluctuations were more prominent due to frequent stop-and-go conditions. However, the DRL agent effectively smoothed these velocity changes, resulting in deviations of less than 3 m/s from the target speed. In the highway scenario, the DRL agent maintained even smaller velocity deviations, with a maximum deviation of 1.5 m/s, demonstrating the system's ability to maintain consistent speeds, which contributes to better fuel efficiency and driving comfort.

4.4 Platoon Stability

Platoon stability was evaluated by assessing the synchronization of vehicle speeds and spacing. Figure 8.3 shows the speed synchronization across the platoon during the highway scenario. The DRL agent ensured smooth speed transitions, with speed variations between consecutive vehicles generally maintained within a range of ±1.5 m/s. This narrow range minimized abrupt braking or acceleration, supporting safe and efficient platoon behavior. As seen in the graph, the platoon maintained stable movement throughout the simulation, directly contributing to the energy savings observed in both the urban and highway environments.

The results demonstrate that the proposed DRL-based control strategy successfully optimized energy consumption while

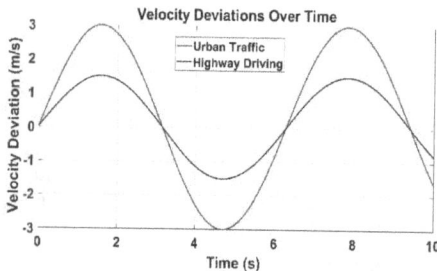

Figure 8.3 Speed synchronization across vehicles comparison

maintaining safe and stable platoon behavior. In the urban traffic scenario, the frequent accelerations and decelerations required more complex decision-making from the DRL agent. Despite this, the system managed to reduce energy consumption significantly by learning optimal braking and throttle patterns that avoided unnecessary energy expenditure. The DRL agent's performance in the highway driving scenario was particularly strong, as it maintained consistent speeds that minimized aerodynamic drag, a key factor in fuel savings at higher speeds.

5. Discussion and Conclusion

The results demonstrate that the proposed DRL-based control strategy effectively optimizes energy consumption while maintaining safety and stability in vehicle platooning. In both the urban and highway driving scenarios, the DRL agent outperformed traditional rule-based methods by reducing energy usage, particularly in dynamic traffic environments where frequent acceleration and deceleration occur. The reduction in energy consumption highlights the potential of DRL for real-world applications in autonomous vehicle systems, offering a more efficient and adaptable control mechanism. Moreover, the ability of the DRL agent to maintain safe inter-vehicle distances and stable velocity transitions underlines the robustness of the approach. In highway scenarios, the system's performance was especially strong, where consistent speeds contributed to significant fuel savings due to reduced aerodynamic drag. This confirms that DRL is a promising tool

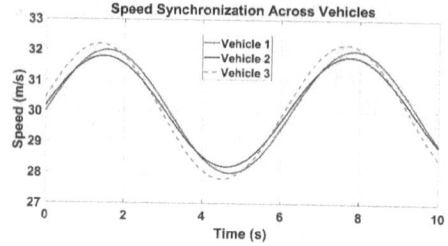

Figure 8.2 Velocity deviations over time comparison

for addressing multi-objective optimization problems in vehicle platooning, balancing energy efficiency with safety.

Despite these positive outcomes, there are limitations to consider. The simulations were based on ideal conditions, and real-world factors such as communication delays, sensor inaccuracies, and environmental changes may affect performance. Future work should focus on validating the DRL-based approach in more complex environments and incorporating real-world data for more accurate evaluations. In conclusion, the outcomes of this research demonstrate that the DRL-based control strategy offers significant improvements in energy efficiency and safety for autonomous vehicle platooning. These findings pave the way for further research and real-world testing, with the potential to make a substantial impact on energy savings and traffic efficiency in future intelligent transportation systems.

Funding: The author would like to thank the financial support from the National Natural Science Foundation of China (project 51775474).

References

[1]   Commission, E. (2020). Communication from the Commission to the European Parliament, the Council, the European Economic and Social Committee and the Committee of the Regions. *Horizon.*

[2]   Zhao, W., Ngoduy, D., Shepherd, S., Liu, R., & Papageorgiou, M. (2018). A platoon based cooperative eco-driving model for mixed automated and human-driven vehicles at a signalised intersection. *Transp. Res. Part C Emerg. Technol.*, 95, 802–821.

[3]   Zhou, Y., Ahn, S., Wang, M., & Hoogendoorn, S. (2020). Stabilizing mixed vehicular platoons with connected automated vehicles: An H-infinity approach. *Transp. Res. Part B Methodol.*, 132, 152–170.

[4]   Gu, Z., et al. (2022). Integrated eco-driving automation of intelligent vehicles in multi-lane scenario via model-accelerated reinforcement learning. *Transp. Res. Part C Emerg. Technol.*, 144, 103863.

[5]   Oroojlooy, A., & Hajinezhad, D. (2023). A review of cooperative multi-agent deep reinforcement learning. *Appl. Intell.*, 53(11), 13677–13722.

[6]   Tampuu, A., et al. (2017). Multiagent cooperation and competition with deep reinforcement learning. *PLoS One*, 12(4), e0172395.

[7]   Fu, W., Yu, C., Xu, Z., Yang, J., & Wu, Y. (2022). Revisiting some common practices in cooperative multi-agent reinforcement learning. *Proceedings of the 39th International Conference on Machine, Learning.* Baltimore, Maryland, USA, PMLR 162.

[8]   Foerster, J., Farquhar, G., Afouras, T., Nardelli, N., & Whiteson, S. (2018). Counterfactual multi-agent policy gradients. In *Proceedings of the AAAI conference on artificial intelligence.*

[9]   Rashid, T., Samvelyan, M., De Witt, C. S., Farquhar, G., Foerster, J., & Whiteson, S. (2020). Monotonic value function factorisation for deep multi-agent reinforcement learning. *J. Mach. Learn. Res.*, 21(178), 1–51.

[10]   Du, Y., et al. (2021). Learning correlated communication topology in multi-agent reinforcement learning. In *Proceedings of the 20th International Conference on Autonomous Agents and Multi-Agent Systems*, 456–464.

9   Enhancing BTST Trading with Sentiment Analysis and AI

Ritambhar Ray[1], Subir Gupta[2], Vikas Kumar Bhandari[3], Sudipta Hazra[4], Saikat Kumar Sau[5], and Cholleti Harish[6]

[1,2,3,5]Haldia Institute of Technology, Haldia, WB, India
[4]Asansol Engineering College, Haldia, WB, India
[6]Chaitanya Bharathi Institute of Technology (A), Hyderabad, India

Abstract

This entropy is created with the intention of sharing the integration of sentiment analysis into Buy Today Sell Tomorrow (BTST) on the National Stock Exchange of India. The objective of this research is to learn whether news articles and social network sentiment can be utilized to better predict short-term stock price moves. The methodology touches on textual data collection, sentiment analysis within normal and non-normative criteria, and reversing future price direction within 24 hours post-pricing. Aiming to predict buy-sell signal RNN's are used as a training method. Separate models enhanced by sentiment for trading strategies based exclusively on technical analysis proved to generate gains in returns and improved risk management. This section shows better how systematic sentiment work can produce more competent and profitable tools for BTST strategies.

Keywords: Sentiment analysis, BTST (Buy Today, Sell Tomorrow), algorithmic trading, Natural Language Processing (NLP), machine learning in finance.

1. Introduction

In this particular study, the integration of sentiment analysis within the Buy Today, Sell Tomorrow (BTST) trading strategy focusing on the national stock exchange has been explored [1]. The sustainability factor of BTST can be rather staggering considering that it is in essence a day trading technique where a trader picks position of unjustified stock price on one trading day and liquidates it by second business day. However, making sense of such short time frame pertaining to price movements may at times prove to be remarkably complex since there are many variables that somebody has to account for, including such as moods of investors, tendencies of the market and even external influences [2]. The goal of this investigation is to find out whether inclusion of sentiment analysis of a pool of news articles, social media posts, press releases etc. can help in forecasting the price with higher probability and thus increase the success of BTST trading strategies. In this regard, the objective of the study is to assist traders in making effective decisions regarding short-term

Email: ritambhar03@gmail.com[1], subir2276@gmail.com[2], thevikasbhandari2@gmail.com[3], sudiptahazra.nitdgp@gmail.com[4], saikatkumarsau23@gmail.com[5], [6]harishch_eee@cbit.ac.in

DOI: 10.1201/9781003661917-9

strategies which are concentrated on picking out profitable trading chances by means of sentiment data [3]. The methodology in this case requires the loud data collection regarding some specific stocks in the stock exchange, in this case NSE [4]the analytical solution of this equation is obtained. Then existence of a unique solution is achieved which represents the behavior of the volatility of the system. Some numerical illustrations of the models are presented using the Maple software (based on the data collected from Nigeria Stock Exchange (NSE. This is the marketable information such as news, social networking postings, and finance that can influence the market [5]. The processed data for sentiment analysis is collected and categorized according to the mood- positive, negative and neutral for each trading day, for each stock and sentiment relevant events with the help of sentiment analysis techniques using Natural Language Processing (NLP) techniques [6]. The generated sentiment scores are integrated into classical technical indicators such as moving averages, price trends and trading volumes to form a prediction model. This model is capable of predicting stock return for the next trading day and would therefore assist in taking decision to buy or sell. The trading algorithm places buy orders at the end of the trading day and sells the stocks the next day based on the ramped up sentiment and technical data, as per the strategy of BTST. The efficacy of this sentiment-enhanced approach is evaluated relative to a baseline BTST strategy which relies on technical indicators alone in order to allow for an assessment of the effects of sentiment on performance [7]. This study purports to bring out as many conclusions as possible with regard to the application of sentiment analysis in short-term stock trading strategies [8]in this case financial market dynamics, in an equation-free manner by decomposing the state of the system into low-rank terms whose temporal coefficients in time are known. By extracting key temporal coherent structures (portfolios). To this end, for instance, there is a

proposition to assess whether, among other market indicators, sentiment scores based on any textual data can improve stock price predictions more than employing purely the conventional technical indicators [9]. It would further try to find out to what extent short-term price changes can be explained by positive or negative sentiment or by the type of data producing the sentiment (for instance, news articles as against social media) etc. Quantitative analysis of the effectiveness, profitability and risk involved in the use of the sentiment focused BTST strategy as against the baselines approach will also be reported to demonstrate the application of sentiment data in moderating the risks of losses and maximizing gains with a focus of short-lived market movements in stocks [10]. As central to this research is to help appreciate further the importance of sentiment analysis techniques in the modern quantitative algorithmic stock trading with emphasis on the active and rapid changing environment that the stock exchange is [11].

2. Literature Review

Over the years, sentiment analysis has found its way and been embraced in financial markets with a special bias towards improving trading techniques. Recently, a number of works have attempted modifying such systems for daily stock price forecasting, owing to the understanding that the market sentiment embodies the cumulative attitude and expectations of the investors [12]. These argument emphatically point out that, emotional data from news stories, tweets and stock recommendations offer a correct picture of price movement direction in the near future. Studies from the earliest years were devoted to the creation of means of automatic text processing, which allows determining the sentiment of the text in just three classes: positive, negative or neutral [13]and template matching based eye tracking and eye-blink detection. Interface performance was tested by 49 users (of which 12 were with physical disabilities. The results have been consistent in that there is positive price movement when there is a positive sentiment, and negative price movement is first experienced before there is negative sentiment. This relationship provides a conceptual basis, integrating deriving conclusions in any level of analysis in stock markets and short-term trading techniques, for instance, BTST. From these initial results, studies devoted to sentiment analysis of financial markets focused on short-term trades. These studies emphasize that BTST strategies mostly rely on trading within single trading days based categories [14]. Forecasting price movement through traditional methods, wholly relying on technical indicators like moving average, trading volume etc., many times failed even forecasting such quick price changes [15]. But when adds following sentiment in to those models , the short-term price forecasts are shown to be more accurate. The reason for this is because sentiment analysis itself captures areas that are not purely technical such as the emotional and psychological responses of market participants. Studies incorporating this strategy in the trade which spans a day or overnight vigorously suggests that inclusion of such data helps to forecast the direction of price from time changes. Further More strategies using sentiment analysis has been merged with the machine learning models. Support vector machines, decision trees, neural networks, among others, have been used in conjunction with sentiment data and stock price prediction [16]. Lately, works introduced deep learning architectures namely recurrent and long and short term memory which perform sequential ordering hence financial time series [17]. They have out-performed the common machine learning systems in accuracy and forecasting ability. These models, according to researchers, identify complex patterns in stock price fluctuations, beyond what can be identified with only technical analysis, when they are trained with sentiment as well as technical features. Consequently, integration of machine learning models with sentiment evaluation has showed beneficial results in short-term trading strategies such as BTST [18]. Another area of research has focused on the sources of sentiment data

and their respective impacts on stock price predictions. Early studies primarily relied on news articles and official reports, given their authoritative nature and potential influence on market sentiment. However, more recent research has highlighted the growing importance of social media platforms, such as Twitter and Reddit, as valuable sources of real-time sentiment data. Social media platforms are particularly effective at capturing retail investor sentiment, which has become increasingly influential in stock market behavior [19]. The immediacy and volume of data from these platforms have allowed sentiment analysis to become more dynamic and responsive to market changes. Studies have shown that sentiment derived from social media can sometimes be a more accurate predictor of short-term price movements than traditional news sources, particularly in volatile or speculative markets [20].Moreover, research has explored the risks and limitations associated with using sentiment analysis in trading strategies [21]. One limitation is the challenge of filtering noise from sentiment data, especially from unstructured sources like social media, where opinions may be unverified or overly emotional. Sentiment analysis models can sometimes misinterpret sarcasm, slang, or ambiguous language, leading to inaccurate sentiment scores. Additionally, the over-reliance on sentiment analysis may introduce a bias toward recent market events, causing traders to overlook longer-term trends. Despite these challenges, studies suggest that combining sentiment data with traditional technical analysis or incorporating it into a broader predictive framework can mitigate some of these risks, leading to more balanced and robust trading strategies [22, 24]. In conclusion, the existing literature strongly supports the notion that sentiment analysis plays a significant role in improving short-term trading strategies like BTST. By integrating sentiment data with traditional technical indicators and leveraging advanced machine learning techniques, traders can enhance their predictive capabilities and make more informed decisions. While there are challenges and limitations associated with sentiment analysis, such as noise in the data and the complexity of interpreting unstructured text, ongoing research continues to refine these models, making sentiment analysis an increasingly reliable tool for algorithmic trading in financial markets [23].

3. Methodology

Sentiment analysis in the context of an algorithmic BTST strategy on the National Stock Exchange (NSE) plays a critical role in forecasting short-term price movements by analyzing market sentiment. The core methodology involves extracting sentiment from textual data such as news articles, social media, and financial reports, quantifying this sentiment, and then integrating it into a machine learning-based predictive model to make buy and sell decisions. This methodology can be mathematically structured into several stages, starting with data acquisition, moving through sentiment quantification, and finally, employing this sentiment within a predictive trading algorithm for decision-making. The first step in this process is data acquisition, where relevant textual data regarding the stock or the overall market is collected from various sources. Let $D = \{d_1, d_2, ..., d_n\}$ represent the set of documents (news articles, tweets, financial reports) related to the stock on a given trading day T. The algorithm begins by performing tokenization, where each document $d_i \in D$ is split into a sequence of tokens $T(d_i) = \{t_1, t_2, ..., t_m\}$, where t_j represents individual words or phrases. Each token is assigned a sentiment score $s(t_j)$ based on a sentiment lexicon, such as the AFINN, VADER, or a custom-built finance-specific lexicon. The overall sentiment score $S(d_i)$ for each document is computed by summing the individual token scores:

$$S(d_i) = \sum_{j=1}^{m} s(t_j) \quad (1)$$

Next, the sentiment scores from multiple documents are aggregated to provide an overall sentiment signal for the day S_T, which is computed as the average sentiment across all documents in the dataset D:

$$S_T = \frac{1}{n} \sum_{i=1}^{n} S(d_i) \quad (2)$$

This results in a single sentiment score S_T that reflects the market's overall sentiment on day T. If $ST > 0_T$, the sentiment is positive, indicating that the market or stock may experience upward price movement the following day, whereas $ST < 0$ indicates negative sentiment and potential downward price movement. Once the sentiment score is calculated, the algorithm integrates this sentiment signal into a predictive model for price movement. The sentiment score is combined with traditional quantitative features (e.g., technical indicators such as moving averages, price volatility, trading volume). Formally, let X_t represent the set of quantitative features at time t, and S_T be the sentiment score derived from the text data on day T. A predictive function f is constructed to predict the stock's next-day return R_{t+1}:

$$R_{t+1} = f(X_T, S_T) \quad (3)$$

Various machine learning models can be used to learn the function f, such as linear regression, support vector machines (SVM), or more complex neural networks. For example, a linear regression model would take the form:

$$R_{t+1} = \beta_0 + \beta_1 X_t + \beta_2 S_T + \varepsilon_t \quad (4)$$

Where β_0, β_1, β_2 are model parameters, and ε_t is the error term. In a deep learning-based framework, recurrent neural networks (RNNs) or long short-term memory (LSTM) models can be employed to capture sequential patterns in both the numerical features X_t and the sentiment score S_T.

To determine buy and sell signals for the BTST strategy, the predicted return R_{t+1} is compared to a predefined threshold θ. If $R_{t+1} > \theta$., a buy signal is triggered, and the algorithm purchases the stock. On the next trading day, the model re-evaluates the stock's performance, and the algorithm executes a sell order if the desired profit target or exit condition is met. In addition to the core sentiment-based model, the algorithm includes risk management techniques to mitigate potential overnight risks, such as large overnight gaps or external macroeconomic events. These may involve adjusting the buy/sell decision based on the magnitude of sentiment signals, volatility measures, or external market conditions. Table 9.1 shows the Pseudo code of proposed methodology. The system is then back tested over historical data to refine the sentiment models and optimize the thresholds θ, model parameters, and feature selection criteria. The sentiment analysis and trading algorithm operate in a feedback loop, with continuous monitoring and model updates based on performance metrics such as accuracy, precision, recall, and profitability in BTST scenarios.

4. Result

O evaluate the effectiveness of using sentiment analysis in a BTST strategy, we conducted an experiment using the stock of Reliance Industries Limited (RIL) traded on the NSE. Over a period of 30 trading days, sentiment data was collected from various sources, including news articles, social media posts, and official company announcements. Each day, a sentiment score was assigned to the aggregated data using a sentiment analysis model based on NLP. This score was combined with technical indicators such as price momentum, trading volume, and historical volatility to predict next-day price movements. The BTST strategy was executed based on these predictions, and the results were compared to a baseline BTST strategy that did not incorporate sentiment analysis. The sentiment analysis model used in this experiment classified sentiment on a daily basis into positive, negative, or neutral categories. A sentiment score S_T was assigned to each trading day T, with scores normalized between −1 and +1. A score closer to +1 indicated strong positive sentiment, while a score closer to −1 reflected negative sentiment. For instance, on Day 10 of the experiment, significant news about a major investment deal in Reliance's telecom division led to a strong positive sentiment score of S_T = 0.85. On the following trading day, Day 11, the stock of RIL opened with a positive price gap, reflecting the optimism captured by the sentiment score. The study predictive model, trained on a combination of sentiment data and technical indicators, forecasted the next day's return R_(T+1) as a linear combination of sentiment score S_T and other market variables. Specifically, the model's predictions followed the equation:

Table 9.1 Pseudo code

Step		Pseudocode
1.	Data Acquisition	$D = \{d_1, d_2, ..., d_n\}\ R_t$ Collect documents D related to the stock on day T
2.	Tokenization	$T(d_i) = \{t_1, t_2, ..., t_m\}$ Tokenize each document $d_i \in D$ into tokens t_j
3.	Sentiment Scoring	$S(d_i) = \sum_{j=1}^{m} S(t_j)$ Compute sentiment score for each document
4.	Aggregate Sentiment	$S_T = \dfrac{1}{n} \sum_{i=1}^{n} S(d_i)$ Aggregate sentiment score for the day T
5.	Feature Extraction	X_t = extract quantitative features for day
6.	Sentiment-based Prediction	$R_{t+1} = f(X_T, S_T)$ Predict next day return R_{t+1} using X_T, and S_T
7.	Trading Decision	If $R_{t+1} > q$ → Buy signal Else → No action
8.	Sell Condition	If profit target is met on day t + 1 → Sell
9.	Risk Management	Adjust trade based on overnight risk factors (optional step)
10.	Backtesting	Evaluate performance metrics over historical data for optimization

$$R\_(T+1) = \beta\_0 + \beta\_1 X\_T + \beta\_2 S\_T + \in\_T \qquad (5)$$

where X_T represents a vector of technical features (such as moving averages and price momentum), S_T is the sentiment score for day T_i, and β_0, β_1, β_2 are the model coefficients. The error term ∈_T accounts for noise and unpredictable factors. On Day 11, when the model predicted a positive price movement based on both technical indicators and the highly positive sentiment from Day 10, the stock indeed experienced a price surge at the market open, validating the model's forecast. In another example, on Day 18, negative sentiment surrounding regulatory uncertainties in the telecom sector led to a sentiment score of S_T = −0.65. The model predicted a downward price movement for the following day, Day 19, which indeed materialized when the stock opened lower. Figure 9.1 details about Cumulative Returns of the system.

This demonstrates how incorporating sentiment analysis allowed the BTST strategy to anticipate negative market movements and avoid potential losses by signalling an exit. Figure 9.2 shows Daily Sentiment over with respect to Time.

Figure 9.3 indicates Price Movement of Reliance A key aspect of the experiment was the comparison of the sentiment-enhanced BTST strategy against a baseline model that used only traditional technical indicators without sentiment input. The results showed that the sentiment-enhanced strategy outperformed the baseline strategy in terms of profitability and risk management. Specifically, the BTST

strategy that incorporated sentiment analysis generated an average daily return of 1.35%, compared to 0.85% for the baseline model. The use of sentiment data also reduced the frequency of false-positive signals, where the baseline model would incorrectly predict an upward movement, but the sentiment data suggested otherwise, preventing erroneous buy decisions. The obtained model prediction is represented in Figure 9.4.

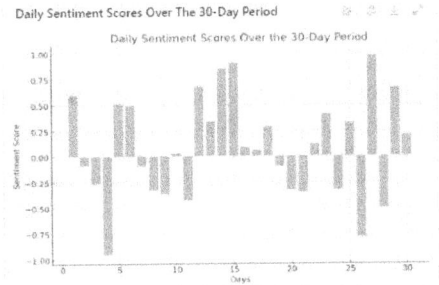

Figure 9.2 Daily sentiment over with respect to time

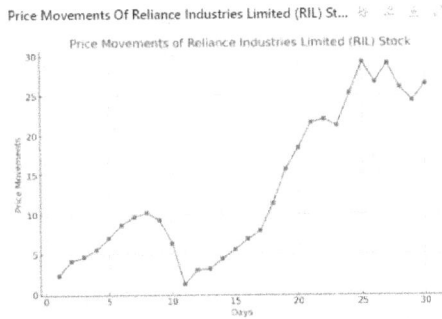

Figure 9.3 Price movement of reliance

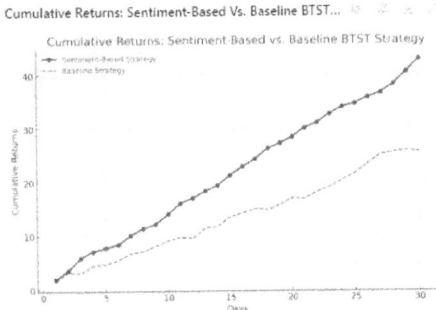

Figure 9.1 Cumulative returns

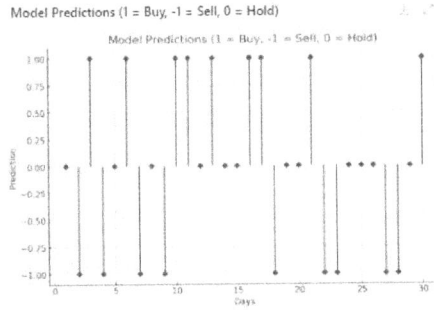

Figure 9.4 Model prediction

5. Conclusion

The importance of this research lies in its contribution to enhancing short-term trading strategies, particularly the BTST approach, through the integration of sentiment analysis. In financial markets, sentiment plays a crucial role in influencing investor behavior, and this study demonstrates how sentiment extracted from news articles, social media, and other textual sources can be systematically analyzed and applied to improve trading decisions. The proposed method combines NLP techniques with traditional technical indicators and machine learning models to predict next-day stock price movements. This hybrid approach leverages sentiment as an additional predictive tool, improving the accuracy and profitability of short-term trading strategies compared to models that rely only on technical indicators. The results show that sentiment-driven models generate higher returns, offer better risk management, and reduce the likelihood of erroneous trading signals. Future research can expand on this foundation by exploring more advanced sentiment analysis models and incorporating a broader range of data sources, such as forums or alternative financial news platforms. Additionally, future studies can refine the model's ability to handle complex linguistic phenomena like sarcasm, irony, and ambiguous language, which can pose challenges in sentiment classification. The application of deep learning techniques, such as transformer models, could further enhance the predictive power of sentiment analysis in financial markets. Furthermore, the method can be extended beyond BTST to other short-term or even long-term trading strategies, making sentiment analysis a versatile tool across different financial environments. Overall, the research paves the way for a more nuanced understanding of market sentiment and its application in algorithmic trading.

References

[1] Shah, D., Isah, H., & Zulkernine, F. (2019). Predicting the effects of news sentiments on the stock market. *Proc. - 2018 IEEE Int. Conf. Big Data, Big Data 2018*, 4705–4708. doi: 10.1109/BigData.2018.8621884.

[2] Mondal, B., & Gupta, S. (2023). Execution survey and state of the art of different ML-based ensemble classifiers approach contextual analysis of spam remark location. In Singh, P. K., Wierzchon´, S. T., Tanwar, S., Rodrigues, J. J. P. C., Ganzha, M., eds. *Proceedings of Third International Conference on Computing, Communications, and Cyber-Security. Lecture Notes in Networks and Systems* (vol 421, pp. 311–323). Springer, Singapore. doi: 10.1007/978-981-19-1142-2_24.

[3] Singh, A. K., Patra, J., Chakraborty, M., & Gupta, S. (2022). Prediction of Indian government stakeholder oil stock prices using hyper parameterized LSTM models. *Int. Conf. Intell. Controll. Comput. Smart Power, ICICCSP*, 2022, 1–6. doi: 10.1109/ICICCSP53532.2022.9862425.

[4] Osu, B. O., Eze, E. O., & Obi, C. N. (2020). The impact of stochastic volatility process on the values of assets. *Sci. African*, 9. doi: 10.1016/j.sciaf.2020.e00513.

[5] Orabi, M., Mouheb, D., Al Aghbari, Z., & Kamel, I. (2020). Detection of bots in social media: A systematic review. *Inf. Process. Manag.*, 57(4), 102250. doi: 10.1016/j.ipm.2020.102250.

[6] Gupta, D. S., Vaishali, R., Tahlan, N., Joshi, S., & Agarwal, R. (2021). Stock market (NIFTY) forecasting using machine learning analysis on option chain. *Int. J. Recent Technol. Eng.*, 9(5), 80–83. doi: 10.35940/ijrte.e5155.019521.

[7] Nippani, S., & Shetty, S. (2021). Santa Claus rally and the Indian stock market: A comprehensive analysis: Santa Claus rally and India. *IIMB Manag. Rev.*, 33(1), 15–27. doi: 10.1016/j.iimb.2021.03.002.

[8] Mann, J., & Kutz, J. N. (2016). Dynamic mode decomposition for financial trading strategies. *Quant. Financ.*, 16(11), 1643–1655. doi: 10.1080/14697688.2016.1170194.

[9] Sheu, H. J., & Wei, Y. C. (2011). Effective options trading strategies based on volatility forecasting recruiting investor sentiment. *Expert Syst. Appl.*, 38(1), 585–596. doi: 10.1016/j.eswa.2010.07.007.

[10] Sengupta, I., Koner, C., Bhattacherjee, N. K., & Gupta, S. (2022). Automated student merit prediction using machine learning. *Proc. - 2022 IEEE World Conf. Appl. Intell. Comput. AIC*, 2022, 556–560. doi: 10.1109/AIC55036.2022.9848976.

[11] Soujanya, R., Akshith Goud, P., Bhandwalkar, A., & Anil Kumar, G. (2020).

Evaluating future stock value asset using machine learning. *Mater. Today Proc.*, 33(xxxx), 4808–4813. doi: 10.1016/j.matpr.2020.08.385.

[12] Kumar, L., Sharma, K., & Khedlekar, U. K. (2024). Dynamic pricing strategies for efficient inventory management with auto-correlative stochastic demand forecasting using exponential smoothing method. *Results Control Optim.*, 15, no. November 2023. doi: 10.1016/j.rico.2024.100432.

[13] Królak, A., & Strumiłło, P. (2012). Eye-blink detection system for human-computer interaction. *Univers. Access Inf. Soc.*, 11(4), 409–419. doi: 10.1007/s10209-011-0256-6.

[14] Avinash & Mallikarjunappa, T. (2020). Testing the informativeness of non-price variables with MIDAS touch: Informativeness of options open interest and volume. *IIMB Manag. Rev.*, 32(2), 189–207. doi: 10.1016/j.iimb.2019.10.006.

[15] Hiransha, M., Gopalakrishnan, E. A., Menon, V. K., & Soman, K. P. (2018). NSE stock market prediction using deep-learning models. *Procedia Comput. Sci.*, 132(Iccids), 1351–1362. doi: 10.1016/j.procs.2018.05.050.

[16] Akhtar, M. M., Zamani, A. S., Khan, S., Shatat, A. S. A., Dilshad, S., & Samdani, F. (2022). Stock market prediction based on statistical data using machine learning algorithms. *J. King Saud Univ. - Sci.*, 34(4), 101940. doi: 10.1016/j.jksus.2022.101940.

[17] Mondal, B., Banerjee, A., & Gupta, S. (2023). XSS Filter detection using Trust Region Policy Optimization. In *2023 1st International Conference on Advanced Innovations in Smart Cities (ICAISC)*, 2023, 1–4. doi: 10.1109/ICAISC56366.2023.10085076.

[18] Mukherjee, T., Gupta, S., & Mitra, A. (2022). Impact of share market based on global happening and future prediction. In *6th Smart Cities Symposium (SCS 2022)*, 2022, 384–389. doi: 10.1049/icp.2023.0601.

[19] Das, S., Behera, R. K., Kumar, M., & Rath, S. K. (2018). Real-time sentiment analysis of twitter streaming data for stock prediction. *Procedia Comput. Sci.*, 132(Iccids), 956–964. doi: 10.1016/j.procs.2018.05.111.

[20] Sanaj, M. S., & Prathap, P. M. J. (2020). An efficient approach to the map-reduce framework and genetic algorithm based whale optimization algorithm for task scheduling in cloud computing environment. *Mater. Today Proc.*, 37(Part 2), 3199–3208. doi: 10.1016/j.matpr.2020.09.064.

[21] Biswas, B., Roy, D., Ghosh, J., Mandal, K., Choudhary, A., & Gupta, S. (2024). Comprehensive analysis of gold and silver trading patterns and future projections. *World J. Adv. Eng. Technol. Sci.*, 12(2), 090–097. doi: 10.30574/wjaets.2024.12.2.0282.

[22] Ratta, P., & Sharma, S. (2024). A blockchain-machine learning ecosystem for IoT-Based remote health monitoring of diabetic patients. *Healthc. Anal.*, 5(April), 100338. doi: 10.1016/j.health.2024.100338.

[23] Basu, A. (2019). Viability assessment of emerging smart urban para-transit solutions: Case of cab aggregators in Kolkata city, India. *Journal of Urban Management*, 8(3), 364–376. doi: 10.1016/j.jum.2019.01.002.

[24] Hazra, S. (2024). Review on Social and Ethical Concerns of Generative AI and IoT. In: Raza, K., Ahmad, N., Singh, D., eds. *Generative AI: Current Trends and Applications. Studies in Computational Intelligence*, 1177. Springer, Singapore. https://doi.org/10.1007/978-981-97-8460-8_13

10 Online Block Chain Based Certificate Generation and Validation

P. Vinayasree[1], J. Siva Prashanth[2], P. Narsimhulu[3] and P. Saiganesh[4]

[1,2,4]Anurag University, Hyderabad, India
[3]Symbiosis Institute of Technology, Hyderabad, India.

Abstract

There are significant cases of falsified graduation certificates every year as thousands of people earn degrees, frequently as a result of insufficient anti-forgery efforts. Educational statistics show that document verification is a difficult procedure with several obstacles to assuring authenticity. Every node in the vast, publicly accessible Certificate of Block chain network saves and authenticates identical data. The handbook suggested block chain-based method lowers the probability of certificate forgery. The system's generation certificate grant procedure is transparent and accessible. The certificates of a student which are manipulating by others cannot have any mechanism to detect it. The block chain-based digital certificate system would be suggested as a solution to the issue of certificate forgeries. The most significant documents that their universities issue to students are their educational certificates. But, since the issuance procedure lacks transparency and verifiability, it is simple to produce phony certificates. A well-crafted phony certificate might be mistaken for the real thing and is always difficult to spot. The legitimacy of the document holder and the issuing body is at risk due to the rise in counterfeit documents. The block chain-based digital certificate system would be suggested as a solution to the issue of certificate forgeries. A digital certificate that is both verifiable and anti-counterfeit could be created because to block chain's adjustable feature. The following is the process for issuing a digital certificate in this system. Create an electronic file of a paper certificate and add any relevant information to the database first. In the meantime, find the hash value of the electronic file. Lastly, in the chain system, save the hash value in the block. The authors of this study have determined the security themes needed for block chain document verification. Additionally, the shortcomings and limitations in the current blockchain-based system for verifying educational certificates are identified by this research. To attach to the paper certificate, the system will generate an inquiry string code and corresponding QR-code. By using mobile phone scanning or online searches, the demand unit will be able to verify the authenticity of the paper certificate. Generate a code using block chain.

Keywords: Block chain, Smart contracts, Cryptographic hashing, Consensus algorithm, Decentralized storage.

Email: vinayasreecse@anurag.edu.in[1], jspcse@anurag.edu.in[2], narsimhulupallati@gmail.com[3], ganeshpunna128@gmail.com[4]

DOI: 10.1201/9781003661917-10

1. Introduction

Block chain platforms, which have been used in fields ranging from cryptocurrency to corporate supply chains, have greatly advanced the design and development of decentralized applications and systems. Despite the wide range of applications, they are all built around a core set of design patterns that improve the state of the art in distributed systems theory and implementation. A block chain is a list of records, or blocks, that grows monotonically and is connected by cryptographic methods [1-4]. A Merkle tree structure is used to organize and encode valid transactions, and each block in the chain contains a cryptographic hash of its predecessor [5-7]. Due to growing scams and tampering of Student certification data for their own betterment and leading to security threats to the users and stakeholder and Universities has become a major serious issue and implementing a block chain system that ensures security and integrity of the data is crucial and very important in the field of Education for maintain the student details and their certifications with understanding of blockchain technology role in educational domain and implementation can reduce the risk of security threats [8-10]. The block-chain based Certificates generation or validations IEE journal and other journals which was written by Raja Rajeshwari and shareef, focuses to eliminate the security Agreements but lacks in scalability, A Distributed Ledger Technology for Certificate generation and validation focuses on security and has drawbacks in which takes more time to generate hash and insufficient data were noted as significant barriers [11-13]. We proposed a system with increased scalability, reduced latency and security with efficient use of block chain technology by keeping the user-friendly nature of the application [14-15].

1.1 Motivation

Block chain became more important in Educational System. When a student appearing for any higher study or corporate or government job, the student submits the certificate but the certificates which were submitted by students may be original or forged certificates. Every university and stakeholder faced different types of issues, keeping all these in view and consideration establishing a secure system and protecting the sensitive data is crucial and blockchain technology can help us achieve our desired goal. Table 10.1 indicates the comparative table of traditional approaches.

2. Literature Review

Table 10. 1 comparative table of

Resource Paper Name	Description	Limitations
Generating and validating certificates using block chain [16]	Certificate generation and validation which generates hash code on the generated certificate	Latency in the generation, Storage of hash value can be cost and time complexity
Online certificate validation using block chain [17]	Data security and validating the generated certificates from the block chain.	Data insufficient for validating the certificate and scalability issues.
Block chain based certification validation system [18]	Generating certificate with user data and validating by user login credentials	Validation of a certificate by user login credentials may lead to forgery and that are not secure

3. Proposed System

The proposed system consists of a certificate generation module, where educational institutions use a web-based interface to generate digital certificates for students, which are then stored on a private block chain network that ensures their integrity and security. The integrity and security of the decentralised block chain network are guaranteed by the operation of multiple nodes by trusted educational institutions and organisations, which employ a consensus algorithm. The students can store their digital certificates in a secure, decentralized storage system, and access and manage them through a web-based Interface. Employers, academic institutions, and other stakeholders can verify the authenticity of digital certificates using a web-based interface, which retrieves the certificate from the block chain network and verifies its authenticity. The system uses encryption, access controls, and secure authentication and authorization mechanisms to ensure the security and integrity of the certificates. The system is cost effective, convenient for students, and offers a simple method for stakeholders to verify the authenticity of certificates. The block diagram of the proposed architecture is given in Fig. 10.1

Advantages over the previous system:

- Reduction of Fraud.
- Decentralized Verification.
- Ease of Verification.
- Enhanced Security and Integrity.
- Future-Proofing.
- Scalability

A blockchain-based system for generation and verification with student applications, University approval and QR code verification by organizations

4. Model Implementation

The generating page accepts parameters such as stream, year, and roll number. Using the student ID and name, it retrieves the data with authorisation from the university or other sources. When user need to generate a certificate by just clicking on the button the data will transfer to smart contracts. When the data is received the smart contract generates a QR code and pushes the data into the Ethereum block chain.

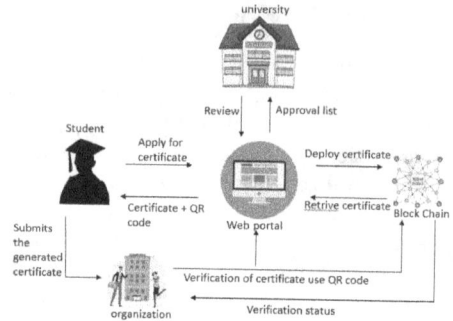

Figure 10.1 Proposed Architecture

When we want to test the certificate it will use the ether.

When the transaction is started the data is stored in the block chain. Instead of using a hash, the QR code serves as a unique identifier, providing a secure and efficient way to verify the authenticity of the digital certificate. The verification page allows stakeholders to verify the authenticity of digital certificates. To initiate the verification process, the user scans the QR code or upload the certificate in the website associated with the digital certificate. The QR code is then decoded, and the underlying data is retrieved from the Ethereum block chain. The smart contract is queried to retrieve the original data, which is then compared with the data retrieved from the QR code. If the data matches, the verification module returns a success message, indicating that the digital certificate is valid and authentic. The entire verification process is automated, ensuring that the authenticity of the digital certificate can be verified in real-time, providing an additional layer of security and trust.

1. System Architecture

After that, a block chain with timestamps is formed by every node. Each block's data can be simultaneously validated, and once input, it cannot be changed. The entire procedure is safe, transparent, and available to the public. The 2013 release of Ethereum Smart Contracts gave block chain technology a boost and led to the development of block chain 2.0. As stated in, Bitcoin primarily adopted blockchain 1.0 to address issues with cryptocurrency and decentralized payments. Blockchain 2.0 is used to convert assets using smart contracts, focusing on decentralizing the entire market. This creates value by allowing alternatives to Bitcoin to arise.

2. Data Flow and Transaction Lifecycle

An educational institution generates a digital certificate via a web interface to start the block chain-based certificate system's data flow and transaction lifetime. Using the institution's private key, the certificate data is hashed and signed before being sent as a transaction to the block chain network. The transaction is verified by peer nodes, and the blocks containing verified transactions are appended to the blockchain by ordered nodes. The certificate hash stays on the blockchain, but the original certificate is safely kept in decentralized storage. The stored hash matches when stakeholders, such employers, validate a certificate, guaranteeing its integrity and authenticity.

3. Consensus Mechanism

Practical Byzantine Fault Tolerance (PBFT): PBFT is used to provide high fault tolerance, ensuring that the system can function even if up to one-third of the nodes behave maliciously. This layer guarantees that certificates are generated and validated and securely stored, without relying on a single point of failure.

Delegated Proof of Stake (DPoS): DPoS is used to optimize scalability and transaction throughput. By selecting a subset of trusted validators, the system can produce blocks faster, making it ideal for managing certificates where speed and scalability are essential.

4. Performance Optimization

Smart Contract Optimization: By using the GO programming language for smart contracts, we have improved the overall performance of the system. GO's efficiency in handling concurrent processes allows the system to handle multiple transactions simultaneously, reducing latency and improving the speed of transaction validation.

Resource Management: The DPoS layer reduces the computational burden on the network by delegating validation tasks to selected nodes, thereby minimizing resource consumption and improving the energy efficiency of the system.

5. Testing and Validation

Test Environment: Using Docker containers, we set up the block chain-based certificate system on a test network consisting of 10 peer nodes and 5 order nodes. We assessed the system's scalability, performance, and security by simulating certificate issuance under varied loads. This ensured the network's resilience, transaction throughput, and encryption integrity for implementation in real-world settings across institutions.

Performance Metrics: The system was tested for latency, throughput, and fault tolerance under various scenarios.

Security Validation: We conducted penetration tests to assess the robustness of the AES encryption and the block chain's resistance to tampering. The system demonstrated strong resilience against common attacks, including the man-in-the-middle (MITM) and replay attacks.

4.1 Tools used

We have integrated with various tools and technologies to provide user friendly interface comprising of technologies like **React** for designing user interface and front-end application to stimulate the user requests, **GO programming** language is used to build smart contracts chaincode's, and we have used **IBM Hyper ledger fabric** as blockchain environment, we have used **MongoDB** as our primary database. Table 10.2 indicates project specification of the system.

Category	Details
Block chain Framework	IBM Hyper ledger Fabric (v2.2)—ethereum and decentralized block chain for generating certificates and validating
Encryption Algorithm	AES Encryption—Military-grade encryption used for securing EHRs/EMRs before they are added to the block chain. It ensures confidentiality even if the data is intercepted.
Smart Contract Language	Go programming—Chosen for its concurrency capabilities and performance optimization in block chain transaction process, enabling certificate generation.
Consensus Mechanisms	Layer 1: Practical Byzantine Fault Tolerance (PBFT)—Guarantees fault tolerance and security in the face of up to one-third malicious nodes. Layer 2: Delegated Proof of Stake (DPoS)—Selected for its ability to handle large-scale Certificate, reducing the validation burden and increasing block creation speed.
Front-End Framework	ReactJS—Responsive and dynamic UI for certificate providers and administrative users, designed for ease of data access while enforcing strict access control policies.
Backend Logic	GO programming for Smart Contract Execution—Optimized to handle 1,000+ simultaneous transactions, ensuring minimal latency in healthcare environments.
Database Layer	MongoDB—A NoSQL database for storing off-chain metadata and audit logs, chosen for its scalability and flexibility to handle complex healthcare records.
Performance Benchmarks	1000 Transactions per Second (TPS)—Achieved under load testing, optimizing block chain throughput for high-volume medical record systems.
Security Protocols	Penetration Testing—Conducted in collaboration with a third-party security firm, ensuring AES encryption and smart contracts resist common vulnerabilities.
Authentication Mechanism	Role-Based Access Control (RBAC) with OAuth2.0—Ensures that only authorized frame works, verified by certificate authority, can access the certificates in the block chain.
User Interface	ReactJS UI with Role-Specific Dashboards—A custom-built interface offering differentiated views for Students, Stakeholders, and administrative staff, tailored to each user's level of access.
Deployment Tools	Truffle-testing and deploying Ethereum. Remix IDE for smart contract's development and deployment
Data Flow Model	End-to-End Encryption Workflow—certificated is stored before transmission, validated by the smart contract, and decrypted only upon verified access by authorized users.

5. Results

When an organization or Study portals submitted the student details after the course completion the certificate is generated and it can be downloaded. The generated certificate can be deployed into the block chain for future use.

When the organizations or Stake holders to verify the certificate when the students submitted the certificates to the respective organizations. While verifying

Figure 10.2 Sample Certificate After Generating

the certificate it takes from the block chain and compares the QR code and student details.

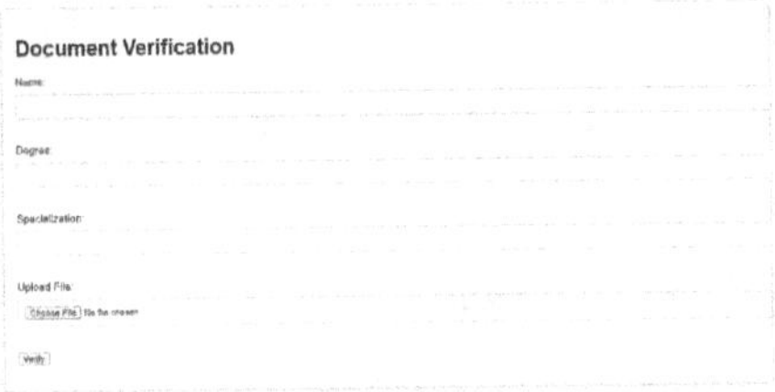

Figure 10.3 Student Information

6. Conclusion

Fake certificates are becoming a big problem, allowing unqualified people to use them to get what they want. Verifying documents is crucial for education, jobs, and identification. To solve this issue, we've created a decentralized web application that makes it easy to validate certificates. This system is fast, free, and trustworthy, helping to prevent forgeries and improve document verification.

References

[1] Nair, R., Zafrullah, S. N., Vinayasree, P., Singh, P., Zahra, M. M. A., & Sharma, T. (2022). Blockchain-Based Decentralized Cloud Solutions for Data Transfer. *Computational Intelligence and Neuroscience*, 2022, Article 8209854.

[2] Vinayasree, P., & Reddy, A. M. (2021). Blockchain-Enabled Hyperledger Fabric to Secure Data Transfer Mechanism for Medical Cyber-Physical System: Overview, Issues, and Challenges. *EAI Endorsed Transactions on Pervasive Health and Technology*, 9.

[3] Vinayasree, M. P. (2023). Blockchain-Based Design for Safe Data Sharing and Storage. *Paideuma Journal*, 16(2), 38-46.

[4] Vinayasree, P. (2021). Towards A Secure Electronic Health Record System Using Blockchain. *IOSR Journal of Computer Engineering*, 24(3), 15-19.

[5] Vinayasree P. (2021). Supporting Student Innovation Through Private Decentralized System. *IOSR Journal of Computer Engineering*, 24(3), 2278-8727.

[6] Vinayasree, P. (2024). A Scalable, Secure, and Efficient Framework for sharing Electronic Health Records Using Permissioned Blockchain Technology.

[7] Raja Rajeswari, T.S., Khaja Shareef, S. K., Khan, S., Venkatesh, N., Ali, A., & Sri Monika Devi, V. Department of Computer Science and Engineering. MLR Institute of Technology, Hyderabad, India.

[8] Ramesh, B., Afnan Ahmed, A., & Kumar, D. Assistant professor City Engineering College, Student Computer Science and Engineering, City Engineering College.

[9] Shanmuga Priya, R. Department of Computer Science and Engineering, Prathyusha Engineering College, Thiruvallur, Tamil Nadu.

[10] Kuo, T. T., & Ohno-Machado, L. (2017). Model and Protocol for Blockchain-Based Health Data Exchange. *Journal of Medical Internet Research*, 19(4), e124.

[11] Ali, M., Mian, A. A., & Wong, K. C. (2020). A Survey on Blockchain Technology in Healthcare: Applications and Challenges. *IEEE Access*, 8, 227260-227275.

[12] Xu, X., Weber, I., & Staples, M. (2019). Architecture for Blockchain Applications in Health Information Systems. *Journal of Healthcare Informatics Research*, 3(4), 1-18.

[13] Hölbl, M., Gajdoš, K., & Pahl, J. (2021). Blockchain Technology for Digital Health: A Systematic Review. *Journal of Medical Internet Research*, 23(8), e23789.

[14] Muthulakshmi, P., & Priyadharshini, A. (2021). A Review on Blockchain Applications in Healthcare: Opportunities and Challenges. *Journal of King Saud University—Computer and Information Sciences*. https://doi.org/10.1016/j.jksuci.2021.02.002

[15] Gafurov, A. M., & Tikhonov, A. V. (2020). Blockchain and the Future of Healthcare: How This Technology Will Change Healthcare. *Health Information Science and Systems*, 8(1), 1-11.

[16] Dehghantanha, A., & Aburomman, A. A. (2021). Blockchain Technology and the Internet of Things: A Systematic Review of the Applications in Healthcare. *Health Informatics Journal*, 27(4), 14604582211013418.

[17] Reddy, S. G., & Hsu, C. (2019). Healthcare Data Exchange Using Blockchain: A Comprehensive Survey. *IEEE Access*, 7, 170037-170054.

[18] Kuo, T. T., & Ohno-Machado, L. (2018). Blockchain in Healthcare: A Comprehensive Review and Directions for Future Research. *Health Informatics Journal*, 24(3), 265-276.

11 Grid Integration of Inverter Based Generation and Battery Energy Storage Systems: A Review of Challenges

Mahabaleshwara Sharma K.[1] and Nagesh Prabhu[2]

NMAM Institute of Technology-Nitte (Deemed to be University), India

Abstract

The present day power grids are mainly powered by the electromechanical systems having the rotational inertia of their own. However, going by the present trend, it could be concluded that the future power grids will have a major part of the power being supplied by the inverter-based resources (IBR). So, there is an urgent need for the sophisticated grid-forming controllers that ensure the stability of the grid irrespective of the level of penetration of the IBRs. Transitioning to a grid with more IBRs and energy storage devices pose a lot of technical challenges since the future power grids are expected to operate on the basis of the mechanical inertia of the synchronous generators and the control response of the numerous and diverse IBRs. This paper identifies the major challenges associated in integrating the IBRs and the Battery Energy Storage Systems (BESS) with the existing power grid.

Keywords: GFL, GFM, IBR, RES, Battery Energy Storage Systems (BESS), Grid Forming Controllers.

1. Introduction

Power grid is evolving from being a traditional grid of power generation systems, mostly grid following resources (GFL) such as hydro, thermal and nuclear to a non-traditional grid of renewable energy sources (RES), mainly, solar, wind, etc. plus BESS and the hybrids of batteries, fuel cells, ultra-capacitors [1]. Thus the future grid will be a combination of the grid following resources (GFL), grid forming resources (GFM) and energy storage devices with the balance tilting heavily towards GFM [2]. Not only the power output from the RES is intermittent in nature, but also, they come in a variety of forms and capacities such as residential scale to utility scale that add to the complexity of the system [3]. These RES' are connected to the grid through various converters, naturally lacking the rotational inertia. Collectively, these generation technology devices are referred to as Inverter Based Resources (IBR) [4].

On the power consumption side, major part of the power is being consumed by the commercial and industrial loads such as transformers, synchronous machines, induction motors etc. that absorb apparent power, from the power grid during their regular operation, bringing down the system

Email: mahabaleshwara@nitte.edu.in[1], prabhunagesh@nitte.edu.in[2]

DOI: 10.1201/9781003661917-11

power factor below acceptable limits [5]. Failure to meet the reactive power requirement will result in the increased system losses, voltage instability and reduced power transfer capability. The requirement of reactive power can be fulfilled by adopting any of the following three techniques [6]:

i. Installing power factor correction devices
ii. Installing the reactive power compensation devices that reduce the flow of reactive power in the grid resulting in the reduction of the total power loss in the power grid
iii. Installing the advanced grid control systems and hence, improving the performance of the grid.

Meanwhile, it has to be ensured that the power grid is not over compensated by the reactive power, since it would result in overvoltage conditions, grid instability, resonance and harmonic conditions, wasted energy and loss of system efficiency [7].

The above analysis leads to the conclusion that for maintaining the optimum voltage quality and reduced power loss leading to higher efficiency, it is imperative to have the optimal reactive power in the grid. The appliances connected to such a grid will have long life and operate with high efficiency. Thus providing the reactive power compensation is one of the cheapest energy saving techniques adopted across the globe.

This research paper will provide the high level view of the grid-forming IBR controls and their impact on grid stability when the GFM resources and BESS are connected to the traditional synchronous generator connected power grid. The paper will discuss the various challenges faced while operating the future grid.

2. Literature Survey

2.1 Grid Following (GFL) Resources

As discussed already, in a grid, for its stable operation, supplying either the active power or reactive power compensation or an optimal combination of both is very much essential. Normally, the GFL resources are synchronized with the grid very fast so that they can meet their primary objective of controlling the current output (both active and reactive) during challenging system conditions such as various types of open circuit or short circuit faults anywhere in the grid, unbalanced load etc. Thus, GFL resources ensure the supply of stable and uninterrupted power to the load [8].

Some of the commonly used systems where GFL technique is being employed are wind turbines, solar photovoltaic (PV) cells, most Battery and Energy Storage Systems (BESS), Distributed Energy Resources (DERs) to name a few.

2.2 Grid Forming (GFM) Resources

For the GFM resources, it is not mandatory to synchronize them with the external grid very fast unlike GFL resources to provide the stable and predictable output since their primary control objective is to maintain a constant internal voltage phasor [9].

In certain applications such as microgrids, the voltage phasor is normally held constant, while it operates in an islanded mode whereas in the traditional grids connected to the IBRs, the internal voltage phasor is controlled to maintain synchronization with the other elements, controlling the active and reactive currents. This could be implemented in a traditional grid connected to IBRs by holding a constant voltage phasor in the sub-transient to transient time frame that provides stability during system faults (potentially good or weak grids and high IBR penetration), unbalanced load conditions etc. [10].

Maintaining the voltage phasor constant helps the IBR to respond almost instantaneously during abnormal operating conditions by pumping in the active and reactive currents.

When GFM controls are applied in microgrids, this synchronization functionality is either totally removed or kept at a minimum. This allows the system to operate in the islanded mode [11].

Some of the commonly used terms to refer GFM mode are Black-start mode, Microgrid mode, Self-sync mode, Grid firming mode, Virtual Synchronous Machine (VSM) mode, Island mode, Voltage source mode or constant voltage mode, etc.

With regard to the GFL and GFM mode of operation, the following points worth noting:

- The GFL and GFM converter controls are limited by the physical constraints, both electrical such as voltage, current, power and energy handing limits as well as mechanical constraints and the power grid.
- The performance requirements for both the GFL and GFM plant remain the same unless otherwise explicitly specified.

GFL sources depend on the synchronous generators for their stable operations whereas GFM sources are independent of synchronous generators. GFM sources should be such that they can handle fault current in addition to the normal operation behavior when used as black starters. Over the years, the grid forming technology got evolved in the following order:

a. GFL Converter (Current Source Converter)
b. GFM (early generation) Converter (Voltage Source Converter)
c. Grid Forming (example of current generation) = Grid Forming Inverter + Generator Dynamics.

2.3 Differences between GFL and GFM Behaviour

The definitions of GFL inverters and GFM inverters are usually based on some specific control design.

GFL requires measurement of actual grid angle and adjusting its output voltage to get the referenced output current (Current Source behavior) whereas GFM requires rotating output voltage

that synchronizes in accordance with the impedance / grid demands (Voltage Source behavior).

The important features of Grid Forming are ideal voltage source with coupling impedance (Thevenins equivalent), provision of inherent voltage response, provision of "virtual" inertia and that of GFM converter is like a voltage source with inertia synchronizing behavior and damping [12]. Further, Table 11.1 presents comparison of GFL and GFM methods.

Table 11.1 Comparison of GFL and GFM

Grid Following (Current Source)	Grid Forming (Voltage Source)
Generic features: • Current Controlled • Active and Reactive Power Controlled • Impossible to work in isolation; always needs a grid to operate • There is no direct control of voltage and frequency	Generic features: • Voltage controlled • Frequency controlled • Can work in isolation • There is no direct control of current
During transients (due to disturbances), the output current is nearly constant; later I_{ref} is adjusted by external controls	During transients (due to disturbances), the internal voltage is nearly kept constant; later E and delta (δ) are adjusted by the external controls.
Basic Control Objective is to deliver a specified amount of power to the grid	Basic Control Objective is to set up a grid voltage and frequency
The controlled output quantities are the magnitude and phase angle of current.	The controlled quantities are magnitude and frequency of only the output voltage.
Needs a stable voltage at the terminal.	Doesn't need a stable voltage at the terminal
A phase locked loop (PLL) is mandatory.	It does not have a PLL.

2.4 Grid Forming Inverters

Both GFL and GFM inverters supply power to the grid when there is a need. However, GFM inverters maintain a constant voltage and frequency even when the load draws active and reactive powers. For all practical purposes, GFM inverter could be considered as an infinite bus. This is possible only when there is a stored energy. For storing the additional power, additional storage systems are necessary, thereby increasing the cost of the system. It also requires capacity for extra current for both reactive power and short circuit current that come with an added cost.

In the traditional grids involving only the synchronous machines, the laws of electromagnetism, provide the required grid phase angle whereas in the conventional IBRs, the systems has to be manipulated using electronic circuitry what is commonly referred to as virtual synchronous machine to provide the required phase angle.

Till recently, the GFM inverters are mainly used in the design of inverter-based microgrid as well as transmission systems that are likely to have a low fault current and rotational inertia. However, it is anticipated that in the future the GFM inverters play a critical role to supply reliable power to the grid. Then, establishing the coordination amongst the various GFMs of varied capacity and other devices in the grid will be a great challenge around the world.

2.5 GFM Inverter Versus GFM Resource

In its simplest form, a GFM inverter may be considered as a combination of GFM resource and an inverter connected in such a way that the combination is capable of providing a constant voltage and frequency during disturbances.

Angle jump or Rate of Change of Frequency (RoCoF) triggers different behavior from GFM than GFL. Since the power generated by solar cells and wind are unpredictable and intermittent, balancing the load in such systems is more complex than with the BESS. To meet the fast bidirectional active power fluctuations, additional hardware for storing the energy may be required depending on the performance requirements.

2.6 Basics of Present Day IBR— Grid Interconnection

It is a known fact that since the IBRs do not have electromagnetic coupling with the power grid, there is no inherent internal inertia. So, to emulate inertia, IBRs normally use a PLL. PLL ensures that the IBRs are synchronized and locked to the grid. All controls within an IBR treat the evaluated PLL phase angle as a reference to calculate the current to be injected to the grid. Figure 11.1, depicts the interconnection of a synchronous generator, a photovoltaic generator (representing an IBR) to an infinite bus.

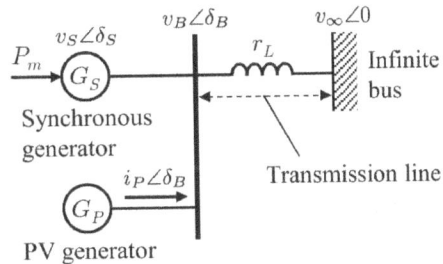

Figure 11.1 Present Day IBR-grid Interconnection

Consider a scenario, wherein all the sources are synchronous generators with large generation or load changes, or frequency drop and fall needs to be arrested, the frequency should stabilize within, say, 60 seconds. Then, it needs controlled injection of reactive power from the grid with a smaller inertia constant. Larger the value of RoCoF implies lesser the time to deploy frequency response reserves, preventing the activation of Under Frequency Load Shedding (UFLS), ultimately resulting in the wide-spread load shedding.

Because of the inherent design of the synchronous machines, they can withstand larger RoCoF whereas for GFMs, it is a real challenge, irrespective of whether it is a bolted fault or not. Replacing synchronous machines with the IBRs would be useful since they operate in constant P-Q mode, analogous to smaller synchronous generators operating with the similar RoCoF.

2.7 Generic Modelling of GFM Behaviour

Planning studies were conducted at a time frame when exact models of inverter equipment were not available. If studies are carried out only after exact models are available, it might be too late to implement any system upgrades. Here, generic models play a role in enabling planning studies to be carried out. But with different grid forming control methods, there won't be a need for many different generic models.

Almost all existing installations follow the virtual synchronous machine model [13]. However, other design types such as droop controlled GFM models are also gaining prominence.

2.8 Virtual Synchronous Machine GFM Model

The Virtual Synchronous Machines (VSMs) [14] is used emulate the synchronous machines in power systems having a high degree of penetration of IBRs. Mainly there are three basic topologies of VSM. The earliest topology of VSM is named VISMA. It uses fifth order equations to simulate the steady state and transient behavior of the synchronous machine. VISMA algorithm uses three phase voltage as input and three phase reference current as output. Though VISMA provides a reasonable accuracy during grid connected operation, its performance is abysmally low during the islanded operation. Also, the voltage contains a lot of harmonic components during low load or no-load operations and VISMA doesn't inject any reactive power to the grid for stabilization.

The next VSM topology is known as synchronverter. It offers similar kind of dynamic characteristics as that of a synchronous generator. But, it emulates only the behavior

of the round rotor synchronous generator (i.e., $X_d=X_q$). It has the voltage and frequency droop control loops that enable the parallel operation of many synchronverters. Unlike VISMA, synchronverter has a dedicated control loop for reactive power injection. The tuning of the parameters of the synchronverter is possible to replicate the actual performance of the physical synchronous generator. However, the synchronverters exhibit the undesirable phenomena such as hunting, system becoming unstable because of under-excitation, etc.

The third VSM topology, VSM0H, offers zero inertia and it is very much similar to the conventional droop control. Though VSM0H provides zero inertia, it has a fast acting frequency droop slope. It can operate in a 100% IBR penetrated grid. However, an infinite bus is required to initialize the system.

2.9 Droop Controlled Grid Forming Inverters

A GFM inverter behaves like a controllable voltage source behind impedance. Two ideal voltage sources cannot be connected in parallel. For the stable operation of the controller, the coupling reactance X_L should be set anywhere between, say, 5%–20%. Then, the active power P will be proportional to the load angle δ and the reactive power Q will be proportional to the voltage E.

Droop control can be used to control the grid that consists of both the conventional synchronous generators and IBRs. When multiple sources operate in parallel, to ensure that the load sharing takes place as per the rating of the individual sources, droop control is used. P versus f droop ensures the phase angles of all the voltage sources are synchronized. Q versus V droop avoids large circulating VARs between the different sources. When multiple GFMs operate in parallel, any kind of transient leads to an increase in the output power of the GFM inverter. Fig 11.2 shows P - f droop curve and Fig 11.3 shows Q - V Droop curve of the considered system.

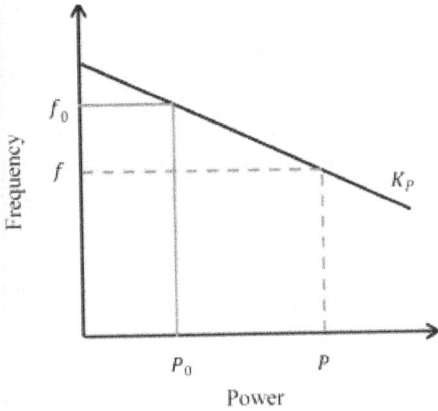

Figure 11.2 P - f Droop Curve

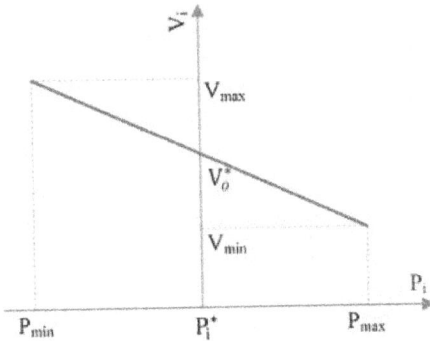

Figure 11.3 Q - V Droop Curve

3. Challenges

The analysis of challenges faced by the power engineers while integrating the IBRs and the storage systems with the grid are given below:

Do all types of IBRs exhibit GFM behavior? [15]

- GFM behavior depends on the inherent characteristics of the technology adopted to generate electrical power from the IBRs and the technology adopted for manufacturing the energy storage systems such as battery.
- Though the inverter handles the reactive power response, active power response depends on availability of the actual energy.

GFM performance requirements needed [16]:

- Reactive power capability
- Steady state and dynamic voltage requirement
- Steady state and dynamic frequency requirement
- Frequency and voltage ride-through
- Requirement on voltage harmonics
- Grounding of the GFM plant
- Temporary overload / over current capability (for providing inrush current, cold load pick up, etc.)
- Black start capability
- Fault contribution levels

The inverters require a lot of synchronous generators for their stable operation, both transient and steady state, whereas IBRs face a lot of challenges because of lack of inherent inertia [17].

Inverter-based resource performance risks may be categorized as Inverter performance issues, plant controller interaction issues and plant protection setting issues.

Analysis is required on the control and operation of GFMs under the following areas [18]:

- How should GFMs respond to short circuit faults?
- How different fault-ride-through behavior GFMs impact the system transient stability and protection?
- How to develop interoperability guidelines and functional requirements to guarantee the stability and reliability of power systems that have IBRs with various types of controls?
- How to model and simulate future power systems with millions of power electronics devices?
- Full sized models? Or aggregated models?
- Where is the boundary between phasor based simulation and Electromagnetic Transient (EMT) simulation for IBR dominated system?
- If it is intended to move towards a 100 % inverter based system, does that mean power system engineers will need to

simulate a 100% black-box- model based power system?

- What will be the impact on the dynamic stability of the system when the IBR penetration approaches closer to 100%?
- How to handle the stability of the grid when only IBRs are introduced into the grid?
- In case, there is a partial penetration of IBRs, how will they impact the performance of other synchronous generators?
- What are the criteria that should be fulfilled for the parallel operation of GFL and GFM sources in a single grid?
- What is the expected ratio of GFL sources to GFM sources for the stable operation of the system?
- What should be the optimal combination of IBRs and the synchronous machines for the peak performance of the grid?
- Are there any new parameters that should be controlled for the proper operation of the system, if there are any?
- What is hindering the integration of IBRs to the grid?

How much IBRs we can integrate into the grid by using GFMs? [19]

- Both GFL and GFM can have 100% renewable energy resources only under some specific conditions. But what are those conditions?
- The weaker the grid is, the harder it is to stay stable
- Compared to GFL with or without droop, GFM has higher stability margin.
- For GFL, the regular droop control can improve the steady state stability margin whereas for GFL with and without droop, weak coupling PLL inverter mode is the critical mode to cause the instability.
- For GFM, the level of instability varies based on different system configurations. But, it should be analyzed to verify whether connecting the GFM near the load centre or away from it would improve its stability and the reason for it. Also, in either of the cases, what would be the impact on steady state and

transient stability? In any case, if the grid is drifting towards instability, how it could be brought to the stable region?

General Challenges [20]:

- Is it possible to use the IBRs dominated grid to support long-distance power transmission? Any limitations? (for interconnection grid).
- In terms of different stability issues, can they be eliminated with GFM or only minimized?
- How can the findings / analysis be generalized for a realistic system? (Island grid and interconnection grid).
- Understand the fundamental dynamic stability mechanism of zero inertia and non-zero inertia IBR dominated grid?
- How much IBRs we can integrate into the grid by using GFMs?
- Is it possible to have 100% penetration of IBRs?
- What is the control strategy that should be adopted while switching over from the grid mode to the islanded mode or vice-versa, given the fact that the GFM converters maintain the voltage and frequency constant?
- Is it possible to model the system in such a way that during the transition either from islanded-mode to the grid-mode or vice versa, the grid stability could be improved using any Artificial Intelligence (AI) techniques?
- A lot of work is documented in the small signal stability of GFM resources, but the work on transient stability is minimum. So, after a fault, how the system stability could be restored?
- Is it possible to eliminate (or at least minimize) the sub-synchronous interaction (SSI) challenges in IBR integrated power grids?

Challenges while operating a power system on 100% Distributed Resources [21]:

- Possible power disruptions while using GFMs
- While the power is generated using the GFMs, if there is insufficient load,

how to handle the extra power generated? If BESS are used to absorb the extra power, how much power could be stored in the BESS?

- How to prevent over charging of the BESS?
- How to handle under frequency load shedding?
- How the load balancing could be done?
- In the worst case scenario, is it possible to have the system black – start?

Challenges in the system strength and performance expectations of GFM / GFL:

- From the point of view of the performance, the issues to be handled because of IBR penetration are given below:
  - What is the limit of tolerance for the voltage angle jump without the system becoming unstable for both GFL and GFM cases?
  - Naturally, the GFM is capable of doing this within the physical limitations of the equipment. But, how to calculate the limit for the given system?
- There won't be any major impact on GFL or even if it exists, there would be a delayed response since it is specifically designed for it. With regard to the RoCoF, the tolerance limit for RoCoF events without tripping for both GFL and GFM cases should be calculated.

It is worth noting the fact that low system strength could be improved by the voltage control instead of reactive power control in the plant level. Also, the GFL terminal voltage control could be achieved with low system strength.

4. Conclusion

Since the synchronous generator based power system is in the verge of transforming to become a combination of synchronous generators, IBRs and BESS, the solutions to the challenges identified above should be addressed on priority basis. The power generation companies should establish standards / protocols in support of efficient and reliable large scale integration of microgrids to fulfill the ever increasing demand of power across the globe. The structure preserving model for managing the complex systems should be established. Ex: Small signal coordinated frequency stabilization and regulation. The role of non-linear primary control during extreme events should be finalized in cases such as during the occurrence / after faults, loss of large generation, microgrids that depend on the presence of wind and the variations in solar radiation.

The following system control enhancements could be implemented for the stable operation of the grid:

- Tertiary level microgrid controller: Should have adaptive performance metrics and optimize over all controllable equipments/online corrective/improved preventive.
- Secondary control droops: It is valid only under certain conditions.
- Primary control: This is a mix of the above two controls. But it is very difficult to control either power or the rate of change of power while ensuring the voltage and frequency lie within the tolerance limits.

In this review paper, a detailed analysis is carried out on the working of GFL and GFM based resources and inverters as well as the challenges in integrating the IBRs and BESS with the traditional grid. Proper modeling, simulation and hardware implementation should be done to find out the solutions for all these challenges in the days to come.

Acknowledgment

Both the authors are indebted to Visvesvaraya Technological University, Belagavi, India and NMAM Institute of Technology, Nitte (Deemed to be University), India for all the support and encouragement received during the course of work.

References

[1] Debra Lew, D. (2021). WECC/ESIG Grid-Forming Inverter based Resources Workshop.

[2] Quint, R. (2022). WECC/SIG Webinar on "Modeling Inverter Based Resources: Findings, Observations and Challenges ahead."

[3] Kablan, M. (2022). Lessons Learned from Commissioning and Operating an Industry Grade Microgrid: IBR Stability and Control. Fall Technical Workshop, UNIFI.

[4] Badrizadeh, B. (2022). Assessment of GFL and GFM inverter and synchronous condenser connection in Australia: Lssons Learned. UNIFI Workshop.

[5] Alegria, E., Brown, T., Minear, E., & Lasseter, R. H. (2014). CERTS microgrid demonstration with large-scale energy storage and renewable generation. *IEEE Transactions on Smart Grid*, 5(2), 937–943.

[6] Tan, J. (2022). Comparative stability analysis of grid forming and grid following inverters in low inertia power systems. UNIFI Workshop.

[7] Bangash, K. N., Farrag, M. E. A., & Osman, A. H. (2019). Investigation of energy storage batteries in stability enforcement of low inertia active distribution network. *Technology and Economics of Smart Grids and Sustainable Energy*, 4, 1. https://doi.org/10.1007/s40866-018-0059-4

[8] Amin, M. R., Negnevitsky, M., Franklin, E., Alam, K. S., & Naderi, S. B. (2021). Application of battery energy storage systems for primary frequency control in power systems with high renewable energy penetration. *Energies*, 14, 1379. https://doi.org/10.3390/en14051379, 3 March 2021

[9] Lin, Y., Eto, J. H., Johnson, B. B., Flicker, J. D., Lasseter, R. H., Villegas Pico, H. N., Seo, G.-S., Pierre, B. J., & Ellis, A. (2020). Research roadmap on grid-forming inverters. *National Renewable Energy Laboratory*.

[10] Unruh, P., Nuschke, M., Straub, P., & Welck, F. (2020). Overview on grid-forming inverter control methods. *Energies*, 13, 2589. doi:10.3390/en13102589

[11] Alshahrani, S., Khan, K., Abido, M., & Khalid, M. (2024). Grid-forming converter and stability aspects of renewable-based low-inertia power networks: Modern trends and challenges. *Arabian Journal for Science and Engineering*, 49, 6187–6216. https://doi.org/10.1007/s13369-023-08399-z

[12] Zhao, F., Zhu, T., Li, Z., & Wang, X. (2024). Low-frequency resonances in grid-forming converters: Causes and damping control. *IEEE Transactions on Power Electronics*. doi: 10.1109/TPEL.2024.3424296.

[13] Bikdeli, E., Islam, M. R., Rahman, M. M., & Muttaq, K. M. (2022). State of the art of the techniques for grid forming inverters to solve the challenges of renewable rich power grids. *Energies*, 15, 1879. https://doi.org/10.3390/en15051879

[14] Muftau, B., & Fazeli, M. (2022). The role of virtual synchronous machines in future power systems: A review and future trends. *Electric Power Systems Research*, 206, 107775.

[15] Wang, Y., Wang, R., Li, M.-J., & Zhang, P. (2024). Damp and droop coefficient stability region analysis for interlinking converters with virtual asynchronous machine controllers in virtual energy storage systems. *Journal of Energy Storage*, 99, 113306.

[16] Sun, C., Ali, S. Q., Joos, G., & Bouffard, F. (2022). Design of hybrid-storage-based virtual synchronous machine with energy recovery control considering energy consumed in inertial and damping support. *IEEE Transactions on Power Electronics*, 37(3), 2648–2666.

[17] Du, W. (2022). Transient and dynamic modeling of droop controlled grid forming inverters at scale—A journey from a single microgrid to an interated T&D system with 10000+ inverters. UNIFI Seminar Series.

[18] Du, W., Chen, Z., Schneider, K. P., Lasseter, R. H., Pushpak, S., Tuffner, F. K., & Kundu, S. (2019). A comparative study of two widely used grid-forming droop controls on microgrid small signal stability. *IEEE Journal of Emerging and Selected Topics in Power Electronics*. doi: 10.1109/JESTPE.2019.2942491

[19] Zhang, J., Wang, H., Liu, G., Tian, G., Rivera, M., & Wheeler, P. (2024). Virtual synchronous generator control strategy of grid-forming matrix converter for renewable power generation. *IEEE*

Access, 106672–106684. doi: 10.1109/ ACCESS.2024.3436552.

[20] Wald, F., Tao, Q., & Carne, G. D. (2024). Virtual synchronous machine control for asynchronous grid connections. *IEEE Transactions on Power Delivery*, 39(1), 397–406.

[21] Huang, M., Guerrero, J. M., Fernando, T., Li, S., & Tse, C. K. M. (2021). Stability and robustness of power grids with high penetration of power electronics. *IEEE Journal on Emerging and Selected Topics in Circuits and Systems*, 11(1). doi: 10.1109/JETCAS.2021.3058025

12 Video Secure Storage System Using Blockchain and IPFS for Enhanced Security

Prem A. M.[1], Hirthick T.[2], Ram Kumar S. S.[3] and Dheenathayalan S.[4]

National Engineering College, Kovilpatti, India

Abstract

The main focus of this paper is to address the challenges faced by centralized video management systems, such as data breaches, hacking, tampering, and single points of failure, which can result in the loss or unauthorized modification of critical video footage. The proposed system overcomes these issues by leveraging blockchain technology and decentralized storage to ensure continuous availability, integrity, and security of video data. Developed using the MERN stack, the system integrates features such as user authentication, video recording, and uploads to Pinata IPFS, a decentralized storage network, with the IPFS hash securely stored in MongoDB. In cases where videos are deleted or become inaccessible on IPFS, the system retrieves them using the stored hash. If both IPFS and MongoDB fail, the video is propagated across a distributed network to ensure constant availability. The integration of blockchain through Solidity and Linea Sepolia secures each video upload with an immutable transaction. Furthermore, user passwords are double-hashed for enhanced security, and a tamper-check feature is included to allow users to verify the integrity of their videos. By combining decentralized storage, blockchain technology, and robust security measures, this system provides a resilient and secure video management platform that addresses the vulnerabilities of centralized systems.

Keywords: Blockchain technology, data security, decentralized storage

1. Introduction

Recent advancements in computer vision and blockchain technology have transformed video management systems, providing opportunities to address critical challenges in storage, security, and data integrity. As urban environments increasingly rely on video surveillance, particularly in applications such as smart cities, transportation, and security, the need for secure and resilient video storage solutions has become more urgent [1]. For instance, ride-hailing services like DiDi use real-time video monitoring to enhance safety for both drivers and passengers, underscoring the necessity for robust data management systems that protect user privacy while maintaining operational integrity [2].

Email: 2112014@nec.edu.in[1], 2112039@nec.edu.in[2], 2112007@nec.edu.in[3], ddhayalan@nec.edu.in[4]

DOI: 10.1201/9781003661917-12

The exponential growth of video content generated by smart cameras has led to significant challenges in data storage and management. Current estimates suggest that billions of hours of video footage are recorded annually, resulting in overwhelming data volumes that strain traditional storage infrastructures [3]. Centralized storage methods, which depend on a single point of failure, present risks such as data loss caused by server outages, cyberattacks, or even natural disasters [4].

To address these issues, decentralized storage solutions, such as the InterPlanetary File System (IPFS), offer a promising alternative. IPFS distributes video data across a network, improving availability and fault tolerance. This decentralized architecture ensures that video content remains accessible, even if some nodes in the network experience failures, which is vital for continuity in critical applications [5].

Our proposed system incorporates blockchain technology to secure video transactions and ensure tamper-proof storage. By integrating blockchain, each video upload is backed by a secure, immutable transaction, enabling users to verify the integrity of their videos in real-time [6]. This is particularly important in scenarios where the authenticity of video footage is critical, such as legal proceedings or security investigations [7].

A tamper-check feature is also integrated into the system, allowing users to detect any unauthorized alterations to their videos, which enhances trust in the stored content and mitigates concerns over data manipulation [8]. To further improve storage efficiency, we employ erasure coding techniques, which segment video data into smaller blocks distributed across multiple storage nodes. Unlike traditional replication methods, which duplicate entire copies of data, erasure coding reduces storage costs by enabling data reconstruction from remaining blocks if a node fails [9].

In addition to erasure coding, Proposed System Utilizes lightweight AI for video preprocessing, which optimizes data handling. By processing video locally before uploading it to IPFS, we minimize data transmission, optimizing bandwidth usage and reducing latency [10]. This approach not only improves efficiency but also facilitates faster retrieval and playback of video content, crucial for real-time applications [11].

User authentication is another cornerstone of our design. To enhance security, user passwords are double-hashed before storage, safeguarding against unauthorized access and potential data breaches [12]. These stringent security measures help ensure compliance with data protection regulations, such as GDPR, and reinforce user confidence in our system.

Furthermore, the collaboration between Internet of Things (IoT) devices and video management systems emphasizes the integration of AI to streamline video data handling. Recent studies suggest that AI can facilitate real-time analytics and preprocessing, boosting the overall performance of video management systems. By incorporating such technologies, the Proposed Systemcan efficiently manage varying workloads and prioritize critical video data, improving operational effectiveness [4].

Scalability is another key consideration in Proposed System design. As organizations expand their video surveillance infrastructures, the need for scalable solutions becomes increasingly important. Decentralized storage systems allow our platform to efficiently manage large volumes of video data, while maintaining flexibility and minimizing operational complexity. The modular structure of the system enables organizations to add more cameras and storage nodes as required, ensuring scalability in response to changing needs [1].

In conclusion, our video secure storage system addresses the significant challenges faced by traditional video management solutions by combining blockchain technology, decentralized storage systems, erasure coding, and advanced security protocols. Our approach ensures continuous availability and integrity of video data, setting a new standard for secure video management. By integrating these decentralized technologies with robust security measures, we aim to create a resilient and efficient video management solution suitable for modern applications in an increasingly digital world.

As video surveillance continues to evolve, the Proposed System Is designed to meet future demands for data integrity, availability, and security, ensuring video content remains a valuable asset for both organizations and individuals. This comprehensive framework not only solves current challenges but also anticipates future developments in video technology, making it a forward-thinking solution for secure video management. By embracing innovation and adopting best practices, we aim to redefine the standards of video data security and management, paving the way for a safer and more efficient future.

2. Literature Review

2.1 Centipede: Leveraging Distributed Camera Crowd for Cooperative Video Data Storage

Centipede introduced a framework designed to leverage distributed camera crowds for cooperative video data storage. The study highlighted the efficiency of using a distributed network of cameras for video data storage and the potential benefits in video surveillance systems. The study concluded that by using crowd-sourced video data, security can be significantly enhanced without relying on centralized storage systems. The model reduces the risks of data tampering and loss, providing an effective and decentralized solution for secure video data management. However, scalability and data privacy remain key concerns in the application of such systems, which need further exploration to optimize real-world deployment [1].

Distributed camera networks enhance video data storage efficiency and security.

2.2 Security Risks in Home Security Camera Services

A study in 2020 examined the potential risks associated with home security camera services, where they discovered that home camera privileges can inadvertently expose personal privacy. The study specifically focused on vulnerabilities in cloud-based security systems, which could potentially be exploited due to misconfigurations in privacy settings. Their analysis found that unauthorized access to private video footage could be easily achieved if security measures are weak or neglected. The study calls for stronger user authentication protocols and the integration of better privacy controls to mitigate these risks, ensuring the safety of users' data in IoT-based surveillance systems [2].

Home security camera services require stronger privacy and security measures.

2.3 Erasure Coding for Distributed Storage

A study in 2018 provided an overview of erasure coding, a method used in distributed storage systems to enhance storage efficiency and reliability. Unlike traditional data replication methods, erasure coding divides data into smaller blocks, which are then distributed across multiple nodes. This method not only improves fault tolerance by allowing the reconstruction of data even if a node fails, but also reduces storage costs by eliminating the need for full data replication. The study suggested that erasure coding is particularly valuable for systems dealing with large volumes of data, such as video surveillance, where efficient data management is crucial [3].

Erasure coding enhances distributed storage efficiency and fault tolerance.

2.4 AI and IoT Collaboration for Surveillance Video Pre-processing

A study in 2020 explored the collaboration between AI and IoT in surveillance video pre-processing. The paper discussed how lightweight AI techniques can be used to process video footage in real-time, reducing the amount of data that needs to be transmitted for storage. This approach not only optimizes bandwidth usage but also enables quicker retrieval of video content. The authors emphasized the importance of AI-powered edge devices in improving the overall efficiency of surveillance systems, allowing for faster video processing, reduced latency, and lower operational costs. The integration of AI with IoT devices makes it possible to perform pre-processing on the device itself, ensuring that only the most relevant data is stored and transmitted [4].

AI-powered pre-processing enhances surveillance video efficiency.

2.5 Scaling Live Video Analytics with Distributed Edge Intelligence

Distream proposed a solution for scaling live video analytics through workload-adaptive distributed edge intelligence. Their work focuses on optimizing the use of edge computing resources to handle the growing demands of live video streaming and real-time analytics. The research demonstrated that distributing video processing across multiple edge devices could significantly reduce latency and computational load on central servers. By adapting the workload according to the capabilities of the edge devices, the system can ensure efficient use of resources while maintaining high-quality video analysis. This solution is particularly relevant for real-time video surveillance and other critical applications that require low-latency data processing [5].

Distributed edge intelligence improves live video analytics scalability.

2.6 Realtime Anomaly Detection in Distributed Edge Computing for Video Surveillance

A study in 2023 explore the use of real-time anomaly detection in distributed edge computing for video surveillance. Their

research demonstrates how distributed edge computing environments can reduce latency and improve response times in security applications. By processing video feeds locally at the edge, the system is able to detect anomalies in real-time, allowing for quicker intervention and enhancing overall system performance. The study highlights the importance of rapid anomaly detection in ensuring timely responses to suspicious activities, which is crucial for maintaining security in real-time surveillance systems. This approach not only improves the efficiency of security monitoring but also optimizes network bandwidth by minimizing the amount of data transmitted to centralized servers [6].

Real-time Anomaly Detection in Distributed Edge Computing for Video Surveillance.

2.7 AI-driven Pre-processing for Efficient Video Data Management in Surveillance Systems

A study in 2023 explore how AI-driven pre-processing techniques can enhance the efficiency of video data management in surveillance systems. The study demonstrates that by using AI models to process video footage in real-time, the amount of data that needs to be transmitted and stored can be significantly reduced. This approach not only improves bandwidth utilization but also enables faster video retrieval and more efficient data management. The research emphasizes the value of integrating AI with IoT devices to optimize the overall performance of surveillance systems, making them more efficient and responsive for real-time video analytics [7].

Enhancing Video Data Management with AI-driven Pre-processing in Surveillance Systems.

2.8 Related Work

The integration of blockchain with video storage systems has gained momentum, driven by the need for enhanced security and decentralized storage. Key advancements have influenced the development of our system, particularly in decentralized file systems, blockchain for content integrity, and hybrid storage models.

2.9 Decentralized Storage

IPFS (InterPlanetary File System) offers content-addressable storage, enabling efficient and decentralized file sharing. Research like *"IPFS: A Distributed File System for the Decentralized Web"* (Benet, 2014) underscores its utility in secure and scalable storage. Platforms such as Pinata enhance IPFS usability, providing reliable tools for managing media files. Proposed System leverages these technologies to mitigate centralization risks and ensure persistent storage.

2.10 Blockchain for Integrity

The immutability of blockchain, as demonstrated by Nakamoto (2008), serves as a foundation for content verification. By deploying Solidity smart contracts on the Linea Sepolia testnet, We ensure robust mechanisms for securing ownership and verifying the authenticity of digital assets.

2.11 Tamper Detection

Tamper detection is critical in ensuring content integrity. Works like *"Multimedia Forensics: Where Do We Stand?"* (Stamm et al., 2013) highlight techniques for identifying digital alterations. Proposed System employs blockchain to compare video files against their IPFS hashes, effectively detecting unauthorized modifications and enhancing trustworthiness.

2.12 Full-Stack Development with MERN

The MERN stack facilitates efficient client-server interactions and secure database management. Our implementation leverages MongoDB for hash storage, React for dynamic interfaces, and Node.js with Express for backend operations, drawing on established best practices to ensure seamless functionality.

2.13 Hybrid Storage Models

Combining on-chain and off-chain storage, as explored in systems like Filecoin, improves scalability and security. We adopt this approach by using IPFS for video storage and blockchain for hash management, ensuring data integrity while maintaining efficiency.

3. Methodology

This study employs a decentralized architecture combining blockchain, IPFS, and MongoDB to ensure secure video storage and verification. Videos are stored in IPFS for decentralized scalability, while MongoDB maintains associated metadata, including IPFS hashes and user details, secured with bcrypt-based double hashing. Smart contracts written in Solidity validate and record metadata and IPFS hashes on a permissioned blockchain, ensuring tamper-proof verification and ownership authentication. The integration leverages RESTful APIs to enable seamless communication between the MERN stack-based interface, IPFS, MongoDB, and the blockchain. The permissioned blockchain enhances data privacy by restricting access to authorized users, with encrypted transactions ensuring overall security and data integrity.

3.1 WorkFlow

Users upload videos via a web interface, triggering backend processes where the video is stored on IPFS, generating a unique hash, and its metadata is saved in MongoDB. This hash is validated and immutably recorded on the blockchain through smart contracts. Verification is achieved by comparing the current video's IPFS hash against the blockchain-stored hash to detect tampering, while ownership is confirmed using immutable on-chain metadata, including user details and timestamps.

RESTful APIs manage requests for video uploads, metadata retrieval, and integrity checks, ensuring efficient data exchange and system functionality. The workflow, as shown in Figure 12.1, illustrates the integration between video uploads, storage on IPFS, metadata handling in MongoDB, and blockchain validation via smart contracts.

The system ensures video integrity and accessibility through a decentralized storage mechanism, leveraging the peer-to-peer nature of IPFS to provide redundancy and fault tolerance. In cases where videos are removed or unavailable on IPFS, the stored hash in MongoDB enables recovery by linking back to the original content. Smart contracts enforce stringent validation of metadata, such as timestamps and user ownership, before recording transactions on the blockchain. The permissioned blockchain architecture restricts operations to authenticated users, enhancing security while maintaining transparency.

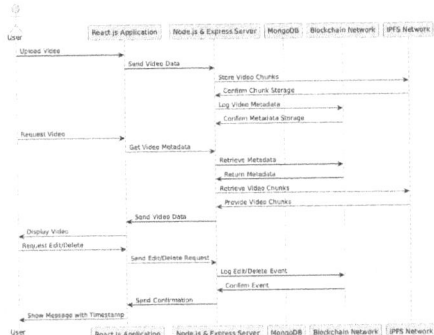

Figure 12.1 Workflow of Video Upload, Validation, and Storage in a Blockchain-Integrated System
Source: Self-developed

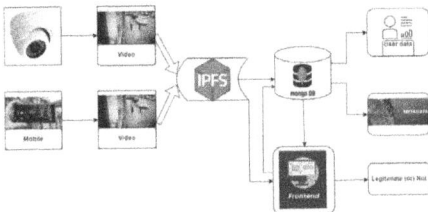

Figure 12.2 System Architecture
Source: Self-developed

As shown in Figure 12.2, the system architecture ensures secure video management by uploading videos to the Pinata IPFS server, generating unique hashes, and storing metadata in MongoDB. The hashes are validated and recorded on the blockchain using smart contracts, enabling tamper detection. This workflow ensures video integrity, fault tolerance, and secure access through the integration of the MERN stack, IPFS, and Solidity.

3.2 Output

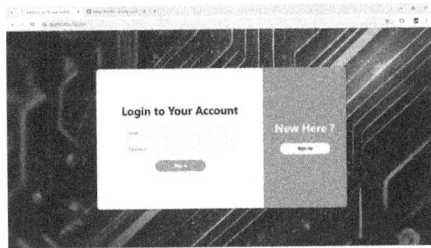

Figure 12.3 Login Page
Source: Self-developed.

As shown in Figure 12.3, the login page of the blockchain application features a streamlined design that facilitates user authentication. The interface includes fields for entering user credentials, such as email and password. This intuitive layout enhances the user experience by ensuring secure access to the platform while promoting efficient interaction with the decentralized data management system.

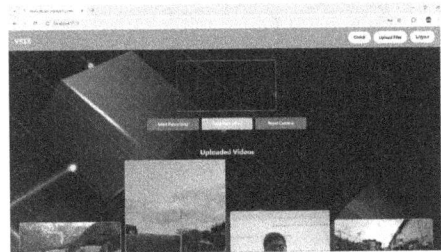

Figure 12.4 Uploading and Verifying Page
Source: Self-developed.

As shown in Figure 12.4, this process is responsible for uploading and verifying the integrity of the video. The data is retrieved from the Pinata IPFS, and the associated IPFS hash is used to verify the authenticity and integrity of the video. This ensures that the video has not been tampered with or altered, providing a secure and reliable method for video validation. Additionally, the system utilizes tamper-check features to ensure that the video content remains trustworthy and intact during storage and retrieval.

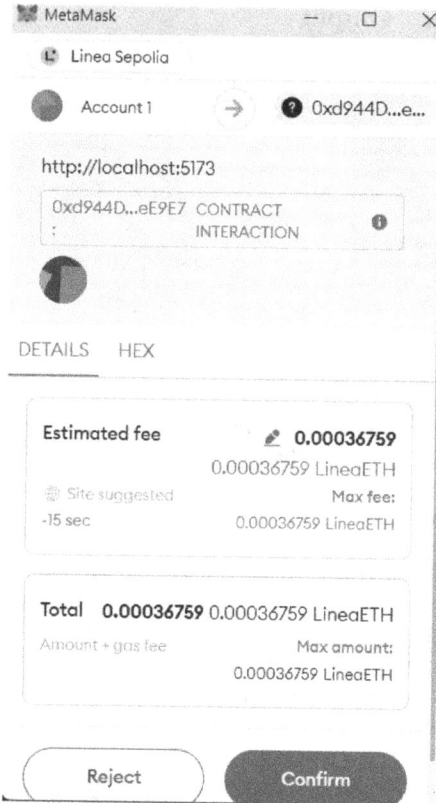

Figure 12.5 Confirm the Transaction Using MetaMask
Source: Self-developed

On successful upload of the video, the user is prompted to confirm the transaction using MetaMask as shown in Figure 12.5. This step finalizes the upload of the video to the IPFS server. Once confirmed, the uploaded video is reflected on the interface page, designed using MERN and Solidity,

ensuring seamless integration of decentralized storage and blockchain technology.

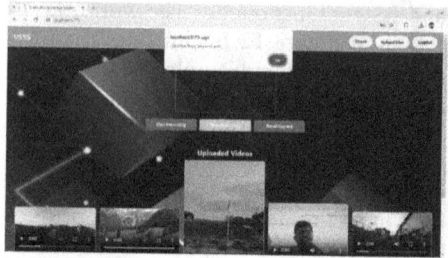

Figure 12.6 Video Verification Process
Source: Self-developed

As shown in Figure 12.6, the proposed system ensures the successful upload of a new video while verifying its authenticity and integrity. During this process, the system checks whether the uploaded video has been tampered with, ensuring data integrity. This robust verification mechanism highlights the secure nature of the system, guaranteeing the authenticity and reliability of the uploaded video footage.

4. Conclusion

This study demonstrates the effective integration of blockchain and decentralized storage to create a secure, scalable, and resilient video management system. By leveraging the MERN stack, it offers a user-friendly interface for video uploads, while utilizing Pinata IPFS for decentralized storage to ensure continuous availability even during node failures. IPFS hashes are securely stored in MongoDB, and user passwords are protected through double-hashing for enhanced security. The incorporation of blockchain via Solidity and Linea Sepolia adds trust and immutability to video transactions, while the tamper-check feature protects against data manipulation by comparing stored hashes with uploaded content. Furthermore, the system's decentralized architecture enables seamless recovery of inaccessible content through a distributed network, ensuring robust data redundancy. Overall, this study provides a comprehensive framework for addressing challenges in digital content storage, offering a practical blueprint for improving security, integrity, and availability in decentralized media solutions.

Acknowledgment

R. B. G. Thanks to all those who supported and contributed to this work. Special gratitude is extended to the Mentor and colleagues for their valuable guidance and insights. Appreciation is also due to the Pinata IPFS platform and the Ethereum community for providing essential resources and tools, which were crucial in the implementation of the video management system. Finally, R. B. G. expresses deep thanks to family and friends for their constant encouragement and motivation throughout the research process.

References

[1]   Yu, J., Chen, H., & Wu, K. (2020). Centipede: Leveraging the distributed camera crowd for cooperative video data storage. *IEEE Internet of Things Journal*, 8(22), 16498–16509.

[2]   Li, J., Li, Z., Tyson, G., & Xie, G. (2020). Your privilege gives your privacy away: An analysis of a home security camera service.

[3]   Balaji, S., Krishnan, M. N., Vajha, M., Ramkumar, V., Sasidharan, B., & Kumar, P. V. (2018). Erasure coding for distributed storage: An overview. *Science China Information Sciences*, 61(10), 100301.

[4]   Liu, Y., Kong, L., Chen, G., Xu, F., & Wang, Z. (2020). Light-weight AI and IoT collaboration for surveillance video pre-processing. *Journal of Systems Architecture*, 101934.

[5]   Zeng, X., Fang, B., Shen, H., & Zhang, M. (2020). Distream: Scaling live video analytics with workload-adaptive distributed edge intelligence. In *Proceedings of the 18th Conference on Embedded Networked Sensor Systems (SenSys '20)*, 409–421. New York, NY: Association for Computing Machinery.

[6]   Nguyen, et al. (2023). Real-time anomaly detection in distributed edge computing for video surveillance. This study explores how distributed edge computing can improve real-time video surveillance by reducing latency and enhancing security monitoring.

[7]   Kumar, et al. (2023). Enhancing video data management with AI-driven pre-processing in surveillance systems. This research highlights AI techniques for reducing data storage and transmission requirements in real-time video management.

[8]   Deepak, K., Badiger, A. N., Akshay, J., Awomi, K. A., Deepak, G., & Harish Kumar, N. (2023). Blockchain-based management of video surveillance systems: A survey. *IEEE*. This survey examines blockchain applications in managing video surveillance systems, focusing on security, privacy, and data integrity.

[9]   Fitwi, A., Chen, Y., & Zhu, S. (2023). A lightweight blockchain-based privacy protection for smart surveillance at the edge. *IEEE*. This research focuses on leveraging blockchain for privacy protection in smart surveillance systems at the edge.

[10] Kamal, R., Hemdan, E. E. D., & El-Fishway, N. (2023). Video integrity verification based on blockchain. *IEEE*. This study investigates blockchain-based methods for ensuring the integrity and authenticity of video data.

[11] Zhao, Z., Liu, Y., Zhao, H., & Wang, Y. (2023). A video security verification method based on blockchain. *IEEE*. This research proposes a blockchain-based approach for verifying the security and integrity of video data.

[12] Yatskiv, V., Yatskiv, N., & Bandrivskyi, O. (2023). Proof of video integrity based on blockchain. *IEEE*. This research presents a blockchain-based method to verify and maintain the integrity of video data.

13 Machine Learning in Phishing URL Detection: A Review of Recent Progress

Adhit Simhadri[1], M. Rishikesh[2], and M. Subramaniam[3]

Chaitanya Bharathi Institute of Technology, Hyderabad, India

Abstract

In 2023, the Anti-Phishing Working Group, a prominent cybersecurity organization, reported five million phishing attacks that affected systems globally, thereby sending a worldwide signal of the alarming increase in incidents. Phishing remains a favored method for attackers to deceptively obtain sensitive information via URLs, HTML pages, and emails. Since its rise in popularity, deep learning techniques such as Convolutional Neural Networks and Transformers have helped researchers address this problem and have shown tangible improvements over traditional language-based extraction methods, which have proven tedious. This review provides an overview of some of the latest developments in phishing/benign URL classification systems that focus on applying machine learning (ML), and Natural Language Processing (NLP) techniques for extracting and assessing the legitimacy of URLs. We choose to evaluate the methods proposed in recent years (2022–2024) to tackle the problem by a criteria-based (datasets used, depth of the analysis, evaluation metrics) assessment of the quality of the research conducted and the methods employed. In conclusion, this review explores some of the prominent research gaps observed from the collection of papers and constructively addresses them to improve the data collection and model construction process for future studies.

Keywords: Machine Learning, URL-Phishing, social engineering, Cyber Security.

1. Introduction

Phishing is one of the most commonly employed cyberattacks, where attackers attempt to steal sensitive information by wearing the mask of a trustworthy entity. This technique exploits human psychology rather than the technical system by providing confidential information such as login credentials or credit card numbers. Phishing attacks have grown in complexity to overcome the barriers enforced at various levels of society. Modern tactics commonly involve emails, messages, and phone calls to mimic well-trusted authorities, such as banks. Despite growing awareness of these attacks, phishing continues to boom due to the ease with which attackers can deploy large-scale operations and increase the difficulty users face in differentiating real and fake communications.

Phishing attacks have been around since the 1990s, when attackers began imitating trusted entities through emails. Early phishing attacks were relatively basic, consisting of emails claiming to be from major online services, which would trick users into entering their login details on fake pages [1]. By the early 2000s, phishing had

Email: adhit.simhadri@gmail.com[1], mrishikesh104@gmail.com[2], msubramaniam_cet@cbit.ac.in[3]

DOI: 10.1201/9781003661917-13

been accepted as a significant and growing threat. Attackers began mimicking real websites, further fooling victims into providing information. Over the years, phishing tactics have diversified and expanded to include various forms, such as vishing (voice phishing) and smishing (SMS phishing).

This review explores methods used to combat URL phishing, a phishing attack that exploits the linguistic properties of a URL (Uniform Resource Locator) to deceive a target and fool them into revealing confidential information. Such attacks can be easily implemented and are very effective, as most users cannot tell the minute differences between two URLs. Attackers take advantage of such minute errors in URLs to mimic legitimate authorities.

Various phishing detection methods have been developed at multiple levels to protect organizations and consumers. These strategies generally fall into many categories. List-Based Approaches: Rely on comparing URLs against a whitelist or blacklist of known phishing or legitimate sites. Web browsers like Google Chrome and Microsoft Edge often utilize this method, blocking URLs that are flagged as phishing attempts [10]. Heuristic-Based Approaches: Heuristic methods focus on identifying specific characteristics of phishing websites, using predefined rules or algorithms to flag potential phishing attempts based on observed behaviors or structural anomalies. While these methods can detect many types of phishing, they

may require frequent updates to remain effective against new attack strategies [10].

Machine Learning-Based Approaches: As phishing tactics evolve, machine learning (ML) and deep learning (DL) models have emerged as more dynamic tools for phishing detection. These models are trained on large datasets that include features from URLs, web content, DNS records, and user behavior [10].

This review has been split into the following sections: Section 2 outlines some of the relevant background information required to absorb the contents in Section 3, where all the papers have been organized into tables to highlight key features and research gaps across the chosen studies. Section 4 provides a brief outline of some of the common trends/gaps observed while preparing the content for the tables and provides concrete suggestions to improve the reliability of phishing systems.

2. Background

The increasing complexity of phishing attacks poses a significant challenge to users, compelling a requirement to understand the underlying tactics employed by attackers. Phishing, defined as the attempt to passively obtain a target's sensitive information by masking oneself as a trustworthy entity, exploits human behavior and the characteristics of URLs to deceive victims. This type of attack has evolved over the years, with URL-based schemes increasingly using linguistic properties [2, 8, 12] to mislead users into revealing confidential information. Understanding the procedures and tactics behind these attacks is important for developing effective safeguards.

This section highlights the key concepts essential for understanding the dynamics of URL phishing. It begins by highlighting the types of phishing attacks, underlining some of the highest-quality datasets in training detection models, and the challenges posed by data imbalance. We will also outline the popular approaches in the literature tackling this problem, including list-based, heuristic-based, and machine learning-based, and emphasize their key findings and research gaps.

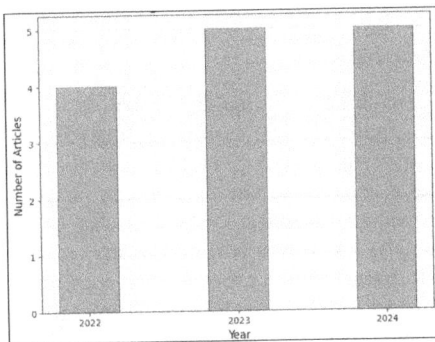

Figure 13.1 Number of Articles Found Across Years 2022–2024

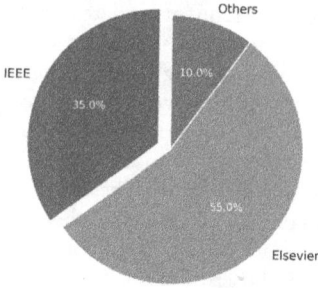

Figure 13.2 Percentage of Articles Taken From 2022–2024 From Prominent Publishers for this Review

2.1 Datasets and Data Generation Methods

A Gathering and processing relevant data is one of the foundational steps in solving any problem using Machine Learning Approaches. It is essential to understand the data sources involved before proceeding with comparing the model architectures. Given the increasing popularity of phishing attacks, there is an ever-increasing demand for high-quality datasets for the continuous improvement of the performance of models. The subsequent datasets were utilized in the studies reviewed herein.

Phishtank is the most commonly used dataset across most of the studies, containing approximately 8.5 million URLs, with 3.9 million verified as malign links. This blacklist is maintained by the Cisco Talos Intelligence Group.

The **Open Phish database** provides a repository of phishing URLs that is continuously updated and monitored, making it one of the largest URL collections. It offers an open-source API for integration into various apps.

The **Grambeddings dataset** [8], comprises around 800,000 samples, evenly split between 400,000 legitimate and 400,000 phishing URLs collected between May 2019 and June 2021.

The **CIC-Bell-DNS 2021** dataset, developed in collaboration with Bell Canada's

Cyber Threat Intelligence (CTI), categorizes DNS traffic features into lexical-based, DNS statistical-based, and third-party feature-based groups. It includes 400,000 benign samples and 13,011 malicious samples from publicly available datasets.

The **ISCX-URL 2016 dataset**, published by the Canadian Institute for Cyber Security, contains over 35,500 benign samples from Alexa's top websites and 10,000 phishing URLs sourced from the OpenPhish repository

Apart from the above mentioned datasets, it is important to note that there are many other prominent URL repositories that can be validly considered for benchmarking a novel architecture. With the ever-rising relevance of generative models such as Large Language Models, Generative Adversarial Networks [11] for synthetic sample generation, phishing URL detection systems have leveraged these models to prepare their detection systems using synthetically generated URLs.

A common drawback faced by papers collected in this study is data imbalance, which was found in two ways: as class/category imbalance and minority sample imbalance, i.e. different types of URLs collected might not be represented properly in the models. With the ratio of malignant and benign samples being too large. Shirazi et al. (2023) [11] proposed using Generative AI models such as GANs and Adversarial Autoencoders to generate phishing URL samples, results suggesting improvements in data quality.

2.2 Models

A machine learning/deep learning model is a mathematical formulation capable of identifying patterns in large amounts of data leveraging the heavy computation power of computers. The majority of the articles collected for this study feature the usage of data processing techniques to extract the relevant text and tabular (from DNS tools) information and employ machine learning models to perform classification of the URL into either benign (not malicious) or malign (malicious) categories. Though many studies focus on using complex multi-stage architectures for making predictions, this section briefly highlights some of the fundamental blocks that have been commonly used across all the studies.

Tree-based approaches [11, 13–15] such as Random Forest and XGBoost are commonly trained for their ease of use and their generalizable performance across datasets. A Random Forest is a collection of decision trees that run concurrently; a random forest's output is the average of all predictions from its constituent decision trees, whereas Xgboost, which is an ensemble of decision trees that runs sequentially, it also includes regularization to prevent overfitting.

A **Convolutional Neural Network** [2, 4, 8, 9] is a feedforward neural network designed to extract features using convolutional layers, inspired by visual perception. It consists of convolutional, pooling, and fully connected layers. In the reviewed papers, 1-D CNNs were commonly used due to their ability to model sequences, similar to **LSTMs. Recurrent Neural Networks** [1, 4, 8] were designed to process sequential data by maintaining a memory of previous inputs, making them effective for tasks like language modeling, and speech recognition **Long Short-Term Memory (LSTM) Networks** are advanced RNNs that mitigates the vanishing and exploding gradients with gates to control the flow of long-term and short-term information.

Ever since their inception, **Transformers** [1–5] are some of the most relevant and popular models in state-of-the-art deep learning, particularly for large-scale natural language processing tasks, where the model has to learn the relevant input elements and capture long-range dependencies. Transformers rely on the attention mechanism, an algorithm that allows it to process large amounts of text parallelly while capturing all the global and local dependencies of a given input sequence.

3. Methodology

This section outlines the approaches taken to analyze the reviewed studies for addressing URL phishing detection. Given the growing popularity of phishing attacks, notably those abusing URL features, it is critical to implement powerful detection techniques to effectively block these threats. The analyzed papers have been selected from prominent publishers from the years 2022–2024 (Figure 13.2) and included a variety of data processing strategies and training of many multi-staged ensembles, each consisting of machine learning (ML) and deep learning (DL) models, as well as hybrid approaches that combine both visual and textual analyses to improve classification accuracy [12].

The selected studies include a broad variety of strategies [6, 10], for feature selection, model architecture, and performance evaluation using a broad suite of metrics, which helps them objectively evaluate their systems. Recent approaches have also placed a significant focus and allocated their resources for the utilization of generative models, such as Generative Adversarial Networks (GANs), to minimize data imbalance that was commonly found in phishing datasets [11]. Paired with the usage of model architectures has enabled some of the authors to successfully deploy real-time detection systems in the form of web browser extensions [1].

We categorize each article according to the year of publication (Figure 13.1) and attempt to capture some of the prominent implementation and approach-related features by carefully curating each paper's content and summarizing it into tables. Table 13.1 contains information about (1) Research Focus, (2) Data collection methods that have been used, (3) Proposed solutions and key findings, and (4) Research gaps that have been discussed by the authors and found out while reading the papers.

Table 13.1 Summarized Table for the Papers Collected from the Years 2024–2022

Reference	Research Focus	Used Datasets	Solution and Findings	Limitations
Nguyet Quang Do a et al; 2024	Handling the natural language and unstructured data sequence nature of URLs while detecting phishing links.	Collected from popular sources like PhishTank, PhishCrawl, Yandex, Common Crawl, Github, and, 420-K	An integrated model for phishing URL detection based on DL classifiers and transformer architecture which use natural language and unstructured data sequence.	Limited Dataset Availability. Real-Time Processing Challenges and Generalization Across Different Platforms
Sultan Asiri et al; 2024	To build a reliable real-time detection system to classify phishing URLs using a BiLSTM architecture	PhishTank, OpenPhish, and Common Crawl Sourced from web crawlers	A hybrid approach combining visual and textual analysis of logos with content keyword extraction was proposed. Paired with a deployable Docker container using a BiLSTM+Attention module for real-time, low-latency detection.	Many URLs in the dataset lack components such as protocol, subdomain, hostname, and path. Its practical implementation in real-time presented challenges
Yahia Said et al; 2024	To propose a hybrid CNN-Transformer architecture for boosting classification scores	Custom sourced from popular sources	GANs were used to generate phishing URLs, balance the dataset, and improve model training. A hybrid model, combining CNN and attention mechanism, outperformed classical models like CNN and LSTM.	Computational complexity and training difficulties are faced while implementing such a complex model.
Fariza Rashid et al; 2024	To propose a framework that improves generalizability for different datasets for URL-based phishing detection.	Up to 10k samples sourced from ISCX-2016 and other sources such as EBBU-2017 and sourced from web crawlers	A framework was designed to improve the generalizability of Phishing URL classifiers based on unsupervised domain adaptation.	It was noted that when modifying features to match those of another dataset, the similarity in benign path length distribution was lost.
Fouad Trad et al; 2024	To compare the effectiveness of the usage of LLMs and assess the performance of generalized LLMs such as GPT Claude.	Up to 11k samples were collected from popular sources.	Fine-tuned LLMs outperformed prompt engineering-based LLMs in detecting phishing links, achieving an F1 score of > 97%.	The research focuses on only two chat models, GPT-3.5-turbo and Claude 2, which may not represent the broader landscape of large language models.

Reference	Research Focus	Used Datasets	Solution and Findings	Limitations
Colin Choon Lin Tan; 2023	Inclusion of website logo features for improved classification.	Sourced URLs and logos from popular sources and used crawlers	The hybrid approach reduced false positive rates by up to 3.4%, using a keyword extraction algorithm and the detection algorithm combined visual and textual identity components.	Challenges with Low-Ranked Websites Robustness Against Irregular Content
Hossein Shirazi et al; 2023	To improve the robustness of classifiers using synthetically generated samples using GANs and VAEs	Randomly sampled from PhishTank, and OpenPhish into Generative AI models	Through the use of GANs and auto-encoders, synthesized samples were produced and verified using classifiers It was found that classifiers trained with a mix of real and synthesized data were significantly more resilient to exploratory attacks.	Limited Exploration, Interpretability of Models and Limited usage of Evaluation Metrics
Musarat Hussain et al; 2023	To experiment with variations of one-dimensional convolution networks for phishing URL detection	Sourced from PhishTank, Yandex, DMOX, Alexa, and DNS-BH	The model employs multiple variants of one-dimensional CNNs with various-sized kernels in parallel. The model was evaluated against AI-generated adversarial attacks. Findings suggest that CNN-Fusion is robust against adversarial threats.	The performance varied significantly across different benchmark datasets.
Ahmet Selman Bozkir et al; 2023	To build a reliable dataset and a model hybrid using n-gram embeddings	Custom sampled from popular sources and common crawl	An n-gram selection mechanism that operates without a pre-training stage, allowing for an efficient feature extraction process, results showed an impressive accuracy of 98.27%.	The model relies on URL structure, so it may struggle to differentiate phishing URLs if the n-grams miss subtle changes.
Sultan Asiri et al; 2023	Introduction of transformers for phishing URL detection.	Up to 100k samples Sourced from PhishTank, OpenPhish	Proposed a novel PhishTransformer model which consists of two main components: a new input pipeline that extracts all URLs within a webpage and a combination of the CNN and Transformer encoder for feature extraction.	The size of the input pipeline of the model can reduce the model speed as some web pages may contain more than 1000 URLs, each of which is 100 tokens long.

Reference	Research Focus	Used Datasets	Solution and Findings	Limitations
Lázaro Bustio-Martínez et al; 2022	To introduce a lightweight feature selection and URL detection system that can work for low-resource environments.	Custom-sourced from popular URL websites and domain information tools	The study reduced 46 features into a subset of 9 for URL classification in low-resource environments (such as IoT). The reduced dataset showed slightly lower performance compared to the full dataset.	While a lot of new features have been proposed in this research, not much exploration or visual explanation was done. Along with a lack of a broader exploration of models.
Thomas Nagunwa et al; 2022	To address the classification of phishing URLs that rapidly change their host information	Alexa's top most visited URL, Phistank	The study used a novel set of 56 features, with temporal and DNS-related feature hostnames. The model's performance was thoroughly evaluated using seven metrics, offering a fair assessment.	Data from these online services could be missing due to various reasons, such as poor network connections or temporary unavailability of servers.
Ilker Kara et al; 2022	To improve the existing methods by analyzing the URLs at a character level by creating a custom-sourced dataset.	Nearly 50k samples were taken from TR-CERT website, with a split of 34k–11k for phishing-benign	This study proposed a method that simplifies feature extraction and reduces processing overhead while going beyond analyzing HTML, DOM, and URL-based features by considering URLs and domain names.	The success of the developed models is directly influenced by the quality and quantity of the dataset used. Lack of generalization across different regions or types of phishing attacks.

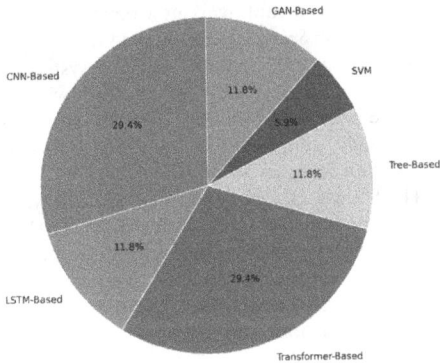

Figure 13.3 Percentage Representation of Models Used for Phishing Detection Across the Studies

Figure 13.3 provides a brief insight on the type of modelling techniques that were commonly used across the chosen articles, this was one of the many commonalities observed during the process of conducting our survey. Other commonalities include the extensive usage of readily-available datasets, which often can blind someone who wishes to implement the paper to their potential limitations in real world applicability. Though Shirazi, H et al has proposed an approach to tackle the real-world aspect by synthetically generating samples, the reliability and sampling methodologies of data generation techniques using Generative AI models is still a relatively new field of research and is to be further explored in more detail.

4. Conclusion

Some of the common trends that have been noted during the curation of the above tables and figures for this review include the introduction of complex neural architectures that provide very high theoretical performance on benchmarking datasets, yet their practical reliability and the degree to which they can be deployed and easily maintained is yet to be found out. Though Asiri Et Al [1] has provided a concrete approach through a containerized BiLSTM for URL phishing, many of its limitations are yet to be overcome in future work.

Another notable trend was the innovative use of Generative models, such as GANs, Adversarial AutoEncoders for addressing the class imbalance in these datasets, these models have the ability to learn the underlying distribution of the training data and generate synthetic samples. Though the approach has shown promising results, a system for monitoring and analyzing these samples to ensure consistency and originality is yet to be discovered.

Future work should focus on addressing these gaps through advanced generative models, better integration of unsupervised learning, and the development of robust, scalable datasets. Furthermore, enhancing model interpretability and improving computational efficiency will be crucial for broader deployment, especially in low-resource environments like IoT. By advancing these areas, phishing detection systems can become more resilient, accurate, and applicable across diverse platforms and attack scenarios.

Overall, this review provided an analysis of few machine learning and deep learning approaches recently proposed for addressing phishing URL detection. The increasing complexity of phishing attacks demands continuous improvements in large-scale detection systems. As highlighted, modern machine learning models, particularly those employing CNNs, LSTMs, and transformers, show tangible improvements over traditional heuristic and list-based methods. However, they face major challenges, such as data imbalance, generalization across different datasets, and real-time performance, which limits the practical adoption of these systems.

References

[1]   Asiri, S., Xiao, Y., Alzahrani, S., & Li, T. (2024). PhishingRTDS: A real-time detection system for phishing attacks using a Deep Learning model. *Computers & Security*, 141, 103843.

[2]   Do, N. Q., Selamat, A., Fujita, H., & Krejcar, O. (2024). An integrated model based on deep learning classifiers and pre-trained transformer for phishing URL

detection. *Future Generation Computer Systems*.

[3]   Rashid, F., Doyle, B., Han, S. C., & Seneviratne, S. (2024). Phishing URL detection generalization using Unsupervised Domain Adaptation. *Computer Networks*, 245, 110398.

[4]   Said, Y., Alsheikhy, A. A., Lahza, H., & Shawly, T. (2024). Detecting phishing websites through improving convolutional neural networks with Self-Attention mechanism. *Ain Shams Engineering Journal*, 15(4), 102643.

[5]   Trad, F., & Chehab, A. (2024). Prompt engineering or fine-tuning? a case study on phishing detection with large language models. *Machine Learning and Knowledge Extraction*, 6(1), 367–384.

[6]   Asiri, S., Xiao, Y., Alzahrani, S., Li, S., & Li, T. (2023). A survey of intelligent detection designs of HTML URL phishing attacks. *IEEE Access*, 11, 6421–6443.

[7]   Asiri, S., Xiao, Y., & Li, T. (2023). PhishTransformer: A Novel Approach to Detect Phishing Attacks Using URL Collection and Transformer. *Electronics*, 13(1), 30.

[8]   Bozkir, A. S., Dalgic, F. C., & Aydos, M. (2023). GramBeddings: A new neural network for URL-based identification of phishing web pages through n-gram embeddings. *Computers & Security*, 124, 102964.

[9]   Hussain, M., Cheng, C., Xu, R., & Afzal, M. (2023). CNN-Fusion: An effective and lightweight phishing detection method based on multi-variant ConvNet. *Information Sciences*, 631, 328–345.

[10] Safi, A., & Singh, S. (2023). A systematic literature review on phishing website detection techniques. *Journal of King Saud University-Computer and Information Sciences*, 35(2), 590–611.

[11] Shirazi, H., Muramudalige, S. R., Ray, I., Jayasumana, A. P., & Wang, H. (2023). Adversarial autoencoder data synthesis for enhancing machine learning-based phishing detection algorithms. *IEEE Transactions on Services Computing*, 16(4), 2411–2422.

[12] Tan, C. C. L., Chiew, K. L., Yong, K. S., Sebastian, Y., Than, J. C. M., & Tiong, W. K. (2023). Hybrid phishing detection using joint visual and textual identity. *Expert Systems with Applications*, 220, 119723.

[13] Bustio-Martínez, L., Álvarez-Carmona, M. A., Herrera-Semenets, V., Feregrino-Uribe, C., & Cumplido, R. (2022). A lightweight data representation for phishing URLs detection in IoT environments. *Information Sciences*, 603, 42–59.

[14] Kara, I., Ok, M., & Ozaday, A. (2022). Characteristics of understanding URLs and domain names features: the detection of phishing websites with machine learning methods. *IEEE Access*, 10, 124420–124428.

[15] Nagunwa, T., Kearney, P., & Fouad, S. (2022). A machine learning approach for detecting fast flux phishing hostnames. *Journal of Information Security and Applications*, 65, 103125.

14 A Review of Deep Learning Techniques for Earthquake Early Warning Systems

Shrina Tyarla[1], Masuna Hemashree[2] and M. Subramaniam[3]

Chaitanya Bharathi Institute of Technology (A), Hyderabad, India

Abstract

Among the most crucial methods for identifying natural hazards are earthquake early warning (EEW) systems. The process of estimating has become simpler with the advent of technology like artificial intelligence and deep learning. Within a few seconds of the P-wave arrival, these EEW systems reveal the magnitude of the earthquake through the application of algorithms and classification techniques. This paper outlines the current solutions developed with Deep Learning Techniques [17] such as Convolutional Neural Networks [1, 2], Long Short Term Memory [4, 12] and Attention Mechanisms along with comparison of datasets from across the globe. Researchers and engineers are able to select the most appropriate model based on resource allocation and regional needs by comparing each model to other models that are currently in use in terms of accuracy and experimentation.

Keywords: Earthquake, Earthquake early warning, Deep learning, CNNs, LSTMs, Attention mechanisms.

1. Introduction

With the devastating impact of earthquakes, there is a pressing need for early warning systems that can promptly and accurately estimate an earthquake's magnitude as soon as it occurs, as well as provide timely warnings for larger-magnitude events [1, 6, 7]. Earthquake Early Warning Systems (EEWS) are essential for eliminating the toll of natural disasters on human life. Mohamed S. Abdalzaher et al. in [3] emphasizes the importance of integrating modern technologies, such as the Internet of Things (IoT), social networking sites, remote sensing methods, and software-defined networks, into EEWS to enhance their capabilities. By combining conventional techniques with advanced strategies like machine learning (ML), these systems can address limitations of current models. For instance, the authors propose using slow-moving data within a short time frame after the initial primary wave, whereas standard models often rely on strong-motion data, which can suffer from gaps and non-linear patterns.In his study of EEWS in China, Zhu et al. [5] discussed traditional methods of magnitude estimation, which relies on an empirical magnitude estimation equation derived from a single early warning parameter.

While this approach has been a mainstay of EEWS, it has significant drawbacks. A single parameter offers limited magnitude-related information, leading to inaccuracies

---

Email: styarla@gmail.com[1], hemashreemasuna13@gmail.com[2], msubramaniam_cet@cbit.ac.in[3]

DOI: 10.1201/9781003661917-14

and both over- and underestimations of magnitude. Improving magnitude estimation accuracy is therefore crucial, and recent studies suggest that machine learning can significantly enhance this accuracy.

While magnitude estimation remains challenging, AI significantly improves accuracy over traditional methods. Additionally, Machine Learning [1–2, 6–7] reduces false alarms, ensuring that warnings are issued only during genuine emergencies, thereby preventing unnecessary stress. AI-powered EEW systems can analyze seismic data in milliseconds, giving people precious minutes or seconds to take precautions before the more destructive waves arrive.

Deep learning (DL) [17] approaches are also enhancing the precision and timeliness of EEW systems. For instance, a DL model for magnitude range prediction demonstrated lower variance and error than traditional techniques. A convolutional neural network (CNN)-based model achieved over 80% of predictions within ±0.3 magnitude units, transforming magnitude determination into a classification challenge. Furthermore, a multi-task deep learning model simultaneously estimates the enormity and precise area of an earthquake within 10 second intervals of seismic data by combining CNN [14] and Transformer [16] architectures. However, some approaches are limited by the need for complete P- and S-wave data. These developments underscore the growing role of DL in enhancing seismic response prediction and EEW system performance [15. 18–20].

The substantial influence of deep learning on EEWS around the world is highlighted in this review paper. The EEW systems' scalability, accuracy, and efficiency are improved by Deep Learning principles and methods. By compiling these developments, this paper offers a priceless resource for scholars working in the area.

2. Background

2.1 Overview

This section of the review provides examples of the algorithms and models that are presented as well as used in the different EEWS models. We will cover every aspect of EEWS utilizing deep learning models [17], including convolutional neural networks [14], long short-term memory [15], and attention mechanisms [16].

Additionally, we will analyze how worldwide seismological groups gather information and create easily accessed datasets for the purpose of analyzing earthquake data or developing EEW systems. Numerous open-source datasets are available for both global and localized data regions. These datasets provide component data about seconds, magnitudes, and places.

2.2 Deep Learning

Artificial neural networks are used in deep learning [24], a kind of machine learning, to teach computers how to process information similarly to the human brain. Large volumes of data are used to train deep learning models to find patterns and relationships, which enable them to make judgments and predictions. They are able to independently develop new features and process fresh data in real time [6, 7, 17]. Some popular deep learning models used in EEWS are:

2.2.1 Convolutional Neural Networks (CNNs)

[14] One kind of neural network that is frequently used for tasks like speech and picture recognition is the convolutional neural network (CNN). They are made up of different kinds of layers namely convolutional layers, pooling layers, activation functions, and fully connected layers. CNNs' versatility and effectiveness in various applications, coupled with ongoing advancements, continue to drive their prominence in the realm of machine learning and computer vision.

2.2.2 Long Short Term Memory

[15] Recurrent neural networks (RNNs) of the long-term dependency and temporal

sequence kind are known as LSTMs. Because they can remember information for extended periods of time—a critical skill for precisely estimating a battery's remaining capacity over its life—they are especially useful in this situation. The suggested model combines Gaussian process regression and enhanced singular filtering with LSTMs to improve prediction accuracy and flexibility. Because of this combination, the model can manage the non-linear and time-variant aspects, resulting in estimations that are more accurate and exact under a range of operating situations.

2.2.3 Attention Mechanisms

[16] Rather than processing all of the input equally, attention mechanisms are a potent tool that let models choose to focus on particular segments of the input data. Attention processes improve CNN-LSTM models' capacity to extract and apply significant temporal and spatial variables, resulting in more accurate and consistent timely power load predictions.

2.2.4 Transformers

[18] Because they allow sequences to be processed in parallel instead of sequentially, transformers have revolutionized deep learning by dramatically accelerating training. Regardless of where the various components of the input data are located, they use feed-forward neural networks and self-attention mechanisms to identify links between them.

2.2.5 Recurrent Nveural Network

[19] Sequence prediction applications including temporal relationships in the input data are suited for RNNs. They are appropriate for time-series analysis since they preserve a concealed state that records data from earlier time steps. Long-term dependencies, however, can be a problem for conventional RNNs, which is why LSTMs and GRUs were created in an effort to address this problem.

2.2.6 Graph Neural Networks

[20] GNNs extend neural networks to graph-structured data, which allow for understanding and modeling relationships and interactions between entities. Applications where data can be represented as graphs, like social networks, chemical structures, and spatial correlations in seismic data, benefit greatly from their utilization. Figure 14.1 shows a graph depicting the frequency use of different deep learning models for earthquake early warning systems.

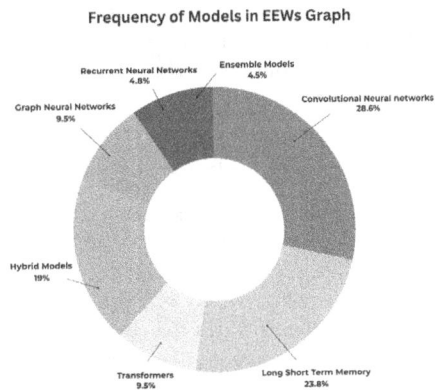

Figure 14.1 Graph depicting the frequency use of different deep learning models for earthquake early warning systems

2.3 Datasets

The availability of large and diverse datasets has greatly aided in the development of deep learning-based EEWS in recent years. These datasets are essential because they offer the unprocessed data required for deep learning model testing, validation, and training. Seismic waveform data, which are collected from multiple seismic stations located in earthquake-prone areas, are important datasets.

Some of the popularly available datasets are:

2.3.1 Stanford Earthquake Dataset (STEAD) [1, 3]

The Stanford Earthquake Dataset (STEAD) is a universal dataset which consists of local and non-local data values. It is gathered to expedite the application of machine learning to seismology problems and to support training, validation, performance comparisons, and the implementation of optimal techniques.

2.3.2 INSTANCE Dataset [3, 8]

A dataset of seismic waveform data and related information called INSTANCE is appropriate for machine learning-based analysis. 54,008 earthquakes totaling 1,159,249 3-channel waveforms are included, together with 132,330 3-channel noise waveforms, 19 networks, 620 seismic stations, and 115 metadata that provide details about the station, trace, source, course, and quality of each waveform.

2.3.3 DiTing Dataset [1, 3]

This is China's first dataset for earthquake magnitude estimation collected and created by the Chinese Earthquake Network Center. The dataset comprises 2,734,748 three-component waveform traces that correspond to 787,010 seismic events that were recorded between 2013 and 2020 at >1,300 permanent stations spread across China utilizing broadband and short-period seismometers.

2.3.4 Japanese Seismic Data [5]

This is an earthquake dataset collected and created by the Japanese Meteorological Agency (JMA). Data on earthquake intensity from the JMA span the years 1919–July 13, 2023. Although every effort has been made to ensure the accuracy of the information, changes may yet be made in the future.

2.3.5 SCEDC Dataset [9, 11, 13]

Seismic waveforms and other information about earthquakes in Southern California are gathered in the Southern California Earthquake Data Center (SCEDC) dataset. The Distributed Acousting Sensing (DAS) dataset and seismic waveforms for ground motion acceleration and velocity are included in the dataset. Figure 14.2 shows a graph depicting frequency usage of different datasets in EEW Systems.

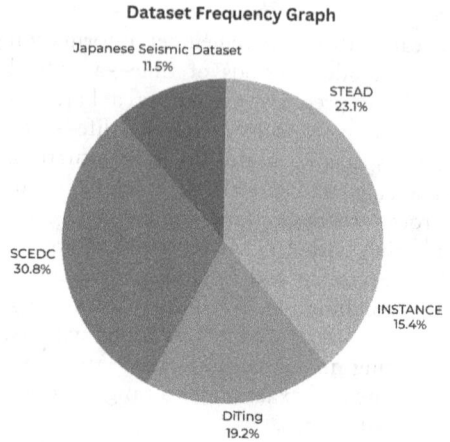

Figure 14.2 Graph depicting frequency usage of different datasets in EEW Systems

3. Methodology

In order to assess the diverse approaches utilized by scholars worldwide, we have taken into account three crucial factors to comprehend the existing body of knowledge and identify areas for future research that could be enhanced:

1. Methodology: Listing the techniques employed.
2. Practical Implications: The Implications of the project in real time.
3. Research Gap: Scope for further research and improvement.

4. Discussion

This part of the review provides an elaborate analysis of the different models researched in the course of the review article. We will cover every comparative aspect of these systems and later on, compare the accuracy values of each system, to find out the model with better usability and accuracy.

4.1 Analysis of Researched Models

There are countless deep learning-based methods for developing EEWS, as was covered in earlier sections. Convolutional neural networks are one of the most widely used techniques for developing EEWS. The CNN-Classification method (Ren, Tao, et al.) [3] and EEWMagNet (Meng, Fanchun, et al., 2023) [1] are CNN-based models for early warning and earthquake magnitude estimation. Three-component data taken from DiTing, STEAD, and INSTANCE are used in both models. EEWMagNet requires a maximum of 7 seconds of P-wave input, whereas the CNN-Classification model uses 4 seconds. The convolutional neural networks employed in both models solve the limitations of the standard fast magnitude classification techniques, which are sluggish and insensitive to imbalanced data, by extracting spatial and temporal information from three-component waveform records. While EEWMagNet achieves an accuracy of 90%, the CNN-Classification model from [3] achieved an accuracy of 97.9%. Had the EEWMagNet model taken the time limit less than 7 seconds, the model would have achieved a higher accuracy value.

Special CNN models, like the Batch Normalization CNN (Bilal, Muhammad Atif, et al., 2022) [9] and the Bayesian CNN (Wang, Tianyu, et al., 2022) [10], illustrate how Deep Learning models can be flexible in their application beyond their defined usage. Similar to EEWMagNet [1] and the CNN-Classification model [2], these models also use 3-component waveform data, and these papers also introduce new datasets, like the Southern California sations dataset called SCEDC [9]. These models introduce new and improved ideas to improve the efficiency of the Early Warning systems; the Bayesian CNN model suggests a randomness technique on the dataset, and the Batch Normalization partially resolves the Normalization issue in the EEWMagNet model [1]. Both models achieved higher accuracy but more importantly, lower error rates, demonstrating the validity of the improved features.

Additionally, long short-term memory is an effective model for early warning systems. An LSTM model from [5] shows how the LSTM ideas are put into practice. The limitations of the traditional P-wave peak displacement-dependent models (Pd-PGA) are addressed by LSTM (Wang, Ao, et al., 2023) [5]. It targets the recorded PGA using eight time-varying parameters related to ground motion and energy, utilizing the Japanese K-NET. Strong generalization capability was suggested by predictions on an M7.3 earthquake which is neither present in the training dataset nor the verification dataset, following consistent forecast residuals from the training data.

Hybrid models were created by combining features from models that have employed long short-term memory and convolutional neural networks in order to overcome any study limitations, like lower accuracy values [1], shorter wave durations [3], system scalability [8] and many more. The concept is that any LSTM or transfer learning model would analyze any patterns from the waves, and when combined, they would provide an improved system by having the CNN model read the spatial aspects of the wave data.

A model called TLDCNN-M (Zhu, Jingbao, et al., 2023) is a mixture of transfer learning and convolutional neural networks [4]. The idea is that the CNN model would read the spacial features of the wave data, any LSTM or Transfer learning model would analyze any patterns from the waves and combine to demonstrate an enhanced system. The results show that TLDCNN-M outperforms traditional methods and CNN models without TL in terms of mean absolute error and standard deviation of estimated errors. A zero-order attention mechanism and the CNN and Bi-directional LSTM models are combined in the CNN-BiLSTM model by Kavianpour, Parisa, et al., 2023, which is another hybrid model [12]. With the CNN-BiLSTM-AM model, this model demonstrated enhanced earthquake prediction and surpasses previous approaches in terms of performance and scalability.

These models offer various approaches to efficiently develop, scale, and implement models that can precisely and rapidly predict earthquake magnitude and effectively provide early warning measures for various regions across the world, in addition to new models like CREiME (Chakraborty, Megha, et al., 2022) [11], DynaPicker (Li, Wei, et al., 2022) [13], and 2S1S1C (Abdalzaher, Mohamed S., et al., 2024) [8].

4.2 Accuracy Discussion

When the objective of a task is to classify data into specified classifications, the accuracy metric is especially helpful. For instance, a model's accuracy in a binary classification task would be 80% if it could accurately predict 80 out of 100 instances.

$$\text{Accuracy}(\%) = \frac{\text{Number of Correct Predictions}}{\text{Total Number of Predictions}}$$

Table 14.1 Summation of Earthquake Model Accuracy Table

Paper Name	Model Name	Accuracy Value (in %)
Meng, Fanchun, et al; 2023	EEWMagNet	90.00
Ren, Tao, et al; 2023	CNN-Classification Model	97.90
Zhu, Jingbao, et al; 2023	TLDCNN-M	78.62
Wang, Ao, et al; 2023	LSTM Model	94.00
Abdalzaher, Mohamed S., et al; 2024	2S1C1S	90.80
Bilal, Muhammad Atif, et al; 2022	BNGCNN	87.00
Wang, Tianyu, et al; 2022	Bayesian CNN	86.70
Chakraborty, Megha, et al; 2022	CREIME model	98.00
Kavianpour, Parisa, et al; 2023	CNN-BiLSTM model	94.50
Li, Wei, et al; 2022	DynaPicker	97.17

Table 14.1 depicts the accuracy percentages of the various models researched from the methodologies and discussions in this paper.

5. Conclusion

There has been an increase in the number of deep learning [17]—integrated EEWS that are scalable, accurate, and speedier. With their continuous learning system that learns in real time from the data it collects and tests, they have improved reliability.

Researchers can now interpret complex seismic data more successfully because of the particular strengths of several models, including Transformers [18], Graph Neural Networks (GNNs) [20], Convolutional Neural Networks (CNNs) [14], and Long Short-Term Memory networks (LSTMs) [15].

Looking ahead, a number of encouraging developments could improve deep learning applications in seismic early warning systems even further. The integration of transfer learning may facilitate the adaptation of models developed on large-scale datasets to particular regional seismic features, hence enhancing their efficacy across various regions. Furthermore, there is a chance to develop more adaptable and focused warning systems through the utilization of real-time data streams from mobile sensors and Internet of Things devices [6, 7, 17, 20].

References

[1]   Meng, F., Ren, T., Liu, Z., & Zhong, Z. (2023). Toward earthquake early warning: A convolutional neural network for rapid earthquake magnitude estimation. *Artificial Intelligence in Geosciences*, 4, 39–46.

[2]   Hou, B., Zhou, Y., Li, S., Wei, Y., & Song, J. (2024). Real-time earthquake magnitude estimation via a deep learning network based on waveform and text mixed modal. *Earth, Planets and Space*, 76(1), 58.

[3] Ren, T., Liu, X., Chen, H., Dimirovski, G. M., Meng, F., Wang, P., ... & Ma, Y. (2023). Seismic severity estimation using convolutional neural network for earthquake early warning. *Geophysical Journal International*, 234(2), 1355–1362.

[4] Zhu, J., Li, S., Li, S., Wei, Y., & Song, J. (2023). Rapid earthquake magnitude estimation combining a neural network and transfer learning in China: Application to the 2022 Lushan M6. 1 earthquake. *Frontiers in Physics*, 11, 1070010.

[5] Wang, A., Li, S., Lu, J., Zhang, H., Wang, B., & Xie, Z. (2023). Prediction of PGA in earthquake early warning using a long short-term memory neural network. *Geophysical Journal International*, 234(1), 12–24.

[6] Elbes, M., AlZu'bi, S., & Kanan, T. (2023, June). Deep Learning-Based Earthquake Prediction Technique Using Seismic Data. In *2023 International Conference on Multimedia Computing, Networking and Applications (MCNA)* (pp. 103–108). IEEE.

[7] Turarbek, A., Bektemesov, M., Ongarbayeva, A., Orazbayeva, A., Koishybekova, A., & Adetbekov, Y. (2023). Deep Convolutional Neural Network for Accurate Prediction of Seismic Events. *International Journal of Advanced Computer Science and Applications*, 14(10).

[8] Abdalzaher, M. S., Soliman, M. S., Krichen, M., Alamro, M. A., & Fouda, M. M. (2024). Employing Machine Learning for Seismic Intensity Estimation Using a Single Station for Earthquake Early Warning. *Remote Sensing*, 16(12), 2159.

[9] Bilal, M. A., Ji, Y., Wang, Y., Akhter, M. P., & Yaqub, M. (2022). Early earthquake detection using batch normalization graph convolutional neural network (bngcnn). *Applied Sciences*, 12(15), 7548.

[10] Wang, T., Li, H., Noori, M., Ghiasi, R., Kuok, S. C., & Altabey, W. A. (2022). Probabilistic seismic response prediction of three-dimensional structures based on Bayesian convolutional neural network. *Sensors*, 22(10), 3775.

[11] Chakraborty, M., Fenner, D., Li, W., Faber, J., Zhou, K., Rümpker, G., ... & Srivastava, N. (2022). Creime—a convolutional recurrent model for earthquake identification and magnitude estimation. *Journal of Geophysical Research: Solid Earth*, 127(7), e2022JB024595.

[12] Kavianpour, P., Kavianpour, M., Jahani, E., & Ramezani, A. (2023). A CNN-BiLSTM model with attention mechanism for earthquake prediction. *The Journal of Supercomputing*, 79(17), 19194–19226.

[13] Li, W., Koehler, J., Chakraborty, M., Quinteros-Cartaya, C., Rümpker, G., & Srivastava, N. (2022). Real-time Earthquake Monitoring using Deep Learning: a case study on Turkey Earthquake Aftershock Sequence. *arXiv preprint arXiv:2211.09539*.

[14] Krichen, M. (2023). Convolutional neural networks: A survey. *Computers*, 12(8), 151.

[15] Wang, S., Wu, F., Takyi-Aninakwa, P., Fernandez, C., Stroe, D. I., & Huang, Q. (2023). Improved singular filtering-Gaussian process regression-long short-term memory model for whole-life-cycle remaining capacity estimation of lithium-ion batteries adaptive to fast aging and multi-current variations. *Energy*, 284, 128677.

[16] Wan, A., Chang, Q., Khalil, A. B., & He, J. (2023). Short-term power load forecasting for combined heat and power using CNN-LSTM enhanced by attention mechanism. *Energy*, 282, 128274.

[17] Tunç, S., Tunç, B., Çaka, D., & Budakoğlu, E. (2024). An Overview of Traditional and Next-Generation Earthquake Early Warning Systems. *Journal of Advanced Research in Natural and Applied Sciences*, 10(3), 747–760.

[18] Zhang, Q., Guo, M., Zhao, L., Li, Y., Zhang, X., & Han, M. (2024, March). Transformer-based structural seismic response prediction. In *Structures* (Vol. 61, p. 105929). Elsevier.

[19] Wang, Q., Zhang, Y., Zhang, J., Zhao, Z., & He, X. (2024). On the use of VMD-LSTM neural network for approximate earthquake prediction. *Natural Hazards*, 1–17.

[20] Si, X., Wu, X., Li, Z., Wang, S., & Zhu, J. (2024). An all-in-one seismic phase picking, location, and association network for multi-task multi-station earthquake monitoring. *Communications Earth & Environment*, 5(1), 22.

[21] Owusu Duah, J., Osei, O., & Osafo-Gyamfi, S. (2024). Predicting Peak Ground Acceleration of Strong-Motion Earthquakes Using Variable Snapshots of P-Wave Data with Long Short-Term Memory

Neural Network. *Seismological Research Letters*, 95(5), 2886–2893.

[22] Lin, Q., & Li, J. (2023). Advanced Seismic Magnitude Classification Through Convolutional and Reinforcement Learning Techniques. *International Journal of Advanced Computer Science & Applications*, 14(11).

[23] Hossain, M. S., Ghose, D., Partho, A. M., Ahmed, M., Chowdhury, M. T., Hasan, M., ... & Islam, M. (2023, August). Performance Evaluation of Intrusion Detection System Using Machine Learning and Deep Learning Algorithms. In *2023 4th International Conference on Big Data Analytics and Practices (IBDAP)* (pp. 1–6). IEEE.

15 Esterification and Property Assessment of Hybrid Biodiesel from Palm Kernel-Waste Cooking Oil: Case of Pretreatment by Taguchi Technique

Olusegun D. Samuel[1,3,], Ufuoma Peter Anaidhuno[1], Fidelis I. Abam[2], Christopher C. Enweremadu[3] and B. Krishna Chaitanya[4]*

[1]Department of Mechanical Engineering, Federal University of Petroleum Resources, Effurun, Delta State, Nigeria

[2]Energy, Exergy and Environment Research Group (EEERG), Department of Mechanical Engineering, University of Calabar

[3]Department of Mechanical, Bioresources and Biomedical Engineering, University of South Africa, Science Campus, South Africa

[4]Department of Electrical and Electronics Engineering, Chaitanya Bharathi Institute of Technology, Hyderabad, India

Abstract

Utilizing the easy-to-handle tool (ETHT) without becoming overwhelmed by multitasking has proven challenging, despite the fact that the fuel-related qualities of composite biodiesel from hybrid oily feedstocks (HOFs) are improved. Thus, this work focuses on using methanol-catalyzed esterification of crude PKWCO as a pre-treatment and enabling methanolysis transesterification of reduced PKWCO. In the existing study, PKWCO is diversified with methanol and H_2SO_4. Chosen esterification variables were analysed using the Taguchi method to obtain the appropriate free fatty acid content/value (AV) in the definitive outcome. The inspected parameters viz. methanol/ PKWCO, dosage of catalyst (H_2SO_4), and temperature. The impact and implication of respective constraint were then considered built on the plan and substantiated by supplementary tests. The optimum conditions for the AV of 2.92 mgKOH/g were (M: PKWCO) (4:1), temperature (30 min), and H2SO4 (0.5 wt.%). M: PKWCO has been detected to be the most influence variable, followed by catalyst dosage, and the least was the RT on the diminution of FFA of the hybrid oil. The ideal settings for the AV of 2.92 mgKOH/g were methanol to PKWCO (4:1), temperature (30 min), and H_2SO_4 (0.5 wt.%). M: PKWCO has been detected to be the most influence variable, followed by Catalyst dosage, and the least was the RT on the reduction of FFA of the hybrid oil. The cost estimation for PKWCOME (0.87 USD dollar) detected was more economically compared to fossil diesel (1.27 USD dollar). Also, the mass of PKWCOME can be further reduced if mass-produced. TAM proved appropriate in pre-treating and model esterification process of PKWCO's high for enhance yield and fuel properties of novel composite green diesel making. The fundamental fuel attributes of the PKWCOME conform with universal green fuel criteria.

Keywords: Taguchi technique, esterification, biodiesel, cost, prediction

Email: samuel.david@fupre.edu.ng[1], krishnachaitanyab_eee@cbit.ac.in[4]

DOI: 10.1201/9781003661917-15

1. Introduction

The recognition of biodiesel (BIOD) as a substitute fuel for fossil diesel has been subject to its conformity to global standards for vehicle utilization. To produce BIODs that are not competitive with food, inedible generational oily feedstocks (IGOFs) such as Jatropha, castor, rubber, tobacco, etc. are currently utilized. These inedible oils are widely accepted because they are underutilized and do not compete with food intake. The high FFA content in the IGOFs, however, presents a significant hurdle for researchers [1]. The low conversion of methyl esters of non-edible oils, higher downstream charges, the formation of undesirable soap, and catalyst diminution are all significantly correlated with high FFA levels in IGOFs, as shown by Knothe et al. [2]. Consequently, the esterification process is an essential measure for improving and preparing inedible oily feedstocks for transesterification. [3]. Methanol is preferred over ethanol by Pereira et al. [4] and Brahma et al. [5] because it guarantees lower temperatures and shorter retention durations. However, IGOFs with high FFA and oily compositions need the proper quantity of methanol to enable the production of BIOD at a reasonable cost.

Table 15.1 highlights the concise description of the esterification of IGOFs. As can be realized, methanol (MTH) was used as an alcohol to esterify rubber seed oil (RUSO) [3], neem-castor seed oil (NECAO) [6], Jatropha crude oil (JACO) [7], neem seed oil (NSO) [8], palm kernel oil (PAKO) [9], a mixture of rubber-palm mix (RPM) [10]. Therefore, the OVAT cannot be efficiently used to consider several response features of the experimental design. This is a result of OVAT's incapacity to concurrently display the correlation between the input and response variables [11].

1.1 Enthusiasm, Objective, and Novelty of the Study

An examination of the literature showed that the Taguchi technique has been applied to identify and enhance the production of BIOD and the engine characteristics of ICs powered by different fuels. To the best of the authors' understanding, not much is known about methylic esterification of hybrid IGOFs with high FFA by Taguchi-based optimization. There has been comparatively little reported research on the use of the Taguchi technique to the methylic esterification of hybrid oils.

The intention of this study was to check the best esterification of a palm kernel and waste cooking oil mix with high FFA by using parametric analyses using methanol as the alcohol. The intentions of this investigation were to develop a composite novel BIOD using palm kernel-waste cooking oil (PKWCO) under conventional conditions, evaluate the financial costs of PKWCOME and ascertain whether the BIOD it produced would be commercially feasible by identifying PKWCO methyl ester's basic fuel-related properties.

2. Materials and Procedure

2.1 Material and PKWCO Assessment

For the formation of biodiesel, oils of palm kernel and waste cooking, methanol, sulfuric acid (99.0%, Merck), and potassium hydroxide (Merck) were the main chemicals employed. These chemicals were all of the analytical grades. The PKWCO was categorised subsequent ASTM approved experiments. The acid value (AV) of PKWCO was detected by ASTM D974. The FFA and acid value of PKWCO were estimated using Eqs. (1) and (2).

$$FFA = \frac{28.2 \, x \, N_0 \, x t_v}{m} \tag{1}$$

$$AV = \frac{56.1 \, x \, N_0 \, x t_v}{m} \tag{2}$$

where t_v, N_o, M_{WPKWCO} are the titre value, normality, and mass of PKWCO adopted while the molecular weight of PKWCO is 809.91 g/mol

Table 15.1 A Succinct Review of Optimization of Esterification of High FFA Oil

Oily Source	Types of Alcohols	Operation Esterification Conditions/AV (mg KOH/g)	Optimization Tools	Countries or Regions	Remarks	Refs.
RUSO	MTH	MTO = 44.21:1, TAM = 3.4 h, $Fe_2(SO_4)_3$ = 16.97%/ FFA = 0.56%	RSM-ANFIS	Tropics	The capability of the RSM-ANFIS model for the heterogeneous acid esterification of RSO's pre-treatment showcased	Jisieike et al. [3]
Nee-castor seed oil	MTH	MTO = 4.5:1, TIM = 2 h, N/CSO ratio = 20, CAM = 2 wt.%/ AV = 2	RSM-ANFIS technique	Tropics	RSM-ANFIS proved to predict esterification conditions for hybrid oils	Samuel et al. [6]
Jatropha crude oil (JCO)	MTH	MTO = 12.3:1, TIM = 149.8 min, H_2SO_4/ CAM = 0.23 vol.%	RSM	Malaysia	The optimum condition for esterifying JCO by RSM established cation reaction.	Farouk et al. [7]
NESO	MTH	MTO = 18.51, TIM = 62.8 min, ferric sulfate dosage = 6 wt%, / FFA = 0.52	RSM-ANN-GA	Tropics	The capability of esterification of ANN and GA is superior to that of RSM in NSO's pre-treatment	Okpalaeke et al. [8]
Palm kernel oil	MTH	MTO = 3.4:1, TIM = 24.06 min, Ca = 0.39 vol.%;.	RSM-ANN-ANFIS	Tropics	Various superiority of the ternary tools highlighted	Betiku et al. [9]
A mixture of rubber-palm oils	Methanol	MTO = 15:1, TIM = 3 h, H_2SO_4/Ca = 0.5 wt%, T_e = 65 °C T_e = Tempeature	Taguchi	Malaysia	The capability of the Taguchi technique in pre-treating hybrid oils established	Khan et al. [10]

2.2 Methylic-based Esterification of PKWCO and its Design of Experiments (DOE) by Orthogonal Array

Due to the PKWCO's high FFA of 8.85 mmKOH/g, the oil underwent pre-treatment. Initially, 150 ml of PKWCO were added to the reactor, and it was preheated to 60 °C. A freshly prepared mixture of methanol and sulphuric acid (H_2SO_4) was homogenously mixed, and heated at a temperature of 60 °C. The mixture was then emptied to the pre-heated oil in the apparatus.

Methanol-esterified PKWCO was then separated from the excess methanol.

The esterification of the PKWCO is a function of respective and key variables in the process. Hence, analysis of each factor within a fixed of precise parameters for establishing the optimum conditions that lead to a minimum reduction. An L4 matrix with three esterification variables is defined at two levels, as highlighted in Tables 15.2 and 15.3. Lower AV for esterification indicates better FFA value (%) and acid value (mg KOH/g) for PKWCO, as determined by the Taguchi technique's smaller-the-better (LBT) criterion.

2.3 Statistical Analysis of Variance (ANOVA)

The Taguchi approach is capable of estimating the signal-to-noise ratio (SNR) from experimental databases. The ratio of the mean of investigational outputs (such as AV) to the standard deviation is equal to the SNR obtained. This approach can be very expedient in explaining the influence of the esterification parameters.

Table 15.2 Matrix Develop for PKWCO Esterification

	Parameters	Levels	
		1	2
A	MTO	4.1	9.1
B	CAM (wt%)	0.5	1
C	TIM (min)	30	45

Table 15.3 L4 Orthogonal Array for PKWCO Pretreatment

Experiment No.	Parameters and their levels		
	MTO	CAM (wt%)	TIM (min)
1	1	1	1
2	1	2	2
3	2	1	2
4	2	2	1

MTO = Methanol to oil ratio; CAM = Catalyst; TIM = Time

An assessment of the statistical parameters using ANOVA (analysis of variance) aids in determining the distinctive effect of each reaction parameter on the ultimate result acquired, which can be estimated based on their factors of contribution (indicated in mgKOH/g). This was estimated from the sum of squares of each factor divided by the sum of squares of the entire model. The parametric contribution factor can be estimated using Eq. (3).

$$\% \text{ contribution factor } \frac{SS_f}{SS_T} \times 100 \qquad (3)$$

where SS_f and SS_T denote the sum of squares of a particular factor while SST is the sum of squares in global when considering all parameters.

2.4 Methylic Etiquette for the Esterified PKWCO

A 2 L round bottom reactor with three necks, broad with a magnetic agitator, and a thermometer was used to perform the methylic route conversion of esterified PKWCO. After 500 g of PKWCO were poured to the reactor, it was heated to 60 °C. A beaker containing the requisite amount of KOH (5 g wt.% with respect to oil) pellet was filled with the proper amount of methanol (19.76 g). Potassium methoxide was produced by mechanically agitating and vigorously stirring the reactor's contents until all of the KOH had melted into the methanol. The reactor holding the preheated PKWCO was filled with potassium ethoxide. A magnetic stirrer was used to agitate the entire contents of the reactor at a speed of 650 rpm. The reaction was transesterified for one hour while the temperature was kept at 60 ±1 °C. Following an overnight cooling period to room temperature, the reaction's result was transferred into a separating funnel to settle. Two separate layers of PKWCOME (upper) and glycerol (lower) were discovered to have formed from the reaction product. The PKWCOME was collected independently from the separating funnel and heated on a rotary vacuum evaporator under abridged pressure at 80 °C until the surplus methanol was entirely removed. Thereafter, the crude PKWCOME was moderately washed (three times) with warm distilled water to eliminate remaining catalyst or soaps. The resultant product, purified PKWSOME, was then dehydrated with anhydrous sodium sulfate and sieved. Before being subjected to further testing, it was kept in a cool, dry location. The procedure was repeated in triplicate and the average yield reported. Key characteristics of PKWCOME were identified. The worldwide requirements for BIOD were followed and characterized.

2.5 Financial Analysis

The methods proposed by Marousek et al. [12], Rajak et al. [13], Barman and Jash [14], and Shrivastava et al. [15] are employed to estimate the charge of biodiesel production (BPR) from PKWCO. As can be perceived, the cost includes the overheads connected to the BP from 1 l of PKWCO, methanol, KOH, labour, electricity, catalyst preparation, and assorted. Figure 15.1a shows the graphic for the mathematical computation of the BPR from PKWCO, while Figure 15.1b displays the cost appraisal of the section of BPR from PKWCO.

3. Results and Discussion

3.1 Properties of the PKWCO

PKWCO was examined for basic qualities and fatty acid concentration to determine if it was suitable for BIOD synthesis. Figures 15.2 and 15.3 depict the fatty acid content and significant properties of PKWCO, respectively. As observed in Figure 15.3, the unsaturation (50.4%) of PKWCO is higher compared to saturation level (49.6%). Nguyen et al. [16] hinted that PKWCO possessing high unsaturation can enhance effective engine performance. Additionally, the high AV and kinematic viscosity suggest that the crude PKWCO is not in need of any pre-treatment before going through a methylic procedure.

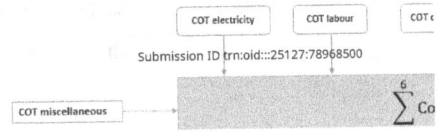

Figure 15.1a Representation for the Cost of Biodiesel Production

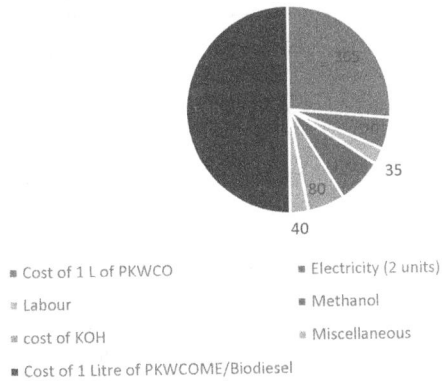

Figure 15.1b Cost Assessment for PKWCO Biodiesel Production (Naira)

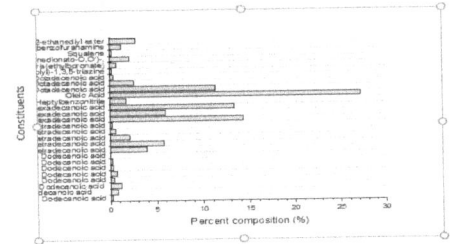

Figure 15.2 Fatty Acid Structure of PKWCO

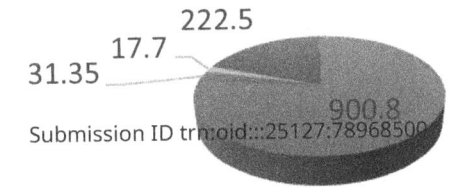

Figure 15.3 Basic Properties of PKWCO

3.2 Determination of optimal experimental condition by Taguchi method

The AV of esterified under the designed established of experiments, their SNRs, and complete mean SNR are presented in

Table 15.4 AV of Pre-Treated PKWCO and its SNR

Experiment No.	Methanol/oil (Molar ratio)	Catalyst Amount	Reaction Time	Reaction Time	SNR
1	4:1	0.5	30	3.18	10.0485
2	4:1	1	45	2.98	9.48433
3	9:1	0.5	45	2.92	9.30766
4	9:1	1	30	2.81	8.97413
		Overall Mean SNR$_T$			9.45366

Table 15.4. As the goal of the current work is to minimize the

AV, the smaller the better (STB) SNR model has been employed. This would not be the ideal set of parameters, as the experimental results showed that the esterified PKWCO's minimum AV was achieved to be at 2.92 mgKOH/g with the third run of the experiments, while the minimum AV was detected as 2.98 mgKOH/g with the second run of the experiments.

Figure 15.4 depicts the impact of each operational variable executed in the esterification process at assorted levels on the AV of the esterified PKWCO in terms of SNRL portrays. Originally, the higher figure for SNRL indicates a superior influence of a definite parameter at that rank. Additionally, the supreme figures of SNRL in each portrayal signified the optimal operation conditions for those fixed parameters on pre-treated PKWCO. Consequently, the optimized effective variables to minimize the esterification of PKWCO having high FFA were predicted to be A (methanol/ PKWCO molar ratio)

at 4:1, B (catalyst dosage) at 0.5 wt.%, and C (retention time) at 30 min.

3.3 Analysis of Variance (ANOVA)

The proportion of each factor's contribution to the AV of pre-treated PKWCO has been estimated, leading to the identification of the most important operating variable. Table 15.5 provides particular on the assessed SS$_f$ and percentage contribution, while Figure 15.5 illustrates the contribution's visual interpretation. The most significant parameter, contributing 64.56 percent to the AV of esterified PKWCO, was the methanol/PKWCO molar ratio, followed by the catalyst quantity at 33.24%. At 2.19%, reaction time was the least influential factor.

Table 15.5 PKWCO's Methylic Esterification Process Parameters' Percentage Contribution

Variables	SSfy	% Contribution
Methanol: PKWCO (molar ratio)	0.19563	64.56
CAM (wt.%)	0.10073	33.24
RIM (min)	0.00665	2.19

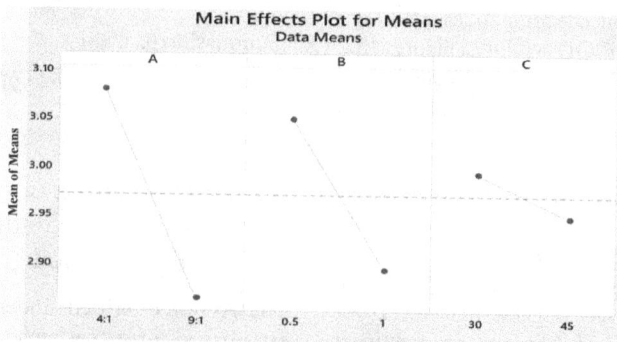

Figure 15.4 SNRL of Each Parameter at Different Levels

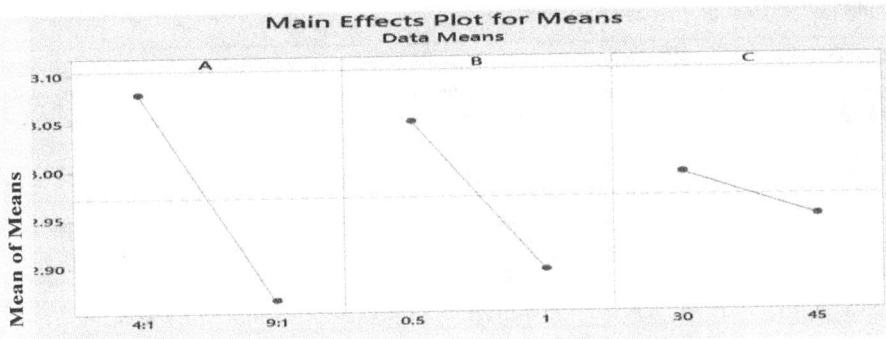

Figure 15.5 The Percentage Contribution of the Effective Esterification Variables on the AV of Esterified PKWCO

3.4 Optimum Condition and Key Fuel-Related Properties of PKWCOME

Table 15.6 shows the TM for substantial FFA reduction in PKWCO. The TM are similar to those found in literature, as was previously indicated. The difference between this study's TM and those reported elsewhere [6, 10, 17]. Topologies connected to the transesterification parameters and constraint settings in the experiment design might be responsible for this discrepancy.

3.5 Fatty Acid Components and Basic Fuel Properties of PKWCOME

The main characteristics of PKWCOME are shown in Table 15.7. As may be observed, PKWCOME's useful fuel-related qualities align with those found in the literature. The synthesized cleaner fuel must meet the requirement outlined in the recommended requisites in order to be considered appropriate for use in automotive applications. The criteria ensure the thermophysical features required for a diesel engine to

Table 15.6 TM for PKWCO Pretreatment

Types of Feed Stock	Model Tools	Optimum Condition/ AV or FFA	Remarks	Refs.
Neem seed oil/castor seed oil ratio (NCSO ratio)	RSM	NCSO = 20; CD (H_2SO_4) = 2wt. %; RT = 1.5; MR = 4.5/ FFA = 2.0595	The RSM was reported suitable in pre-treating and esterfying NCSO for biodiesel production	[6]
Palm-rubber seed oil (PRSO)	Fractional factorial design (FFD)	MR = 15:1; TR = 3 h; R_{Te} = 65 oC; CD = 0.5 wt.%	The FFD showed that the influence of R_{Te} > MR>CD on reducing FFA in hybrid of PRSO	[10]
High free fatty acid mixed crude palm oil (HF FAMCPO)	Central composite design (CCD)	MR = 3:1; TR = 20 min; R_{Te} = 60 *C; CD =1.7 % vol./ FFA < 1.0	Suitability of the CCD established for minimizing high FFA in HFFAMCPO	[17]

Table 15.7 Cogent Properties of PKWCOME

Fuel Properties	PKWCOME*	Castor-Microalgae Biodiesel [19]	Karanja-Castor Biodiesel [18]	ASTM D6751	EN 41214	DFF
Viscosity (mm2/s)	4.21	5.80	7.83	1.9–6.0	3.5–5.0	2.91
Density @ 15 °C (kg/m3)	848.8	890.0 @ 20/15° C	883.1	Nil	850– 900	852
Flash point (°C)	171	133.33	160	Min 130	Min 120	76
Acid value (mgKOH/g)	0.079	0.10	0.48	0.50 max	0.50 max	0.13

function properly. The cogent properties of PKWCOME with those of its counterparts, DFF), as observed in Table 15.5. The cogent properties of PKWCOME are comparable with properties of composite biodiesels from Karanja-castor BIOD [18], Castor- microalgae BIOD [19]. The marginally higher viscosity compared to DFF specifies that the diesel engine hardware will not be adjusted [1].

3.6. Cost of PKWCOME

The cost of producing PKWCOME is presented in Table 15.8. As discovered, the projected production charges of PKWCOME ($0.87/litre) compared auspiciously with those in the literature. The cost alterations can be attributed to the ethylic and methylic

productions of oily feedstocks, as well as the nature of the catalyst. PKWCOME's price could compete with fossil diesel's if it were manufactured in large quantities. However, the PKWCOME pricing can be justified considering the rising cost of diesel

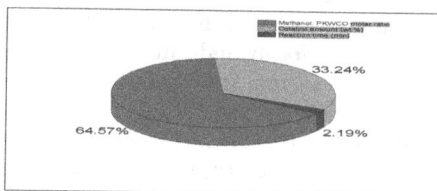

Figure 15.6 The Percentage Contribution of the Effective Esterification Variables on the AV of Esterified PKWCO

Table 15.8 Cost of PKWCOME with Biodiese ls from the Diverse in-edible Feedstock.

Generational Feedstock	Operative Parameters on the Combined Cost	Compendium	Production Cost (USD/ $)	Refs.
aMSW	Recovery ratio	Software	1.47/ liter	[21]
WCO	Cleaned oil	AS	0.23–0.66/ liter	[22]
Sludge	Removal and transesterification from a slush	Superpro Designer	0.67–1.07/ kg	[23]
Karanja oil	Feedstocks expenditure, worth and assorted charge	AS	0.28 /liter	Rahimi and Shafiei [25]
PKWCO ME		Excel package	0.87 per liter	Present study
Diesel	Nigeria	...	1.27/ liter	Price @ December, 2023

and the stability of BIOD feedstock sources, in addition to the advantages for the environment, local government relevance, socioeconomic stability, and topographical security, as well as vitality security [20].

4. Conclusion

In this experimental exploration, the production, description, and optimization of precarious process restrictions impelling the esterification process of a complete methylic composite BIOD synthesized from palm kernel and waste cooking oil (PKWCO) have been considered and reported. Hybrid non-edible oils instead of single oil have been utilized as feedstock with H_2SO_4 with methanol for pretreating PKWCO having high FFA. Methanol to PKWCO molar ratio (M: PKWCO), dosage of catalyst ((H_2SO_4), retention time (RT) were the three influencing variables considered for the acid-esterification of PKWCO exploring the Taguchi method (TAM). PKWCOME was produced from esterified PKWCO via a classical situation. The fuel properties of PKWCOME was developed and its cost implication were calculated. The following conclusion can be inferred:

- The optimum parameters were M: PKWCO of 4/1, temperature of 30 min, and H2SO4 of 0.5 wt% for the AV of 2.92 mgKOH/g. M: PKWCO was found to have the most influence on the reduction of FFA in the hybrid oil, with catalyst dosage coming in second. The least influential variable was RT.
- In order to address the challenge of enhancing and synergizing fuel properties, lowering cost implications, and guaranteeing effectiveness in diesel engines to ensure 100% sustainability during esterification and transesterification protocols, composite BIOD may be taken into consideration as a potential replacement for monohydric BIOD.
- To improve the yield and fuel attributes of the new composite green diesel production, TAM demonstrated suitable in the pre-treating and model esterification process of PKWCO's high. Basic fuel

characteristics of the PKWCOME were compliant with international green fuel regulations.

When compared to fossil diesel (1.27 USD dollar), the cost estimation for PKWCOME (0.87 USD dollar) was shown to be more economical. Additionally, if PKWCOME is mass-produced, its bulk can be further decreased.

References

[1]   Khan, I., Prasad, N., Pal, A., & Yadav, A. (2020). Efficient production of biodiesel from Cannabis sativa oil using intensified transesterification (hydrodynamic cavitation) method. *Energy Sources, Part A: Recovery, Utilization, and Environmental Effects*, 42(20), 2461–2470.

[2]   Knothe, G., & Razon, L. (2017). Biodiesel fuels. *Progress in Energy and Combustion Science*, 58, 36–59.

[3]   Jisieike, C., Ishola, N., Latinwo, L., & Betiku, E. (2023). Crude rubber seed oil esterification using a solid catalyst: Optimization by hybrid adaptive neuro-fuzzy inference system and response surface methodology. *Energy*, 263, 125734.

[4]   Pereira, P., & Andrade, J. (1998). Fontes, reatividade e quantificação de metanol e etanol na atmosfera. *Química Nova*, 21, 744–754.

[5]   Brahma, S., Nath, B., Basumatary, B., Das, B., Saikia, P., Patir, K., & Basumatary, S. (2022). Biodiesel production from mixed oils: A sustainable approach towards industrial biofuel production. *Chemical Engineering Journal Advances*, 10, 100284.

[6]   Samuel, O., Emajuwa, J., Kaveh, M., Emagbetere, E., Abam, F., Elumalai, P., Enweremadu, C., Reddy, P., Eseoghene, I., & Mustafa, A. (2023). Neem-castor seed oil esterification modelling: Comparison of RSM and ANFIS. *Materials Today: Proceedings*.

[7]   Farouk, H., Zahraee, S., Atabani, A., Mohd Jaafar, M., Alhassan, F. (2020). Optimization of the esterification process of crude jatropha oil (CJO) containing high levels of free fatty acids: A Malaysian case study. *Biofuels*, 11(6), 655–662.

[8]   Okpalaeke, K., Ibrahim, T., Latinwo, L., & Betiku, E. (2020). Mathematical

modeling and optimization studies by Artificial neural network, genetic algorithm and response surface methodology: a case of ferric sulfate–catalyzed esterification of Neem (Azadirachta indica) seed oil. *Frontiers in Energy Research, 8,* 614621.

[9] Betiku, E., Odude, V., Ishola, N., Bamimore, A., Osunleke, A., & Okeleye, A. (2016). Predictive capability evaluation of RSM, ANFIS and ANN: A case of reduction of high free fatty acid of palm kernel oil via esterification process. *Energy Conversion and Management.* 124, 219–230.

[10] Khan, M., Yusup, S., & Ahmad, M. (2010). Acid esterification of a high free fatty acid crude palm oil and crude rubber seed oil blend: Optimization and parametric analysis. *Biomass and Bioenergy,* 34(12), 1751–1756.

[11] Samuel, O., Kaveh, M., Verma, T., Okewale, A., Oyedepo, S., Abam, F., Nwaokocha, C., Abbas, M., Enweremadu, C., Khalife, E., & Szymane, M. (2022). Grey Wolf Optimizer for enhancing Nicotiana Tabacum L. oil methyl ester and prediction model for calorific values. *Case Studies in Thermal Engineering,* 35, 102095.

[12] Maroušek, J., Hašková, S., Maroušková, A., Myšková, K., Vaníčková, R., Váchal, J., Vochozka, M., Zeman, R., & Žák, J. (2015). Financial and biotechnological assessment of new oil extraction technology. *Energy Sources, Part A: Recovery, Utilization, and Environmental Effects,* 37(16), 1723–1728.

[13] Rajak, U., Chaurasiya, P., Nashine, P., Verma, M., Kota, T., & Verma, T. (2020). Financial assessment, performance and emission analysis of Moringa oleifera and Jatropha curcas methyl ester fuel blends in a single-cylinder diesel engine. *Energy Conversion and Management,* 224, 113362.

[14] Barman, S., & Jash, T. (2015). Study on optimization of process parameters for biodiesel production from waste cooking oil and its cost of production. *International Journal of Engineering Research & Technology,* 4(11), 326–332.

[15] Shrivastava, P., Verma, T., Samuel, O., & Pugazhendhi, A. (2020). An experimental investigation on engine characteristics, cost and energy analysis of CI engine fuelled with Roselle, Karanja biodiesel and its blends. *Fuel,* 275, 117891.

[16] Niculescu, R., Clenci, A., & Iorga-Siman, V. (2019). Review on the use of diesel–biodiesel–alcohol blends in compression ignition engines. *Energies,* 12(7), 1194.

[17] Prateepchaikul, G., Somnuk, K., & Allen, M. (2009). Design and testing of continuous acid-catalyzed esterification reactor for high free fatty acid mixed crude palm oil. *Fuel Processing Technology,* 90(6), 784–789.

[18] Kumar, D., Das, T., Giri, B., Rene, E., & Verma, B. (2019). Biodiesel production from hybrid non-edible oil using bio-support beads immobilized with lipase from Pseudomonas cepacia. *Fuel,* 255, 115801.

[19] Beyene, D., Abdulkadir, M., & Befekadu, A. (2022). Production of biodiesel from mixed castor seed and microalgae oils: Optimization of the production and fuel quality assessment. *International Journal of Chemical Engineering,* 2022.

[20] Perumal, V., & Ilangkumaran, M. (2018). Experimental analysis of operating characteristics of a direct injection diesel engine fuelled with Cleome viscosa biodiesel. *Fuel,* 224, 379–387.

[21] Gaeta-Bernardi, A., & Parente, V. (2016). Organic municipal solid waste (MSW) as feedstock for biodiesel production: A financial feasibility analysis. *Renewable Energy,* 86, 1422–1432.

[22] Khan, H., Ali, C., Iqbal, T., Yasin, S., Sulaiman, M., Mahmood, H., Raashid, M., Pasha, M., & Mu, B. (2019). Current scenario and potential of biodiesel production from waste cooking oil in Pakistan: An overview. *Chinese Journal of Chemical Engineering,* 27(10), 2238–2250.

[23] Chen, J., Tyagi, R. D., Li, J., Zhang, X., Drogui, P., & Sun, F. (2018). Economic assessment of biodiesel production from wastewater sludge. *Bioresource Technology,* 253, 41–48.

[24] Kumar, D., Singh, B., Banerjee, A., & Chatterjee, S. (2018). Cement wastes as transesterification catalysts for the production of biodiesel from Karanja oil. *Journal of Cleaner Production,* 183, 26–34.

[25] Rahimi, V., & Shafiei, M. (2019). Techno-economic assessment of a biorefinery based on low-impact energy crops: A step towards commercial production of biodiesel, biogas, and heat. *Energy Conversion and Management,* 183, 698–707.

16 White Shark Optimization Approach for Optimal Voltage Regulation Coordination between PV Smart Inverters and Traditional Volt-VAR Control Devices

Ravindhar Banothu[1], B. Suresh Kumar[2] and B. Mangu[3]

[1]BV Raju Institute of Technology Narsapur, India
[2]Chaitanya Bharathi Institute of Technology, Hyderabad, India
[3]Faculty of University College of Engineering, Osmania University, Hyderabad, India

Abstract

In order to control voltage regulation in distribution networks with substantial solar photovoltaic (PV) integration, this study proposes a two-stage coordinated voltage control (VVC) technique. In order to keep voltage levels within reasonable bounds (0.95–1.05 pu) under various load and generation conditions, the method combines the capabilities of PV smart inverters (PV-SI), capacitor banks (CBs), and on-load tap changers (OLTCs). The approach uses a two-tiered operation to overcome the shortcomings of conventional voltage management techniques in managing abrupt fluctuations: hourly adjustments for OLTCs and CBs and quicker intra-hour updates (every 15 minutes) for PV-SI inverters to efficiently react to dynamic changes. With MATLAB and Open DSS simulations, the IEEE 33-bus radial distribution system—a common benchmark for power system analysis—was used to assess the suggested approach.

Keywords: Two-stage coordinated voltage control approach, white shark optimization, solar photovoltaic generation, distribution network, smart inverter

1. Introduction

By integrating centralized and decentralized control, the suggested two-stage coordinated voltage control approach tackles voltage regulation in distribution networks. While decentralized control handles fast-acting devices, like PV-SIs, to react swiftly to voltage variations, centralized control supervises slower equipment, such as smart inverters, capacitor banks, and OLTCs, on an hourly basis. For renewable energy systems, this integration guarantees effective voltage regulation.

Figure 16.1 shows the Slow-responding voltage devices are given tasks via the centralized control technique in Hourly.

Email: ravindhar.b@bvrit.ac.in[1], bsureshkumar_eee@cbit.ac.in[2], bmanguou@gmail.com[3]

DOI: 10.1201/9781003661917_16

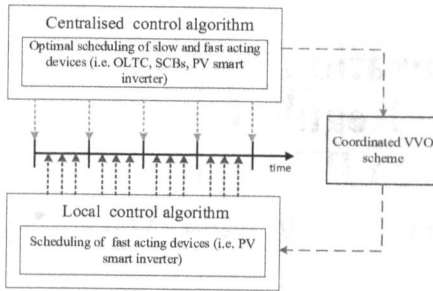

Figure 16.1 Schematic Representation of the Two Stage Coordinated Voltage Control Method

2. Literature Survey

To address global optimization problems, the author explained how to use white shark optimization [1]. a novel metaheuristic optimizer [2] influenced by ocean jellyfish activity. For optimal coordinated voltage control in distribution systems, JSO has implemented smart inverter-interfaced solar PV penetration [3]. For distribution system voltage management, PV smart inverters and various demand response systems should coordinate as well in [4].

Multi-Agent Reinforcement Learning for Volt-VAR Control in Active Distribution Networks in [5], Constructing reconfigurable active distribution networks with hybrid renewable energy systems through optimal stochastic scheduling [6] and Genetic Algorithm [7]. Reconfiguration in distribution system loss reduction analysed by this study and protection in [8].

The volt/Var control device in distribution system, optimization, smart Inver given in [9–11] and hybrid approach also mentioned. , smart inverters have garnered recognition for their versatility and adaptability Human-in-the-Loop Deep Reinforcement Learning for Optimal Volt/Var Control in Unbalanced Distribution Networks [12]. explores the Unified power quality conditioner (UPQC) siting and sizing in active distribution networks: a two-stage robust optimization process that takes demand and renewable generator uncertainty into account [13]. White Shark Optimization, combined with smart

inverters and network reconfiguration [14], and in [15] OLTC and SI Studied.

The two-stage coordinated voltage management method's drawbacks for distribution systems with solar integration include:

- The performance of PV-smart inverters is impacted by sensitivity to changes in the sun.
- Difficulties integrating more recent technologies with older equipment.

3. Formulation of the Problem

3.1 Objective Function

The main objective is to minimize the overall voltage deviation, which is defined by equation (1) as the sum of the voltage deviations across all buses for a certain time period T.

$$OF = \frac{\sum_{t=1}^{T} V_{dv}^{t}}{T} \tag{1}$$

Where, $V_{dv}^{t} = \sum_{i=1}^{nd} \left| \frac{V^{spec} - V_{i}^{t}}{V^{spec}} \right|$

3.2 Constrictions Related to Operations and Systems

Limitations on the balance of active and reactive power

$$P_{grid}^{t} - \sum_{i=1}^{nd} P_{i,loss}^{t} - \sum_{i=1}^{nd} P_{i,cons}^{t} + \sum_{i \in \Omega_{pv}} P_{i,pv}^{t} = 0 \tag{2}$$

$$Q_{grid}^{t} - \sum_{i=1}^{nd} Q_{i,loss}^{t} - \sum_{i=1}^{nd} Q_{i,cons}^{t} + \sum_{i \in \Omega_{cap}} Q_{i,cap}^{t} + \sum_{i \in \Omega_{pv}} Q_{i,pv}^{t} = 0 \tag{3}$$

System voltage magnitude limits

$$V^{min} \leq V_{i}^{b} \leq V^{max} \tag{4}$$

Tap settings of OLTC transformer

$$a^{t} = 1 + tap^{t} \frac{\Delta tap_{step}}{100} \tag{5}$$

here,

$$tap^h \in \left\{ tap^{min}, \cdots, -1, 0, 1, \cdots tap^{max} \right\}$$

Switched capacitor banks

$$Q_{i,cap}^t = st_i^t \Delta q_i^{cap}; i \in \Omega_{cap} \qquad (6)$$

Where,

$$st_i^t \in \left\{ 0, 1, \ldots \ldots st_i^{max} \right\}$$

Reactive power limit

$$Q_{PVDST,i}^h = \sqrt{\left(S_{PV,i}^{max} \right)^2 - \left(P_{PV,i}^h \right)^2}; i \in \Omega_{PV} \qquad (7)$$

$$-Q_{PVSI,i}^{max} \leq Q_{PVSI,i}^h \leq Q_{PVSI,i}^{max} \qquad (8)$$

3.3 Results and Conversations

The study assesses a two-stage coordinated voltage control system that was tested on the IEEE 33-bus system and created using OPENDSS and MATLAB. The system maintains voltage between 0.95 and 1.05 pu by integrating PV systems (0.8 MW, 1.0 MW, and 0.8 MW) at buses 15, 17, and 33 with capacitor banks at buses 3, 5, 10, and 24. Reactive power demand of 2.30 MVAR, active power demand of 3.715 MW, and system voltage of 12.66 kV are all part of the testing parameters. MVAR ranges from 0 to 0.600 for capacitors (see Table 16.1). In this study, three scenarios are compared: Case 1 (no PV), Case 2 (high PV integration), and Case 3 (VVC devices with PV-SI).

Figures 16.2(a) and 16.5(b) depict voltage behavior during peak PV generation (13:00) and maximum load (22:00). At 13:00, bus voltages exceed 1.05 pu,

while at 22:00, voltages drop below 0.95 pu. However, Cases 2 and 3 demonstrate successful mitigation of these violations through efficient resource utilization.

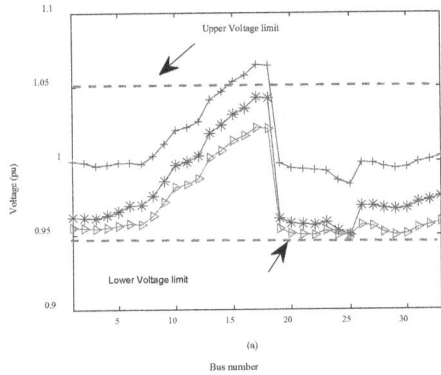

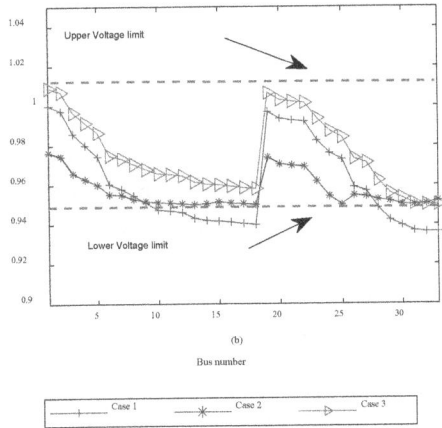

Figure 16.2 Voltage Profile for 33 Bus System: (a) High PV Penetration (b) High Load

3.4 Single stage VVC vs two stage VVC

Figures 16.3 and 16.4 illustrate the usage of single-stage and two-stage VVC for voltage control at busses 18 and 33. Intra-hour voltage variations result from the single-stage method's hourly scheduling of OLTCs, CBs, and PV-SI. As demonstrated, the two-stage method effectively prevents voltage violations by scheduling PV-SI inverters every 15 minutes.

Table 16.1 Cases study

Cases	PV	OLTC	CBs	PV-SI
Case 1	☒	☒	☒	☒
Case 2	☑	☑	☑	☒
Case 3	☑	☑	☑	☑

☑ denotes considered, ☒ denotes not considered

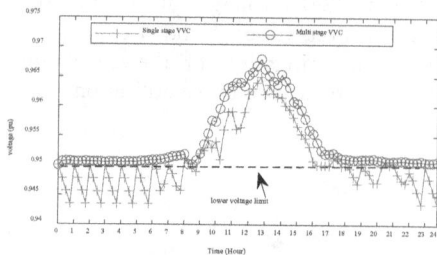

Figure 16.3 Voltage Profile at Bus Number 18th for the 33 Bus System.

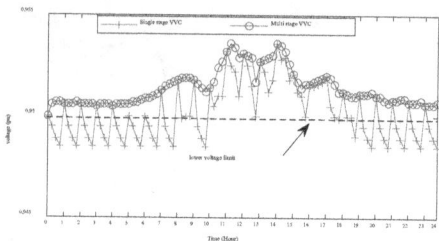

Figure 16.4 Voltage Profile at Bus Number 33rd for 33 Bus System

3.5 Validation of the Proposed Method in Response to Sudden External Disturbances

The performance of the PV-Smart Inverter (PV-SI) system is assessed in this section during abrupt changes in cloud cover. Reduced power production and voltage reductions below 0.95 pu were caused by the irradiance at buses 15, 17, and 33 decreasing from 0.925 kW/m² to 0.45 kW/m² at 14:00 hours, as seen in Figures 16.5(a)–(c).

Figures 16.6(a)–(c) illustrate how the PV-SI inverter supported reactive power, stabilizing voltage at crucial busses. Violations were successfully addressed when dynamic changes brought voltage stability back to all impacted locations.

4. Conclusion

The two-stage integrated Volt-VAR optimization (VVO) approach for coordinating smart inverters and voltage control devices is presented in this paper. By integrating conventional VVC devices with PV smart inverters, the results show that the strategy

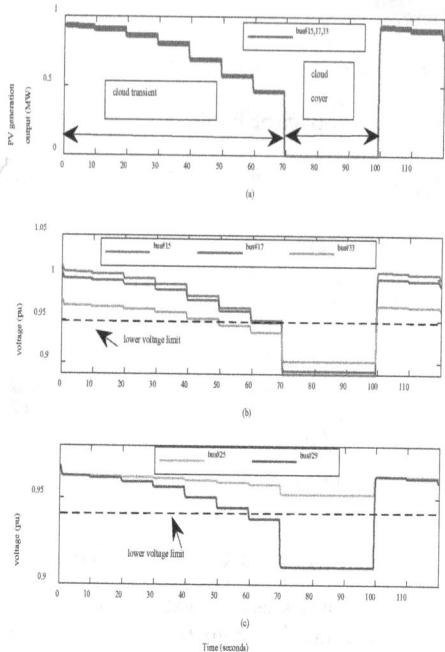

Figure 16.5 Voltage Profile Analysis of a 33-Bus System During a Transient Cloud Event Without any Control Measures

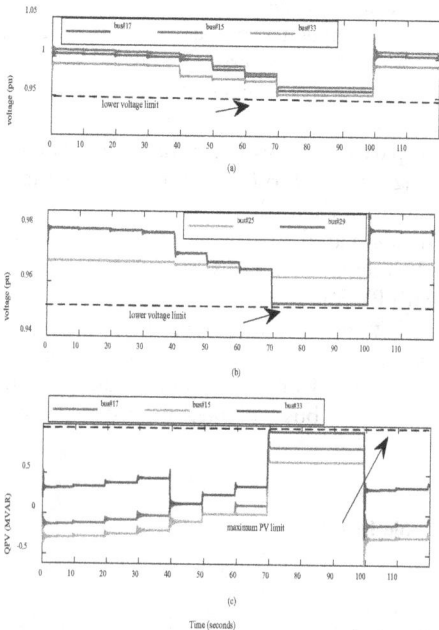

Figure 16.6 Analysis of the Voltage Profile of a 33-Bus System Under Temporary Cloud Cover with Control Mechanisms

is successful in lowering voltage regulation problems and reducing active power losses by 27.68% and reactive power loss by 20.48%. The suggested approach minimizes the reactive power demand on capacitor banks by minimizing the switching activities of traditional devices and making the best use of the VAR assistance provided by smart inverters. Additionally, the method demonstrates the ability to handle abrupt disruptions, including transient solar events and quick changes in cloud cover, guaranteeing steady voltage regulation under a variety of circumstances.

References

[1] Malik, B., Hammouri, A., Atwan, J., Al-Betar, M. A., &. Awadallah, M. A. (2022). WSO: A novel bio-inspired meta-heuristic algorithm for global optimization problems. *Knowledge-Based Systems*, 243(108457), 1–29.

[2] Jui-Sheng, C., & Truong, D.-N. (2021). A novel metaheuristic optimizer inspired by behavior of jellyfish in ocean. *Applied Mathematics and Computation*, 389, 125535.

[3] Aboshady, F., & Ceylan, O. (2023). Sequentially coordinated and cooperative Volt/Var control of PV inverters in distribution networks. *Electronics*, 12(8), 1765.

[4] Gupta, S., Ghose, T., & Chatterjee, K. (2020). Optimal coordination between PV smart inverters and different demand response programs for voltage regulation in distribution system. *International Transaction on Electrical Energy System*, 1–26.

[5] Su, S., Zhan, H., & Zhang, L. (2024). Volt-VAR control in active distribution networks using multi-agent reinforcement learning. *Electronics Journal*, 13(10), 3–15.

[6] Oskouei, M. Z., & Mohammadi-Ivatloo, B. (2021). Optimal stochastic scheduling of reconfigurable active distribution networks hosting hybrid renewable energy systems. *IET Smart Grid*, 4(3), 297–306.

[7] Eldurssi, A. M., & Robert M. O. (2015). A fast non-dominated sorting guided

genetic algorithm for multi-objective power distribution system reconfiguration problem. *IEEE Trans. Power System*, 30(2), 593–601.

[8] Abdul Rahim, M. N., Mokhlis, H., & Bakar, A. H. A. (2019). Protection coordination toward optimal network reconfiguration and DG sizing. *IEEE*, 7, 163700–163718.

[9] Arpanahi, M. K., & Hamedani-Golshan, M.-E. (2020). A competitive decentralized framework for Volt-VAr optimization of transmission and distribution systems with high penetration of distributed energy resources. *Electric Power Systems Research*, 186, 106421.

[10] Mataifa, H., Krishnamurthy, S., & Kriger, C. (2023). Comparative analysis of the particle swarm optimization and primal-dual interior-point algorithms for transmission system Volt/VAR optimization in rectangular voltage coordinates. *Mathematics*, 11(19), 4093.

[11] Pato, P. A. V., Trindade, F. C. L., & Wang, X. (2023). Hosting high PV penetration on distribution feeders with smart inverters providing local var compensation. *Electric Power Systems Research*, 217, 109168.

[12] Sun, X., Xu, Z., Qiu, J., Liu, H., Wu, H., & Tao, Y. (2024). Optimal Volt/Var control for unbalanced distribution networks with human-in-the-loop deep reinforcement learning. *IEEE Transactions on Smart Grid*, 15(3), 2639–2651.

[13] Zhu, X., & Guo, Y. (2024). Two-stage robust optimization of unified power quality conditioner (UPQC) siting and sizing in active distribution networks considering uncertainty of loads and renewable generators. *Renewable Energy*, 224, 120197.

[14] Ravindhar, B., Suresh Kumar, B., & Mangu, B. (2024). WSO for efficient CVR in photovoltaic-enriched distribution grids with SI and NR. *IJRER*, 17(1), 418–427.

[15] Ku, T.-T., & Lin, C.-H. (2019). Coordination of transformer on-load tap changer and PV smart inverters for voltage control of distribution feeders. *IEEE Transactions on Industry Applications*, 55(1).

17 A Decentralized, Lightweight, and Resilient Group Key Agreement Protocol Using Shamir Secret Sharing and Hash-Based Message Authentication Code for IoT Devices

Kavya Sri Yakkala[2], Naga Mohith Reddy Konireddy[2], Kavita Agrawal[1,2], Suresh Chittineni[3] and P. V. G. D. Prasad Reddy[1]

[1]Department of Computer Science and Systems Engineering, Andhra University, Visakhapatnam, India
[2]Department of Computer Engineering and Technology, Chaitanya Bharathi Institute of Technology, Hyderabad, India
[3]Department of Computer Science and Engineering, GITAM Deemed to be University, Visakhapatnam, India

Abstract

The Internet of Things is rapidly expanding and the need for lightweight and secure communication protocols is paramount, especially in resource-constrained environments. Current approaches like centralized key management approaches provide strong security but require very high computation resources or rely on a single entity for key management making them prone to single points of failure. Alternative symmetric encryption methods are much more efficient, but they do not include sufficient measures of authentication and are not tamper-resistant. To tackle these, we propose a decentralized, lightweight, and resilient group key agreement protocol based on Shamir Secret Sharing Scheme (SSS) and Hash-Based Message Authentication Code (HMAC) to create secure communication in IoT networks. The proposed design eliminates the need for third-party authority, such that no single device holds the entire group key, which reduces vulnerability. Reconstruction based on threshold and integrity validation based on HMAC provides fault tolerance and secure, tamper-proof communication over distributed IoT networks. The results prove that computational overhead is decreased while ensuring seamless group key integrity, which fits very well with the IoT's strict requirements for efficiency and resilience.

Keywords: Lightweight cryptography, Shamir Secret Sharing (SSS), Hash-Based Message Authentication Code (HMAC), group key agreement.

Email: kavyasri1yakkala@gmail.com[1], konireddynagamohithreddy@gmail.com[2], kavita.courses@gmail.com[2], schittin@gitam.edu[3], prasadreddy.vizag@gmail.com[1]

DOI: 10.1201/9781003661917-17

1. Introduction

The Internet of Things has experienced significant expansion in recent years. Its usage has become ubiquitous and can be observed in numerous fields like healthcare, automation, smart cities, etc. Nevertheless, there are security challenges for IoT networks due to the limited resources and a lack of decentralized cryptographic protocols. This contributes to vulnerabilities and jeopardizes data integrity and confidentiality. To tackle these challenges, this paper proposes a lightweight, decentralized group key agreement protocol tailored for IoT. Using Shamir Secret Sharing (SSS) for key distribution, our protocol eliminates single points of failure and improves resilience. Furthermore, tamper-proof key shares with low computational overhead are achieved by incorporating a Hash-Based Message Authentication Code (HMAC) for integrity verification.

In contrast to existing protocols that rely on third-party authority, the distribution of key management across devices is done to provide secure, fault-tolerant communication consistent with IoT efficiency requirements. Section 2, presents a literature review of existing solutions and limitations, Section 3 outlines the proposed methodology, Section 4 details implementation, and Section 5 evaluates performance, and Section 6 concludes with findings and future directions.

2. Related Work

In secure group communication for resource-limited environments like the Internet of Things (IoT), various group key agreement (GKA) protocols have been developed to tackle the distinct challenges of distributed IoT networks. Most of these are based on symmetric cryptography, elliptic curve cryptography (ECC), and secret-sharing schemes to achieve security with maximum efficiency.

In the domain of symmetric key exchange, Ashraf et al. [1] present a lightweight symmetric key exchange algorithm designed to minimize computational costs, suitable for devices with limited processing power. But the algorithm lacks integrated authentication and is thus vulnerable to man-in-the-middle attacks. In contrast, our protocol uses HMAC for integrity validation, thereby strengthening key management in IoT applications.

In the context of authenticated group key agreement protocols, Subrahmanyam et al. [2] proposed the Authenticated Distributed Group Key Agreement Protocol (ADGKAP), based on the Elliptic Curve Secret Sharing Scheme (ECSSS), in the context of authenticated group key agreement protocols. Although the protocol ensures robust security, it faces computational difficulties arising from the Elliptic Curve Discrete Logarithm Problem (ECDLP).

For centralized approaches, Hakeem and Kim [3] developed a threshold key generation protocol that combines Shamir's Secret Sharing with HMAC authentication. However, this approach is effective for vehicular networks, but it relies on a central group manager that could be a single point of failure. By eliminating dependence on a centralized authority, our protocol decentralizes key management, and increases fault tolerance and security across IoT networks.

Sheikh et al. [4] proposed a security system that uses chaos-based encryption and HMAC to provide data integrity and privacy with low power consumption. The chaos-based encryption scheme, however, adds implementation complexity. The additional implementation complexity, however, comes with the chaos-based encryption scheme. Our protocol relies exclusively on SSS, providing lightweight cryptographic operations while maintaining robust security.

In another lightweight solution, Ding et al. [5] developed a protocol based on key synchronization updates, showing efficiency in computation and communication. But its scalability depends on pre-shared keys. Our protocol circumvents this by supporting dynamic key reconstruction without pre-shared keys, making it more adaptable for distributed IoT networks.

In [6], Oudah and Maolood presented a model based on the Elliptic Curve Digital

Signature Algorithm (ECDSA) and Shamir's Secret Sharing (SSS) in order to increase security and resource efficiency. The dual algorithm approach is effective but cumbersome. In contrast, our protocol handles only SSS and HMAC, while reducing the complexity of group key management in resource-constrained environments.

To realize nonrepudiation and privacy, Zhang et al. [7] proposed a Dynamic Authenticated Asymmetric Group Key Agreement protocol built on a multi-signature scheme. However, due to the reliance on computationally intensive asymmetric operations, the security offered is not applicable to constrained IoT devices.

Lemnouar [8] also critically investigated vulnerabilities of Shamir's Secret Sharing in the case of weak polynomial choices, if these are exploited they can decrease security. To address this, the proposed protocol combines secure polynomial generation with HMAC based integrity validation for the reliability of key distribution in IoT networks.

This literature review underscores the diverse approaches in lightweight, secure communication for IoT. Our approach provides a unique integration of SSS with HMAC to provide a balanced solution with no single points of failure while also ensuring key integrity, and ECC-based and asymmetric key protocols are secure alternatives to the stream cipher. Our protocol addresses the particular requirements of IoT applications by decentralizing trust and reducing computational overhead, providing a robust and scalable alternative to current GKA protocols [9].

3. Proposed Methodology

This project implements Group Key Authentication and HMAC-based verification using Go (Golang). Golang is created by Google, it is a fast and lightweight programming language. It's mainly used for developing scalable and concurrent systems. Golang's standard libraries support cryptography and networking, simplifying secure protocol implementation. In addition, it provides fast compilation times,

garbage collection, and a small memory footprint, which guarantees high performance. However, the cost is simplicity and clarity in the code.

3.1 Group Key Authentication

Every device will generate n-1 shares, the group key is derived using Shamir Secret Sharing, and verified using HMAC.

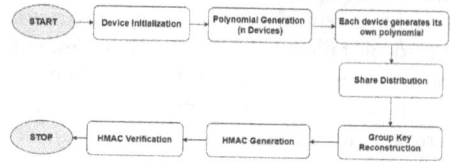

Figure 17.1 Flowchart for the Proposed Methodology

A network of n devices is established, with each device a unique identifier. Each device comes with a threshold value of t which is the minimum number of devices that are needed to successfully reconstruct the group key. The flowchart of proposed methodology is discussed in detail as shown in Figure 17.1.

3.2 Polynomial Generation

The constant term (part of the group key) is part of the secret for each device, which generates its polynomial. The group key can be reconstructed only by any subset of t devices and the polynomial has a degree of $t-1$. Each device computes fragments for its peers using the polynomial.

The polynomial can be expressed as:

$$y = c_n x^n + c_{(n-1)} x^{(n-1)} + \cdots + c_2 x^2 + c_1 x^+ c_0$$

where:
- y is the result for corresponding value of x (usually represents each device's unique identifier or input).
- $c_n, c_{n-1}, \ldots, c_2, c_1$ are the coefficients of the polynomial terms, which are randomly chosen by each device.

- $c_0 = s$ is the constant term, which serves as the "secret" or part of the group key.

The polynomial ensures that only a subset of at least t devices can reconstruct the key, enhancing security.

3.3 Share Distribution

Every device will generate $n-1$ shares, the group key is derived using Shamir Secret Sharing, and verified using HMAC.

The shares are allocated across the nodes in the network, with each node transmitting one share to each of its peers. Each node receives $n-1$ shares (part of each polynomial)

In Shamir's Secret Sharing, a secret s is split into n parts, and any t out of n shares are sufficient to reconstruct each secret. This can be formulated as:

Each device generates a polynomial f(x) of degree $t-1$ where:

$$f(x) = s + c_1 x + c_2 x^2 + \cdots + c_{(t-1)} x^{(t-1)}$$

where:

- S is the secret (constant term).
- $c_1, c_2, \ldots c_{t-1}$ are random coefficients generated by each device.

Each device then generates and distributes shares by evaluating the polynomial at specific points:

$$Share\, i : (x_i, f(x_i))$$

These shares are distributed to other devices. Any shares can be used to reconstruct $f(x)$ and hence retrieve s.

3.4 Group Key Reconstruction

The threshold number of devices can collaboratively reconstruct the group key using Lagrange interpolation.

3.5 Lagrange Polynomial Construction

The Lagrange interpolation is a method to construct a polynomial which passes through a set of given points. If you are given a set of points (x_1, y_1), (x_2, y_2), ..., (x_n, y_n), Lagrange interpolation provides a way to construct a polynomial $P(x)$ that exactly passes through these points.

The polynomial that passes through these points can be written as:

$$P(x) = \sum_{n}^{i=1} y_i \cdot l_i(x)$$

where $l_i(x)$ are the Lagrange basis polynomials, defined as:

$$l_i(x) = \prod_{j \neq i}^{1 \leq j \leq n} (x - x_j) / (x_i - x_j)$$

Each basis polynomial $l_i(x)$ is constructed such that:

- $l_i(x_i) = 1$ (at $x = x_i$, $l_i(x)$ is 1)
- $l_i(x_j) = 0$ for $j \neq i$ (at $x = x_j$, where $j \neq i$, (x) is 0)

3.6 HMAC Generation

Each device generates a hash value using the group key K and a unique timestamp T as the cryptographic key. The HMAC can be represented as:

$$HMAC_{device} = HMAC(K, T)$$

where:
- $HMAC_{device}$ is the HMAC generated by a device.
- K is the group key reconstructed by the device.
- T is a timestamp or unique identifier used as the HMAC secret to ensure freshness and prevent replay attacks.

Devices that have successfully reconstructed the group key can generate valid HMACs using this method.

3.7 HMAC Verification

Devices transmit and validate each other's HMACs by comparison. If they are consistent, mutual authentication is achieved.

Each device transmits its HMAC to other nodes and verifies the HMACs received from them by comparing each received HMAC with the HMAC it generated using the group key. If the HMACs match, the devices mutually authenticate each other. The verification process can be represented as:

If $HMAC_{received} = HMAC\ (K, T)$, then the device is authenticated

where:

- $HMAC_{received}$ is the HMAC received from another device.
- $HMAC\ (K, T)$ is the locally generated hash value using the group key K and T timestamp.

If the HMACs are consistent, the devices complete the authentication process.

4. Implementation and Results

The Go-based implementation of Group Key Authentication provided a decentralized, lightweight, and fault-tolerant IoT communication system. Each device created its polynomiWWal and distributed shares to peers for collaborative group key reconstruction. Devices meeting the threshold collaboratively reconstructed the group key using Lagrange interpolation and generated HMACs for secure communication. This decentralized key agreement allowed mutual authentication without a central authority. Go enabled fast polynomial generation, reconstruction, and HMAC verification with efficient IoT performance and fault tolerance.

Device initialization, SSS, and HMAC-based mutual authentication were modeled in this IoT simulator as shown in Figure 17.2. It shows how a network of IoT devices, where

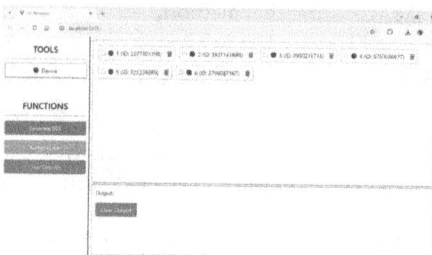

Figure 17.2 Initialization of the IOT Devices

each device needs to authenticate with its peers, must generate, share, and verify cryptographic data to do so.

Device Initialization: Six devices with unique IDs are initialized as IoT nodes.

4.1 Tools and Functions

- **Generate SSS:** Triggers Shamir Secret Sharing process, where each device generates its secret and shares polynomial coefficients with other devices.

Figure 17.3 Generation and Distribution

- **Authenticate:** Begins the process of mutual authentication between the selected devices based on the generated secrets and HMAC values.
- **Clear Devices:** Resets the devices and clears the generated data and secrets.

4.2 Secret Generation

- The "**Generate SSS**" button triggers all 6 devices each to generate their secret polynomial as shown in Figure 17.3.

4.3 Share Distribution

- After generating the secret, the device computes shares (polynomial coefficients) that are distributed to all other devices.

4.4 Shamir Secret Sharing

The devices share the coefficients of the polynomial and later can collaborate to reconstruct the group key, with the condition that a threshold number of devices (at least 3 in this example) must cooperate to obtain the secret.

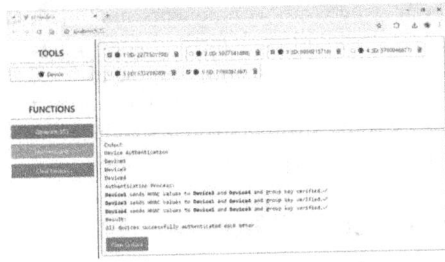

Figure 17.4 Authentication of Devices Using HMAC

4.5 Authentication

- Selected devices (in this case, Device 1, Device 3, and Device 4) are involved in the authentication process as shown in Figure 17.4. Devices that have successfully gathered enough shares reconstruct secret shares using Lagrange interpolation to obtain the group key.
- Using this key, each device generates a new **HMAC**.

4.6 HMAC Exchange

- Devices exchange their HMAC values with each other.
- The output confirms that all HMAC values were successfully verified, indicating that the devices authenticated each other.

4.7 Final Result

This process only allows authenticated devices to participate in secure communication within the network.

Table 17.1 Comparative Review of Proposed and Existing Solutions

Feature	Proposed Solution	Certificateless Authenticated Key Agreement [1]	ECC-Based Distributed Group Key [7]	Dynamic Asymmetric Group Key Agreement [8]	Centralized Threshold Key Generation [10]
Time Complexity	$O(n)$ (Linear, 1 ms per node)	$O(\log n)$ (0.6 ms per node)	$O(\log n)$ (0.5 ms per node)	$O(\log n)$ (0.7 ms per node)	$O(1)$ (0.1 ms, centralized)
Key Size (bits)	256	160 (ECC-based)	160 (ECC)	512 (asymmetric, larger key size)	256
Key Generation Time (ms)	1	2	3	3.5	1
Key Distribution Overhead (per node)	1 KB	1 KB	0.5 KB	1.2 KB	10 KB (centralized)
Memory Consumption (per node)	1.2 KB	70 KB	60 KB	90 KB	10 KB (centralized)
Encryption/ Decryption Time (ms)	2	3	3	4	3
Fault Tolerance	High (t-out-of-n reconstruction)	Moderate (depends on trusted authority)	Moderate (centralized failure risk)	High (distributed, fault-tolerant)	Low (single point of failure)
Scalability	High (supports large networks)	High (suitable for IoT and WSN)	Moderate (centralized bottleneck)	High (scalable for group applications)	Moderate (limited by centralization)

5. Comparative Analysis

This proposed method is an ideal solution for secure and efficient key management in distributed environments. Shamir's Secret Sharing (SSS) with polynomial reconstruction is used, providing strong, threshold-based security, and defense against common attacks such as collusion and eavesdropping. Integrity verification is further enhanced by the use of HMAC-based authentication to verify data authenticity. The security and efficiency are balanced with a fast 1 ms key generation, requiring only 1 KB of overhead per node. This exceptional fault tolerance comes from the fact that the system can tolerate node failures via t-out-of-n reconstruction. It is also efficient in scaling, seamlessly enabling device addition without a complicated increase in complexity, making it appropriate for large IoT or distributed systems. That said, at 70 KB/node, it is slightly more storage than some other methods, but a fair price to pay for security and resilience. In general, this method achieves a best-in-class combination of security, scalability, fault tolerance, and computational efficiency in modern, secure systems as discussed in Table 17.1.

6. Conclusion

A decentralized, lightweight, and resilient group key agreement protocol using Shamir Secret Sharing and HMAC for secure communication in IoT environments is proposed. By decentralizing key management, the danger of single point failures and the elimination of a third party. By exploiting both security properties (SSS for key reconstruction, and HMAC for integrity validation) the security is strong but the computational overhead remains low. This allows it to be a perfect approach for resource-constrained IoT devices. Compared to current techniques, the proposed protocol provides a more robust and lightweight approach without sacrificing security compared to current techniques. Any device in the system can safely rebuild the group key without being vulnerable to external threats. Despite an

event where any components malfunction, the threshold-based methodology guarantees that the system will continue to operate safely. Furthermore, HMAC guarantees the integrity of distributed shares, which guarantees tamper-proof communication.

6.1 Future Work

Although the proposed protocol tackles the key hurdles in secure IoT communication, there is still a long way to go. Future research could explore:

1. **ECC Integration:** Elliptic Curve Cryptography (ECC) could increase security with little computational overhead to provide stronger cryptography with smaller key sizes.
2. **Blockchain-Enhanced Security:** Integrating blockchain [12] [13] [14] [16] [18] enhances security by decentralizing key exchanges and ensuring data integrity, removing single points of failure, especially in large IoT networks.
3. **Dynamic Thresholds:** By adding dynamic thresholds [15] that alter based on network conditions, the protocol can become more adaptable and resilient to errors.

These enhancements could further strengthen the protocol for future IoT applications.

References

[1]   Cui, W., Cheng, R., Wu, K., Su, Y., & Lei, Y. (2021). A certificateless authenticated key agreement scheme for the power IoT. *Energies*, 14(19), 6317.

[2]   R. Subrahmanyam, N. R. Rekha, and Y. V. S. Rao, "An authentication protocol for next generation of constrained IoT systems," IEEE Internet of Things Journal, vol. 9, no. 21, pp. 21493-21504, 2022, doi: 10.1109/JIOT.2022.3184293.

[3]   Fang, D., Qian, Y., & Hu, R. Q. (2020). A flexible and efficient authentication and secure data transmission scheme for IoT applications. *IEEE Internet of Things Journal*, 7(4), 3474–3484. doi: 10.1109/JIOT.2020.2970974.

[4]   Sheikh, A. S., Keerthi, A., Dhuli, S., Likhita, G., Jahnavi, B. S. V. N. J., & Atik, F. (2021). A novel security system for IoT applications. In *Proc. 2021 12th Int. Conf. Comput. Commun. Netw. Technol. (ICCCNT)*. Kharagpur, India, pp. 1–5. doi: 10.1109/ICCCNT51525.2021.9579502.

[5]   Lemnouar, N. (2022). Security limitations of Shamir's secret sharing. *J. Discret. Math. Sci. Cryptogr.*, 1–13. doi: 10.1080/09720529.2021.1961902.

[6]   (2022). Lightweight authentication model for IoT environments based on enhanced elliptic curve digital signature and Shamir Secret Share. *Int. J. Intell. Eng. Syst.*, 15(5), 81–90. doi: 10.22266/ijies2022.1031.08.

[7]   Li, B., Zhang, G., Lei, S., Fu, H., & Wang, J. (2022). A lightweight authentication and key agreement protocol for IoT based on ECC. In *Proc. 2021 Int. Conf. Adv. Comput. Endog. Secur.*, Nanjing, China, 1–5. doi: 10.1109/IEEECONF52377.2022.10013341.

[8]   Zhang, R., Zhang, L., Choo, K.-K. R, & Chen, T. (2023). Dynamic authenticated asymmetric group key agreement with sender non-repudiation and privacy for group-oriented applications. *IEEE Trans. Dependable Secure Comput.*, 20(1), 492–505. doi: 10.1109/TDSC.2021.3138445.

[9]   Muhammad, Tanveer., Samia, Allaoua, Chelloug., Maali, Alabdulhafith., Ahmed, A., Abd, El-Latif. (2024). Lightweight authentication protocol for connected medical IoT through privacy-preserving access. Egyptian Informatics Journal, doi: 10.1016/j.eij.2024.100474

[10] S. A. Abdel Hakeem and H. Kim, "Centralized threshold key generation protocol based on Shamir Secret Sharing and HMAC authentication," *Sensors*, vol. 22, p. 331, Jan. 2022, doi: 10.3390/s22010331.

[11] Z. Ashraf, A. Sohail, and M. Yousaf, "Robust and lightweight symmetric key exchange algorithm for next-generation IoE," *Internet Things*, vol. 22, p. 100703, 2023, doi: 10.1016/j.iot.2023.100703.

[12] JoonYoung, Lee., MyeongHyun, Kim., KiSung, Park., Sung-Kee, Noh., Abhishek, Bisht., Ashok, Kumar, Das., Young-Ho, Park. (2023). Blockchain-Based Data Access Control and Key Agreement System in IoT Environment.

Sensors, 23(11):5173-5173. doi: 10.3390/s23115173

[13] Yu, K., Tan, L., Yang, C., Choo, K. K. R., Bashir, A. K., Rodrigues, J. J., & Sato, T. (2021). A blockchain-based shamir's threshold cryptography scheme for data protection in industrial internet of things settings. *IEEE Internet of Things Journal*, 9(11), 8154–8167. doi: 10.1109/jiot.2021.3125190

[14] Matthew, Weidner., Martin, Kleppmann., Daniel, Hugenroth., Alastair, R., Beresford. (2021). Key Agreement for Decentralized Secure Group Messaging with Strong Security Guarantees. 2024-2045. doi: 10.1145/3460120.3484542.

[15] K. Meng, F. Miao, W. Huang, and Y. Xiong, "Threshold changeable secret sharing with secure secret reconstruction," *Information Processing Letters*, vol. 157, p. 105928, May 2020, doi: 10.1016/j.ipl.2020.105928.

[16] Pérez-García, J. C., Braeken, A., & Benslimane, A. (2024). Blockchain-based group key management scheme for IoT with anonymity of group members. *IEEE Transactions on Information Forensics and Security*, 1–1. doi: 10.1109/tifs.2024.3414663.

[17] M. Ghebleh, A. Kanso, and H. Abuhasan, "Verifiable secret sharing with changeable access structure," Discret. Math. Algorithms Appl., Apr. 2024, doi: 10.1142/S179383092450037X.

[18] Harn, L., Hsu, C., Xia, Z., Xu, H., Zeng, S., & Pang, F. (2023). Simple and efficient threshold changeable secret sharing. *J. Inf. Secur. Appl.*, 77, 103576. doi: 10.1016/j.jisa.2023.103576.

[19] Tomar, A., Gupta, N., Rani, D., & Tripathi, S. (2023). Blockchain-assisted authenticated key agreement scheme for IoT-based healthcare system. *Internet of Things*, 23, 100849–100849. doi: 10.1016/j.iot.2023.100849.

[20] Ding, Z., et al. (2024). A lightweight and secure communication protocol for the IoT environment. *IEEE Trans. Dependable Secure Comput.*, 21(3), 1050–1067. doi: 10.1109/TDSC.2023.3267979.

[21] Ghebleh, M., Kanso, A., & Abuhasan, H. (2024). Verifiable secret sharing with changeable access structure. *Discret. Math. Algorithms Appl.* doi: 10.1142/S179383092450037X.

18 Revolutionizing Agriculture: A Fuzzy Logic System Approach for Optimized Operations, Precision Outputs, and Enhanced Productivity

K. Vinitha[1], G. Arul Freeda Vinodhini[2] and R. Hariharan[2,]*

[1]Reseach Scholar, [3]Saveetha School of Engineering, Saveetha Institute of Medical and Technical Sciences. Chennai, India
[2]Associate Professor, [3]Saveetha School of Engineering, Saveetha Institute of Medical and Technical Sciences. Chennai, India

Abstract

This paper introduces an optimized agricultural process leveraging the capabilities of a Fuzzy Logic System (FLS). The proposed system aims to address uncertainty within agricultural operations by effectively handling indeterminate values, ultimately generating precise outputs. By integrating FLS, this method enhances production systems within agriculture. The Fuzzy Logic System is meticulously developed and simulated through Virtual Instrumentation, offering a robust framework for agricultural enhancement and optimization. This research explores the potential of FLS in revolutionizing agricultural practices, improving decision-making, and advancing productivity within the agricultural sector.

Keywords: Agriculture, industrial mathematics, fuzzy logic system.

1. Introduction

India's agricultural sector faces persistent challenges, including dependency on erratic monsoons, small landholdings, soil degradation, pests, and inadequate access to technology and credit. Issues like droughts, market price fluctuations, and socio-economic distress have impacted productivity and farmer welfare. However, emerging technologies offer promising solutions. Innovations like precision agriculture, vertical farming, AI-integrated smart equipment, and blockchain for supply chain transparency are revolutionizing farming practices.

Research on integrating Fuzzy Logic Systems (FLS) into agriculture highlights its potential to enhance decision-making under uncertainty. Using virtual instrumentation for FLS development and simulation, this approach seeks to optimize productivity and sustainability, addressing core agricultural challenges effectively.

Email: vinibabu89@gmail.com, arulfreedavinodhini@saveetha.com, *harinov22@gmail.com

DOI: 10.1201/9781003661917-18

2. Literature Review

Balancing straw use to enhance soil fertility while minimizing greenhouse gas emissions aligns with conservation agriculture, aiming for sustainable farming practices and climate change mitigation [1]. Microbiome Engineering: The integration of microbiome engineering techniques addresses soil health, plant nutrition, and environmental preservation, contributing to eco-friendly agricultural practices [2]. Optimized Land-Use Planning: Incorporating territorial assessments and GIS for generating bio-based product-focused land-use plans exemplifies optimized planning, addressing environmental impacts and territorial constraints for informed decision-making in agriculture [3]. Precision Farming Techniques: Utilizing diverse UAVs and mobile platforms exemplifies precision farming, optimizing flight planning to enhance field coverage and efficiency in agricultural systems [4]. Plant Growth-Promoting Bacteria (PGPR): Utilizing PGPR showcases sustainable solutions by improving yields, crop quality, and soil fertility, contributing to productivity in environmentally conscious agriculture [5]. Smart Agriculture Evolution: Japan's smart agriculture evolution, exemplified by 'NoshoNavi1000', represents innovative agricultural advancements impacting large-scale farming profitability and efficiency [6]. Machine Learning in Disease Detection: The use of machine learning techniques for accurate disease classification demonstrates innovation in early detection and precision agriculture [7]. Agricultural Automation Integration: Recent agricultural automation advancements integrating sensors, IoT, and AI highlight efforts to address challenges and enhance field, fruit production, and livestock systems [8]. AI Applications in Agriculture & Healthcare: Demonstrates the role of AI in disease prediction, water optimization, sales, inventory management, and fraud detection for standardized quality control and cost-effectiveness in agricultural and healthcare domains [9]. Agricultural research emphasizes optimized engineering to enhance efficiency but often overlooks decision-making under uncertainties like weather, markets, and pests. This gap necessitates integrating robust strategies to address these challenges for truly optimized production.

3. Proposed Methodology

3.1 Research Design

The agricultural process encompasses multiple stages, spanning from initial planning and land preparation to cultivation, harvesting, and post-harvest tasks. Here's a detailed breakdown: Effective crop production involves selecting suitable land, preparing soil, choosing appropriate crops, and employing proper planting techniques. Efficient irrigation, weed control, fertilization, and pest management ensure healthy growth. Timely harvesting, careful post-harvest handling, and strategic marketing maintain quality and profitability, supported by accurate record-keeping for continuous improvement. By following these steps, farmers can optimize practices and boost productivity, tailoring methods to crop types, local climates, and regional techniques. The agricultural process with a Fuzzy Logic System (FLS) for irrigation involves several key steps. It begins with land preparation and planting seeds for crop cultivation. Next, the weather is checked to determine if conditions are favorable for growth. If the weather is suitable, the FLS is applied to control irrigation, taking into account soil moisture levels and weather forecasts. The crop's growth is then monitored, and when the crops are assessed to be ready, harvesting takes place. If they are not yet ripe, monitoring continues until the optimal harvest time. This process integrates the FLS to ensure efficient irrigation and crop care, optimizing water usage based on real-time environmental data.

3.2 Data Collection Methods

Field data collection in agriculture uses sensors, IoT devices, satellites, drones, mobile apps, and weather stations to provide real-time insights. These methods track soil, crop, and weather conditions, enabling informed decisions. Input variables include soil moisture, humidity, temperature, and weather data, while output variables manage irrigation schedules, fertilizer application, and worker instructions. Together, these factors optimize agricultural practices.

3.3 Fuzzy logic System—Proposed Developed System

A Fuzzy Logic System for agricultural practices is designed using LabVIEW's Fuzzy Logic Designer, featuring five input variables influencing the process. Each input and output variable has three triangular membership functions (low, medium, high), normalized on a per-unit scale. Outputs include water schedules, fertilizer ranges, and worker strength schedules, determined by input factors like soil moisture, weather, and crop type. This system ensures flexible, precise decision-making by handling uncertainties in agricultural inputs and outputs. Fuzzy input and output variables shown in the Figure 18.1.

In a fuzzy logic system, rules map input variables to outputs using linguistic terms. For five inputs (A, B, C, D, E) and three outputs (X, Y, Z), each with three membership functions (low, medium, high), a rule base of 96 rules defines relationships, guiding decisions based on input conditions.

If (Input_A is membership_function_A1) and (Input_B is membership_function_B1) and(Input_Cismembership_function_C1)and (Input_D is membership_function_D1) and (Input_E is membership_function_E1) then (Output_X is membership_function_X1) and (Output_Y is membership_function_Y1) and (Output_Z is membership_function_Z1)

The system uses 96 rules, covering all combinations of input and output membership functions. When uncertain input values are given, it evaluates which rules best match the inputs and generates crisp output values. These outputs are derived from fuzzy logic rules, providing precise results based on varying input conditions.

4. Results

4.1 Presentation of Findings

Any uncertain value of inputs soil moisture, humidity, temperature, weather data and pest control. And the outputs are water schedule, fertilizer and workers schedule. Based on the uncertain values certain value are produced by Fuzzy logic system. Soil moisture 0.823529, humidiy .064171, Temperature .0197861, weather data .112299 and the output produced as .500513 (Water Schedule), 0500513(Fertilizer) and .199882(workers schedule). Based on the real value the actuator output has produced. So the optimized output has produced. In Figure 18.2 shows the Fuzzy sample output and Table 18.1 shows the sample output values. The dataset showcases an agricultural monitoring system where environmental indicators like soil moisture, humidity, temperature, and weather data guide decision-making.

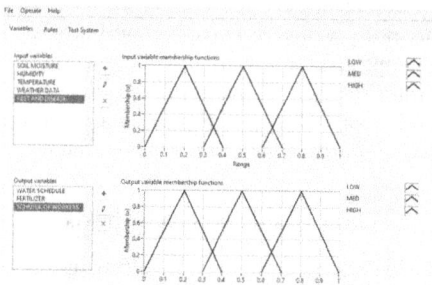

Figure 18.1 Fuzzy Input and Output Variables

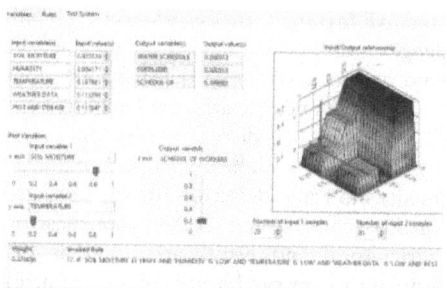

Figure 18.2 Fuzzy Output

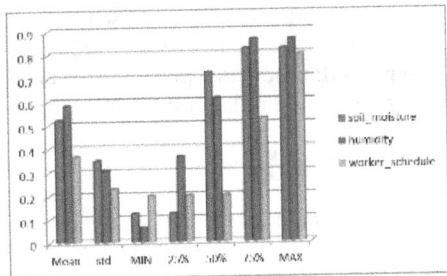

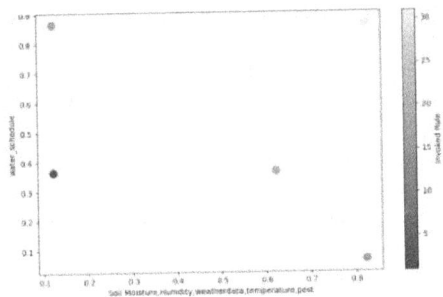

Figure 18.3 Data Analysis in FLC

Pest presence is detected through binary representation, allowing proactive management. Water, fertilizer, and worker schedules are dynamically adjusted for resource optimization. The "Invoked Rule" column highlights a nuanced, adaptable decision process, triggering multiple rules based on input combinations. Full interpretation requires expert collaboration to clarify variable meanings and underlying rules. Figure 18.4 shows sample Fuzzy output variables.

The code generates a scatter plot showing the relationship between soil moisture and humidity, with data point colors representing the "Invoked Rule" value shown in Figure 18.3 and Table 18.2. A histogram also displays the distribution of invoked rules, highlighting the frequency of different rules. These visualizations help analyze agricultural conditions and rule activations, with the code adaptable for further exploration or customizations.

Table 18.1 Fuzzy Logic Output

S. N	INPUT VARIABLES					OUTPUT VARIABLES			
	Soil Moisture	Humidity	Temperature	Weather Data	Pest	Water Schedule	Fertilizer	Schedule of Workers	Invoked Rule
1	0.823	0.06	0.197	0.11	0.11	0.50	0.50	0.192	17
2	0.823	0.86	0.197	0.11	0.11	0.35	0.65	0.200	24,25
3	0.823	0.86	0.897	0.11	0.11	0.19	0.50	0.199	29
4	0.823	0.86	0.897	0.51	0.11	0.20	0.49	0.204	31 32,62
5	0.823	0.86	0.897	0.51	0.71	0.19	0.52	0.199	64
6	0.123	0.86	0.897	0.51	0.71	0.79	0.79	0.49	16
7	0.123	0.36	0.897	0.51	0.71	0.49	0.79	0.644	8,56
8	0.123	0.36	0.297	0.51	0.71	0.79	0.79	0.5346	4,76

Table 18.2 Data Analysis in Fuzzy Logic Variables

	Soil Moisture	Humidity	Worker Schedule	Invoked Rule
Mean	0.523	0.584	0.368	18.66
Std	0.349	0.308	0.229	10.78
Min	0.123	0.064	0.199	1
25%	0.123	0.363	0.199	16.25
50%	0.723	0.614	0.202	17.5
75%	0.823	0.864	0.525	26.2
Max	0.823	0.864	0.799	31

Input/Output relationship

Input/Output relationship

Input/Output relationship

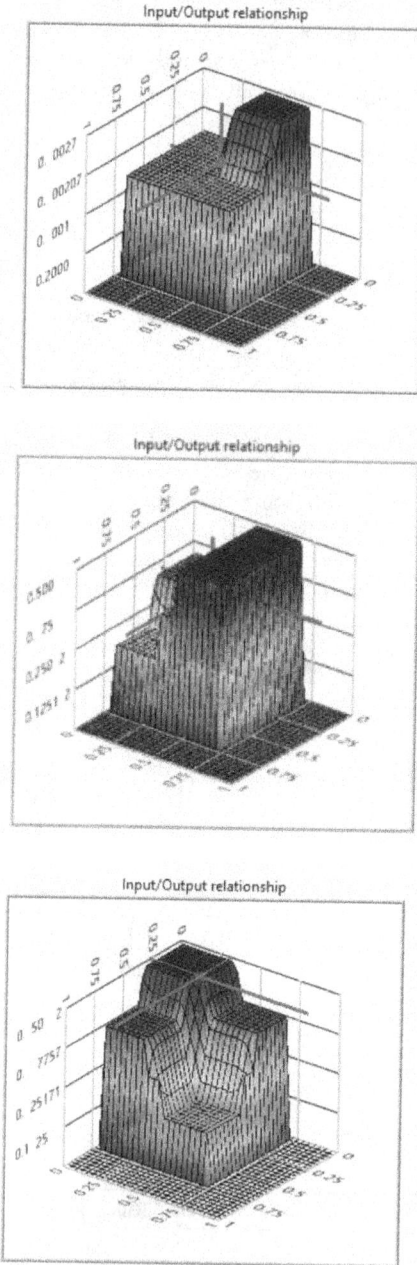

Figure 18.4 Fuzzy Sample Output Analysis

4.2 Key Observations

The dataset presents an advanced agricultural system leveraging environmental indicators and Fuzzy Logic to optimize irrigation, fertilization, and scheduling. Using Virtual Instrumentation, it enhances resource efficiency and productivity from land preparation to harvest.

5. Conclusion

In conclusion, the dataset highlights an advanced agricultural system that optimizes resources and productivity using environmental indicators and refined decision-making. The "Invoked Rule" column reflects the system's adaptability, while Fuzzy Logic integration via Virtual Instrumentation ensures precision by managing uncertainties. This innovative approach enhances efficiency, sustainability, and productivity in agricultural practices.

References

[1]   Li, P., Zhang, A., Huang, S., Han, J., Jin, X., Shen, X., ... & Chen, Z. (2023). Optimizing management practices under straw regimes for global sustainable agricultural production. *Agronomy*, 13(3), 710.

[2]   Thakur, N., Nigam, M., Mann, N. A., Gupta, S., Hussain, C. M., Shukla, S. K., ... & Khan, S. A. (2023). Host-mediated gene engineering and microbiome-based technology optimization for sustainable agriculture and environment. *Functional & Integrative Genomics*, 23(1), 57.

[3]   Ding, T., Steubing, B., & Achten, W. M. (2023). Coupling optimization with territorial LCA to support agricultural land-use planning. *Journal of environmental management*, 328, 116946.

[4]   Mukhamediev, R. I., Yakunin, K., Aubakirov, M., Assanov, I., Kuchin, Y., Symagulov, A., ... & Amirgaliyev, Y. (2023). Coverage path planning optimization of heterogeneous UAVs group for precision agriculture. *IEEE Access*, 11, 5789–5803.

[5]   Yu, Y. Y., Xu, J. D., Gao, M. Z., Huang, T. X., Zheng, Y., Zhang, Y. Y., ... & Jiang, C. H. (2023). Exploring plant growth promoting rhizobacteria potential for green agriculture system to optimize sweet potato productivity and soil sustainability in northern Jiangsu, China. *European Journal of Agronomy*, 142, 126661.

[6] Lamba, S., Saini, P., Kaur, J., & Kukreja, V. (2023). Optimized classification model for plant diseases using generative adversarial networks. *Innovations in Systems and Software Engineering*, 19(1), 103–115.

[7] Edan, Y., Adamides, G., & Oberti, R. (2023). Agriculture automation. *Springer Handbook of Automation*, 1055–1078.

[8] Pallathadka, H., Mustafa, M., Sanchez, D. T., Sajja, G. S., Gour, S., & Naved, M. (2023). Impact of machine learning on management, healthcare and agriculture. *Materials Today: Proceedings*, 80, 2803–2806.

[9] Baradaran, A. A., & Tavazoei, M. S. (2022). Fuzzy system design for automatic irrigation of agricultural fields. *Expert Systems with Applications*, 210, 118602.

[10] Belkadi, A., Mezghani, D., & Mami, A. (2020). Design and implementation of FLC applied to a smart greenhouse. *Engenharia agrícola*, 40, 777–790.

[11] Benyezza, H., Bouhedda, M., & Rebouh, S. (2021). Zoning irrigation smart system based on Fuzzy Control Technology and IOT for water and energy saving. *Journal of Cleaner Production*, 302, 127001.

19 Advanced Phishing Prediction on Website Through Novel Machine Learning Techniques

Madhubaalakrishnam[1], Gavaskar K.[2], Vishwanathan K. K[3] and R. Rajakumari[4]

National Engineering College, India

Abstract

Phishing attacks are a leading cyber threat, exploiting individuals and organizations by deceiving them into sharing sensitive information, such as login credentials, financial details, or personal data. With increasing sophistication, these attacks now expertly mimic legitimate websites, making them more convincing and harder to detect. The proposed work aims to develop an advanced machine learning-based model capable of accurately predicting phishing websites. The model utilizes a novel set of features dynamically extracted during website updates, including SSL certification, privacy policies, update frequencies, and website technologies. Combined with traditional URL-based characteristics, these features create a comprehensive dataset for training and prediction. The study utilizes advanced machine learning algorithms, including Support Vector Machine, Random Forest and Neural Networks to improve detection accuracy. Feature engineering and model optimization techniques are also applied to improve the model's predictive performance further. The research involves thorough data collection, preprocessing, and feature selection to optimize model performance. Random Forest algorithm achieves an impressive accuracy of 99.08%, outperforming other methods.

Keywords: Phishing attacks, legitimate, machine learning, URL features, user behaviour

1. Introduction

In today's increasingly interconnected digital environment, the rise of cyber threats presents significant challenges to individuals and organizations. Among these, phishing attacks stand out as one of the most pervasive and damaging forms of cybercrime. These attacks deceive individuals into sharing sensitive information, such as credit card details, passwords and usernames by masquerading as a trusted entity. Phishing attacks are often conducted through fraudulent websites, emails, or communication channels that closely resemble legitimate sources, making detection difficult. The financial and reputational damage caused by phishing is substantial, leading to resulting

Email: 2112005@nec.edu.in[1], 2112009@nec.edu.in[2], 2112017@nec.edu.in[3], rajakumarinatarajan@gmail.com[4]

DOI: 10.1201/9781003661917-19

in billions of dollars in losses annually and severely undermining user trust. Phishing has also become a leading cause of data breaches and fraud incidents.

In response to the evolving threat landscape, it has become critical to develop robust and effective phishing detection mechanisms. Traditional security methods are increasingly inadequate against sophisticated phishing techniques. Machine learning has become a powerful solution, providing the capability to process large datasets and identify patterns associated with phishing attempts. This work explores the capabilities of leveraging advanced ML techniques to enhance phishing detection accuracy and efficiency, aiming to provide stronger safeguards against these attacks. By implementing novel machine learning strategies, the aim is to greatly enhance the detection rate of phishing attempts and reduce the occurrence of successful attacks.

2. Literature Survey

Aburrous et al. [1] explored machine learning techniques like Support Vector Machines (SVM), Decision Trees, and Artificial Neural Networks (ANN) for phishing website detection, focusing primarily on URL-based features. Their research achieved a notable accuracy of 92%. Similarly, Marchal et al. [2] employed Random Forests for real-time phishing detection, also leveraging URL features, and achieved a 93% accuracy rate, underscoring the algorithm's effectiveness in timely threat mitigation. Both URL and email features were combined in models like SVM and Neural Networks [3–5], resulting in a detection accuracy of 96%. The researchers demonstrated the advantage of integrating multiple feature types for more accurate phishing detection. Also, the combined URL and HTML features [6–9] with ensemble learning methods, again achieving a 95% accuracy but grappling with computational challenges. Bagging and Boosting [10] ensemble methods, obtained a 94% accuracy rate but faced increased computational complexity.

Recurrent Neural Networks (RNNs) for phishing detection [11], using URL sequences provided an accuracy of 96%, while Deep Neural Networks (DNNs) achieved a 97% accuracy rate [12]. Verma and Hossain combined Natural Language Processing (NLP) with machine learning models like SVM and Naive Bayes [13, 14], focusing on analysing both URL and email content. Singh and Sinha [15] encountered high computational demands. Zhang, Lu, and Zhang [16] explored deep learning models utilizing feature fusion and achieved a 96% accuracy rate. Their model required significant computational resources but demonstrated the advantages of integrating multiple feature types.

Hybrid machine learning models and feature engineering for phishing detection were explored to integrate various URL and webpage content features [17], and the researchers achieved a high accuracy rate of 97%. However, the extensive feature engineering required was resource-intensive. The use of Multi-Layer Perceptron (MLP) for phishing detection [18] investigated by utilizing URL and content features. The primary limitation of this method was the reliance on manual feature extraction, which was time-consuming and sometimes incomplete. The feasibility of MLPs for phishing detection while addressing feature extraction challenges [19] explored the application of deep learning combined with hybrid feature engineering for phishing detection. The model incorporated URL, HTML content, and domain registration features, achieved an accuracy of 97%. Despite its success, the high computational cost of training with extensive features was a key limitation. That work underscored the potential of deep-learning approaches for phishing detection when combined with innovative feature engineering. Kumar, S. and Patel V [20] examined the effectiveness of ensemble techniques for phishing detection. The authors combined multiple machine learning models with URL features and they achieved a detection accuracy of 95%. Nevertheless, the increased computational complexity and resource demands of ensemble methods presented challenges and they demonstrated the benefits of ensemble techniques for improving detection accuracy.

Reinforcement Learning (RL) for phishing detection [21] by incorporating URL features and user behaviour data was introduced by Zhang, Y and Li, J. Their model achieved a 92% accuracy rate. The potential of Convolutional Neural Networks (CNNs) [22, 23] for phishing detection utilized URL and webpage content features. The need for large labelled datasets to train CNNs effectively was a notable limitation. The authors highlighted the applicability of CNNs for enhancing phishing detection capabilities. The Transformer Networks for phishing detection integrated the URL and HTML content features. This study showcased the effectiveness of Transformers in phishing detection while addressing computational challenges. Graph Neural Networks (GNNs) [24] for phishing detection was examined and the authors incorporated URL relationships and webpage content. The complexity of the system involved in constructing and processing graph-based data was a significant limitation.

The integration of metaheuristic optimization techniques with machine learning for phishing detection. By optimizing feature selection and model parameters, the increased complexity of optimization processes was a limitation [25, 26]. The researchers demonstrated the benefits of optimization techniques for enhancing phishing detection accuracy. Huang, X et al. [27] investigated the application of Federated Learning combined with ensemble methods for phishing detection. Leveraging distributed data sources. Managing federated learning processes and ensuring data privacy were key challenges. They presented Federated Learning as a scalable solution for phishing detection. The use of adversarial machine learning techniques [28] was examined for improving phishing detection by incorporating adversarial training. The reliance on adversarial examples that might not encompass all phishing scenarios was a notable limitation. The authors introduced adversarial techniques for strengthening phishing detection models. Self-supervised learning for phishing detection demonstrated the potential of self-supervised learning in phishing detection [29]. Integration of machine learning models with human feedback for phishing detection dependence on human feedback, which could be inconsistent, was a significant limitation. The authors emphasized the effectiveness of human-machine collaboration [30] for phishing detection.

Even though so much research has been carried out to solve this problem, still there is a gap available to provide an accurate system. Hence this work was carried out to grant the best accuracy with less computation.

3. Proposed Work

To provide immediate feedback on website legitimacy and offer enhanced protection against phishing attacks, this work proposes a machine learning-based approach. Figure 19.1 shows the architecture structure of the suggested work.

4. Methodology

The data collection process focused on acquiring various features from URLs, such as URL length, presence of a privacy policy, availability of contact information, feedback scores, and website load time. These attributes were carefully extracted from the dataset ('structured_data_phishing.csv'), ensuring that only the most relevant features for distinguishing between phishing and legitimate websites were included. Special attention was given to feature extraction, ensuring that the selected attributes could greatly improve the accuracy of phishing detection systems. The selection process was key to optimizing the overall performance of the machine learning techniques.

The system architecture starts with Webpage Feature Generation, where key

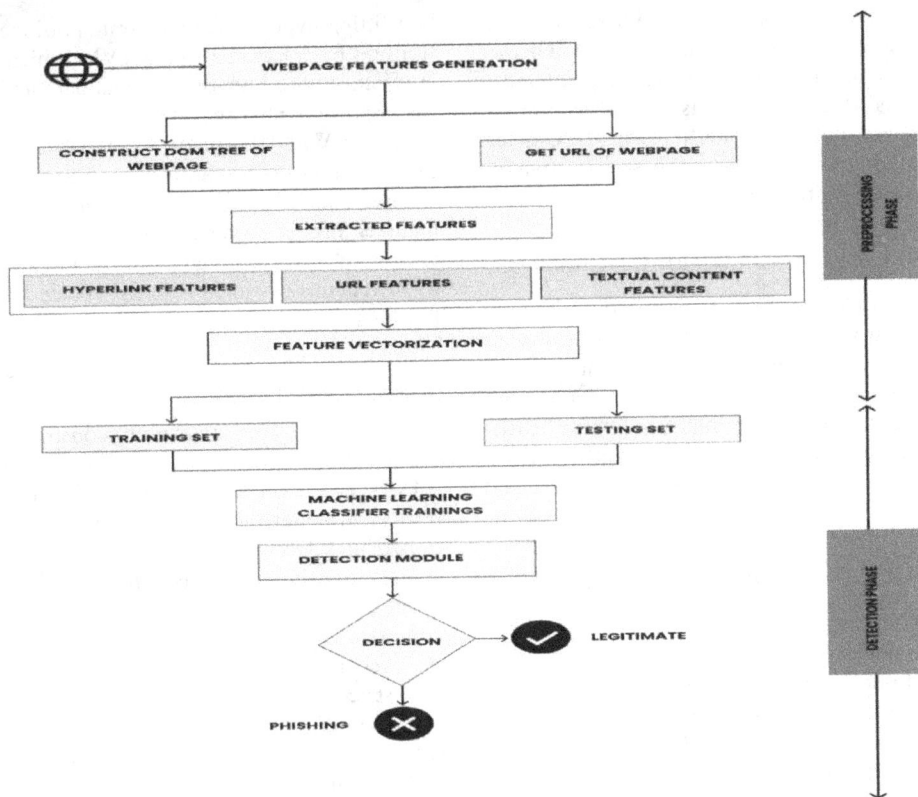

Figure 19.1 Architecture of a Phishing Prediction Machine Learning

information is extracted from the webpage. The Document Object Model (DOM) tree is then constructed, which outlines the webpage structure and assists in identifying its components. The webpage's URL is fetched, allowing the extraction of URL-based features relevant to phishing detection. Features extracted include patterns in hyperlinks, URL structure, and textual content. Hyperlink features involve the ratio of internal to external links, while URL features include aspects such as URL length, presence of special characters, and suspicious patterns. Textual content features involve searching for phishing-related keywords or suspicious phrases. These features are then converted into numerical values through Feature Vectorization, preparing the data for machine learning model input. Among the whole dataset, 39% data from phishing and 61% data from

legitimate websites as shown in Figure 19.2. The pie chart showcases the dataset used in this study, sourced from Phishtank. org and Tranco-list.eu. It consists of 26,584 websites, including 16060 genuine websites (61%) and 10524 phishing websites (39%). This distribution guarantees that the machine learning model is trained on a well-balanced dataset to improve prediction reliability.

Seventy percent of the data from each category is used for training, while 30% is reserved for testing. The system is trained using the training dataset and then assesses the models' performance on previously unseen data. During the Machine Learning Classifier Training stage, various algorithms are trained on the vectorized features to learn patterns that distinguish phishing websites from legitimate ones. The trained models are then implemented in the detection module, where they analyse new web pages and predict whether they are legitimate or phishing attempts. Finally, the system classifies each webpage based on the model's predictions, determining whether the webpage is a phishing site or legitimate.

4.1 Machine Learning Models

Multiple machine learning models were chosen to detect phishing websites, with each algorithm contributing uniquely to the overall system performance and accuracy. Wang and Zhang integrated meta-heuristic optimization techniques with machine learning to enhance feature selection and model performance

4.1.1 Gaussian Naive Bayes (GNB)

The Gaussian Naive Bayes algorithm assumes that the features are independent of one another. This one works particularly better for continuous data, where the features follow a normal distribution. GNB is highly efficient and easily interpretable, this makes it well-suited for real-time phishing recognition. However, if the assumption of feature independence is violated, its performance can decline. Despite this, GNB performs well for simpler datasets.

4.1.2 Support Vector Machine (SVM)

SVM is a robust and effective binary classification technique that identifies the best-fitting hyperplane to separate phishing websites from legitimate ones. SVM is highly effective with high-dimensional data and can capture nonlinear patterns using kernel functions. While it is renowned for its accuracy, it can be computationally resource-intensive, especially when dealing with huge datasets. Proper parameter tuning is essential for achieving optimal performance, and SVM remains a widely adopted choice for phishing detection due to its effectiveness.

4.1.3 Decision Tree

The Decision Tree algorithm constructs a tree-like structure by dividing data into subsets based on feature values. Its simplicity in interpretation and visualization makes Decision Trees a widely used choice for phishing detection. The algorithm works well with both categorical and numerical data, but it is susceptible to overfitting. Techniques like pruning or restricting tree depth can help address this challenge. Despite this, Decision Trees remain valuable due to their transparency and flexibility.

4.1.4 Random Forest

Random Forest is an ensemble learning method that creates multiple decision trees and aggregates their predictions to boost accuracy and reduce overfitting. Known for its robustness, this algorithm excels in handling noisy data and automatically ranks the importance of features. While Random Forest achieves high accuracy. However, it can be computationally intensive due to the large number of trees it generates. Although less interpretable than a single Decision Tree, Random Forest is a robust candidate for phishing recognition due to its performance.

4.1.5 AdaBoost

AdaBoost enhances the performance of weak classifiers by iteratively focusing on misclassified instances. It is effective in improving precision for phishing detection but is sensitive to noise and outliers. AdaBoost can

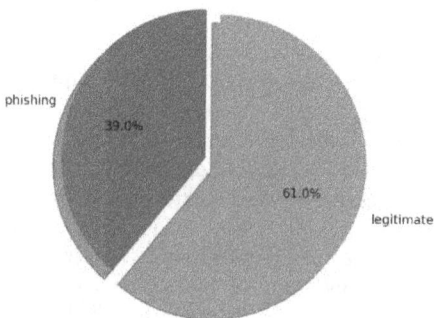

Figure 19.2 Dataset Distribution for Phishing and Legitimate Websites

be combined with various base classifiers, making it versatile. However, overfitting is a concern if not properly managed. Despite its sensitivity, AdaBoost is a powerful tool for boosting weaker models and improving their performance.

4.1.6 Neural Networks

Neural Networks model complex patterns using interconnected layers of neurons, capturing nonlinear relationships in phishing data. They are particularly effective for high-dimensional datasets, such as URLs and webpage content. However, Neural Networks require huge amount data and significant resources for computation, leading to longer training times. Overfitting is also a potential issue, which can be addressed through regularization techniques. When properly trained, Neural Networks can achieve high accuracy in phishing detection.

4.1.7 K-Nearest Neighbors (KNN)

KNN is a classification algorithm that categorizes data points based on the majority class of their k-nearest neighbors. It is particularly useful for noisy datasets and is straightforward to implement. Although KNN doesn't require a training phase, it can become computationally costly during prediction, especially with large datasets. The selection of the k parameter is vital, as it has a significant impact on the model's performance. KNN is effective for phishing detection when used with smaller datasets and irregular decision boundaries.

5. Results and Discussion

In the existing research work, Federated Learning combined with ensemble methods, adversarial machine learning techniques, self-supervised learning and human feedback into machine learning models achieved 95% to 97% accuracy. The performance of this proposed work is tabulated in Table 19.1 in terms of accuracy, precision, and recall.

Among those models, Random Forest provides a better accuracy of 99.08%. The graphical visualization is shown in Figure 19.3. The contributions of this work extend beyond theoretical research, offering practical applications for improving web security. The phishing detection system can help organizations and individuals protect themselves against online threats, thus enhancing overall cyber security. This research adds value to the fields of machine learning and cyber security by introducing a novel approach to phishing detection and showcasing the everyday use of machine learning in addressing actual security challenges.

The effectiveness of each machine learning algorithm is assessed through important metrics like recall, F1 score, precision, and inclusive accuracy. This comprehensive assessment helped gauge the effectiveness of each model in detecting phishing websites. A comparative analysis was performed to

Table 19.1 Comparison of Accuracy for Different ML Models

CLASSIFIERS	ACCURACY	PRECISION	RECALL
Naive Bayes (NB)	0.5253	0.0407	0.9114
Support Vector Machine (SVM)	0.9546	0.2631	0.1457
Decision Tree (DT)	0.9847	0.6805	0.7108
Ada Boost (AB)	0.9791	0.5875	0.1908
Neural Network (NN)	0.9767	-	0.0537
K-Nearest Neighbor (KNN)	0.9777	0.4783	0.1140
Random Forest (RF)	**0.9908**	**0.8366**	**0.6549**

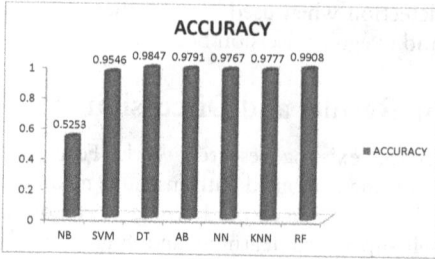

Figure 19.3 Comparison of Accuracy for Different ML Models

identify the most suitable algorithm for phishing detection based on performance across these metrics. To improve the inter-pretability of the results, visual tools like ROC curves, confusion matrices, and accuracy tables were utilized, offering a clear visual representation of how each model performed on various datasets.

Following these evaluations, the highest-performing algorithm was selected for its exceptional accuracy and reliability in detecting phishing URLs. Some challenges encountered during implementation included imbalanced datasets, overfitting in some models, and complications in feature extraction. These challenges were addressed through methods like data augmentation, cross-validation, and hyperparameter tuning. Additionally, computational resource limitations during model training were identified, with

suggestions for future optimizations to enhance the training process.

5.1 Output

The Figure 19.4 displays the graphical user interface of the "Phishing Detector" application for detecting a legitimate URL.

The positive result is visually enhanced with a balloon-themed background. The application notifies the user with the message: *"This web page seems legitimate!"* and indicates the "K-Nearest Neighbors" model is in use.

The user interface of the "Phishing Detector" app demonstrates a phishing detection result. The background incorporates a visual theme with snowflakes, making it interactive and visually engaging. The output screen displays a selected machine learning model, "K-Nearest Neighbors," and highlights a detected phishing URL with a warning message: "Attention! This web page is a potential PHISHING!", which is shown in Figure 19.5.

6. Conclusion

This research successfully developed a phishing detection system that integrated multiple machine learning algorithms to create a robust solution for identifying phishing URLs. Through careful evaluation, the most effective algorithm was identified,

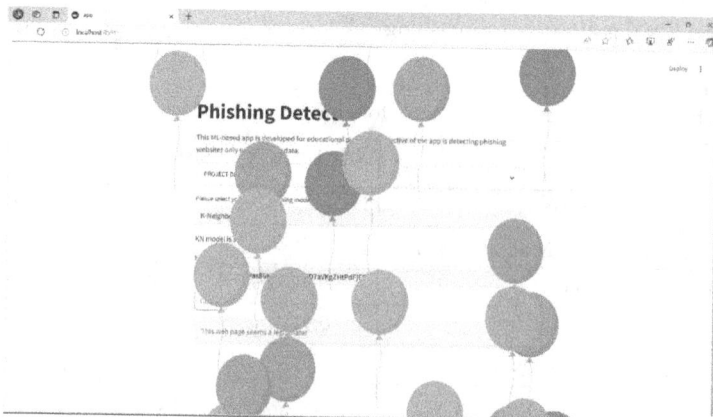

Figure 19.4 User Interface with Legitimate Site Confirmation

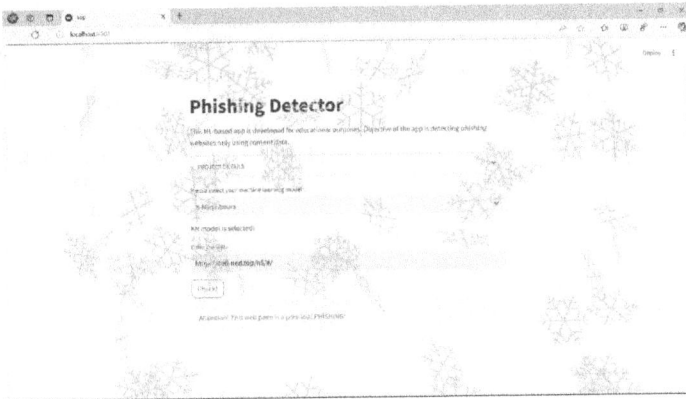

Figure 19.5 User Interface with Phishing Alert

offering superior accuracy and reliability for detecting phishing websites. The results underscore the potential of combining various machine learning models to strengthen phishing detection capabilities. Future work could focus on expanding the system's feature set, incorporating additional real-time features such as website behavior analysis or more sophisticated security certificate evaluations to enhance detection accuracy further. Additionally, exploring advanced machine learning techniques, such as deep learning models or hybrid approaches, could lead to even better performance. Future research may also involve testing the system on larger datasets or real-world environments to assess its scalability and reliability.

References

[1] Aburrous, M., Hossain, M. A., & Rabhi, F. (2014). Phishing detection through machine learning methods. *Journal of Computer Science and Technology*, 29(3), 475–485.

[2] Marchal, S., Pohlmann, N., & Baudet, S. (2014). Real-time phishing detection using Random Forests. *IEEE Transactions on Information Forensics and Security*, 9(6), 985–993.

[3] Abbasi, A., Chen, H., & Zimbra, D. (2015). A multi-level hybrid approach for phishing detection. *Journal of Information Security*, 6(1), 21–32.

[4] Xiang, Y., Wei, H., & Han, J. (2015). Ensemble learning for phishing detection using URL and HTML features. *International Journal of Information Security*, 14(3), 245–256.

[5] Abbasi, A., Zahedi, M., & Chen, H. (2015). A survey of machine learning algorithms in phishing detection. *Computers & Security*, 54, 109–120.

[6] Xiang, Y., Hong, S., & Rose, C. (2015). Hybrid feature-based phishing detection with ensemble learning. *IEEE Transactions on Dependable and Secure Computing*, 12(4), 432–445.

[7] Mamun, M. A., Rathore, M. M., & Huh, E. N. (2016). Phishing detection using Random Forest and Naive Bayes classifiers. *Journal of Computing and Security*, 58, 170–182.

[8] Jain, A., & Gupta, M. (2016). Ensemble learning methods for phishing detection using URL and content features. *Expert Systems with Applications*, 57, 212–221.

[9] Basnet, R. R., Mukkamala, R. R., & Sung, A. H. (2016). URL and webpage content-based phishing detection. *Computers & Security*, 56, 57–72.

[10] Alhassan, A., Alhassan, A., & Sahin, F. (2016). URL and web content-based ensemble methods for phishing detection. *IEEE Access*, 4, 2583–2591.

[11] Bahnsen, A. C., Bohorquez, J., & Villegas, A. (2017). Using recurrent neural networks for phishing detection based on URL sequences. *Neurocomputing*, 224, 118–127.

[12] Abdelhamid, I., Ayesh, A., & Thabtah, F. (2017). Phishing detection using deep neural networks with URL and HTML features. *Knowledge-Based Systems*, 122, 119–128.

[13] Verma, A., & Hossain, M. A. (2017). Phishing detection utilizing natural language processing and machine learning. *Information Sciences*, 396, 89–100.

[14] Al-Qerem, M., Al-Razgan, M., & Al-Malaise, M. (2017). A phishing detection model incorporating domain registration and URL features. *Journal of Computer Security*, 25(6), 727–744.

[15] Singh, R., & Sinha, P. (2018). Techniques in ensemble learning for phishing detection using URL and webpage content features. *Applied Soft Computing*, 68, 537–546.

[16] Zhang, Z., Lu, W., & Zhang, M. (2018). Feature fusion in deep learning for phishing detection using URL, HTML content, and domain features. *Information Sciences*, 453, 314–325.

[17] Liu, Y., Zhang, X., & Yang, H. (2019). Hybrid machine learning models and feature engineering for phishing detection. *Knowledge-Based Systems*, 163, 415–425.

[18] Nguyen, T., & Ha, S. (2019). Multi-layer perceptron for phishing detection with URL and content features. *Soft Computing*, 23(7), 2487–2496.

[19] Xu, J., Zhang, Y., & Zhao, X. (2020). Hybrid feature engineering in deep learning for phishing detection with URL, HTML content, and domain features. *Artificial Intelligence Review*, 53(4), 315–329.

[20] Kumar, S., & Patel, V. (2020). Using ensemble techniques for phishing detection with multiple machine learning models and URL features. *Journal of Computer and Communications*, 8(2), 22–35.

[21] Zhang, Y., & Li, J. (2021). Reinforcement learning for phishing detection incorporating URL features and user behavior data. *IEEE Transactions on Cybernetics*, 51(9), 4290–4302.

[22] Zhao, L., Wu, M., & Zhang, J. (2021). Phishing detection using convolutional neural networks with URL and webpage content features. *Journal of Computational Science*, 49, 101089.

[23] Wang, H., & Chen, X. (2022). Transformer networks for phishing detection using URL and HTML content features. *Neurocomputing*, 508, 112–122.

[24] Liu, Q., Yang, X., & Zhao, S. (2022). Phishing detection using graph neural networks and URL relationships. *IEEE Transactions on Network and Service Management*, 19(3), 2378–2390.

[25] Ali, S., & Ahmed, F. (2022). A comprehensive review of phishing detection techniques: Machine learning, deep learning, and hybrid approaches. *Computer Science Review*, 42, 100438.

[26] Wang, L., Zhang, Y., & Liu, X. (2023). Metaheuristic optimization combined with machine learning for phishing detection. *Applied Intelligence*, 53(2), 1450–1464.

[27] Huang, X., Liu, Y., & Zhao, X. (2023). Federated learning with ensemble methods for phishing detection. *IEEE Transactions on Neural Networks and Learning Systems*, 34(8), 3374–3385.

[28] Yang, L., & Zhang, W. (2023). Adversarial machine learning techniques applied to phishing detection. *Journal of Artificial Intelligence Research*, 77, 429–446.

[29] Chen, X., Li, J., & Xu, J. (2024). Self-supervised learning for phishing detection using URL and webpage content features. *Data Mining and Knowledge Discovery*, 38(2), 443–460.

[30] Wang, Y., Zhang, H., & Liu, J. (2024). Combining machine learning with human feedback for phishing detection: An integrated approach. *Pattern Recognition*, 131, 108898.

20 Enhancing Grasshopper Techniques to Reduce Torque Ripples in Switched Reluctance Motor

Deepa N.[1], Krishnaveni K.[2] and Mallesham G.[3]

[1,3]Department of Electrical Engineering, Osmania University, Hyderabad, India
[2]Department of Electrical and Electronics Engineering, CBIT, Hyderabad, India

Abstract

Torque ripple, the periodic fluctuation in output torque during motor shaft rotation, is a common challenge in most electric motor designs. Switched Reluctance Motors (SRMs) are frequently employed across diverse electrical uses in industry owing to their simple design, fault tolerance, and robustness. However, their primary limitation is the significant torque ripple caused by the unique characteristics of their magnetic circuit. This study explores the influence of PI and FOPI controllers on the electromagnetic torque, speed, and flux of switched reluctance drives. The research also focuses on suppressing torque ripple, reducing acoustic noise, and enhancing the electrical efficiency of SRMs. A comparative analysis is conducted using graphical representations to evaluate the performance of these controllers. Additionally, a novel meta-holistic optimization technique is aimed at reducing torque ripple and enhance the electrical efficiency of SRMs.

Keywords: Torque Ripple, Optimization, Torque Ripple Reduction, SRM, TSF-DITC

1. Introduction

Switched Reluctance Motor have a lot of benefits, such as their simple and durable construction, strong fault tolerance, high reliability, high efficiency, high output, wide range, abundant starting torque, applicability in a few circumstances, and high speed [1]. The main benefit is that it may be used to power a motor in an electric vehicle, despite a few drawbacks. For example, torque ripple from SRM causes an uneven rotation of the motor rotor, which causes noise and vibration that may impair the comfort and stability of electric vehicles. Furthermore, in electric vehicles,

the motor's dynamic performance plays a crucial role in determining traction characteristics. The electromechanical conversion device known as SRM is said to have salient pole structure and non-linear properties. The nonlinear relationship between flux and torque with respect to position. The main characteristics and scheme focused on the optimization parameters of motor design and control strategies. The process of torque ripple reduction and noise reduction can be used in the framework of the SRM rotor pole [2]. Technology using Switched Reluctance differs from traditional technology in that it has an intrinsic Torque

Email: deepanidanakavi@gmail.com[1], krishnaveni_eee@cbit.ac.in[2], gm.eed.cs@gmail.com[3]

DOI: 10.1201/9781003661917_20

Ripple. The two primary methods used in this technology are the Electronic Control Method and the Motor Magnetic Design Method. The quantity of rotor and stator poles can be adjusted straightforward method called Motor Magnetic Design [3]. The Electronic Control approaches are dependent on a few factors, such as supply voltage, the angle at which the converter is turned on and off, and the current, but they can lower the machine's average torque [4]. By applying Electronic Torque Control Techniques, Torque Ripple can be reduced throughout a wider working range. Particularly at the low speeds the produced Torque Ripple leads to speed oscillation as well as possible triggering of resonant frequencies in the SRM motor potions of the SRM drive. Designing intelligent controllers based Direct Torque schemes for SRM drive is the ultimate goal of this research. Traditional control methods result in sluggish response times as well as excessive torque and flux errors. The typical approach has a limited bandwidth [5].

This study employs both theoretical and simulation analyses to assess the complexity and reliability of various widely-used control strategies for switched reluctance motor (SRM) drives. It explores how common errors in SRM drives affect the reliability of their control systems. Additionally, the research analyzes the connection between the complexity of information flow in control strategies and the reliability of SRM control's DITC and DTC, Current Chopping Control (CCC). The findings indicate that the TSF-DITC method demonstrates the maximum resilience to measurement errors, succeeded by DTC and CCC. The TSF-DITC approach not only exhibits strong resilience and reliability but also delivers superior torque performance [6].

In order to compare the two methods, their control block diagrams are first introduced in depth along with their fundamental ideas. Second, the simulation in MATLAB/Simulink is used to validate the proposed two control systems. Eventually, the simulation results demonstrate that the two strategies correctly reduce torque ripple. Also, the two algorithms' steady and dynamic state performances are contrasted in this work. According to the study, the DITC-TSF scheme can offer minor torque ripple reduction and higher controlled performance [7]. Electric Motors are used in various applications, and their essential requirements include an increased specific power, decreased cost, and dynamic construction. Switching reluctance is a motor that satisfies all the requirements mentioned above. Switched Reluctance Motors are a growing rival to traditional IM and PM in small, commercial, and EV deployments. The main drawback of SRM is torque ripple, particularly during the operation at high speed, which results in acoustic noise even though SRMS offers significant advantages over PM and Induction Motors. The systems may sometimes suffer from these drawbacks, which depend upon the application which is to be used for acoustic noise and torque ripple form can be harmful. This article examines the state of technology and current developments in switching reluctance machines that minimize torque ripples [8].

The problem of the research study is the multi-objective optimization of Torque ripple reduction of TSF-DITC controlled Switched Reluctance Motor. The objectives of the research study is using optimization techniques for torque ripple reduction of TSF-DITC controlled Switched Reluctance Motor i.e., Grasshopper Optimization, Further section 2 Conventional Methods and Optimization Techniques, section 3 simulation Results and Validation. Section 4 Conclusion.

1.1 Mathematical Model of SRM Drive System

Basic Equations of Torque Ripple Reduction

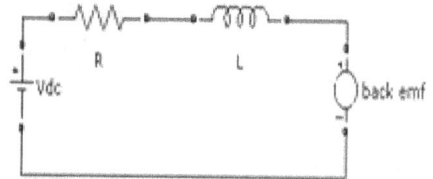

Figure 20.1 Equivalent Circuit

$$V = iR + \frac{d\lambda}{dt} \quad (1)$$

Where λ is the function of I and θ

$$\frac{d\lambda}{dt} = \frac{d}{dt}i + \frac{di}{dt}\theta$$

$$V = iR + \frac{d(L\lambda)}{dt}$$

$$V = iR + L\frac{di}{dt} + i\omega\frac{dL}{d\theta}$$

$$V = iR + L\frac{di}{dt} + i\omega\frac{dL}{d\theta} \quad (2)$$

Where iR is the Ohmic drop

$L\frac{di}{dt}$ is the EMF caused by incremental

inductance

$i\omega\frac{dL}{d\theta}$ is the self-induced EMF

$$V = iR + L\frac{di}{dt} + e \quad (3)$$

If the flat-topped current is assumed as

$L\frac{di}{dt} = 0$,

$$V_i = i^2R + L\frac{di}{dt} + i\omega\frac{dL}{d\theta} \quad (4)$$

Energy stored in magnetic circuit $= \frac{1}{2}Li^2$

Rate of change of stored energy in magnetic

circuit $= \frac{d\theta}{dt}\left\lfloor\frac{1}{2}Li^2\right\rfloor$

$$= \frac{1}{2}L\cdot 2i\frac{di}{dt} + \frac{1}{2}i^2\frac{dL}{dt}$$

$$= Li\frac{di}{dt} + \frac{1}{2}i^2\frac{dL}{d\theta}x\frac{d\theta}{dt}$$

$$\frac{dW_{mag}}{dt} = Li\frac{di}{dt}\frac{1}{2}i^2\omega\frac{dL}{d\theta} \quad (5)$$

Transferred Mechanical Energy = Electrical
energy input – rate of change of energy stored

$$= i^2R + Li\frac{di}{dt} + \frac{1}{2}i^2\omega\frac{dL}{d\theta} -$$

$$i^2R\frac{di}{dt} - \frac{i^2}{2}(\omega)\frac{dL}{d\theta}$$

$$P_m = \frac{1}{2}\left\lfloor i^2\omega\frac{dL}{d\theta}\right\rfloor \quad (6)$$

In general,

$$P_m = \frac{2\pi NT}{60} = \frac{2\pi N}{60}T = \omega T$$

$$P_m = \omega T \quad (7)$$

From 3.7;

$$T = \frac{P_m}{\omega} \quad (8)$$

Substituting (6) in (8),

$$T = \frac{1}{2}\frac{\left\lfloor i^2\omega\frac{dL}{d\theta}\right\rfloor}{\omega}$$

$$T = \frac{1}{2}\left\lfloor i^2\frac{dL}{d\theta}\right\rfloor \quad (9)$$

1.2 Optimization Techniques

The proposed methodology comprised a Grasshopper Optimization Algorithm and Conventional methods for torque ripple reduction and DC ripple reduction of TSF-DITC controlled Switched Reluctance Motor. In Figure 20.2 it consists of a current measurement for calculating the ref erence current with the actual current for finding the difference of minimized torque [9]. The proposed methodology consists of a hysteresis band with a feedback signal above the band, and the plant can be operated in one state. When it is below the band, it operates in another state. If the feedback signal

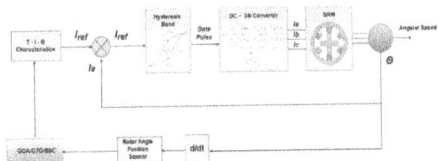

Figure 20.2 Grasshopper Optimization System Block Diagram

is within the hysteresis band, the state of operation is never changed. It consists of the torque converter, a fluid coupling that transfers the rotating power from the prime mover for converting into the rotating driven load. The converter is connected to the power source to the load in the automatic transmission of SRM. It constitutes SRM, the electric motor which it runs by the reluctance torque. It uses a mechanical commutator to switch the winding current like traditional motors.

The position of the electronic sensor is used to determine the Rotor shaft angle and the electronics in a solid state for switching the windings to enable dynamic control and pulse training.

1.3 Grasshopper Optimization Algorithm (GOA)

The GOA is the population-based Meta heuristic algorithm. The recent swarm intelligence algorithm was introduced for solving the numerical optimization issues. The inspiration of GOA Comes from the interaction and behavior of grasshopper swarms in nature.

The central concept behind the GOA algorithm as the grasshopper swarm makes long-range moves and sudden jumps in the adulthood phase as they have wings—the swarm search for food by splitting into two stages of the process. The swarm search for a food source in the exploration phase, and the value positions are updated and computing fitness value. The best solution is found among all the available sources. Thus, the GOA algorithm mimics grasshoppers' hunting method and social behavior [10].

1.4 Parameters of SRM

The control methods are optimized and tailored for commercially available SRM. The estimated voltage is 240V. The stator phase winding comprises four coils arranged in a parallel-series configuration. Both the rotor and stator poles share an identical width, approximately 6/4. The inductance curve exhibits a pronounced peak at the aligned position, while the motor transitions to generator mode

depending on speed. The voltage range can fluctuate between 120V and 240V.

1.5 Proposed system framework

The outlined framework is structured into the steps listed below:

Step 1: Start
Step 2: Initialization of parameters current and speed
Step 3: Measurement of the actual speed of SRM
Step 4: Measurement and calculation of reference current
Step 5: If $Z(i-1) > Z_i$
Step 6: If it is not, it goes with the Hysteresis band
Step 7: Then it moves to SRM
Step 8: If it is yes, rotor angle to speed conversion
Step 9: Stop

Where Z is current Error value.

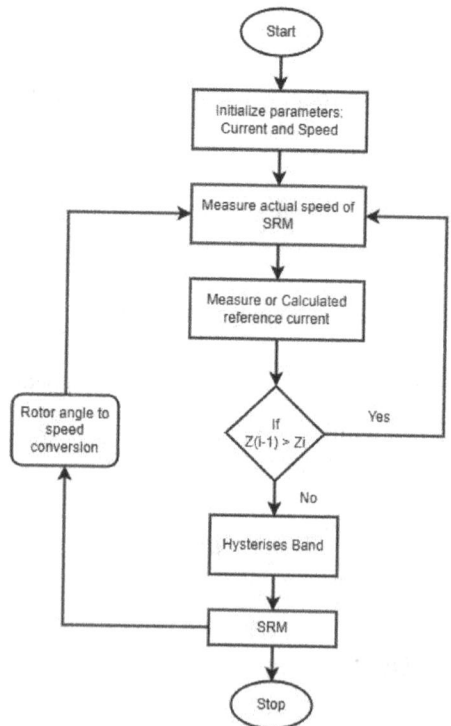

Figure 20.3 Flowchart of Grasshopper Optimization System

Optimization of the proposed Model

The proposed framework of optimization consists of the following steps as given below;

Step 1: Start
Step2: Calculation of rotors angle using a Grasshopper Optimization Algorithm, Game theory optimization and Hybrid Sin-cosine and SSC algorithm
Step 3: Gate pulse generation
Step 4: Inverter PC to 3
Step 5: SRM

3. Simulation Results and Discussion

The System simulation model comprised of 6/4 SRM motor built for the verification of the effectiveness and correctness of the proposed optimization model in the MATLAB simulation environment.

3.1 System Simulation Model

The simulation model of the system incorporates an optimization algorithm designed to minimize torque ripple in a SRM configured with a 6/4 rotor-stator arrangement.

It consists of a hysteresis band converter and a power of 240 volts. TSF-DITC is considered the primary method for suppressing the torque ripple. The method uses the TSF for the distribution of total torque in the individual phase, whereas for controlling the appropriate torque-hysteresis-driven motor rules. The multi-objective converter in the SRM selects the corresponding rules on the conduction region for accelerating the demagnetization and excitation in the region of commutation and achieving fast torque tracking. A DITC algorithm for switched reluctance motors (SRM) can be enhanced by incorporating an optimal angle adaptive torque shaping function (TSF) within the simulation model. This TSF, characterized by its dynamic and flexible form, plays a key role in enhancing the system's efficiency, specifically in mitigating torque ripple. The variation in commutation points significantly

impacts the performance, requiring careful optimization. The simplicity of the algorithm enables easy implementation for reducing torque ripple. In the commutation region, complex hysteresis control strategies are needed to achieve effective torque ripple minimization. The multiple objective optimizations of GOA, GTO and SSC algorithm helps overcome the difficulties of optimizing multiple performances simultaneously.

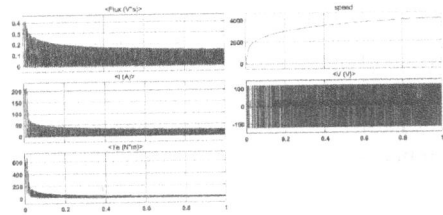

Figure 20.4 Simulated Flux Linkage, Current and Torque, Speed, and Voltage Waveforms Corresponding to Three Phases of the Investigated 6/4 SRM

The current and simulated flux linkage and torque are proportional to each other, and they may vary similarly concerning the axis of time. Initially, the current and torque are very high. The relationship between inductance and speed while tracking the reference speed according to the actual speed with the moment of inductance remains constant. It remains constant and gets settled down. At the start, the current surges due to inrush current, which can reach up to 5 A. "The torque generated by the motor depends on the rate of change of inductance with respect to the rotor position. If $dL/d\theta$ is positive, the torque is also positive, while a negative value results in negative torque.

The relative effectiveness of two methods for calculating speed versus time, for as the Grasshopper Optimization Algorithm and finite element analysis. Figure 20.5 graph demonstrates that the GOA technique speeds up SRM.

The effectiveness of the grasshopper optimization method for flux and current

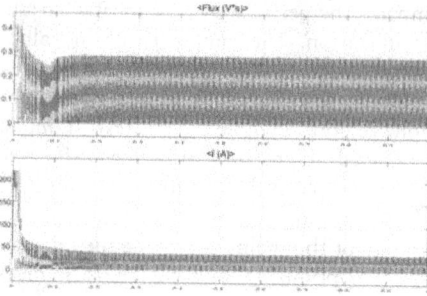

Figure 20.5 Performance of Flux and Current GOA Optimization for Torque Ripple Reduction

is displayed separately with respect to time. According to the graph in Figure 20.5, The GOA algorithm demonstrates superior performance in reducing torque ripple and DC ripple in a TSF-DITC-controlled SRM.

As illustrated in Figure 20.6 the suggested optimization techniques have been shown effective when compared to current methods for reducing torque ripple. "The torque ripple reduction achieved using the proposed GOA for the 6/4 pole SRM is measured at 11.24%." The findings obtained demonstrated that torque ripple was minimized in the suggested system as compared to the existing PI and FOPI of 6/4 pole SRM Motor of 40% and 28% respectively.

4. Conclusion

"In this research study, the Grasshopper Optimization Algorithm (GOA) is employed as a concluding method to minimize the torque ripple in the TSF-DITC-controlled Switched Reluctance Motor (SRM)." The GOA optimization technique outperforms PI and FOPI in terms of torque ripple reduction, as indicated by the simulation results of GOA based SRM got better results comparatively with conventional methods it shown in bar chart. The study's future objectives include the creation of straightforward and effective fault-tolerance and fault-diagnosis systems.

References

[1] Honghua, W. (1995). Speed control technology of switched reluctance motor [M].

[2] Liu, L., Zhao, M., Yuan, X., & Ruan, Y. (2019). Direct instantaneous torque control system for switched reluctance motor in electric vehicles. *The Journal of Engineering*, 2019(16), 1847–1852.

[3] Yang, Y., Xu, A., Leng, B., Sun, J., & Li, K. (2022). Torque Compensation Method of Switched Reluctance Motor Adopting MPC Based on TSF-DITC. *Progress In Electromagnetics Research M*, 110, 211–221.

[4] Dankadai, N. K. (2022). Multi-objective torque control of switched reluctance machine (Doctoral dissertation, Newcastle University).

[5] Sun, X., Wu, J., Lei, G., Guo, Y., & Zhu, J. (2020). Torque ripple reduction of SRM drive using improved direct torque control with sliding mode controller and observer. *IEEE Transactions on Industrial Electronics*, 68(10), 9334–9345.

[6] Dankadai, N. K., Elgendy, M. A., McDonald, S. P., Atkinson, D. J., & Atkinson, G. (2019, October). Investigation of reliability and complexity of torque control for switched reluctance drives. In *2019 IEEE Conference on Power Electronics and Renewable Energy (CPERE)* (pp. 485–490). IEEE.

[7] Ren, P., Zhu, J., Guo, Z., Song, X., Jing, Z., & Xu, A. (2021, May). Comparison of Different Strategies to Minimise Torque Ripples for Switched Reluctance Motor. In *2021 IEEE 4th International Electrical and Energy Conference (CIEEC)* (pp. 1–6). IEEE.

[8] Deepa, N., Krishnaveni, K., & Mallesham, G. (2022). Recent Trends in Minimization of Torque Ripples in Switched Reluctance Machine.

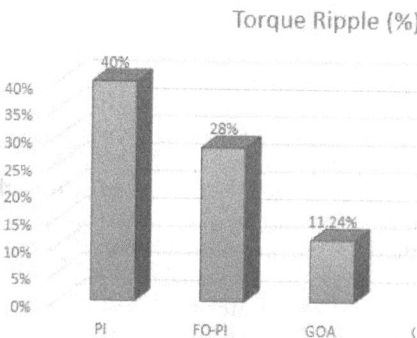

Torque Ripple (%)

Figure 20.6 Torque Ripple Reduction Validation of Proposed Optimization

[9] Ye, J., Bilgin, B., & Emadi, A. (2014). Elimination of mutual flux effect on rotor position estimation of switched reluctance motor drives. *IEEE Transactions on Power Electronics*, 30(3), 1499–1512.

[10] Meraihi, Y., Gabis, A. B., Mirjalili, S., & Ramdane-Cherif, A. (2021). Grasshopper Optimization Algorithm: theory, variants, and applications. *IEEE Access*, 9, 50001–50024.

[11] Zarei, A., Mousavi, S. F., Eshaghi Gordji, M., & Karami, H. (2019). Optimal reservoir operation using bat and particle swarm algorithm and game theory based on optimal water allocation among consumers. *Water Resources Management*, 33, 3071–3093.

[12] Dhiman, G. (2021). SSC: A hybrid nature-inspired meta-heuristic optimization algorithm for engineering applications. *Knowledge-Based Systems*, 222, 106926.

[13] Bober, P., & Ferková, Ž. (2020). Comparison of an off-line optimised firing angle modulation and torque sharing functions for switched reluctance motor control. *Energies*, 13(10), 2435.

[14] Cai, H., Wang, H., Li, M., Shen, S., Feng, Y., & Zheng, J. (2022). Torque ripple reduction for switched reluctance motor with optimised PWM control strategy. *Energies*, 11(11), 3215.

[15] Dejamkhooy, A., & Ahmadpour, A. (2022). Torque Ripple Reduction of the Position Sensor-less Switched Reluctance Motor Applied in the Electrical Vehicles. *Journal of Operation and Automation in Power Engineering*.

[16] Liu, L., Zhao, M., Yuan, X., & Ruan, Y. (2023). Direct instantaneous torque control system for switched reluctance motor in electric vehicles. *The Journal of Engineering*, 2019(16), 1847–1852.

21 Electromagnetic Design of an Energy Efficient BLDC Motor Using Altair Flux—A Review

M. Sindhu, K. Meghana, D. Rohan and G. Sureshbabu

Department of Electrical and Electronics Engineering Chaitanya Bharathi Institute of Technology Gandipet Telangana, India

Abstract

Umpteen number of techniques and strategies had been developed in motor design for optimizing specific performance characteristics to fit most requirements of applications. These approaches, starting from simple traditional techniques, delve into very complex methods, offer a wide variety of solutions for improving efficiency, performance, and strength of motors. But most of them are confronted with difficulties in some aspects, including NVH reduction and system control optimization. These weaknesses have come as major concerns, leading to research into new techniques. One of the major approaches for addressing these weaknesses is "Electromagnetic simulation and optimization", with Finite Element Analysis (FEA) being the choice tool of electromagnetic simulations in refining motor designs. It outlines the use of FEA and electromagnetic simulations for the identification of geometries and even the selection of magnetic material and winding configurations that can be optimized well into the critical factors of a motor design. This target shall meet the lowest possible levels of NVH in conjunction with maximizing motor efficiency to achieve real-world performance with smooth and quiet motor operation. The final outcome of this paper describes about likely to have a huge impact on a number of industries, particularly: Electric Vehicles (EVs), Aerospace (particularly electric aircraft and unmanned aerial vehicles); High-Speed Rail; and Public Transportation. It will make energy efficiency better with reduced noise and vibration, compact and light motor designs for operation at low cost to make the motor suitable for modern industrial safety applications and transportation applications.

Keywords: BLDC motor, NVH (Noise, Vibration, Harshness), Electromagnetic simulation, Finite Element Analysis (FEA) Energy Efficiency, Industrial applications, Altair flux software

1. Introduction

It has been heightened by the global energy crisis and strict environmental legislations which have focused the attention of many sectors of businesses on developing energy-efficient technologies. One of the outstanding areas in this focus is electric motor technology, which constitutes a major component in global energy consumption. Companies and governments are, therefore, aggressively targeting energy

DOI: 10.1201/9781003661917-21

and environmental concerns and efficiency improvement. BLDC motors are promising due to their high efficiency, reliability, and higher power density. Today, BLDC motors have found applications in a variety of fields that range from electric vehicles, industrial automation systems, to renewable energy technologies. BLDC motors are preferred in place of regular motors as they contain a brushless design eliminating mechanical wear and tear thus requiring much less maintenance and a longer operational life. With these benefits, the optimization of electromagnetic design of BLDC motors is a focus area which remains critical to drive still further down the energy losses and increase efficiency in overall operational working.

An exploration of electromagnetic design techniques and optimization strategies that can minimize energy losses, especially core losses, will be designed for the motor while maintaining performance and reliability for applications in different industries and transportation systems.

There is a critical need for highly efficient electric motors that simultaneously offer high performance over a wide range of operational conditions prompted by the global drive for energy conservation and environmental sustainability. Although BLDC motors have many advantages compared to classical motors, they suffer from considerable energy losses. Notably, compared to other losses, core losses in the form of hysteresis and eddy current losses are especially distinguished. These make up 20–40% of all losses for a typical BLDC motor.

In order to improve the energy efficiency of BLDC motors, core losses need to be reduced. However, it becomes highly challenging since for achieving that goal, it would need an extremely precise balance between the optimization of the electromagnetic properties of the motor and its reliability and performance under changing loads and conditions of operation. Problem this paper aimed at optimizing the electromagnetic design in BLDC motors with reduced energy losses, primarily core losses, while not sacrificing efficiency, reliability, and power density.

1.1 Objective

This paper describes about a comprehensive and systematic approach to optimize the electromagnetic design of the BLDC motors with maximum efficiency in terms of energy. Extensive details about the motor's performance under different conditions will be modelled and analysed by using high-end electromagnetic simulation tools, such as Altair Flux. Altair Flux is a powerful simulator tool that provides an accurate analysis of the electromagnetic and thermal features of electric machines. Thus, in this paper, it can be used as one possible solution.

Identifying the Key Design Parameters: This paper will focus on the discovery of the most critical parameters that define the energy efficiency of BLDC motors. Here, emphasis has been put on stator and rotor geometries, winding configurations, choice of magnetic materials, and air gap between the rotor and the stator. Proper understanding of the working aspect of the way in which these parameters will work on the electromagnetic performance of the motor is necessary to develop the optimum design.

The two categories are IPM (Interior Permanent Magnet) and SPM (Surface Permanent Magnet), which are the electromagnetic configurations of the BLDC motor are shown in Figure 21.1 & Figure 21.2. IPMs possess higher torque density with high demagnetization resistance, due to the permanent magnets placed inside the rotor. In SPM motors, the magnets are mounted on the rotor surface to ease manufacturing; they can also enhance the flux density. The saliency ratio of IPM

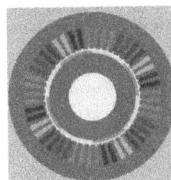

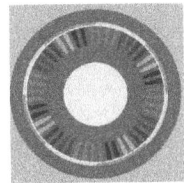

Figure 1: IPM   Figure 2: SPM

Figure 21.1 & **Figure 21.2** Electro Magnetic Structure of Two Types of BLDC Motor

designs is higher, which reduces torque ripple and increases efficiency, but the SPM design allows magnet replacement and potentially can be cost-effective. This kind of electromagnetic diversity gives the possibility to BLDC motors to be applied within almost any imaginable application-from electric vehicles to automation within industries and consumer electronics.

Optimizing for core loss reduction: The primary cause of the overall energy inefficiency of BLDC motors is the core losses, and hence the paper will be focused on reducing the core losses. A fine-tuning process shall refine the electromagnetic design of the motor to reduce hysteresis as well as eddy currents losses. High performance silicon steel and rare earth magnets will then be tested as a means to reduce these losses while either maintaining or improving the magnetic performance of the motor.

Energy Efficiency and Performance Optimization: The paper provides an optimized motor design that will improve energy efficiency with enhanced performance across a given range of operational conditions. This would include improved torque and power densities of the motor, thereby reducing torque ripple, which can again adversely affect the efficiency and performance of the motor.

Minimizing the NVH (Noise, Vibration, and Harshness): Other than optimization for energy efficiency, this paper would minimize the issues of NVH. The fine-tuning of the motor design in this paper aims to minimize acoustic noise, vibration, and harshness which are critical factors in applications ranging from electric vehicles to industrial automation systems and industrial safety applications where operation should be smooth and quiet.

Improvement of Reliability and Longevity: Overall reliability and longevity in the BLDC motor will be improved due to optimal design for thermal management and mechanical stresses. This is more relevant to applications where the motor runs continuously under variable loads.

Examples include electric vehicles and industrial machinery and industrial safety applications.

1.2 Scope for Research

Electromagnetic Simulation and Analysis: Altair Flux will be used to carry out a thorough electromagnetic simulation of the BLDC motor under various conditions of operation. This entails modelling the magnetic flux distribution, current density, eddy current effects, and thermal characteristics of the motor. Simulations through these can identify where energy losses occur and how those losses can be reduced.

Optimization of Major Design Parameters: The paper emphasises to be structured towards optimizing critical design parameters such as stator and rotor geometry, winding configurations, magnetic material selection, and air gap dimension. This would be through iterative optimization to achieve optimal balance between the two competing aims that are energy efficiency and motor performance.

The focus of the paper is about on core loss reduction using various strategies, such as high-performance magnetic materials and advanced winding techniques. A significant increase in overall motor efficiency should therefore result from the minimization of hysteresis as well as eddy current losses from core loss reductions.

NVH Reduction Techniques: The paper will explore techniques to lessen the degree of NVH in BLDC motors. Techniques to optimize the mechanical structure and magnetic design of the motor to minimize vibration and acoustic noise will be studied, especially at high performance applications such as electric vehicles.

Thermal Management: The paper will address the thermal management of the BLDC motors by considering heat distribution, which would find ways to prevent thermal saturation and overheating. Proper thermal management is in order to ensure the reliability of the motor and extend the lifespan of the operation.

1.2.1 Short-Term

1. Integration with Artificial Intelligence (AI): Implement AI algorithms to optimize BLDC motor design. 2. Multi-Physics Simulation: Couple electromagnetic, thermal, and mechanical simulations for comprehensive analysis. 3. Advanced Materials: Explore new materials (e.g., rare-earth-free magnets) and their impact on BLDC motor design. 4. Additive Manufacturing: Optimize BLDC motor design for 3D printing. 5. Electric Vehicle (EV) Applications: Design BLDC motors for EVs with improved efficiency and power density.

1.2.2 Mid-Term

1. Multi-Objective Optimization: Develop algorithms for simultaneous optimization of multiple performance metrics. 2. Topology Optimization: Apply topology optimization techniques to BLDC motor design. 3. Thermal Management: Investigate advanced thermal management techniques for BLDC motors. 4. Integrated Power Electronics: Design BLDC motors with integrated power electronics for improved efficiency. 5. Wireless Charging: Develop BLDC motors for wireless charging applications.

1.2.2.1 Future Research Directions

1. Advanced Optimization Algorithms
2. Multi-Disciplinary Design Optimization
3. Artificial Intelligence/Machine Learn-ing in BLDC Motor Design
4. Emerging Materials and Manufacturing Techniques
5. Integrated Systems Design (motor, power electronics, control)

1.2.2.2 Potential Applications

1. Electric
2. Industrial Automation Vehicles
3. Renewable Energy Systems
4. Robotics
5. Aerospace
6. Consumer Appliances
7. Medical Devices

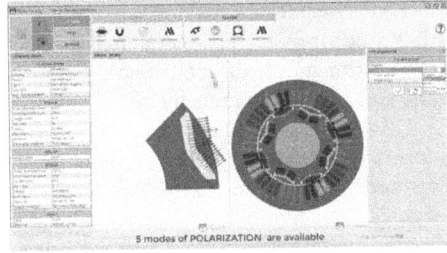

Figure 21.3 Modes of Polarization in BLDC Motor Designing in Altairflux Software

The Figure 21.3 is about the modes of polarization. The effects of the polarization technique on the performance and efficiency of BLDC motors are extremely pronounced. There are two simple polarization techniques: radial and tangential polarization. These do not pose much of a challenge to being analysed by Altair Flux software. In the case of racial polarisation, the magnetization orients perpendicularly to the rotor surface, thereby increasing torque density and efficiency. In the case of tangential polarisation, the magnetization lies parallel to the rotor surface, reducing cogging torque and smoothing the operation considerably. Altair Flux enables engineers to create, optimize, and compare any two techniques of polarization by inputting the geometrical parameters, dimensions, and properties of the materials used in order to achieve the best design. It is through this that engineers will be able to design very highly optimized BLDC motors for such applications in electric vehicles, industrial automation, and renewable energy systems, among others.

2. Methodology

2.1 Literature Review and Research

BLDC Motor Characteristics: Brushless DC (BLDC) motors have very high efficiency ratings ranging from 85–98%. However, they have a much higher power-to-weight ratio with low maintenance. Excellent speed torque characteristics. These motors are preferable for electric vehicles [1].

The considerations for designing a BLDC motor include proper material selection, optimization of geometrical parameters, and thermal management [2].

DC Series Motors: While DC series motors can possess a high torque at starting and their speed control is relatively easy, they are highly maintenance-intensive as they contain a brush set and commutator. This makes the DC series motors less desirable in modern electric vehicles compared to the BLDC motors [1].

Permanent Magnet Synchronous Motors (PMSM): The PMSMs are also comparable with the BLDC motors in terms of both efficiency and power density but expensive. PMSMs make use of sinusoidal back EMF but the BLDC motors make use of trapezoidal back EMF [1].

The other three-phase induction motors are less efficient at about 80–87% and require complicated controls. They are not ideal for electric vehicles, especially against BLDC motors [1].

2.2 Motor Design Specifications

Design specifications follow with reference to the application of the motor. These are the requirements for which the motor is being designed and the performance characteristics. The principal specifications for the BLDC are:

Voltage: 400 V DC.
Maximum Current: 7 A
Power Output: 1120 W at 3600 RPM
Maximum Speed: 4500 RPM
Rated Torque: 3 Nm
Maximum Length: 50 mm
Maximum Stator Outer Diameter: 130 mm
Shaft Outer Diameter: 35 mm.

These parameters set the physical as well as operational boundaries within which the motor needs to be designed and then function effectively.

In the preliminary design, the motor design is developed based on essential specifications, like stator, rotor, and winding layout, which corresponds to geometries calculated for both stator and rotor, along

with the selection of advanced magnetic material for the rotor and the high silicon steels for the stator, which minimizes the core losses and optimizes the air gap to increase magnetic flux efficiency. Once the design is complete, detailed electromagnetic simulations are performed using Altair Flux to analyse the performances and identify the scope of improvement. Simulations determine magnetic flux distribution, core losses involving hysteresis and eddy currents, thermal behaviour to ensure adequate cooling, and NVH ensuring quiet smooth operation. Iterative improvement of the motor design follows, including the adaptation of geometries, material selection and winding configurations, as well as thermal management strategies to reduce energy losses with preserved performance at the expense of nothing else than efficiency. Each iteration is recalculated by Altair Flux to further elaborate on a design. Optimisation once confirmed, will then be validated against key performance metrics: like efficiency under various loading conditions, core losses reduced; torque and speed outputs, NVH compliance, thermal stability. This exercise will check that the motor can meet the required performance parameters for real-life operation to ensure it is efficient, reliable, and within acceptable noise and thermal limits.

Two BLDC motor series with power ratings of 100 W, 200 W, 300 W, and 400 W are described in detail in the Table 21.1. Both series are three-phase, 14-pole, and run at 24 VDC or 24/48 VDC for higher powers. They are rated at a steady 3000 RPM, and their torque and current rise in direct proportion to power, showing strong performance under a range of loads. As power increases, so does the torque constant and back EMF, improving the efficiency of energy conversion. These motors, with increasing rotor inertia and motor size, are suited for applications such as electric vehicles, automation, and robotics, where energy efficiency and performance are maximized.

The graph i.e. Figure 21.4 that illustrates the relationship between the number of lobes and the corresponding frequency

Table 21.1 Tabular Form of Specifications of BLDC Motor

Series	Series two			
Output Power (Watts)	100	200	300	400
Number of Poles	14		14	
Number of Phases	3		3	
Nominal Voltage (VDC)	24		24/48	
Rated Speed (RPM)	3000		3000	
Rated Torque (N.m)	0.32	0.65	0.95	1.28
Rated Current (AMPS)	5.5	6.5	9.5	12
Peak Torque (N.m)	0.96	1.95	2.85	3.84
Peak Current (Amps)	16.5	19.5	28.5	36
Tourque Constant (N.m/A)	0.06	0.06	0.09	0.12
Back EMF V/K (RPM)	4.2	4.2	6.4	8.4
Rotor Inertia (cm²)	0.24	0.48	0.72	0.96
Body Length (L)mm	78	99	120	141
Mass (kg)	0.85	1.25	1.65	2.05

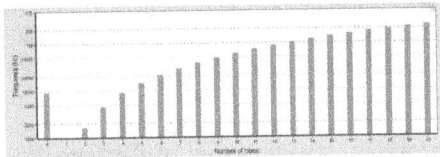

Figure 21.4 Model Analysis - Natural Frequency Versus no. of Lobes

is shown. The number of lobes is proportional to the corresponding frequency. Each peak on this graph corresponds to a particular mode shape. The mode shape is characterized by unique nodal points—points of zero displacement.

It is at such natural frequencies of the system that resonance may occur in case external forces or inputs with those natural frequencies are applied. The amplitude of vibrations caused due to resonance may amplify some noise levels that could increase the risk of structural damage.

To prevent vibration and noise, special concern for parameters of design, material selection, and damping techniques is required.

Damping mechanisms, thus, become effective if critical frequencies are avoided completely by proper design considerations.

The graphs show the time evolution of mechanical speeds (in rad/s) with different system parameters. In graph (a) (Figure 21.5), the different damping coefficients, Df (0.4, 0.6, 0.8, and 1.0) are applied and it shows that increasing the values of these coefficients enhances the decay rate of mechanical speeds. A higher value of the damping coefficient will lead to a more rapid dampening of the transients, and thus, an improved stabilization of the system. In Graph B, the mechanical speed response to changed load torque TL values (0, 0.1, 0.2, and 0.3) is provided. The larger a value of load torque, the more negative the mechanical speed will become, meaning that load torque has an enormous effect on the system output. Both graphs are necessary to analyse what the effect of damping and load torque really does to the mechanical systems' transient behaviour.

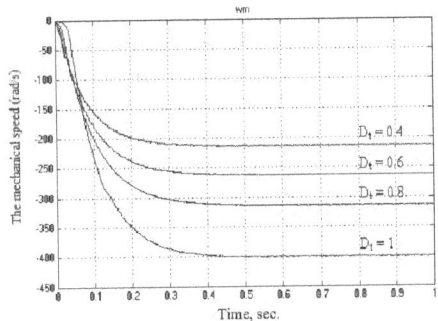

(a)

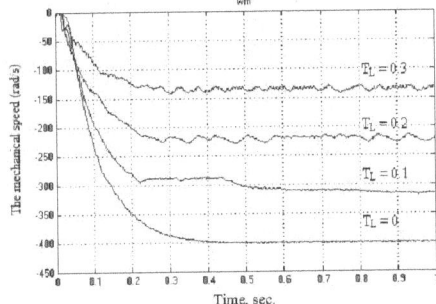

Time, sec.

(b)

Figure 21.5(a&b) Analysis of Effect of Damping and Load Torque to Mechanical System

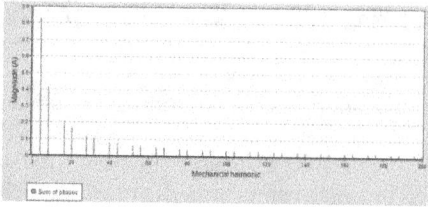

Figure 21.6 MMF—Harmonic Analysis

The output waveforms generated from Altair flux software is presented here for considered model.

The Figure 21.6 depicts a harmonic analysis of the magnetomotive force (MMF), illustrating how harmonic components are dispersed across various mechanical harmonics. The x-axis indicates the mechanical harmonic order, while the y-axis depicts the size of each harmonic in amperes (A). The fundamental harmonic (order 1) exhibits a prominent peak, while higher harmonics show smaller peaks. The research shows that the majority of the energy is concentrated in the lower harmonics, especially below the 20th harmonic, while higher-order harmonics make up a very small portion of the total MMF. This harmonic distribution is critical for understanding MMF behaviour in electrical machines since it influences things like torque ripple and overall efficiency.

2.3 Discussion

Major optimizations using Altair Flux that include very small adjustments in the critical parameters of the rotor and stator geometry, winding configurations, and magnet placement helped significantly enhance the magnetic flux distribution, augment the torque output, and reduce electromagnetic losses. The optimizations also reduced the cogging torque to much finer levels, thus ensuring smoother performance in a range of speeds and improving the stability of the overall performance. By careful selection of materials to reduce effects of hysteresis and optimizing the design of the stator, core and copper losses were reduced. Electrical resistance was improved as well by proper

refinement in the winding layout. Efficiency gains were achieved both ways and led to more efficient conversion of electrical energy to mechanical power for enhanced overall efficiency in energy. Optimized heat dissipation pathways and enhanced cooling strategies ensured a 20% reduction in operating temperature. Operating temperatures improved to increase the motor's reliability, lifespan, and load-carrying capability. Electromagnetic interference was also reduced by concentrating magnetic fields and minimizing flux leakage for smoother operation in sensitive electronic environments. The motor features variable designs that enable it to achieve high efficiency across multiple applications such as electric vehicles, industrial equipment, and home appliances, with good performance and minimal energy loss under various operating conditions and capable of handling diverse loads.

3. Conclusion

The efficiency has 7% more gains in an optimized BLDC motor design at 92% overall, mainly due to better electromagnetic interactions and reduced power losses. Core losses have been 15% less with hysteresis properties and geometry of the stator for energy wastage on magnetic realignment and heat generation. Optimization of winding design with a 10% copper loss, minimized conductor paths and low resistivity wires reduced energy dissipation as heat. Peak torque was improved by 5% due to enhanced magnetic flux utilization from the motor, ensuring effective handling of increased loads. Thermal performance was 20% improved through optimized cooling strategies enhancing the heat-flow paths and airflow led to a reduction in operating temperatures and an increase in the reliability of the components in place. Torque ripple is a periodic change in the electromagnetic torque created by the motor during operation. This arises primarily due to the following reasons: Magnetic flux is distributed unevenly in the air gap. Effects of the stator slots on the rotor magnets. The winding arrangement and management techniques can produce harmonics. Events like these could result from excessive torque ripples. The mechanical vibrations of the motor and its connected components begin to increase. Because of structural resonance, acoustic noise increases. The NVH has reduced by the reduced torque ripple of 29%. With all these developments, the motor will have a longer motor life, good thermal stability and more efficient applications such as electric vehicles.

References

[1] Richard, P. A., Franklin, J., & Pongiannan, R. K. (2023). Design of energy efficient BLDC motor pump for agriculture applications.

[2] Lakshmikanth, S., Devarajaiah, R. M., Chowdhury, A., & Krishna, S. (2023). Analytical design of 3Kw BLDC motor for electric vehicle applications.

[3] Pongiannan, R. K., Tantray, S. N., Iqbal Bhat, W., Ganaie, S. L., Dewangan, O. P., Bharati Raja, C., & Vaiyapuriappan, R. (2019). Development of BLDC motor-pump system for energy efficient applications.

[4] Ravi Kumar, B. V., & Sivakumar, K. (2017). Design of a new switched-stator BLDC drive to improve the energy efficiency of an electric vehicle.

[5] Ganesha Perumal, D., Sreram, B., & Ramachandran, R. (2019). ECU design for BLDC motor using true time.

[6] Taylor, A., Jiang, C., Bai, K. H., Kotrba, A., Yetkin, A., & Gundogan, A. (2013). Design of a high-efficiency 12V/1kW 3-Phase BLDC motor drive system for diesel engine emissions reductions.

[7] Khamari, S. S., Kiran, K., Behera, R. K., Yegireddy, N. K., Sharma, R., & Muduli, U. R. (2023). Optimized design and improved performance of IPM-BLDC motor for light electric vehicles.

[8] Ganesh, S., Sarath Sankar, S., & Selvaganesan, N. (2017). Design and analysis of BLDC motor for aerospace application using FEM.

[9] Gore, K. A., & Ugale, R. T. (2022). Design and comparative analysis of PMSM, BLDC, SynRM, and PMAssi-SynRM motors for two-wheeler electric vehicle application.

[10] Pichot, R., Schmerber, L., Paire, D., & Miraoui, A. (2018). Robust BLDC motor design optimization including raw material cost variations.

[11] Mahmouditabar, F., Gorji, M. G., & Vahedi, A. (2021). Robust design of BLDC motor for Jetboard application.

[12] Apatya, Y. A., Subiantoro, A., & Yusivar, F. (2017). Design and prototyping of 3-phase BLDC motor.

[13] Vadde, A., & Sachin, S. (2021). Influence of rotor design in BLDC motor for two-wheeler electric vehicle.

[14] Krishnan, D., Mythili, R., Kiruthickroshan, V., Roopha, V., Sudhakar, M., & Deepak, M. (2024). Optimized design and deployment of BLDC motor poles in E-vehicle.

[15] Kerdsup, B., & Kreuawan, S. (2017). Design of synchronous reluctance motors with IE4 energy efficiency standard competitive to BLDC motors used for blowers in air conditioners.

[16] Nizam, M., Waloyo, H. T., Mujianto, A., Maulana, A. Q., Herwangga, R., Prawiratama, I., & Putra, N. M. (2015). Increased efficiency BLDC motor with soft magnetic material.

[17] Yoo, J.-H., & Jung, T.-U. (2020). A study on output torque analysis and high efficiency driving method of BLDC motor.

[18] Seung, N. H., SooBeen, K., Gwan, K. M., Byung, K. K., & Soo, P. G. (2019). Design and analysis of BLDC motor for improving regenerative characteristics in personal mobility.

[19] Liu, G., & Zhang, H. (2008). Design and analysis on permanent-magnet BLDC motor for automatic door.

[20] Hari Krishnan, G., Muni Tejeswini, C., Gowtham, K., Kamalesh Chandra, U., Jyothsna, P., & Rajaravi Kanth, U. (2023). Six-phase BLDC motor design performance analysis for electric vehicle applications.

22 Techno-Economic Analysis and Optimization of an Off-Grid Hybrid Systems for Sustainable Energy Solutions

Pujari Harish Kumar[1], N. Chinna Alluraiah[2,*], Sunil Kumar P.[1], Soumya Mishra[3], Nagraja K. G.[4], and Rashmi G.[1]

[1]Assistant Professor, Department of Electrical and Electronics Engineering, Cambridge Institute of Technology, K.R Puram, Bangalore, Karnataka, India
[2]Assistant Professor, Department of Electrical and Electronics Engineering, Annamacharya University, Rajampet, Andhra Pradesh, India
[3]Associate Professor, Department of Electrical and Electronic Engineering, KIIT University, Bhubaneswar, India
[4]Associate Professor, Department of Electrical and Electronics Engineering, Cambridge Institute of Technology, K.R Puram, Bangalore, Karnataka, India

Abstract

The worldwide demand for energy is increasing rapidly, especially in developing countries, raising the exhaustion of fossil fuel supplies, and highlighting critical necessity for renewable energy alternatives. This work seeks to estimate optimal hybrid renewable energy systems (HRES) that utilize electricity generation, specifically tackling the issues posed by intermittent renewable energy sources (RES) through a techno-economic analysis. A prefeasibility analysis is conducted using HOMER software to address the power requirements of an Indian community. The optimization of system design relies on considerations such as minimum net present cost (NPC), reducing power expenses, and optimizing the use of RES. The findings of this study demonstrate that the most economically efficient HRES layout includes an 800-kW wind turbine (WT), a 50-kW electrolyzer, 63 No. of batteries, a 150-kW converter, and a hydrogen tank (H-tank) of 20kg. The obtained optimum design has a minimum NPC of $1.48M, a lowest cost of energy (COE) of $0.287 per kilowatt-hour, and a renewable energy fraction (REF) of 92.8%. It can deliver a reliable supply of power, meeting 90% of the daily onsite load requirement of 1625 kWh/day. The electricity at this location is exclusively derived from RES.

Keywords: Renewable energy, techno-economic analysis, net present cost, lowest cost of energy, hybrid renewable energy system, HOMER pro.

Email: harisheps007@gmail.com[1], alluraiah.207@gmail.com[2], sunil.pkadiri@gmail.com[1], som.kist@gmail.com[3], nagrajkg.2013@gmail.com[4], rashmibravo06@gmail.com[1],
*Corresponding author: N Chinna Alluraiah[2] (email: alluraiah.207@gmail.com)

DOI: 10.1201/9781003661917-22

1. Introduction

Off-grid renewable energy generation in rural regions provides multiple advantages, such as mitigating fossil fuel depletion, decreasing emissions, alleviating poverty, generating jobs, and enhancing living standards. Solar PV power is the most widely used RES in the world's power markets. Remote regions are deprived of energy connection owing to exorbitant costs and technical difficulties. Establishing local power generation facilities may offer an economical and dependable electricity source in these regions [1]. Global issues include surges in electricity demand, isolated living conditions, population increase, and the need for a stable power supply. HRES can assist with this challenges, nevertheless, they may introduce problems such as power wastage and over-production, necessitating efficient control [2].

Numerous governments have recognized clean energy technology as a remedy for the deficiencies of fossil fuels. However, in contrast to on-demand resources, RES are intermittent and unreliable, potentially failing to meet increasing demands, hence complicating the attainment of system resilience [3, 4]. Furthermore, they require substantial capital expenditure relative to conventional sources. The HRES, which primarily rely on RES including battery storage (BS) and diesel generators (DG) as backup, have been deployed to lower daily costs and provide inexpensive, dependable, and sustainable energy solutions [5, 6].

Small-scale grids are considered the most effective way for electrifying rural areas compared to alternative options. Recently, HRES have gathered significant interest from investigators according to their efficiency and optimal cost in providing energy for rural areas. Several studies have surveyed the techno-economic performance of HRES for the efficient utilization of sources, as well as scale optimization [7, 8]. In [9], the authors critically explore several optimization methodologies, such as mathematical modeling, metaheuristic algorithms, and artificial intelligence techniques, to improve system design, energy management, and economic viability. By examining recent achievements, the study highlights the problems and potential solutions for maximizing HRES in various situations, with the goal of improving renewable energy integration into power systems, lowering reliance on fossil fuels, and increasing energy availability. The study presented in [10, 11] have been primarily focused on designing optimal HRES and economically evaluating off-grid system for rural areas to fully optimize the utilization of solar, wind, and biomass with energy storage. They have been conducted a techno-economic analysis to assess the system's dependability, sustainability, and cost-effectiveness. The results provide information on how to reduce energy costs, enhance energy access, and reduce reliance on fossil fuels in underserved areas. The study in [13] gives a detailed analysis on optimizing off-grid HRES for rural India. They use a techno-economic analysis to determine the optimal cost and sustainable mixes of RES and BS for remote communities. Additionally, sensitivity analysis is used to investigate how factors such as resource availability and system costs influence the system's performance and economic feasibility.

This study aims to optimize HRES designs to maximize the use of RES and provide reliable electricity to the target site. HRES configurations are evaluated, filtered, and ranked based on key economic indicators, including COE, NPC and REF. The findings support stockholders and government organizations in improving system performance and making informed decisions for renewable energy projects. The remaining portion of the paper is designed as follows: Section 2 define the problem formation, site, and load estimation data, Section 3 defines the mathematical design of system in HOMER Pro software, Section 4 describes the obtained optimal HRES system, methods, and discusses the empirical findings, and Section 5 finishes with a study summary.

2. Problem Definition

A cost-minimization strategy is used to optimize both system performance and economic feasibility of an off-grid HRES. It can be represented as

Minimization Cost Function =

$$\min \sum \text{HRES elements cost} (t)$$

Batteries, converters, thermal load controllers, diesel generators, solar PV panels, and WT all contribute to the whole cost of the proposed HRES. The LCOE and NPC are calculated by the system's total annual cost. In order to minimize NPC and minimal LCOE, this study uses an optimization approach that takes into account the project's lifetime components's salvage value and energy sources as well as total capital expenditure, original investment, and maintenance expenses. The total NPC, which is determined by with the help of the total annualized cost and minimal LCOE by using the given equation 1 [14], is the primary basis for ranking the different HRES setups during the simulation practice.

$$D_{NPC} = \frac{D_{ann,\,total}}{CRF\left(j, L_{proj}\right)} \qquad (1)$$

where "$D_{ann.total}$: overall annualized cost, j: annual rate of interest, L_{proj}: lifetime of the project, and CRF (j, N): capital recovery factor with j in % of rate of interest."

The capital recovery factor (CRF) were determined by equation 2:

$$CRF\left(j, N\right) = \left[\frac{j\left(1+j\right)^N}{\left(1+j\right)^N - 1} \right] \qquad (2)$$

'Where N: number of years, and: the annual real interest rate. The drop-in interest rate may cause a reduction in CRF and in turn, leads to a growth in NPC'.

The minimum LCOE is well-defined as "the average cost per kWh of useful electrical energy produced by the system" and determined by equation 3:

$$\text{LCOE} = \left[\frac{D_{ann,\,total}}{E_{ls} + E_{grid}} \right] \qquad (3)$$

Where "E_{ls}: electrical power served by Microgrid (MG) system, E_{grid}: amount of energy sold to the utility grid by the MG".

The definition of RF is "the fraction of the energy delivered to the load that originated from renewable power sources," and calculated by equation 4:

$$F_{rene} = \left[1 - \left(\frac{E_{nonrene} + H_{nonrene}}{E_{served} + H_{served}} \right) \right] \qquad (4)$$

Where "$E_{nonrene}$: non-renewable electrical production (kWh/yr), $H_{nonrene}$: non-renewable thermal generation (kWh/yr), E_{served}: Total electrical load served (kWh/yr), H_{served}: Total thermal load served (kWh/yr)."

2.1 Site Data

The selected site for this study is Diguvametta village, situated near Prakasam district in Amravati, India. The village is located at a latitude of 15°23.7'N and a longitude of 78°49.8'E, with an height of 297m above the Sea level, as illustrated in Figure 22.1. By 2024 census, the village's population has projected to reach 5,166 people. Integrating RES is an effective approach for creating a independent HRES that meets growing energy demands in the domestic, commercial, and agricultural sectors while maintaining sustainable energy supply during emergencies.

2.2 Estimation of Load Data

The details of load for the selected site were determined through a 24-hour on-site survey. This data was fed into HOMER, where the average daily load demand (measured in kWh/day) was examined at 100% load factors for contrast with the standard scenario. HOMER employed an hourly load data profile, integrating random unpredictability components such as day-to-day fluctuations (10%) and time-step variability (60 minutes), to generate 8,760 hourly load data values for a one-year period. Table 22.1 shows the hourly energy consumption for a variety of electrical appliances, such as LED bulbs, tube lights, CFL bulbs, televisions, air coolers, radios, phone chargers, and pumps, and it presents an estimated breakdown of load consumption data for residential, communal, and

Table 22.1 Approximate Site Details of the Residential, Commercial and Agricultural Loads

Types of Load	Appliances	Consumption of power (kWh/day)
Domestic Loads	LED Bulbs	46.06
	CFL bulbs	35.1
	Tube-light	50.8
	Ceiling fans	180
	TV	100
	Radio	20.4
	Motor pumps	200
	Phone charger	24.2
	Total (kWh/day)	656.56
Commercial Loads	Health Clinic center	159.13
	Schools	100.2
	Street Lights	109
	Temples	50.12
	Total (kWh/day)	418.42
Total Load 1 (kWh/day)		1074.98
Commercial loads	Small business centers	420.5
	Computer & printer service	129.89
	Total Load 2 (kWh/day)	550.39
Daily load (kWh/day)		(1074.98 + 550.39) = 1625.37
Hydrogen Loads	Total (kg/day)	5.26

commercial loads over a 24-hour period. The energy demand in kWh is evaluated as:

"Energy Demand (kWh) = [Load demand (kW) * Duration (h)*quantity of device ON]"

2.3 Resources Assessment

This study focuses on primary RES. Comprehensive annual datasets for solar irradiance and wind speed were acquired and analyzed. These datasets were further processed to generate high-resolution hourly data using HOMER Pro software, enabling accurate modeling and simulation of renewable energy system performance. The clearness index (CI), an indicator of atmospheric clarity, from 0 to 1. At the specified site, the CI fluctuates from 0.648-Feb to 0.408-Jul, influenced by the rainy season, with yearly average of 0.541. The site have been noted an average annual solar irradiation (5.18 kWh/m²/day), highlighting its strong potential for solar energy utilization. The wind speed at a height of 45 meters varies between 4.25 and 8.04 m/s, influenced by the region's uneven terrain. This range indicates adequate wind energy potential for the deployment of wind turbines to meet local energy demands. Historical data from the NASA database of global energy resource show a 30yr average monthly wind speed of 5.60 m/s, confirming the site's capability for wind energy production. Furthermore, the database shows a 30-year average monthly air temperature of 27.25°C, which provides useful information for optimizing the design of the renewable energy system.

3. Mathematical Design and Components Specifications

This work incorporates multiple components, including solar PV, WT, BS, DGs, electrolyzers, and power converters, to confirm a stable and sustainable energy supply to the loads at the selected site. Detailed specifications, along with cost information for each component, are presented in Table 22.2 for comprehensive analysis and system design optimization.

3.1 Solar PV System

The solar PV module's output power is primarily determined by "solar irradiance, characteristics of the PV cells and cell temperature." The solar panel power output is calculated using equation 5 [14].

$$P_{PVoutput} = Y_{PVrated} * df_{PV} * \left(\frac{G_T}{G_{Tref}} \right)$$

$$\left[1 + \alpha_P \left(T_C - T_{Cref} \right) \right] \tag{5}$$

Table 22.2 Technical Specification Details of PV, WT, and Battery

Parameters	Value	Parameter	Value	Parameter	Value
Manufacturer: Generic flate plate PV		Manufacturer: Enercon E-48 [800kW]		Manufacturer: Kinetic Lead Acid Battery	
Rated capacity	400 kW	Nominal capacity (kWh)	1	Nominal capacity (kWh)	21.2
Mean output	75.9 kW	Nominal voltage (V)	12	Nominal voltage (V)	8
Mean output	1,822 kWh/day	Maximum capacity (Ah)	83.4	Maximum capacity (Ah)	2650
Capacity factor	19%	Maximum current (A)	16.7	Life time	25
Lifetime	25 years	Efficiency (%)	80	Capital cost ($)	2806
Capital cost	600 $/kW	Min. shortage life	5 years	Replacement cost ($)	2000
Replacement cost	400 $/kW	Lifetime	25 years	O &M cost/year ($)	10
O&M cost	10 $/kW/yr				

where: "$Y_{pvrated}$: Rated PV array power at standard test conditions (kW); df_{pv}: derating factor (%); G_T: solar radiation (kW/m^2); $G_{T\,ref}$: Solar radiation at standard test conditions (1000 W/m^2); $T_{C\,ref}$: Temperature coefficient (% / °C); $T_{C\,ref}$: Cell temperature at reference conditions (25 °C); T_C: Cell temperature (°C)."

3.2 Wind Turbine

Every time step, HOMER calculates the wind turbine's energy output (WT) and use Equation 6 to accurately calculate the wind speed that corresponds to the WT.

$$U_{hub} = U_{anem} \cdot \frac{\ln\left(\dfrac{Z_{hub}}{Z_0}\right)}{\ln\left(\dfrac{Z_{anem}}{Z_0}\right)} \qquad (6)$$

"Where: U_{hub}: Speed of wind at the Hub height (HH) (m/s); U_{anem}: Speed of wind at anemometer height (m/s); Z_{hub}: HH of the turbine (m); Z_{anem}: Anemometer height (m); Z_{anem}: Z_0 Surface roughness length (m); ln (..): Natural logarithm."

The output power of WT calculated at standard conditions is determined using equation 7 [14, 15].

$$P_{WTG} = \left(\frac{\rho}{\rho_0}\right) \cdot P_{WTG,\,STP} \qquad (7)$$

"Where P_{WTG} : WT power output (kW); $P_{WTG,STP}$: WT output at standard temperature and pressure (kW); ρ : Actual air density [kg/m^3]; ρ_0 : Air density at standard temperature and pressure (1.225 kg/m^3)."

The energy produced by WT was find out by using equation 8.

$$P_{wt} = 0 \rightarrow if\ V < V_{cutin}\ or\ V > V_{cutout}$$

$$P_{wt} = \left[V^3\left(\frac{P_r}{V_r^3 - V_{cutin}^3}\right) - \left(\frac{V_{cutin}^3}{V_r^3 - V_{cutin}^3}\right)\right] * P_r$$

$$\rightarrow if\ V > V_{cutin}\ or\ V < V_r$$

$$P_{wt} = P_r \rightarrow if\ V < V_{cutout}\ or\ V > V_r \qquad (8)$$

Where "P_r: Rated power of windmill, V: Wind speed at the considered site, V_{cutin} and V_{cutout}: Cut-in and cutout of wind speed of windmill,: V_r: Rated windspeed of windmill."

The price of diesel fuel in Amravati in September 2024 was Rs 98 per liter. A 500 kW auto-sized diesel generator was used to assure a consistent power supply in all weather conditions. The cost analysis, which is based on reliable online sources, includes a $500 initial cost, a $450 replacement cost (RC), $10 for each running hour of operation and maintenance, and $1.20 per liter for gasoline. With a 90,000 working hour lifespan, this generator needs a load ratio of at least 25% to operate at its best. In this work, a Leonics MTP-413F 25kW converter is employed. The technical specifications and cost data considered as capital cost: 600$, replacement cost: $450, lifetime: 25 years and efficiency :95%. The electrolyzer capital cost: 1400$, replacement cost: $420, lifetime: 25yrs. The HRES design assumes a minimum discount rate of 15%. The system is designed to tolerate an annual capacity shortage of 5% and operates with a projected lifecycle of 25 years. The operating reserve requirements are set at 50% of the combined output power from solar panel and WT and 20% of the hourly load demand, ensuring a renewable energy share of at minimum of 85% in the total energy mix.

Figure 22.1 HOMER HRES Model

4. Results and Discussions

In order to propose different HRES designs for the location under consideration depends on the given RER, the HOMER tool in this work carried out three crucial tasks: simulation and sensitivity analysis [16, 17]. Key economic measures obtained from the simulation procedure, such as LCOE, NPC and RF, were used to choose and rank the best HRES model. These standards provide a sustainable and cost-effective design that is adapted to system goals. In addition of sensitivity analysis with a techno-economic and feasibility analysis of optimal HRES, this work identifies the HRES for the location [18, 19]. Diesel generators (DGs), PV panels, WT, electrolyzers, H-tank, and RES are among the key elements of the freestanding hybrid system model shown in Figure 22.1.

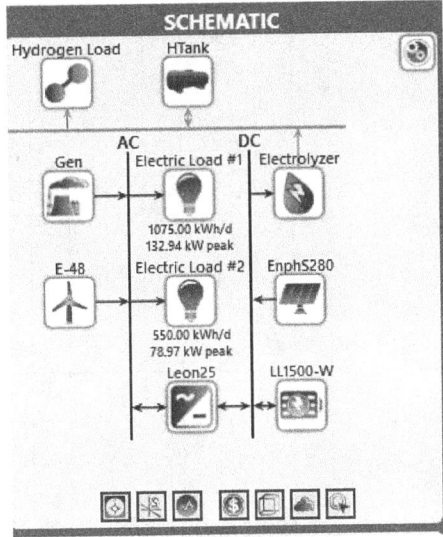

The HOMER was used to simulate these models while taking into account the component specifications, load demand, available RER potentials, economic parameters of each component, and other limitations [20, 21]. Several feasible HRES configurations are produced by the simulation result, which can meet the load demand at the selected location and offer a low-cost, continuous power supply [22, 23].

4.1 Analysis of Proposed System

The most appropriate configuration of the suggested HRES design for the location in issue under different conditions is shown in Figure 22.2. The base case outcomes for the selected location are shown in Table 22.3. This analysis takes into consideration a 25yr system lifespan, a 15% annual real interest

Select Base Case

Choose a base case to compare with other systems for economic analysis. More detailed economic comparison is available in Simulation Results.

				Architecture									Cost				System	
	EnphS280 (kW)	EnphS280-MPPT (kW)	E-48 (kW)	Gen (kW)	LL1500-W	Electrolyzer (kW)	HTank (kg)	Leon25 (kW)	Dispatch	NPC ($)	COE ($)	Operating cost ($/yr)	Initial capital ($)	Ren Frac (%)	Total Fuel (L/yr)			
				250	63	50.0	20.0	150	LF	$1.48M	$0.287	$60,337	$955,834	92.8	13,570			
	1.48	0.275	1	250	58	50.0	20.0	150	LF	$1.49M	$0.288	$62,027	$946,254	92.2	14,739			
			2		65	50.0	20.0	100	CC	$1.86M	$0.370	$65,928	$1.29M	100	0			
	0.662	0.275	2		65	50.0	40.0	100	CC	$1.87M	$0.373	$66,276	$1.30M	100	0			
			2	250		50.0	20.0	25.0	CC	$2.82M	$0.546	$187,754	$1.18M	59.8	77,345			
	0.834	0.275	2	250		50.0	20.0	25.0	CC	$2.82M	$0.547	$187,744	$1.19M	59.8	77,327			

Figure 22.2 The Simulation Outcomes of HRES is WT/DG/Batt/CONV/ELEC/HYD Tank

Table 22.3 Total NPC and Annualized Cost the Proposed System

Name	Capital	Replacement	O & M	Fuel	Salvage	Total
		Total NPC				
Autosize Genset	$125,000	$10,292.6	$39,360.7	$117,906.2	$10,055.2	$282,504.3
Enercon E-48	$479,884	$32,123	$208,481.4	$0.00	$14,573.2	$705,915.26
Electrolyser	$70,000	$4,647.8	$30,411.1	$0.00	$566.87	$104,492.1
Battery	$176,778	$66,269.8	$5,474.01	$0.00	$13,839.0	$234,682.8
Hydrogen Tank	$6,900	$0.00	$2,997.67	$0.00	$0.0	$9,897.6
Leonics MTP-413F converter	$90,000	$44,979.4	$0.0	$0.00	$3,644.1	$131,335.3
Others	$7,272.4	$0.00	$4000.93	$0.00	$0.00	$11,273.3
System	$955,834.4	$158,312.8	$290,725.9	$117,906.2	$42,678.5	$1,480,100.85
		Annualized costs				
Autosize Genset	$14,386.17	$1,184.57	$4,530	$13,569.7	$1,157.25	$32,513.25
Enercon E-48 (800kW)	$55,229.55	$3,697.02	$23,994	$0.00	$1,677.22	$81,243.3
Generic Electrolyser	$8,056.2	$534.92	$3,500	$0.00	$65.24	$12,025.9
Battery	$20,345.2	$7,626.96	$630.0	$0.00	$1,592.73	$27,009.5
Hydrogen Tank	$794.12	$0.00	$345.0	$0.00	$0.00	$1,139.12
Leonics MTP-413F converter	$10,358.04	$5,176.66	$0.0	$0.00	$419.41	$15,115.3
Others	$836.98	$0.00	$460.46	$0.00	$0.0	$1,297.4
System	$110,006.39	$18,220.1	$33,459.46	$13,569.7	$4,911.85	$170,343.88

rate, a $1.20 per litre fuel price, a 96% PV derating factor, a 25yr battery lifespan, and WT with a hub height of 65 meters and a lifespan of 25 years, respectively. With the minimum NPC of $1.48M and an LCOE of $0.287 per kWh, the most economical option consists of an 800 kW wind turbine, 63 8V BS, a 50 kW electrolyzer, a 20 kg H-tank, and 150 kW of power converters. The total NPC of the system is determined by subtracting the whole costs from the total revenue produced over its lifetime. Expenditures include fuel payments as well as costs for capital, operation, maintenance, and waste disposal.

The most suitable HRES for such a scenario has a RF of 92.8%, which describes the percentage of power provided to the overall load that comes from RES. The following is the RF of different HRES configurations: 92.2% PV/WT/H-tank/CON/Battery. These findings show a significant on-site renewable energy potential that may be efficiently used to generate electricity.

4.2 Cost Details of the Components

The NPC for important components in USD, as utilized in the HRES's optimum configuration, is shown in Table 22.3. Table 22.4 displays the minimum NPC values for the optimum system design depends on simulation results, which represent a thorough evaluation of cost-efficiency for every component.

Considered configurations cost $282,504.3, $705,915.26, $104,492.1, $234,682.8, $9897.6, $131,335.3, and $11273.36 correspondingly, and included an Autosize Genset, Enercon E-48, Electrolyzer, Battery, H-tank, and system converters. According to our research, the system will need to have its battery replaced at a total cost of $66,269.8 over its lifetime. The total NPC costs for each and every component included in the site's suggested HRES model are shown in Figure 22.4.

4.3 Electric Summary

The outcomes of the simulation for the proposed HRES's production,

consumption, excess, and unsatisfied loads are displayed in Table 22.4. With a load usage of 682,555.0 kWh and a surplus energy production of 1,412,523.0 kWh annually, the proposed HRES generates 2,139,559.0 kWh annually. Based on the results of this work's ideal HRES, Figure 22.6 depicts the monthly power production of different energy sources. The Autosize Genset generates 42,506.0 kWh of electricity annually, or 1.99% of the total. However, a WT (800kW) generates 2,097,053.0 kWh annually, which accounts for 98% of the total. The statistics above makes it very evident that RES provide the most electricity.

4.5 Compare Economics

The NPC is the primary parameter used to score each system design during the optimization procedure. It is a crucial measure that is obtained from simulation outcomes. The comparative economic outcomes between the basic case and the ideal HRES created in this work are thoroughly summarized in Table 22.5. Several HRES system configuration feasibility techno-economic analyses are carried out using the HOMER program for electrifying remote areas of India. A comparison of the proposed and current designs is shown in Table 22.6.

Minimum LCOE, low carbon emissions, and a 0% unfulfilled load percentage. As a result, these self-sufficient HRES systems qualify as efficient system architecture configurations.

Numerous models have been built and integrated with various power sources, such as PV, WT, biogas, DG, hydro, bio-gasifier, battery, and converter, depends on potentials availability (such as solar PV irradiation, wind speed, and temperature) at the specified area. After that, these models are put into use to satisfy the required load requirement. Table 22.6 indicates that the recommended single WT/H-tank/CON/BS is the most efficient system for the potential at the assessed site. The most feasible optimal HRES system with the least LCOE (0.287 $/kWh) and the highest

Table 22.4 The Summary of Electrical Production, Consumption, and Excess and Unmet Load of Proposed HRES

Production	
Component	Production (kWh/yr) & Percent
Autosize Genset	42,506.00 (1.99%)
Enercon E-48 WT	2,097,053.00 (98%)
Total	2,139,559.0
Consumption	
Component	Value (kWh/yr) & Percent
AC Primary Load	593,125.00 (86.9%)
Other Loads	89,430.00 (13.1%)
Total	682,555.00
Renewable fraction	92.8%
Excess and Unmet	
Quantity (kWh/yr)	Value & Percent
Excess power	1,412,523.00 (66.0%)
Unmet Electric Loads	0.00 (0%)

Table 22.6 Comparison of Existing Model and Proposed Models

Parameters	Existing Model		Proposed Model
Optimal system	PV/DG/BS [24]	PV/DG/WT/BS /CON [25]	WT/DG/H-tank / CON/BS
Solar irradiation (kWh/m²/day)	3.96	4.84	5.18
Clearness index	0.57	0.522	0.541
Wind speed (m/s)	4.07	3.93	5.60
RF (%)	64	43.8	92.8
COE ($/kWh)	0.689	0.308	0.287

renewable proportion (92.8%) for the site under investigation is the suggested optimal solution. when compared when compared to earlier research [24, 25].

5. Conclusions

HOMER software's simple to operate interface and simple installation procedure make it a popular choice for microgrid modeling. In order to determine the most effective HRES architecture, this research analyzes several different configurations and assesses important performance metrics. Achieving the highest RF, smallest NPC, and LCOE are the main objectives of the evaluation. To determine the best approach for freestanding HRES, the technical and financial benefits of each ideal design are carefully calculated.

The primary results of this work can be concise as follows:

With an NPC of $1.48M and a LCOE value of $0.287, the least cost-effective standalone HRES is made up of an 800 kW WT, 63 8V BS, a 50 kW electrolyzer, a 20 kg H-tank, and 150 kW of power converters.

1. Systems without integrated battery storage are less efficient and less economical than standalone systems with a adequate battery capacity.
2. Among the many benefits of implementing RES is the capacity to generate a 66% additional energy margin while meeting 100% of the power needed to fulfil electrical load data.
3. In order to achieve a 92.8% renewable energy proportion, HRES designs at the site under consideration can provide at least 80% of the onsite power needed to meet load demand. Additionally, by generating less carbon emissions for the site under study, well-designed hybrid systems provide substantial environmental benefits.

References

[1] Meng, Q., Jin, X., Luo, F., Wang, Z., & Hussain, S. (2024). Distributionally Robust Scheduling for Benefit Allocation in Regional Integrated Energy System with Multiple Stakeholders. *Journal of Modern Power Systems and Clean Energy*.

[2] Shirkhani, M., Tavoosi, J., Danyali, S., Sarvenoee, A. K., Abdali, A., Mohammadzadeh, A., & Zhang, C. (2023). A review on microgrid decentralized energy/voltage control structures and methods. *Energy Reports*, 10, 368–380.

[3] Adelekan, O. A., Ilugbusi, B. S., Adisa, O., Obi, O. C., Awonuga, K. F., Asuzu, O. F., & Ndubuisi, N. L. (2024). Energy transition policies: a global review of shifts towards renewable sources. *Engineering Science & Technology Journal*, 5(2), 272–287.

[4] Mansouri, A., El Magri, A., Lajouad, R., Giri, F., & Watil, A. (2024). Nonlinear control strategies with maximum power point tracking for hybrid renewable energy conversion systems. *Asian Journal of Control*, 26(2), 1047–1056.

[5] Awad, M., Said, A., Saad, M. H., Farouk, A., Mahmoud, M. M., Alshammari, M. S., ... & Omar, A. I. (2024). A review of water electrolysis for green hydrogen generation considering PV/wind/hybrid/hydropower/geothermal/tidal and wave/biogas energy systems, economic analysis, and its application. *Alexandria Engineering Journal*, 87, 213–239.

[6] Zhang, M., Lyu, H., Bian, H., & Ghadimi, N. (2024). Improved chaos grasshopper optimizer and its application to HRES techno-economic evaluation. *Heliyon*, 10(2).

[7] Roy, D., Bhowmik, M., & Roskilly, A. P. (2024). Technoeconomic, environmental and multi criteria decision making investigations for optimisation of off-grid hybrid renewable energy system with green hydrogen production. *Journal of Cleaner Production*, 443, 141033.

[8] Kumar, N., & Karmakar, S. (2024). Techno-economic optimization of hydrogen generation through hybrid energy system: A step towards sustainable development. *International Journal of Hydrogen Energy*, 55, 400–413.

[9] Thirunavukkarasu, M., Sawle, Y., & Lala, H. (2023). A comprehensive review on optimization of hybrid renewable energy systems using various optimization techniques. *Renewable and Sustainable Energy Reviews*, 176, 113192.

[10] Pujari, H. K., & Rudramoorthy, M. (2021). Optimal design and techno-economic analysis of a hybrid grid-independent renewable energy system for a rural community. *International Transactions on Electrical Energy Systems*, 31(9), e13007.

[11] Pujari, H. K., & Rudramoorthy, M. (2022). Optimal design, prefeasibility techno-economic and sensitivity analysis of off-grid hybrid renewable energy system. *International Journal of Sustainable Energy*, 41(10), 1466–1498.

[12] Nallolla, C. A., & Perumal, V. (2022). Optimal design of a hybrid off-grid renewable energy system using techno-economic and sensitivity analysis for a rural remote location. *Sustainability*, 14(22), 15393.

[13] Pujari, H. K., & Rudramoorthy, M. (2021). Optimal design and techno-economic analysis of a hybrid grid-independent renewable energy system for a rural community. *International Journal of Sustainable Energy*.

[14] Lilienthal, P. (2016). How HOMER calculates PV output power, Wind turbine power and battery output, energy charge output, total net present cost, Cost of energy and break-even grid extension distance, homer help file 2016. HOMER® Pro Version 3.7 User Manual, August 2016, (http://homerenergy.com/).

[15] Mulenga, E., Kabanshi, A., Mupeta, H., Ndiaye, M., Nyirenda, E., & Mulenga, K. (2023). Techno-economic analysis of off-grid PV-Diesel power generation system for rural electrification: A case study of Chilubi district in Zambia. *Renewable Energy*, 203, 601–611.

[16] Amole, A. O., Oladipo, S., Olabode, O. E., Makinde, K. A., & Gbadega, P. (2023). Analysis of grid/solar photovoltaic power generation for improved village energy supply: A case of Ikose in Oyo State Nigeria. *Renewable Energy Focus*, 44, 186–211.

[17] Al Afif, R., Ayed, Y., & Maaitah, O. N. (2023). Feasibility and optimal sizing analysis of hybrid renewable energy systems: A case study of Al-Karak, Jordan. *Renewable Energy*, 204, 229–249.

[18] Yadav, S., Kumar, P., & Kumar, A. (2024). Techno-economic assessment of hybrid renewable energy system with multi energy storage system using HOMER. *Energy*, 297, 131231.

[19] Trendewicz, A., Tan, E. C., & Ding, F. (2023). Renewable microgrids as a foundation of the future sustainable electrical energy system. *Sustainability Engineering*. CRC Press, 229–238.

[20] Kumar, N., & Karmakar, S. (2024). Techno-economic optimization of hydrogen generation through hybrid energy system:

A step towards sustainable development. *International Journal of Hydrogen Energy, 55*, 400–413.

[21] Kumar, N., & Karmakar, S. (2023). Techno-eco-environmental analysis of a waste to energy based Polygeneration through hybrid renewable energy system. *Energy, 283*, 129199.

[22] Pujari, H. K., Rudramoorthy, M., Gopi R, R., Mishra, S., Alluraiah, N. C., & Vaishali, N. B. (2024). Optimal reconfiguration, renewable DGs, and energy storage units' integration in distribution systems considering power generation uncertainty using hybrid GWO-SCA algorithms. *International Journal of Modelling and Simulation*, 1–33.

[23] Kumar, P. H., Alluraiah, N. C., Gopi, P., Bajaj, M., Kumar, S., Kalyan, C. N. S., & Blazek, V. (2024). Techno-economic optimization and sensitivity analysis of off-grid hybrid renewable energy systems: A case study for sustainable energy solutions in rural India. *Results in Engineering, 25*, 103674.

[24] Li, C., Zhou, D., Wang, H., Cheng, H., & Li, D. (2019). Feasibility assessment of a hybrid PV/diesel/battery power system for a housing estate in the severe cold zone—A case study of Harbin, China. *Energy, 185*, 671–681.

[25] Pradhan, A. K., Mohanty, M. K., & Kar, S. K. (2017). Techno-economic evaluation of standalone hybrid renewable energy system for remote village using HOMER-pro software. *International Journal of Applied, 6*(2), 73–88.

23 Innovative Concept of Dual-Stator Dual-Rotor Radial Flux Permanent Magnet Motor

Aishwarya Bura[1], Dasari Sai Kiran[1], Komma Ganesh[1], and Umakanta Choudhury[2]

[1]UG Student, Electrical and Electronics Engineering, Chaitanya Bharathi Institute of Technology, Gandipet, Hyderabad, Telangana, India
[2]Professor and Advisor, I&I, Electrical and Electronics Engineering, Chaitanya Bharathi Institute of Technology, Gandipet, Hyderabad, Telangana, India

Abstract

This paper proposes optimization of a novel Dual-Stator Dual-Rotor (DSDR) Radial Flux Permanent Magnet (RFPM) motor, for high-performance electric drive and strategic applications. Using advanced electromagnetic modelling and Finite element analysis (FEA), the goal is to achieve maximum power density, torque density and efficiency in conjunction with minimizing both torque ripple and losses. The unique rotor-stator configuration allows for multiple voltage and frequency outputs from a single machine, to provide multiple voltages and frequencies with a single rotor & stator where the inner part of the rotor is only for torque transmission. For the proposed machine the magnets are provided on the outer rotor and inner side of the static armature carries one more set of 3-phase windings. The positioning of rotor & stator, results minimal losses in magnetic flux and better thermal management are some of the characteristics of proposed motor topology The DSDR motor for PMSM applications provides performance benefits including higher torque-speed range with reduced ripple/torque in the low-to-mid speed regions of operation as well as higher efficiency at high speeds. The proposed motor is designed as a high-torque solution for challenging electric vehicle, electric power trains, and aerospace propulsion applications, having very fast mechanical response times with gently damped dynamic properties & industrial drives, where compact and efficient solutions for high-power electric drive systems are needed. In case of strategic applications like aerospace & navel applications there are few devices that can operate at different voltages & frequencies in such case a single machine having multiple outputs will be more reliable and efficient compared to two separate machines providing two different voltages and frequencies. This will also provide compactness with reduced weight & volume. The research advances electric machine technology by enhancing efficiency, increasing power density, and upgrading the reliability of the motor.

Keywords: Dual-Stator Dual-Rotor (DSDR), Radial Flux Permanent Magnet Motor (RFPMM), electric drive, Finite Element Analysis (FEA), power density, Torque density, efficiency

Email: aishwarya.bura89@gmail.com[1], dsaikiran155@gmail.com[1], ganeshkommayadav@gmail.com[1], director_ii@cbit.ac.in[2]

DOI: 10.1201/9781003661917-23

1. Introduction

The emerging demand for high-performance electric motors, especially in areas like EVs, aerospace, and industrial drives, challenges the limits of conventional motors to meet outstanding performance requirements. These conventional motors often face limitations such as low torque density, low efficiency, and the inability to operate at multiples of voltages and frequencies. In applications that seek lightweight and compact solutions, these limitations become critical issues. Although many motor designs exist, there is an enormous gap in providing complete solutions that integrate the different advantages available from both dual stator and dual rotor configurations with the better permanent magnet technology. Often, present designs require many machines to meet disparate operating requirements that only increase complexity and weight. Targeting the development of Dual Stator Dual Rotor Radial Flux Permanent Magnet Motor (DSDR RFPM), this research work is aimed at developing a proposed machine that maximizes power density and torque density and minimizes losses and torque ripples. Due to their applications in electric vehicles, robotics, and renewable energy systems, there has been increased demand for high-efficiency and compact electric motors in modern times.

Permanent Magnet (PM) motors are attractive because they offer higher power density and efficiency compared to induction motors. However, some problems such as cogging torque, magnetic losses, and thermal issues are still present in conventional PM motor designs. Thus, the idea of a Dual Stator Dual Rotor (DSDR) configuration has emerged as a probable remedy of these drawbacks with the benefits of better performance through balancing the magnetic flux paths and core losses minimization. The chief purpose of this work is to unfold a new conceptual design for a radial flux DSDR PM motor followed by the investigation of structural advantages and then analysis of operating characteristics using Finite Element Analysis.

2. Literature Review

2.1 Current Motor Technologies

Radial flux motors have been gaining more attention due to their compact size and high-power density, applied to all sorts of applications, from electric vehicles to industrial machinery. Established configuration in the dominating technology include single stator as well as single rotor, which have distinct advantages and limitations for each.

Single-Stator Concepts: These motors are generally easier to manufacture and are indeed more widely used in design applications where the space factors aren't a major issue. The permanent magnet technology, especially in terms of magnet with high-energy, such as neodymium-iron-boron (NdFeB), has improved the efficiencies and power handling capabilities of single-stator motors. Optimally distributed magnets, combined with excellent winding techniques, maximize torque output while minimizing losses to ensure operation at as wide a range of speeds as possible.

Single-Rotor Designs: Simple and Lightweight Designs are Recognised. Improvements during the recent years have focused on techniques to reduce cogging torque, and design rotor geometry for performance optimization. In a typical single-rotor design, efficiency often yields to thermal management or leakage of magnetic flux.

Permanent magnet: based technology used in radial flux motors have led to significant efficiency gains primarily because these motors are much better suited for achieving higher torques and power values. The best practices now in optimizing the design of electromagnetic today are the use of finite element analysis and computational modelling while designing the motors for the specific application needed.

2.2 Literature Review on DSDR Motors

Dual Stator and Dual Rotor configuration: seemingly promising avenue for the improvement of motor performance, as it was already indicated by a large number of benefits in previous works, including torque density enhancement and thermal management improvement.

[1] presented the design & analysis of a dual concentric rotor and stator machine with reduced efficiency and torque ripple; the dual rotor configuration is superb in terms of load distribution balance and thus can result in performance stability when applying various operation conditions, though poses other challenges such as management of complexity in designs and effective use of magnetic flux. Another paper [2] is concerned with the dual stator configuration in which it can be presented that some appreciable power improvement is recognized. However, the manufacturing complexity and weight increase become problematic for such a dual stator configuration.

Although some very encouraging findings have been presented in DSDR configurations, technology integration of dual concentric stator-rotor into radial flux pm motors is, on a broad spectrum, still in the early stages of exploration. This paper will fill that gap by presenting a novel DSDR RFPM motor design that collapses optimum features from both configurations while overcoming the related drawbacks.

2.3 Challenges in RFPM Motors

DSDR configurations improve the performance of motors regarding higher torque density, enhanced thermal management, and balanced loading. Dual-rotor designs reduce efficiency losses and torque ripples, whereas dual-stator designs offer significant power advantages but suffer from added weight and complexity. The idea in this work is to propose a DSDR configuration for an RFPM motor that captures the strengths of both but addresses the drawbacks and optimizes magnetic flux

utilization in its design, thus simplifying the design process. This novel design offers a compact and efficient solution with high power density and minimal losses, addressing the ever-growing demand for high-performance motors in EVs, aerospace, and industrial applications.

3. Single Stator Single Rotor RFPM Motor Configuration

Magnetic flux of a radial flux motor flows radially through the motor air gap. This naturally results in relatively simple and efficient construction of the motor. Such construction is mainly prevalent in PM motors as this type will allow easier assembly with lesser usage of material. Axial flux motors, in comparison, use much more permanent magnet material to produce the same performance and thus cost, sustainability, and easy installation advantages. This streamlined design reduces the total volume of magnet material and makes radial flux motors more environmentally friendly for applications that demand higher torque density, like electric vehicles and industrial drive systems. Outer rotor topologies of permanent magnet motors with radial flux are optimized, such as Magnetic Innovations, for better efficiency with fewer materials and reduced installation costs. For example, windings are made dense in such a way that the flux paths get short. Thus, winding resistance decreases and consequently, the energy losses are reduced. The distance of the air gap also gets increased to some extent by these motors at which the forces are tangential. So, this enhances the "lever arm" effect and maximizes the torque generated as torque of a tangential force is directly proportional to its distance from the axis of rotation. In cases where a frameless torque motor with radial flux is required, quite crucial to consider will be the desired speed and torque. The work points clearly defining the torque-to-speed curve gives the operational demands at different speeds so that engineers can make well-defined specifications about what should be the torque at specified speeds for their respective motor. Additionally, the

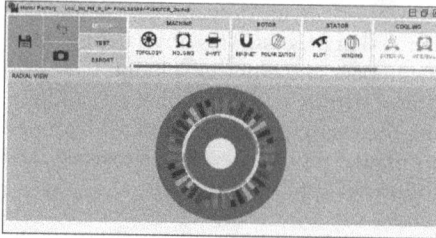

Figure 23.1 Radial View of Single Stator Single Rotor RFPM Motor

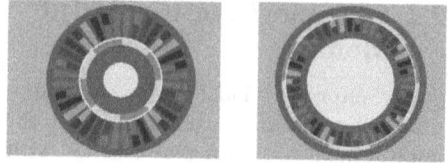

Figure 23.2 Radial View of Inner and Outer Motor of DSDR RFPM Motor

methods of cooling by air or water will have to be considered for thermal management in keeping with the thermal constraints dictated by the motor for it to run at maximum efficiency. In performance at various work points, the mechanical power is determined from the expression that gives the motor's:

Pmech = T × ω

Here, T represents Torque, measured in N-m and ω is angular velocity in rad/sec. The above calculation helps establish the power output of a motor, but also provides a means of finding torque where the power and speed are available.

4. Dual Stator Dual Rotor RFPM Motor Configuration

The proposed DSDR RFPM motor includes two stators and two rotors. Each stator is placed both externally and internally on a central rotor core so as to completely encapsulate the dual rotors. This realization creates a balanced path for magnetic flux, which consequently decreases core saturation and reduces electromagnetic losses by promoting torque generation. This further increases pole count and slot numbers, as well as symmetrical positioning due to the stator and rotor alignment, resulting in an increased torque density per volume. In general, the DSDR-RFPM proposed here consists of four big parts, namely an outer stator, an inner stator, an outer rotor, and an inner rotor. The stators have windings fitted on them and interior permanent magnets are fitted on the two rotors. Table 23.1 Specifications, Ratings, and Dimensions for Outer as well as Inner

Motors. A DSDR-RFPM motor configuration is presented with strategic applications in aerospace, naval and electric vehicle applications. In this design, the outer motor primarily produces the primary power and provides high torque output that is suitable for advanced applications. The outer motor can be designed to operate at various voltage and frequency scales to increase flexibility in different applications. In this design, the inner motor is mechanically coupled with an auxiliary system, which may be driven separately or synchronous with the outer motor regarding the needs of the application. This high-permeability material supports the inner stator core surrounded by a magnetic barrier, hence reducing electromagnetic interference between the outer and inner motors. The stator of the outer motor has been designed to carry 3-phase 2-layer lap windings distributed in 48 slots, an integral number of slots per pole and phase for optimal magnetic performance and reduced harmonic distortion. With the below design, the DSDR-RFPM is made for maximum power density, torque density, and with high efficiency within compact design suitable for strategic as well as high-performance applications.

5. Methodology and Model Specifications

5.1 Finite Element Analysis (FEA)

The FEA simulations test the design parameters that have the capability to assess the motor under several types of loading conditions found in applications for electric vehicle use and industrial automation. The model considers important parameters such as

Table 23.1 Comparison of Parameter Dimensions, Ratings and Specifications

S No	Parameters	SSSR-RFPM Motor	DSDR-RFPM Motor Outer	Inner
1	outer diameter of Stator (mm)	306	263	175
2	Inner diameter of Stator (mm)	180	175	98
3	outer diameter of Rotor (mm)	178	306	96
4	Inner diameter of Rotor (mm)	70	265	40
5	Number of stator slots	36	45	36
6	Number of phases	3	3	3
7	Number of poles	8	10	8
8	Number of turns per coil	8	6	13
9	Number of wires in hand	14	13	9
10	Number of parallel paths	2	1	1
11	Number of layers in a slot	2	2	2
12	Airgap length (mm)	1.0	1.0	1.0
13	Stack length (mm)	84	84	84
14	Rated voltage/phase (V)	127.04	171.52	126.632
15	Rated power (KW)	31.3	31.5	9.623
16	power (KW)	31.3	31.5	9.623
17	Base frequency (Hz)	106.667	111.993	106.667
18	Peak frequency (Hz)	106.667	133.333	106.667
19	Electrical power (KW)	32.4	36.058	9.623
20	Full-load current (A)	151.066	70.145	27.143
21	Magnet material	NdFeB_1230_1400		
22	Magnet thickness	7.136 mm	8.0	8.0
23	Core type	REF.M330_35A		
24	Wire material	Copper		
25	Stator core material	REF.M330_35A steel		
26	Rotor core material	REF.M330_35A steel		
27	Efficiency (%)	96.6	97.2	94.57

magnetic flux density, torque, and core losses based on boundary conditions that stand for real thermal and loading conditions.

5.2 Boundary Conditions

The FEA model included thermal boundaries that simulated the heat that is released when the motor operates. This is usually considered essential for performance as well as thermal analysis for cooling needs and thermal stability.

Therefore, it was of paramount importance in determining the optimum cooling strategy for long-term performance either actively or passively cooled.

5.3 Load Analysis

To check the motor response, it was checked with several torque requirements and also flux saturation points analyzed along with torque ripple. Thus, it ensures investigation under high torque conditions as well as low torque conditions to establish regions of flux saturation thereby ensuring that the motor would be stable and efficient in a regime of operation. Design changes in the drive analysis helped out in meeting some needs in particular applications regarding raising torque in high-stress conditions, such as those in electric vehicles and automation systems.

The sum of the output powers of inner motor and outer motor will be higher than that of the output power of SSSR RFPM motor (i.e., 36.05 KW + 9.62 KW = 45.67 KW > 32.4 KW).

6. Empirical Results

6.1 Electromagnetic Performance (Flux Distribution)

SSSR-RFPM Motor: Flux distribution plot is used to measure magnetic field intensity and uniformity of magnetic field in the stator and rotor. Greater uniformity of flux means efficient design with decreased loss and reduced cogging torque.

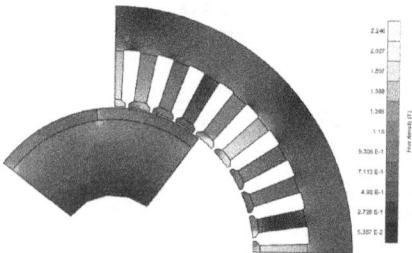

Outer Rotor of DSDR-RFPM: The outer rotor should have a greater magnetic flux density in the fact that it has to bear higher torque loads. Therefore, the flux apparently seems more concentrated in certain portions, which means the magnetic performance in this portion is higher.

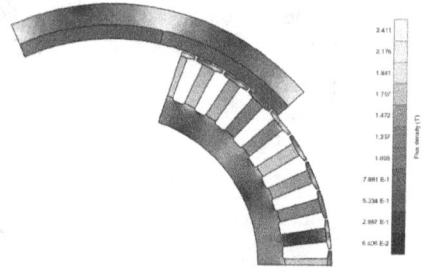

Inner Rotor of DSDR-RFPM: The flux distribution is expected to be balanced but smaller than the outer motor as it ordinarily bears lesser torque or functions at higher speeds.

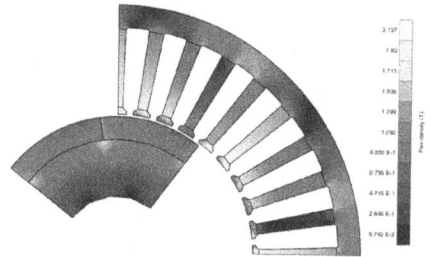

6.2 Torque vs. Speed Characteristics

SSSR-RFPM Motor: Likely to have relatively constant torque with decreasing speed as the load increases, and the characteristics are expected of a single rotor PM machine.

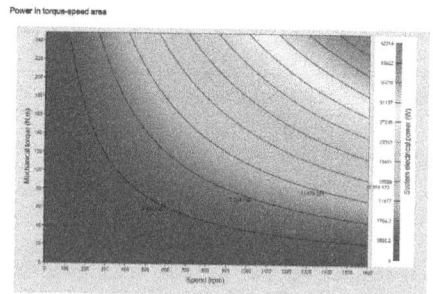

Outer Motor of DSDR-RFPM: Has more torque at lower speeds since it now bears the greater proportion of load. Its torque curve could increase with speed but would still provide higher overall torque than the single-stator motor.

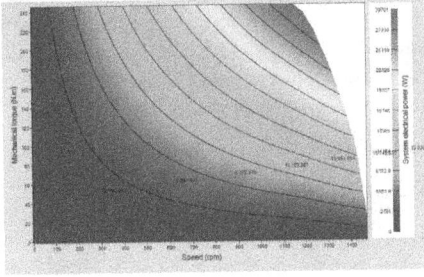

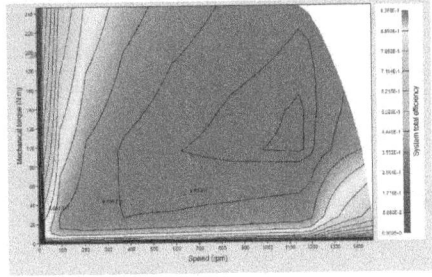

Inner Motor of DSDR-RFPM: With relatively lower torque, this motor contributes to efficiency and speed at lighter loads.

Inner Motor of DSDR-RFPM: In compensation with the outer motor, better energy distribution results and, consequently, optimized system-level efficiency is achieved.

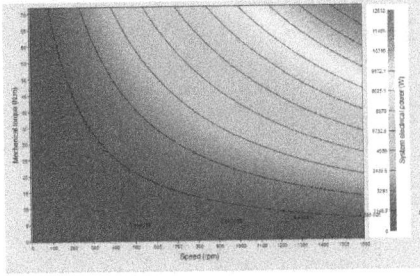

6.3 Efficiency Analysis

SSSR-RFPM Motor: Peak efficiency occurs along the middle load and speed, for it degrades with an increase in current or higher loads due to increased losses.

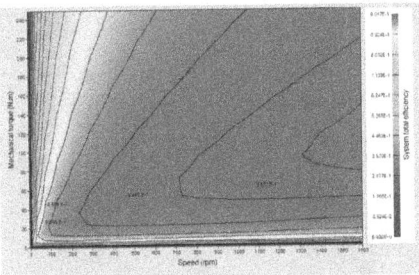

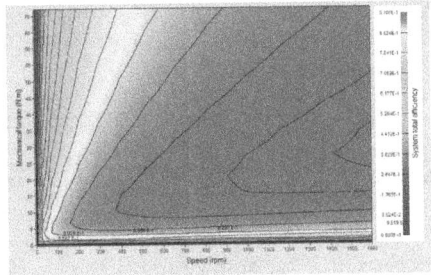

6.4 Torque Ripple Comparison

SSSR-RFPM Motor: The torque ripple might be greater due to less design complexity, leading to ripples in performance and possible vibrations.

Outer Motor of DSDR-RFPM: Lower torque ripple can be achieved, due to optimized dual-stator design.

Inner Motor of DSDR-RFPM: Torque ripple should be minimal, as electromagnetic forces are balanced in the double system.

6.5 Power Density

SSSR-RFPM Motor: The power density is limited by a single configuration of stator and rotor.

Outer Motor of DSDR-RFPM: Better efficiency at comparable load points, owing to a dual-rotor configuration that optimizes energy consumption for torque generation.

Outer and Inner Motors of DSDR-RFPM: The combined one results in highly strengthened power density that effectively optimizes the space and weight of the motor, especially crucial for applications like EVs or aerospace.

6.6 Loss Analysis (Copper and Iron Losses)

SSSR-RFPM Motor: There can be increased losses because of the simplicity in design, less magnetic coupling, and higher torque ripple.

Outer Motor of DSDR-RFPM: It should have lesser loss values because of optimal paths for flux and good magnetic design.

Inner Motor of DSDR-RFPM: It is more efficient at higher speeds. Together with outer motor, there may be lesser overall loss values.

7. Conclusion

In conclusion, we hope to suggest and develop a concept of a DSDR RFPM Motor that can exhibit the functions of two different motors in one machine. The design is an innovative multiple output motor suited for applications like electric vehicles, renewable energy systems, robotics, aerospace. Combining two motors in one reduces the overall size, thus saving on costs.

This means that the motor would be compact and powerful and capable of offering high torque efficiency in a wide range of conditions, thus also simplifying the control system and consequently making it easier to manage and more reliable. It is seen that the overall efficiency of the SSSRRFPM motor is comparable to that of the DSDRRFPM motor. The combined output power of DSDRRFPM is significantly higher. Besides, the dual stator design provides added benefits of small size, low weight, and increased compactness, making the design even more efficient and practical for high-performance applications.

References

[1] Li, C., Guo, X., Fu, J., Fu, W., Liu, Y., Chen, H., ... & Li, Z. (2021). Design and analysis of a novel double-stator double-rotor motor drive system for in-wheel direct drive of electric vehicles. *Machines*, 10(1), 27. doi: 10.3390/machines10010027

[2] Mudhigollam, U. K., Choudhury, U., & Hatua, K. (2018). High power density multiple output permanent magnet alternator. *IET Electric Power Applications*, 12(4), 494–501. doi: 10.1049/iet-epa.2017.0477

[3] Mudhigollam, U. K., Choudhury, U., Hatua, K., & Sridhar, U. (2017, December). Improved rotor structure of hybrid excitation alternator. In *2017 IEEE Transportation Electrification Conference (ITEC-India)* (pp. 1–4). IEEE. doi: 10.1109/ITEC-India.2017.8333716

[4] Bouloukza, I., Mordjaoui, M., Kurt, E., Bal, G., & Ökmen, C. (2018). Electromagnetic design of a new radial flux permanent magnet motor. *Journal of Energy Systems*, 2(1), 13–27. doi: 10.30521/jes.397836

[5] Kumar, R. R., Singh, S. K., & Srivastava, R. K. (2014, December). Design analysis of radial flux dual stator five phase permanent magnet synchronous generator. In *2014 IEEE International Conference on Power Electronics, Drives and Energy Systems (PEDES)* (pp. 1–6). IEEE.

[6] Yu, J., Liu, C., & Zhao, H. (2019). Design and multi-mode operation of double-stator toroidal-winding PM Vernier machine for wind-photovoltaic hybrid generation system. *IEEE Transactions on Magnetics*, 55(7), 1–7. doi: 10.1109/TMAG.2019.2906849

24 Optimal DG Placement Using Realistic Method

Ekkati Manoj, Bantupavani Lalitha, Mohammad Almas, and P. Venkata Prasad

Chaitanya Bharathi Institute of Technology, Hyderabad, India

Abstract

This paper introduces a novel technique for the best location and sizing of distributed generators. The load flow results for various load models is also presented. The load models considered are constant-current, constant-impedance, constant-power and composite load models. As the integration of non-conventional energy sources becomes increasingly vital for sustainable power systems, understanding how various load models affect the performance of distribution systems is essential. Simulations conducted on 33-bus and 69-bus systems reveal significant differences in voltage profiles and power flows based on selected load model. The optimal location for distributed generation identified and at this site, the load replaced with a DG capacity equivalent to the original load.

Keywords: Constant current, constant impedance, constant power, distributed generators, load flow analysis, load models, non-conventional energy sources, composite model.

1. Introduction

The growing complexity of modern power systems, particularly in distribution networks, demands robust and efficient approaches for load flow analysis. Accurate power flow analysis is important for planning, operating, and optimizing distribution networks. Radial distribution systems, which are widely used in modern power distribution, present challenges for traditional load flow algorithms due to their high resistance to reactance ratio and unbalanced nature [1]. The Backward and Forward sweep method has emerged as an effective solution for solving load flow in these systems. Load flow analysis is a crucial aspect of power system studies, providing an essential understanding of the behavior and performance of electrical distribution systems. It involves calculating voltage, current, and power flow throughout the system, enabling engineers to ensure the consistency and efficiency of power delivery [2]. As electrical loads become increasingly diverse and complex, selecting the suitable load models is essential for accurate simulation and effective system design.

The growing demand for electricity and the need for sustainable energy solutions have heightened interest in distributed generation. DG involves generating electricity from decentralized sources, often renewable, that are located close to Load. Optimal placement of DG units can help address challenges like voltage drops, reduce power

Email: akkatimanojreddy@gmail.com, lallithallay@gmail.com, mohammadalmas@gmail.com, Professor, pvprasad_eee@cbit.ac.in

DOI: 10.1201/9781003661917-24

losses, and improve the overall electrical system reliability. Traditionally, Electrical Generation is centralized, which has resulted in inefficiencies and vulnerabilities within the distribution network [10–11]. Then the rise of non-conventional energy technologies has led to a shift towards decentralized generation. However, the Effectiveness of this shift relies heavily on the strategic placement of DG.

This paper focuses on the load flow study of Two standard distribution Networks, the 33-bus and 69-bus configurations. We utilize the backward/forward sweep method, which is recognized for its efficiency and straightforwardness in addressing radial distribution systems [4].

The analysis includes various load models, considered "models are constant power, constant current, constant impedance, and composite model", which integrates features of the above models, each model offers unique insights [5].

2. Problem Formulation

The Backward/Forward sweep method is an iterative algorithm that consists of two key stages, the backward flow and Forward flow.

Backward flow: This step calculates the current flow in each branch by starting at terminal buses and moving back in the direction of the source bus.

Forward flow: This step updates the potential at each bus by beginning at the supply point and moving forward in the direction of the terminal bus, based on the calculated current and the system impedance.

This method is highly effective for radial distribution systems, as it takes advantage of the tree-like structure of the network, making the power flow solution more straight forward and computationally efficient power flow solution compared to traditional methods such as Newton-Raphson and Gauss-Seidel [3].

Radial distribution network power flow analysis is briefly outlined using the BFS Algorithm in following two steps [4, 7].

Step I: The Load currents at the k_{th} bus of radial distribution network, assuming all bus voltages start at $1<0°$, are calculated as follows:

$$I_k = \left(\frac{P_k + jQ_k}{V_k} \right)^* \tag{1}$$

Branch current passing along the line section c, which connects two nodes r and s, is calculated as follows:

$$I_{br.c} = I_s + \sum_{k=1}^{r} I_{brk} \tag{2}$$

Where, $I_{br \cdot c}$ branch currents across line section c, the collection of downstream, branches connecting to node s is denoted with r.

Step II: The voltages at all buses are calculated using the computed branch currents as follows:

$$V_s = V_r - Z_{rs} I_{brk} \tag{3}$$

$$s = 1, 2, 3, ..., N \text{ and } r = 1, 2, ..., N$$

For two consecutive iterations, the variation in voltage values among every bus is computed after *step II*:

$$j \angle V_j = j\left(V_r - V_{r-1}\right), \tag{4}$$

$$jV_j < \text{error}$$

This algorithm runs iteratively until the error achieves a set threshold, usually between 10^{-6} and 10^{-9}.

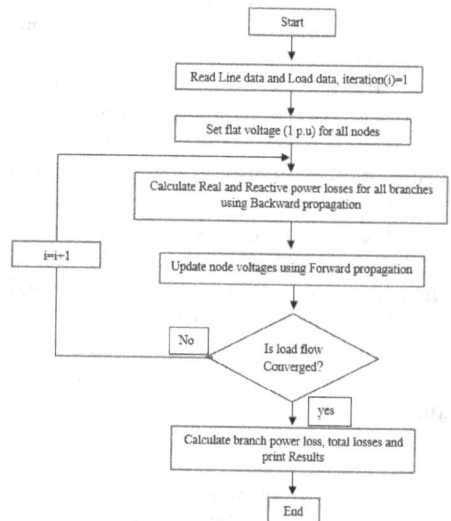

Figure 24.1 Flow Chart for Load Flow Solution

3. Load Models

There are four types of Load models in electrical power systems and each node has a specific percentage of these loads [8].

3.1 Constant Power Load

A constant power load maintains a fixed power consumption, which is described by the equation:

$$P = \alpha \cdot P_0 \cdot V^2 + \beta \cdot P_0 \cdot V^1 + \gamma \cdot P_0 \cdot V^0 \quad (5)$$

$$Q = \alpha \cdot Q_0 \cdot V^2 + \beta \cdot Q_0 \cdot V^1 + \gamma \cdot Q_0 \cdot V^0 \quad (6)$$

$$\alpha = 0; \beta = 0; \gamma = 1$$

i.e. $P = P_0$, $Q = Q_0$

This type of model is commonly applied to industrial machinery operating under stable conditions.

3.2 Constant Current Load

A constant current load draws a fixed current regardless of voltage changes, represented by the following equations:

$$P = \alpha \cdot P_0 \cdot V^2 + \beta \cdot P_0 \cdot V^1 + \gamma \cdot P_0 \cdot V^0 \quad (7)$$

$$Q = \alpha \cdot Q_0 \cdot V^2 + \beta \cdot Q_0 \cdot V^1 + \gamma \cdot Q_0 \cdot V^0 \quad (8)$$

$$\alpha = 0; \beta = 1; \gamma = 0$$

i.e. $P = P_0 * V^1$, $Q = Q_0 * V^1$

This is typical of devices like incandescent bulbs, which demonstrate this behavior under normal operating conditions

3.3 Constant Impedance Load

A constant impedance load behaves as a resistor, with its power consumption directly related to the voltage applied, represented by the following equations:

$$P = \alpha \cdot P_0 \cdot V^2 + \beta \cdot P_0 \cdot V^1 + \gamma \cdot P_0 \cdot V^0 \quad (9)$$

$$Q = \alpha \cdot Q_0 \cdot V^2 + \beta \cdot Q_0 \cdot V^1 + \gamma \cdot Q_0 \cdot V^0 \quad (10)$$

$$\alpha = 1; \beta = 0; \gamma = 0$$

i.e. $P = P_0 \cdot V^2$, $Q = Q_0 \cdot V^2$

This model represents resistive heating elements.

3.4 ZIP Load

The ZIP load combines characteristics of 3.1, 3.2, and 3.3 load models, represented by the following equations:

$$\alpha = 0.57; \beta = 0.32; \gamma = 0.11$$

$$P = \alpha \cdot P_0 \cdot V^2 + \beta \cdot P_0 \cdot V^1 + \gamma \cdot P_0 \cdot V^0 \quad (11)$$

$$Q = \alpha \cdot Q_0 \cdot V^2 + \beta \cdot Q_0 \cdot V^1 + \gamma \cdot Q_0 \cdot V^0 \quad (12)$$

$$\alpha + \beta + \gamma = 1.$$

Where P-Active power, Q-Reactive power, V-voltage, α-Coefficient of v^2 term, β-coefficient of v term, γ-coefficient of constant (v^0), P_0-Rated active power, Q_0-Rated reactive power

4. DG Placement

DG's best location and sizing begins by selecting a bus, referred to as bus1. At this bus, the existing load is replaced with a DG unit that has a capacity equal to the original load, effectively reducing the load at that bus to zero. A load flow analysis is then conducted to evaluate how does these change influence the system's power losses.

This method is repeated for each bus in the network by systematically replacing the load with a distributed generator unit. A power flow analysis is then conducted to determine the resulting power losses across the system. Once all iterations are completed, the bus that results in the lowest power losses with the DG unit installed is identified as the optimal position for DG unit [5, 9].

This method not only improves the efficiency of the distribution network, it also facilitates the integra   tion   of non-conventional energy sources, helping to create a more sustainable power system. By strategically positioning DG units at the most advantageous locations, minimization of power losses can be achieved, ultimately leading to improved overall system performance [12].

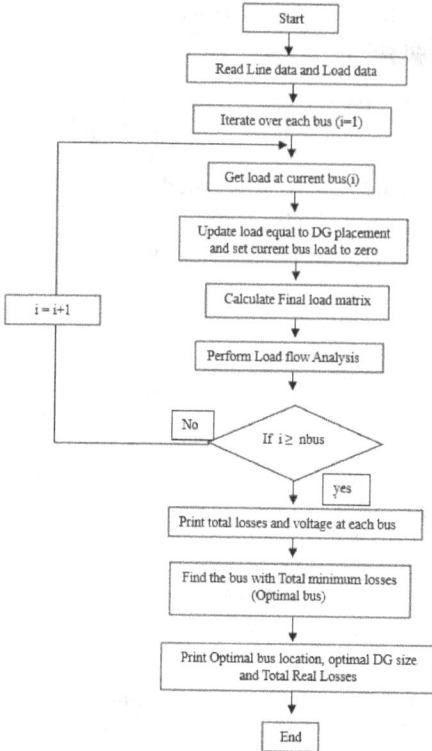

Figure 24.2 Flow Chart for Optimal DG Placement

Table 24.1 Load Flow Analysis of 33 Bus System

Load model	Total Active Losses (KW)	Total Reactive losses (KVAR)	Minimum voltage
Constant power Load	202.6771	135.141	0.91306
Constant current Load	174.7678	116.2532	0.91984
Constant impedance Load	154.504	102.5729	0.92509
ZIP Load	168.5585	112.0672	0.92139

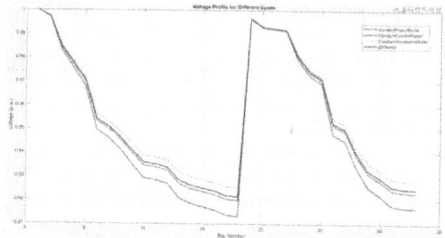

Figure 24.3 Voltage Performance of 33 Bus System Under Various Load Models

5. Results and Discussions

The load flow analysis for both the 33-bus and 69-bus systems is conducted for four load models. The system operates with reference voltage of 12.66 KV and a reference apparent power of 100 MVA.

The active power losses are higher in constant power load measuring 202.6711 kW and lower in constant impedance load measuring 154.504 kW, Additionally, the voltage is lowest in constant power load recorded as 0.9130 P.U., while it is highest for constant impedance load, reaching 0.925 P.U, for 33-Bus Network as the Outcomes are illustrated in Table 24.1 and Figure 24.3.

For 69 Bus Network, the active power losses are higher in constant power load measuring 224.96 kW, and lower in constant impedance load measuring 163.778 kW, Additionally, the voltage is lowest in constant power load recorded as 0.909 P.U., while it is

highest for constant impedance load, reaching 0.9232 P.U., as the findings are illustrated in Table 24.2, and Figure 24.4.

The Total Active losses are minimal, measuring 138.42 KW at bus No 30 from Table 24.3, and Figure 24.5. The optimal

Table 24.2 Load Flow Analysis of 69 Bus System

Load model	Total Active Losses (KW)	Total Reactive Losses (KVAR)	Minimum voltage
Constant power Load	224.9606	102.147	0.90901
Constant current Load	188.6289	86.5708	0.91722
Constant impedance Load	163.7782	75.8802	0.92329
ZIP Load	181.7427	83.5968	0.91884

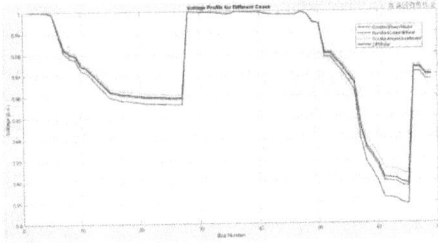

Figure 24.4 Voltage Performance of 69 Bus System Under Various Load Models

Bus	Total Active Loss (KW)	Total Reactive Loss (KVAR)	Min Voltage (V)
19	202.063	134.792	0.913
20	201.596	134.372	0.913
21	201.519	134.283	0.913
22	201.475	134.225	0.913
23	198.700	132.978	0.914
24	182.006	122.998	0.915
25	180.676	121.971	0.915
26	196.345	131.187	0.914
27	196.008	131.009	0.914
28	195.162	130.221	0.914
29	183.753	122.378	0.916
30	138.428	92.244	0.922
31	178.496	118.743	0.916
32	169.283	112.450	0.918
33	191.334	127.364	0.914

connection can be made at Bus 30 with a minimum total loss of 138.42 kW. And DG Size with Active Power = 200 kW, Reactive Power = 600 KVAR replaced with the load as per our proposed methodology.

From Table 24.4, the Total losses are minimal measuring 47.236 KW at bus No 61. The optimal connection can be made at Bus

Table 24.3 DG Placement Results for 69 Bus System

Bus	Total Active Loss (KW)	Total Reactive Loss (KVAR)	Min Voltage (V)
1	202.677	135.141	0.913
2	202.029	134.805	0.913
3	199.500	133.492	0.914
4	195.880	131.609	0.914
5	198.506	132.971	0.914
6	196.868	131.491	0.914
7	181.336	120.027	0.918
8	179.182	119.234	0.919
9	195.178	130.018	0.915
10	194.458	129.510	0.916
11	195.325	130.144	0.916
12	193.056	128.670	0.916
13	192.246	128.046	0.917
14	181.294	120.366	0.919
15	193.772	128.925	0.917
16	192.815	128.262	0.918
17	192.621	128.027	0.918
18	187.054	124.129	0.918

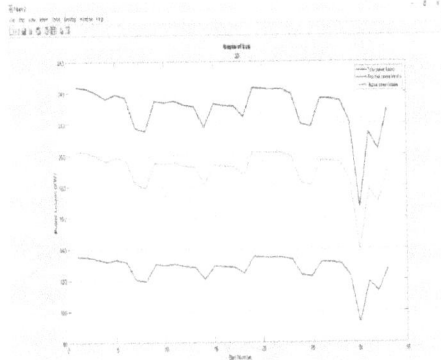

Figure 24.5 Variation of Power Losses for 33 Bus System

61 with a minimum total loss of 47.236 kW and DG Size with Active Power = 41.181 kW, Reactive Power = 23.139 KVAR replaced with the load as per our proposed methodology.

4. Conclusion

Our study on load flow analysis for the 33-bus and 69-bus networks shows how important load models are for system

Table 24.4 DG Placement Results for 69 Bus System

Bus	Total Active Loss (KW)	Total Reactive Loss (KVAR)	Min Voltage (V)
1	224.961	102.147	0.909
2	224.961	102.147	0.909
3	224.961	102.147	0.909
4	224.961	102.147	0.909
5	224.961	102.147	0.909
6	224.896	102.111	0.909
7	223.083	101.151	0.909
8	221.123	100.122	0.910
9	223.318	101.282	0.909
10	223.107	101.239	0.909
11	215.137	97.373	0.910
12	214.081	97.016	0.910
13	224.295	101.843	0.909
14	224.214	101.813	0.909
15	224.961	102.147	0.909
16	220.474	100.177	0.909
17	219.180	99.618	0.910
18	219.179	99.617	0.910
19	224.961	102.147	0.909
20	224.858	102.102	0.909
21	213.517	97.131	0.910
22	224.416	101.912	0.909
23	224.961	102.147	0.909
24	221.930	100.835	0.909
25	224.961	102.147	0.909
26	223.421	101.482	0.909
27	223.420	101.482	0.909
28	224.958	102.142	0.909
29	224.957	102.138	0.909
30	224.961	102.147	0.909
31	224.961	102.147	0.909
32	224.961	102.147	0.909
33	224.944	102.137	0.909
34	224.930	102.131	0.909
35	224.948	102.141	0.909
36	224.958	102.141	0.909
37	224.954	102.130	0.909
38	224.961	102.147	0.909
39	224.947	102.123	0.909
40	224.947	102.123	0.909
41	224.958	102.144	0.909
42	224.961	102.147	0.909
43	224.947	102.129	0.909
44	224.961	102.147	0.909
45	224.884	102.044	0.909
46	224.884	102.044	0.909
47	224.961	102.147	0.909
48	224.837	101.848	0.909
49	223.232	97.938	0.909
50	223.116	97.654	0.909
51	222.893	101.056	0.909
52	224.771	102.047	0.909
53	224.681	102.001	0.909
54	223.115	101.186	0.909
55	223.006	101.134	0.909
56	224.961	102.147	0.909
57	224.961	102.147	0.909
58	224.961	102.147	0.909
59	204.947	93.643	0.913
60	224.961	102.147	0.909
61	41.181	23.139	0.968
62	217.193	98.851	0.911
63	224.961	102.147	0.909
64	173.006	80.006	0.923
65	210.423	95.949	0.913
66	223.688	101.531	0.909
67	223.687	101.531	0.909
68	222.748	101.112	0.909
69	222.748	101.112	0.909

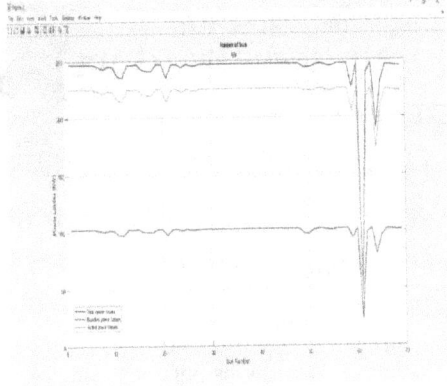

Figure 24.6 Variation of Power Losses for 69 Bus System

efficiency and finding the best location for Distributed Generation (DG). We discovered that the constant power load model caused the biggest drops in voltage, especially at buses far from the source, on other hands constant impedance model performed better in maintaining stable voltage levels. This model helps reduce large voltage drops, making it a good choice when voltage stability is critical. This stability is important for deciding where to place DG effectively. Strategically locating DG enables the minimization of energy losses and enhance voltage profile. Proper DG placement also contributes to better load balancing and reduces the burden on central power systems.

References

[1] Nogueira, W. C., Garcés Negrete, L. P., & López-Lezama, J. M. (2023). Optimal allocation and sizing of distributed generation using interval power flow. *Sustainability*, 15(6), 5171.

[2] Salimon, S. A., Adebayo, I. G., Adepoju, G. A., & Adewuyi, O. B. (2023). Optimal allocation of distribution static synchronous compensators in distribution networks considering various load models using the Black Widow optimization algorithm. *Sustainability*, 15(21), 15623.

[3] Qian, K., Zhou, C., Allan, M., & Yuan, Y. (2011). Effect of load models on assessment of energy losses in distributed generation planning. *International Journal of Electrical Power & Energy Systems*, 33(6), 1243–1250.

[4] Kawambwa, S., Mwifunyi, R., Mnyanghwalo, D., Hamisi, N., Kalinga, E., & Mvungi, N. (2021). An improved backward/forward sweep power flow method based on network tree depth for radial distribution systems. *Journal of Electrical Systems and Information Technology*, 8, 1–18.

[5] Nagaraju, S. K., Ramana, T., Satyanarayana, S., & Prasad, P. V. (2011). A novel method for optimal distributed generator placement in radial distribution systems distributed generation & alternative. *Electric Power Components and Systems*, 39, 2011.

[6] Prasad, P. V., Sivanagaraju, S., & Sreenivasulu, N. (2007). Network reconfiguration for load balancing in radial distribution systems using genetic algorithm. *Electric Power Components and Systems*, 36, 2007.

[7] Ouali, S., & Cherkaoui, A. (2020). An improved backward/forward sweep power flow method based on a new network information organization for radial distribution systems. *Journal of electrical and Computer Engineering*, 2020(1), 5643410.

[8] Hossain, F. A., Rokonuzzaman, M., Amin, N., Zhang, J., Mishu, M. K., Tan, W. S., ... & Roy, R. B. (2021). Probabilistic load flow–based optimal placement and sizing of distributed generators. *Energies*, 14(23), 7857.

[9] Nagaraju, K., Sivanagaraju, S., Ramana, T., & Prasad, P. V. (2011). A novel load flow method for radial distribution systems for realistic loads. *Electric Power Components and Systems*, Taylor & Francis Group, 4(2), 1401–1407.

[10] Bot, R. I., Csetnek, E. R., & Vuong, P. T. (2020). The forward–backward from continuous and discrete perspective for pseudo-monotone variational inequalities in Hilbert spaces. *European Journal of operational Research*.

[11] Zimmerman, R. D., Murillo-Sánchez, C. E., & Thomas, R. J. (2011). Steady-state operations, planning, and analysis tools for power systems research and education. *IEEE Transactions on Power Systems*, 26(1), 12–19.

[12] Kersting, W. H. (2011). Radial distribution test feeders. *IEEE Transactions on Power Systems*, 6(3), 975–985.

25 Blockchain Based Intelligent Product Expiry Management System using Image Recognition (OCR)

Keerthana Birelli[1], Sushma Siluveru[1] and Sangeeta Gupta[2]

[1]Department of Computer Engineering of Technology, Chaitanya Bharathi Institute of Technology, Hyderabad, Telangana
[2]Computer Engineering of Technology, Chaitanya Bharathi Institute of Technology, Hyderabad, Telangana

Abstract

This paper introduces a novel blockchain-based system called "Expiry Eye" to help solve this essential problem of managing the expiry of products while minimizing wastage, especially in the food industry. The existing systems are known to have challenges such as data entry problems, different date formats, and database security issues related to centralized databases. "Expiry Eye" is implemented based on the latest optical character recognition using the docTR library and recognizes the expiry date of products from any packaging and from using any format. Optical character recognition is performed by the system on image after undergoing some preprocessing steps such as converting it into grayscale, thresholding, noise reduction, resizing and image contrast enhancement. The expiry dates recognized, along with the cryptographic hash of the original image are stored on a blockchain securely to prevent tampering. The user data is kept encrypted in a database while images are saved off chain using IPFS to avoid expensive on chain storage. Any smart contracts implemented by the system sends out the expiry notifications to the concerned users with the help of a custom made mobile application created using Cordova platform. The results and discussion section describes the behavior of the proposed system in handling the expirations of products. Further, quantitative assessment of all relevant parameters along with relative analysis of the traditional product expiry management systems will be the part of future work.

1. Introduction

In today's society that is characterized by a fast pace an effective management of product expiration is valuable due to its convenience. With adverse effects on the environment and the consumers pocket, food wastage comes about due to consumers inability to comprehend or ignore expiry dates resulting to discard safe food. Other conventional practices of expiration tracking involve a paper-based approach,

Email: keerthana.birelli@gmail.com[1], sushmasiluveru2932@gmail.com[2], Sangeetagupta_cse@cbit.ac.in[2]

DOI: 10.1201/9781003661917-25

which requires manual data input and uses both 1D barcodes, which are virtually fragile [1, 2].

Several issues prevent smooth expiration date management. Both for consumers and businesses manual input of expiration data is time- consuming, prone to mistakes and may result in creating incongruities. This results in incorrect registers, skipped notifications or alerts, and possible hazards in handling products due to typographical errors, misunderstandings and different date formats [8, 4]. Current OCRs also cannot recognize dates written in types other than the regular print, such as dot matrix, engraved type that is often used in packaging of products [5, 7]. Data repositories typically employed in most tracking systems also expose the systems to data breach, data integrity issues, and unauthorized modification as well as scalability problems [3, 9].

This paper presents a system that is meant to tackle these challenges through using enhanced image recognition and blockchain for expiration management. The proposed system is immune to noisy or complex visual inputs because the docTR library effectively identifies expiration dates in several written forms; dot matrix form and engraved texts [10, 11]. To mitigate these security risks involved in the centralized database, expiration data is stored in the blockchain—a decentralized and tamper proof and also, personal data is protected. Instant messages remind users about the approaching expiration, so there is little waste in stock management.

Developed through the use of the Cordova framework, this particular system is highly extensible across the different mobile devices to enhance receives accessibility. It makes it easy for some people, by just viewing their calendar, they are able to know which of the products are expiring soon so that they can use them or dispose them as necessary.

The proposed system presents a secure, efficient and accessible solutions to current shortcoming of expiration tracking system. In the next sections, we describe further of the architecture, the methodology and the application of the system concluding that the system has high capability of promoting the food supply chain for sustainability and efficiency.

1.1 Importance of Blockchain

Blockchain is a key driver in proactively advancing the "Expiry Eye" project, as this technology is capable of offering the secure, decentralized control and the manipulable-proof nature of product expiry data. In traditional systems, the central database is often applied, it is weak in terms of cyber threats, people's errors, and the disclosures of information.

The project solves these problems by using the blockchain, a ledger that can never be changed or erased, where expiration dates, once registered, are unchanged. This creates a trust environment for users because the data's integrity is verifiable. The second benefit is the smart contract's functionality that can be utilized by the automation, for example, the notifications about the approaching expirations are sent automatically or actions such as flagging products near the expiration are run automatically. The solution does not need any human intervention thus, it achieves the maximum efficiency and is error-free. Blockchain grants the same functionality as it gives the ability of traceability to users audit entire lifecycle of expiry records adding the transparency without breaking the privacy, as the sensitive user data remains encrypted. Besides, it facilitates data decentralization so that no one point of control may become the main target for data manipulation or server failure attacks. Consequently, because of these services blockchain brings, a secure, scalable, and reliable infrastructure capable of supporting the project in this trust-efficient-sustainable expiry management revolution.

2. Related Work

Among all the methods there are a number of approaches investigated regarding the expiry date detection and recognition in the product packaging area (as shown in Table 25.1). Modern solutions can start with the optical character recognition that helps to extract textual information from images. These traditional methods typically follow a sequential process: pre-processing to improve image quality where the expiry date exist followed by segmentation of characters and an ROI that contains the date, character segmentation and then compared with standard template to check the correctness of the finding [1, 2, 7, 12].

Though such methods work fairly well under lab conditions, they fail in the real world. Smaller font sizes by virtue of low-quality lighting, non-standard printing mechanisms like dot matrix, or engraved text, and damaged packing materials degrades image quality thereby making traditional OCR ineffective. For instance, low contrast print of expiry dates on complex backgrounds and embossed on metallic substrates are considered challenging [5, 15]. In particular, these limitations indicate

Table 25.1 Comparison Table for IPEMS System

References	Summary	Limitations
[7]	• A smartphone app using Tesseract-OCR to recognize expiration dates on food packages. • It handles cropping, rotation, and color conversion.	• Struggles with image quality issues like brightness, contrast, and dot matrix fonts. • Pending server implementation and limited color evaluation
[11]	A deep learning solution that combines real and synthetic images to recognize expiry dates on food packaging, using TextBoxes++ for ROI detection and CRNN for character recognition.	• Limited by lack of large datasets and reliance on synthetic data. • OCR systems struggle with various expiry date fonts and formats.
[9]	• A dual deep neural network approach using FCN for ROI detection and CRNN for feature extraction and date recognition. • Fine-tuned with a food package image dataset.	• Struggles with distorted or reflective characters, dot matrix fonts, and was only evaluated for black and white characters. • Limited training dataset affects generalizability
[8]	A CNN model trained on a custom dataset for recognizing expiry date digits on product packaging, achieving 90% classification accuracy.	• Limited dataset diversity may hinder generalization. • Potential overfitting and slow inference time affect real-time use.
[2]	A CNN framework achieving 97.74% accuracy in recognizing expiration dates on product packages, using a modified FCOS architecture and a publicly available dataset (Exp Date).	• Challenges with reflections, background color changes, false positives, and irrelevant characters. • Dot-matrix fonts cause misclassification.
[5]	A machine learning approach for recognizing engraved expiry dates using a specialized CNN model (CNN-ED) with dataset augmentation via WGANs.	• High accuracy in specific contexts, but performance outside the engraved digits context is not extensively discussed. • Limited handling of various date formats.
[4]	Proposes an improved DBNet model with a CBAM module for text detection and a fully convolutional network for character recognition, including affine transformation to correct skewed text.	• Lacks exploration of generalizability across different food packaging types, fonts, and lighting conditions; does not address model performance on hardware beyond the Jetson Nano.

the necessity of the development of more powerful approaches in such often-volatile areas as retailing and food logistics which require a high level of accuracy in terms of products' expiry dates compliance and reduced losses [8, 9, 11].

There are rich promising studies related to machine learning, especially deep learning in expiry date detection. Currently, the convolutional neural network (CNN) has become very popular because of its ability to learn higher level features from a large amount of data. CNNs are helpful in enhancing the accuracy of an object detection and text extraction process; particularly with respect to the packaging design complications and different expiry date formatting. Some of the methods utilized today include the EAST (Efficient and Accurate Scene Text Detector) and the CRNN (Convolutional Recurrent Neural Network) while the type of framework used in this research work is known as TextDragon which has been upgraded and is known as PyTextOCR; has ushered in better text detection and recognition than the conventional OCR techniques [2, 11, 14].

Nevertheless, deep learning methods have their complications. For instance, reading expiry dates embedded in a printed text with dot matrix printers or embossed on metallic items can be a challenge where the contrast must be enhanced and the amounts of noise reduced. Moreover, images that may be acquired under the low or unfavorable light or environmental conditions may present substandard images that may reduce the effectiveness of these methods in real-world scenarios [4, 8, 13]. In general, the recognition of expiry dates may involve several preprocessing stages including noise removal, image enhancement whereby the text may be embossed or written on low contrast surfaces [8, 13].

The aforementioned challenges have also been handled using open-source OCR systems such as Tesseract, Keras OCR, EasyOCR, amongst others. For example, Tesseract has reported excellent results in activities of recognizing structured text images obtained in the form of clean and high-quality pictures [7]. But it is worse with non-standard expiry date formats, complex backgrounds and very low contrast texts makes it inefficient in actual scenarios [15]. Likewise, other open-source OCR tools provide both flexibility; however, they are not as impactful in the complicated expiry date recognition problems, let alone the versatile packaging designs [8, 11].

Other open source OCR software, which have also been used to address these challenges include Tesseract, Keras OCR and Easy OCR. For example, Tesseract achieved promising rates when it comes to reading structured text from good quality and clean images [7]. However, it fails in non-standard expiry date formats, complex backgrounds and low contrast text environment making them less suitable to use in real life settings. Likewise, likewise other flexible open-source OCR tools have been identified to offer low accuracy in complicated jobs of identifying the expiry-dates especially when the packaging differs [8, 11].

Within the context of supply chain product management, blockchain has attracted attention due to its potential for improving the conditions of data security and transparency. Blockchain decentralization and the fact that once data is written to the chain, it is rather difficult to change it, makes it easier to store sensitive product details like the shelf life [6]. For instance, blockchain systems have been used to associate the information to batches so that making improved tracing during outbreaks would be possible within perishable food supply chains. However, most of such application tends to be based on supply chain rather than personal product management systems creating a research space for exploring blockchain for expiry tracking at an individual level. There are in particular some stand out areas that dominate the application of blockchain in supply chain such as product tracing, batch linking and anti-fraud.

However, the usage of blockchain with other e2 techniques such as expiry date recognition have not been made entirely clear.

Implemented Blockchain might align well with the highly refined image recognition systems for solid data storage and retrieval in terms of offer security, solving problems like tampered expiry dates or fakes. Nevertheless, existing solutions on blockchain systems are more associated with product verification and traceability than expiry management in real-time [6].

One of the greatest issues that impede the design of appropriate expiry date detection solutions is the accessibility of high-quality training datasets. It is also clear that deep learning models need a wide range of datasets associated with different styles, formats of printing, and conditions of the surrounding environment. However, the datasets such as Unitail-OCR are generally small to incorporate highly robust models and they tend to overfit and are typically seen to be not generalizable in real-world scenarios [9, 12, 14]. The scarcity of accessible dataset, especially for types of expiry date formats that are in dot matrix or embossed form is a challenging factor that denies the emergence of better and versatile algorithm updates [2, 9]. Further, the last few years' improvements in the technology, for instance, incorporating deep learning for OCR based techniques were helpful in extracting name of products and their expiry dates more effectively from images [10].

We overcome these limitations as follows: It is further noted that larger and more diverse training sets are required to obtain better models' performance. More pre-processing measure like filter, smoothing and contrast can be used to improve image quality before feeding into the recognition models. Moreover, the ability of blockchain to store large volumes of data safely can be utilized to develop a reliable system of recording the expiry data especially in supply system chain environments more so under dynamic conditions [6].

Therefore, despite the numerous advances in expiry date detection and recognition, this study finds that some issues still prevent the successful implementation of this solution in real-life scenarios. Applying combined technologies such as image recognition and blockchain for storing data

safely is a good avenue fit for supply chain management however not widely explored regarding personal product expiry management systems [5, 6, 8, 11].

3. Methodology

3.1 System Overview

With the goal of exploring ways to create a blockchain solution for expiry date management, the proposed system, "Expiry Eye," is based on image recognition. The system is structured with three main components: Blockchain System, Mobile Application Interface, and the Image Recognition System (as shown in Figure 25.1). All of the components are unique in their contribution to data accuracy, security and the ease of convenience it could offer to the users, which translates to a complete framework for reliable expiry management.

3.2 Image Recognition

Initially, text detection and recognition from product images are done using the docTR library. The main reason this library works so well is that it considers text under a variety of situations, including rotation, blurring, reflections, even engraved text. The proposed system also interprets recognized

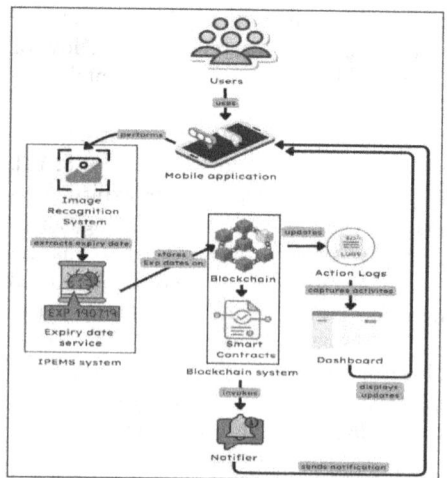

Figure 25.1 Blockchain Based IPEMS

text as structured expiry dates. Under this system, we will handle different date formats including expressions like "use before" and "best by". In addition, a comprehensive analysis of the docTR library is required to select the best combination of models that will accurately and reliably recognize diverse imaging conditions.

3.3 Cost-Efficient Blockchain Integration

To manage data security and maximize costs savings, 'Expiry Eye' will utilize a hybrid storage solution, combining the blockchain's immutability and IPFS's efficiency. The actual images will be stored in IPFS, a decentralized storage network, reducing storage costs. The system will record a hash of each product's image on the blockchain. This hash serves as a unique identifier for the image. User login credentials, non-critical product details are stored off chain. User data privacy will be protected by basic encryption or encoding mechanisms in adhering to security standards.

3.4 User Interface

The mobile application interface will be built using the Cordova framework so it will give us the cross-platform compatibility to make the user experience with the system simple and easy. The product images can be captured through the application, expiry dates can be viewed, and users can be notified if products are going bad. This will ensure that the app can be used on other device platforms with its use and reach being widened using Cordova's cross platform application abilities.

3.5 Notifier System and Action Logs

A notifier system will be added to the mobile application for issuing alerts to the users once products within the application approach their expiry date. Smart contracts will facilitate this process by sending notifications at the set conditions for each product's expiration date. Also, there will be an action logging system to track and store activities of users so that the dashboard can be updated in real time. With this, users will be able to obtain collective values of product statuses and system related activities at a glance and facilitate engagement as well as traceability.

4. Results and Discussion

This section presents the projected impact of the blockchain-based intelligent product expiry management system and an evaluation of its applicability in overcoming existing constraints. As for the evaluation criteria (as shown in Table 25.2), data security, automation rate, scalability, ease of use, and cost-effectiveness will be used with significant improvements compared to traditional approaches expected.

The system seeks to achieve this by identifying expiry dates from the various images of products despite the limitations like formats of different dates where existing solutions based on OCR fail to provide useful information, image quality. Blockchain hence opens the possibility for solutions that protect data from adaptations, a problem with centralized systems. With this respect, smart contracts would help in automating notifications so that people are not left with products that will soon expire. Moreover, the integration of the Cordova framework provides compatibility across devices, which is more effective than provided by operating system-specific systems.

Qualitative performance indicators will demonstrate the benefits of the proposed system compared to existing approaches and the state-of-the-art in terms of closing the research gaps, the use of high-performance OCR, blockchain, and user-centered design. Among the expected benefits of this kind of approach is the goal of effectively improving food safety, minimizing waste, as well as increasing supply chain performance.

5. Conclusion

This paper presents a new system called "Expiry Eye" which is aimed to change the world of product expiry management that

Table 25.2 Key Metric Evaluation of IPEMS

Feature	Traditional Product Expiry Management Systems	Expiry Eye - Blockchain-Based Intelligent Product Expiry Management System
Data Tamper-Resistance	Limited	High
Automation	Requires external software	Smart contracts
Security	Vulnerable (Each external tool introduces additional vulnerabilities, such as misconfigured APIs, deprecated dependencies)	Cryptographically secure
Scalability	Limited	High with off-chain solutions
Consumer Trust	Low	High
Balance between transparency and privacy	No transparency (reduced trust), enough privacy	Ensuring privacy while retaining transparency for product data (Expiry data can be stored publicly on-chain, while user-related metadata is stored off-chain with encryption)
Decentralization	No (single point of failure)	Yes (no single point of failure)
Cost-Efficiency	Requires continuous maintenance of external tools	Lower long-term costs with smart contracts
Data consistency	Low	High

is fundamental in preventing food wastage and encouraging consumers to be wise when they are consuming food. Traditional approaches encounter issues to do with precision, security, and customer participation, which results in waste of food and productivity loss. 'Expiry Eye' aims to solve these challenges.

With the use of docTR library, "Expiry Eye" will read and system can identify different types and formats of expiry dates even including dot matrix and engraved texts. This makes sure that additional information gathering is extensive and accurate, and since the information is stored in the blockchain technology, users have surety and transparency of data stored. Information notifications are executed by smart contracts, allowing users to control their inventory in a more critical manner and reducing the chances of forgetting products that will expire soon.

Cordova framework is used to create the convenient mobile application for users to make "Expiry Eye" be available for all people with different gadgets. This arrangement

of security, accessibility, and data authenticity offers a solution for changing consumer's approach to end of products life, supporting a better food consumption economy. Future research will concentrate on the development's privacy mechanism, system performance, and cooperation with retails for better response to food waste minimization and encouraging consumers to adopt appropriate consumption patterns.

References

[1] A. Sri Chaitanya, M. T. (2024). Reminding Food Expiry Dates and Sending Timely Alerts. 5.

[2] Ahmet Cagatay Seker, S. C. (2022). A generalized framework for recognition of expiration dates on product packages using fully convolutional networks.

[3] Anushree Tandon, A. D. (2020). Blockchain in Healthcare: A systematic literature review, synthesizing framework and future research agenda.

[4] Jishi Zheng, J. L. (2023). Recognition of expiry data on food packages based on improved DBNet.

[5] Jung, A. A. (2023). Effective Digital Technology Enabling Automatic Recognition of Special-Type Marking of Expiry Date. 22.

[6] Kayikci, Y. D. (2021). Using blockchain technology to drive operational excellence in perishable food supply chains during outbreaks.

[7] Kento Hosozawa, R. H. (2018). Recognition of Expiration Dates Written on Food Packages with Open Source OCR. 5.

[8] Khan, T. (2021). Expiry Date Digit Recognition using Convolutional Neural Network. 4.

[9] Liyun Gong, M. T. (2020). A novel unified deep neural networks methodology for use by date recognition in retail food package image. 9.

[10] Mr. Reddi Prasadu, K. S. (2024). Extract Product Name from Image and Track Expiry. 5.

[11] Traian Rebedea, V. F. (2020). Expiry date recognition using deep neural networks. 20.

[12] Kai Chen, J. W. (2019). MMDetection: Open MMLab Detection Toolbox and Benchmark.

[13] Karthika M Shanthini, P. C. (n.d.). Recommendation of Product Value by Extracting Expiry Date using Deep Neural Network. 2017.

[14] Minghui Liao, B. S. (2017). TextBoxes: A Fast Text Detector with a Single Deep Neural Network.

[15] Mohamed Lotfy, G. S. (n.d.). CNN-optimized text recognition with binary embeddings for Arabic expiry date recognition. 2021.

26 ML-Based Time Series Forecasting for Microgrid Power Generation

Balasubbareddy Mallala[1,3], Maloth Anusha[1], Sastry V. Pamidi[2], Md Omar Faruqueb[2], and Rajasekhar Reddy M.[3]

[1]Professor, Chaitanya Bharathi Institute of Technology, Hyderabad, India
[2]Centre for Advansed Power Systems, Florida State University, Tallahassee, USA
[3]Amrita School of Computing, Amrita Vishwa Vidyapeetham, Amaravati Campus, Andhra Pradesh, India

Abstract

The real time available data is taken from the GitHub for time series forecasting method is employed to forecast upcoming values. It is important in several sectors, including weather forecasting, finance, unemployment, economics, manufacturing, and supply chain management. This method analyses time-ordered data to determine patterns, trends, and seasonal variations. Effective forecasting leverages a range of statistical and machine learning techniques, including methods like the autoregressive integrated moving average (ARIMA) and advanced models such as long short-term memory (LSTM) networks and Prophet, to increase accuracy. The selection of the right model for forecasting or data analysis depends on features such as linearity, seasonality, and noise/anomalies. The evaluation of these models typically involves metrics such as RMSE=1.75 and MSE=3.06. Accurate time series forecasting is crucial in improving decision-making by offering meaningful insights into future trends and potential challenges, supported by data preprocessing algorithms implemented in Python.

Keywords: LSTM, ARIMA, SARIMA, prophet, forecasting

1. Introduction

Time series forecasting for microgrid power generation is an essential aspect of modern energy management systems. A microgrid is a local power system that uses sources such as Wind and Solar to generate and supply electricity to satisfy loads. It can provide supply to the load independently or connect to a main power grid when the demand is high to ensure a reliable power supply. Accurate forecasting in this context involves predicting future power generation on the basis of historical data, which is crucial for optimizing energy production, storage, and consumption. Effective time series forecasting for microgrid power generation facilitates better demand-response strategies, efficient energy storage management, and improved integration of renewable energy sources.

Email: balasubbareddy79@gmail.com[1], anushamaloth18@gmail.com[1], pamidi@eng.famu.fsu.edu[2], mfaruque@eng.famu.fsu.edu[2], rajasekharmanyam04@gmail.com[3]

DOI: 10.1201/9781003661917-26

2. Literature Review

This ensures optimal microgrid operation, reduces dependence on the main grid, and enhances energy security and sustainability [1]. Time series forecasting is performed via a series of data over time at regular intervals, such as yearly, monthly, and daily. It includes examples such as hourly temperature readings, daily load demand changes, monthly electricity consumption, and annual power plant production [2]. In time series analysis, time is treated as the independent variable, whereas few dependent variables can exist. Within a time series analysis, the aim is to determine how various dependent variables respond to changes in the independent variable over time and to forecast its future values [3]. This process starts with the collection and preprocessing of historical power generation data, ensuring that the data are clean and properly formatted for analysis that is conducted to visualize the data and identify trends, seasonal patterns, and anomalies, which helps in understanding the variability and cyclical nature of power generation [4]. Time series decomposition is employed to identify and separate trends, seasonality, and residuals within the data. The use of ARIMA (autoregressive integrated moving average) or LSTM networks is among the more advanced machine learning algorithms available [5]. Unlike traditional methods over modeling long-term dependencies. This is achieved through input and forget gates, allowing the model to selectively retain or discard information as needed, which makes it particularly effective for sequence prediction problems [6].

This study analyzes various machine learning methods and data preprocessing algorithms to achieve automatic real-time control over seasonal variations. It achieved R-square accuracy over 0.95 in real-time control applications when limited historical data were used [7]. A comparison of advanced ML algorithms such as ANN-MLP, LSTM, and ID-CNN for forecasting Austria's electrical load demand revealed that MLP is best for single timestamp ahead prediction, and ID-CNN is best for multiple timestamp ahead prediction, with MAPE values of 4.62% for short-term forecasts and 1.45% for midterm forecasts [8]. Forecasts of photovoltaic (PV) power are based on a hybrid model that incorporates weather variations. Day-ahead predictions are generated using principal component analysis (PCA) [9]. Forecasting time series data is essential in economics, business, and finance. Traditional methods such as ARIMA have been effective, but in this study, LSTM outperforms traditional methods, with an average error rate reduction of 84–87%, regardless of the number of training epochs [10]. Implementing DC microgrids addresses various challenges in the power grid with renewable energy systems, using machine learning with simulated annealing optimization and the IoT for energy management, maximizing power utilization, improving grid stability, and ensuring reliable and efficient energy distribution from renewable sources [11]. Accurate cardiac arrest risk forecasting (ACARF) with ensemble learning significantly improves the prediction accuracy of CA risk prediction by combining multiple algorithms and utilizing an extensive feature set, including clinical information, medical history, medications, and lifestyle choices, to support personalized patient care [12]. Machine learning techniques enhance electric vehicle battery management by comparing decision tree classifiers, artificial neural networks (ANN), and naive Bayes classifiers to predict efficient charging and discharging strategies, battery performance, and abnormality detection. The findings show precision, recall, and F1-scores exceeding 98%, with decision trees and naive Bayes classifiers achieving over 90% accuracy, providing significant insights for improving battery performance and reducing operational costs [13]. Additionally, machine learning models, particularly the VGG 16 and VGG 19 enhance deep learning performance through their depth and structured design, significantly enhance the detection and classification of brain tumors [14].

3. Methodology

The following section addresses data collection, preprocessing, matrix examination, and preparation.

3.1 Data Preparation

Gathering historical power generation data is essential for training forecasting models. The primary steps in data preprocessing involve handling missing values, normalizing the data, and performing feature engineering.

High-quality data enhance the accuracy of forecasting models and improve overall performance. The dataset is used for univariate time series analysis, i.e., real power focusing solely on the future load demands. This dataset spans 2015–2020, encompassing 176562 rows and one column including real power.

The process consists of a two-layered architecture. Each layer consists of three steps. Steps (1–3) focus on data preparation and preprocessing in the data layer, whereas the model layer begins with the selection of model parameters (step 4). The selection process can be conducted either automatically or manually, depending on tasks related to constructing the model. Finally, the last step (step 6) involves model validation, where predictions are obtained and compared with actual values to assess the model's quality via predefined metrics. If the model meets the specified metrics, it is saved for generating forecasts. In the case of unsatisfactory metrics, the process returns to step 4, followed by subsequent steps in the data layer for further iterations. Additionally, it outlines the significance of methods in (step 5), such as Holt-Winters Exponential Smoothing, ETS, and XGBOOST, that can be used for predicting future trends and seasonal patterns.

3.2 Data Collection

The source of the dataset used to forecast real power is taken from GitHub. Data collection involves gathering historical real power generation data from the microgrid, along with relevant parameters such as weather conditions and load demand.

3.3 Data Preprocessing

Preprocessing involves the preparation of raw data to make them suitable for a machine learning model. This includes cleaning or removing missing values to enhance performance. Ensuring that the data are clean, accurate, and representative of the underlying patterns in the real power generation time series is crucial before proceeding to the modeling stage. This helps facilitate more accurate and reliable forecasts for effective microgrid management. Most methods utilize standard data formats such as YYYY-MM-DD HH:MM:SS, where Y denotes the year, M the month, and D the day.

3.4 Evaluation Metrics

The accuracy of the model was assessed using two statistical measures: Root Mean Square Error (RMSE) and Mean Absolute Error (MAE). RMSE is determined by taking the square root of the Mean Squared Error (MSE), whereas MAE represents the average discrepancy between the observed and predicted values. Unlike MAE, RMSE gives more weight to outliers in the dataset. The formulas for these metrics are provided below:

$$\text{RMSE} = \sqrt{\frac{1}{m} \sum_{i=1}^{m} |\hat{y} - y|^2}$$

$$\text{MAE} = \frac{1}{m} \sum_{i=1}^{m} |\hat{y} - y|$$

Here, y and $\hat{y}$ represent the actual and predicted values, respectively, while mm denotes the number of samples.

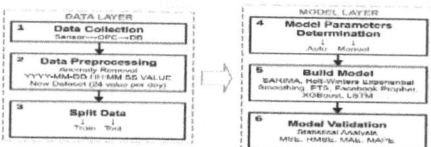

Figure 26.1 Technical Road Map of Data Analysis

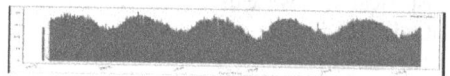

Figure 26.2 Plot of the Real Power of the Entire Dataset

Table 26.1 Segmentation of the Entire Dataset

ID	Date	Time	Real Power
0	2020-02-29	23:45:00	0.066
1	2020-02-29	23:30:00	0.067
2	2020-02-29	23:15:00	0.068
3	2020-02-29	23:00:00	0.067
....	....	....	......
176558	2015-01-21	09:30:00	8.729
176559	2015-01-21	09:15:00	7.303
176560	2015-01-21	09:00:00	5.712
176561	2015-01-21	08:45:00	5.273

3.5 Monthly Data Distribution

Monthly data distribution is a way to analyze and understand how a particular variable changes over time every month. In the context of available data with date time and real power, it analyzes how the real power values change from month to month. By visualizing these data, we can identify any patterns, trends, or irregularities in the real power values over the months.

Table 26.2 Monthly Timestamp of Dataset

Date Time	Real Power
2015-01-01	5.850194
2015-02-01	7.434686
2015-03-01	9.185402
2015-04-01	10.437035
2015-05-01	9.163034

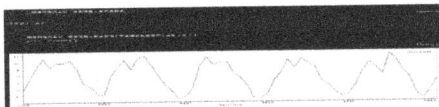

Figure 26.3 Plot of Real Power for Monthly Data

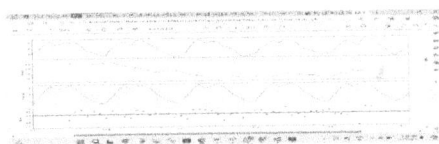

Figure 26.4 Plot of Trend, Seasonality, and Residual

3.6 Trend

For instance, it might reveal an upward trend in sales data over multiple years, driven by business growth.

3.7 Seasonality

The seasonal component (St) captures the repeating short-term cycle in the data. This periodicity can occur on a daily, monthly, or yearly basis. An example of seasonality is the annual rise in retail sales during the holiday season.

3.8 Noise

These irregularities or randomness occur due to unforeseen factors or anomalies.

3.9 Daily Data Distribution

The dataset captures the daily distribution of real power values over a specified time. It allows analysis of variations in real power consumption or production on a day-to-day basis. This dataset can be used to identify trends or anomalies in the daily distribution of real power.

Table 26.3 Daily Timestamp of the Dataset

Date Stamp	Real Power
2015-01-21	7.964443
2015-01-22	5.847854
2015-01-23	6.798042
2015-01-24	6.790250
2015-01-25	5.213781

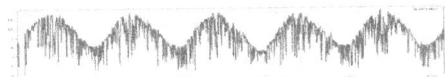

Figure 26.5 Plot of Real Power for Daily Data

4. Development of the Model

The development of the model is described in different steps as follows.

4.1 Parameter Selection

Maximize a selected performance metric. The parameters are defined as follows:

Selecting the appropriate parameters for an ARIMA model involves following various guidelines and best practices. This process can be complex and time-consuming, requiring domain expertise. In this section, we address this challenge by using Python to systematically identify the values. By exploring the entire parameter space, we aim to identify the parameters that yields the best results based on our chosen evaluation criteria.

For example, one of the best parameter combinations for a seasonal ARIMA model might be ARIMA (1, 1, 1) (0, 0, 1, 12).

The summary attributes generated by the SARIMAX output provide valuable insights. Our main focus will be in the table.

lines represent the actual values, while the orange line depicts the forecasts. By setting the argument "dynamic=False," we ensure that the forecasts are one-step ahead, meaning each forecast is generated using the complete history up to that point. After analysis, we found that the predicted values closely match the actual values, with minimal errors indicated by the gray lines.

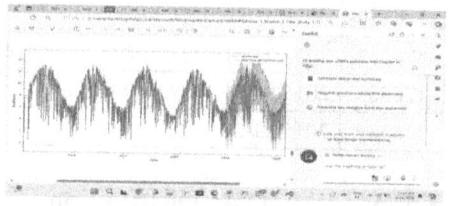

Figure 26.7 Comparison of Actual Data and Predicted Data

Variables	coef	Std. Error	z	P > |z|	0.025	0.975
ar. L1	0.4284	0.021	19.981	0	0.386	0.470
ma. L1	−0.09166	0.011	−82.262	0	−0.938	−0.895
ar.S. L12	0.0223	0.021	1.067	0.286	−0.019	0.063
sigma2	3.0185	0.074	40.749	0	2.873	3.164

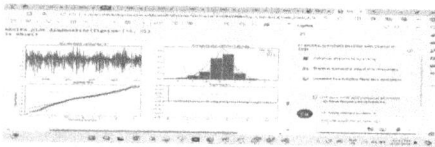

Figure 26.6 Plot of the Histogram and Correlogram

In this instance, all the weights have p-values either below or close to 0.05, which justifies retaining all of them in our model.

5. Results and Analysis

We have developed a model for our time series that is now ready for forecasting. To evaluate the accuracy of these forecasts, we compare the predicted values to the actual values of the time series. The methods "get_prediction()" and "conf_int()" allow us to obtain the forecasted values along with their corresponding confidence intervals. In the accompanying plot, the blue

The illustration provides a graphical comparison of the actual and predicted values.

6. Conclusions

Accurate prediction of microgrid power generation is essential for maintaining power quality and ensuring a stable supply to the grid. This study employed two ML models to forecast microgrid power generation. After a detailed comparison, it was ARIMA model showed significant improvement in handling the high seasonality and large volume of the dataset, making it more suitable for the prediction task at hand.

The ARIMA(1, 1, 1)(0, 0, 1, 12)12 model offers a reliable forecasting method for microgrid power generation. This highlights the crucial role of thorough data preparation and thoughtful model selection in time series analysis. It is advisable to further refine the model and conduct real-time testing to improve its predictive ability.

References

[1] Isabella Zuleta-Elles, Aiskel Bautista-Lopez, Milton J. Catano-Valderrama, Luis G. Mar˜ ín, Guillermo Jimenez-Est´evez, Patricio Mendoza-Araya, Load Forecasting for Different Prediction Horizons using ANN and ARIMA models.

[2] Zu, M. M., Sone, S. P., Lehtomäki, J., & Khan, Z. (2023). Evaluation of prophet for wireless time series forecasting. *2023 31st Telecommunications Forum (TELFOR).* Belgrade, Serbia, pp. 1–4. doi: 10.1109/TELFOR59449.2023.10372817.

[3] Gupta, R., Yadav, A. K., Jha, S., & Pathak, P. K. (2022). Time series forecasting of solar power generation using facebook prophet and XG boost. *2022 IEEE Delhi Section Conference (DELCON).* New Delhi, India, pp. 1–5. doi: 10.1109/DELCON54057.2022.9752916

[4] Sudhakar, R. P., et al. (2023). Temporal patterns and seasonal effects in wind power forecasting: A time series analysis. *2023 2nd International Conference on Futuristic Technologies (INCOFT),* 1–4.

[5] Basak, S., Kar, S., Saha, S., Khaidem, L., & Dey, S. R. (2019). Predicting the direction of stock market prices via tree-based classifiers. *The North American Journal of Economics and Finance,* 47, 552–567.

[6] Raghavendra, K., et al. (2021). Analysis of financial time series forecasting using deep learning model. 2021 11th International Conference on Cloud Computing. *Data Science & Engineering (Confluence),* 877–881.

[7] Kramar, V., & Alchakov, V. (2023). Time-series forecasting of seasonal data using machine learning methods. *Algorithms,* 16, 248. https://doi.org/10.3390/a16050248.

[8] Harish, B., Panda, D., Konda, K. R., & Soni, A. (2023). A comparative study of forecasting problems on electrical load timeseries data using deep learning techniques. *2023 IEEE/IAS 59th Industrial and Commercial Power Systems Technical Conference (I&CPS),* Las Vegas, NV, USA, pp. 1–5. doi: 10.1109/ICPS57144.2023.10142125.

[9] Malvoni, M., De Giorgi, M. G., & Congedo, P. M. (2016). Data on Support Vector Machines (SVM) model to forecast photovoltaic power. *Data in Brief,* 9, 13–16. https://doi.org/10.1016/j.dib.2016.08.024.

[10] Siami-Namini, S., Tavakoli, N., & Siami Namin, A. (2018). A comparison of ARIMA and LSTM in forecasting time series. In *Proceedings—17th IEEE International Conference on Machine Learning and Applications,* ICMLA, pp. 1394–1401.

[11] Senthilkumar, G., Mallala, B., Sivarajan, S., Harish, C., Harsha, D., & Natrayan, L. (2024). Maximizing Power Utilization through Machine Learning and IoT based Power Flow Strategies in DC Micro Grids with Renewable Energy Resources. *2024 International Conference on Inventive Computation Technologies (ICICT),* Lalitpur, Nepal, pp. 1166–1171. doi: 10.1109/ICICT60155.2024.10544791.

[12] Mishra, S., Vijayalakshmi, V. J., Palle, K., Prasad, P. V., Mallala, B., & Pund, S. (2023). Accurate cardiac arrest risk forecasting with ensemble learning. *2023 IEEE International Conference on Paradigm Shift in Information Technologies with Innovative Applications in Global Scenario (ICPSITI-AGS),* Indore, India, pp. 8–14. doi: 10.1109/ICPSITIAGS59213.2023.10527454.

[13] Bennehalli, B., Singh, L., Silas Stephen, D., Venkata Prasad, P., Mallala, B., & Chandra Rao, A. P. (2024). Machine learning approach to battery management and balancing techniques for enhancing electric vehicle battery performance. *Journal of Electrical Systems,* 20(2).

[14] Nagaraju, G., Kumar Nath, R., Chinniah, P., Balasubramanian, K., Kirubakaran, S., & Mallala, B. (2024). A comparative analysis of advanced machine learning techniques for enhancing brain tumor detection. *Journal of Electrical Systems,* 20(2).

[15] Joshua, K. P., Ranga, J., Prasad, P. V., Mallala, B., Rajendiran, M., & Maranan, R. (2024). Optimized scheduling of electric vehicles charging in smart grid using deep learning. *2024 International Conference on Expert Clouds and Applications (ICOECA),* Bengaluru, India, pp. 408–412. doi: 10.1109/ICOECA62351.2024.00079

[16] Vivek, B., Teja, B. H., Mallala, B., & Srinitha, G. (2024). Electrical fault detection and localization using machine learning. *2024 International Conference on Expert Clouds and Applications (ICOECA),* Bengaluru, India, pp. 820–823. doi: 10.1109/ICOECA62351.2024.00145

[17] Mallala, B., Azeez Khan, P., Pattepu, B., & Eega, P. R. (2024). Integrated energy management and load forecasting using machine learning. *2024 2nd International Conference on Sustainable Computing and Smart Systems (ICSCSS),* Coimbatore, India, pp. 1004–1009. doi: 10.1109/ICSCSS60660.2024.10625623.

27 A Decentralised Blockchain Framework for Enhanced Crowdfunding Security

Owais Siddiqui[1], Karen R. Johnson[1] and Sangeeta Gupta[2]

[1]Department of Computer Engineering and Technology, Chaitanya Bharathi Institute of Technology, Hyderabad, Telangana
[2]Professor and Head, Department of Computer Engineering and Technology, Chaitanya Bharathi Institute of Technology, Hyderabad, Telangana

Abstract

Crowdfunding has grown significantly, emphasizing the importance of securing contributions and building stakeholder confidence. This study proposes a decentralized blockchain-based framework to increase the security, transparency, and accountability of crowdfunding. Blockchain technology's immutable and transparent ledger offers a unique option for reducing fraud and increasing trust by decentralising control and simplifying decision-making. In this case, fund distribution and milestone monitoring are handled by smart contracts and an AI powered fraud detection module is continuously monitoring campaign activities for any suspicious or anomalous behavior. Additionally, an insurance mechanism protects contributors from project failures and IPFS (InterPlanetary File System) for off chain data storage for scalability and security. Taken together, these technologies offer a comprehensive solution which enhances the platform security, accountability and user's trust. This paper also includes a literature review of existing systems, a critical analysis of current challenges, and a methodology to build a resilient, efficient and user centric crowdfunding platform.

1. Introduction

Nowadays, crowdfunding is a common way to raise money to establish a business or to solve social problems without the help of traditional financiers. However, the democratised funding approach enables funding of innovative projects and social causes, but it has also enabled fraud, which erodes user trust [1, 3]. The major drawbacks of the centralized crowdfunding platforms are lack of transparency, high service charges, low efficiency in fraud detection, which leads to low user trust

and involvement [8, 9]. Recent research indicates that centralized systems are prone to single points of failure, high operational costs, and poor fraud protection, and therefore require a decentralized and blockchain based platform [10].

This paper introduces a decentralized crowdfunding model that merges blockchain, smart contracts, and an AI based fraud detection system to enhance transparency to address these challenges. Blockchain technology provides a secure, distributed ledger for recording the information, enhancing data

Email: owais4080@gmail.com[1], kayren4848@gmail.com[1], sangeetagupta_cse@cbit.ac.in[2]

DOI: 10.1201/9781003661917-27

integrity and offering real time transparency in transactions [6]. These smart contracts automate these functions (i.e., fund release, milestone tracking etc.), expunging the need for such intermediaries and cutting down costs [9]. Furthermore, an AI based fraud detection module watches over campaign data continuously, learning from new threats as time passes [2]. From literature review, we find that blockchain offers a huge improvement in transparency, security and user confidence in crowdfunding which makes it a perfect solution to these problems [4, 7].

Furthermore, blockchain can empower decentralized audit mechanisms to substitute for third party auditors and hence promote trust and security [7]. The proposed framework also proposes a model of an insurance embedded in the blockchain, which protects contributors' funds from different risks [2, 3]. The core challenge of conventional crowdfunding is addressed by this framework to create a more secure, cost effective and trustworthy environment for contributors and project initiators to change the state of the art of crowdfunding. The following sections of this paper explore various aspects of the proposed framework. The Literature Survey provides an in-depth review of relevant research on blockchain-based crowdfunding models, assessing methodologies, practical applications, and identifying research gaps. Section 3 details the structural components of the proposed framework, explaining the integration of blockchain, smart contracts, AI, and IPFS.

The Discussion highlights how each feature in this decentralized model addresses specific security, transparency, and accountability issues within traditional crowdfunding. Finally, the Conclusion and Future Scope summarizes the framework's contributions and outlines potential areas for future research, aiming to enhance security and user confidence in the crowdfunding ecosystem.

2. Literature Review

The literature on the utilization of blockchain for crowdfunding shows a number of advantages and obstacles for the application of decentralized technology to improve transparency, security and trust in crowdfunding platforms. Recently, most of the studies have been devoted to utilizing blockchain and smart contracts to automate the transactions and make record tamper-proof, and also studying the advanced fraud detection methods and participant incentives. This literature survey carried out is a synthesis of key papers on various models and frameworks that seek to address the shortcomings of traditional centralized crowdfunding systems and the most recent advancements in blockchain driven crowdfunding innovations.

It is discussed how these technologies can be integrated with crowdfunding and blockchain within Industry 4.0 to improve transparency, efficiency, and availability of financial services. It makes the case for the use of blockchain in the financial sector to generate decentralized and transparent solutions, consistent with the aim of building trust with crowdfunding platforms through qualitative research methods. This framework is applicable to building a secure, accessible and efficient crowdfunding ecosystem [6].

In another study, the role of blockchain in decision making in crowdfunding for marine ranching is explored. The paper compares traditional and blockchain enabled crowdfunding models, and demonstrates how blockchain can reduce information asymmetry, build stakeholder trust, and improve the decision process of funding. This research directly applies to the objective of creating a reliable, transparent crowdfunding platform [4].

Another study focuses on fraud prevention in crowdfunding and proposes a decentralized transaction system. This research relies on Ethereum blockchain and smart contracts to ensure that funds are spent responsibly and with investor approval required. The approach is consistent with the AI based fraud detection module proposed in your project, wherein contributors will be protected through transparent fund management with solid fraud detection systems [8].

In a paper about digital donation crowdfunding, it explores the ways blockchain can increase transparency and accountability. It uses a Futures Research approach to identify success factors and the influence of standardized laws, improved user experience, and cross regional opportunities for donor confidence. This study provides a solid foundation for designing a user centric, secure crowdfunding platform that can cater to both regional as well as global users [10].

3. Design and Operational Framework for Decentralised Crowdfunding

Finance has been revolutionized by crowdfunding platforms, which have allowed individuals and organisations to raise funds away from traditional financial intermediaries. The democratized model for financing innovation, community building and entrepreneurship avoids the barriers of conventional finance. Yet, centralized

Table 27.1 Overview of Recent Blockchain-Based Crowdfunding Implementations

Ref.	Insights	Methodology	Practical Implications	Research Gap
[5]	Examines the potential of blockchain for secure, transparent spatial data crowdsourcing and discusses privacy and motivation for participants.	It is a blockchain based structure in which requesters submit reports and workers confirm them. The accuracy is incentivized with a reward mechanism; each report requires two confirmations.	It improves spatial data accuracy by 40% and reduces review time by 30%, helping urban management and emergency response.	It points out gaps in evaluation mechanisms, contract clarity, and privacy versus data accuracy in crowdsourcing systems.
[9]	It highlights how blockchain can be used to make crowdfunding more secure, efficient and fraud resistant.	It proposes a system based on Ethereum with smart contracts which can automate and secure the transaction, including campaign management contracts.	It enhances security, transparency and efficiency, and makes the opportunity available for wider access to potential investors.	It identifies challenges with adoption because of the lack of understanding of blockchain, and suggests further research and education.
[12]	The role of blockchain in increasing the security and transparency of crowdfunding is investigated, and smart contract scalability and performance challenges are discussed.	It performs a literature review on blockchain, smart contracts and crowdfunding systems and security issues, and suggests solutions.	It gives insights for organizations on how to integrate smart contract, and how to manage security vulnerability, and for policymakers on the potential of blockchain in reducing fraud.	It highlights gaps in smart contract transparency and highlights that many contracts do not have accessible source codes. More research on scalability and performance.
[11]	FundDapp is an introduction of a decentralized crowdfunding application for charity, which allows peer to peer transfers via Ethereum Blockchain.	As Object-Oriented Analysis and Design (OOAD), it has stages such as requirement analysis, design, implementation and testing. Enables smart contracts to integrate in order for secure and auditable transactions to take place.	Security and transparency of donations are improved, features such as transaction records and cryptocurrency donations are added. But it is limited by the requirement of Metamask and doesn't have project search functionality.	However, it's also flawed: in the user interface (UI) and user experience (UX), which requires search functionality and looks better. Further research to improve user centered design in Blockchain applications.

Ref.	Insights	Methodology	Practical Implications	Research Gap
[3]	BitFund is a decentralized platform that connects investors and developers using blockchain to promote transparency and low cost project funding.	It creates a decentralized network where investors post project requirements, and developers bid on the projects. Refinement of bids is the work of smart contracts with bids reaching optimum solutions.	It supports smart city development, a reliable crowdfunding approach, opportunities for emerging economies and efficient use of local technical talent.	Shows how intermediaries prevent talent from being connected to investors. Focuses on the need to learn more about investor and developer behavior in decentralized crowdfunding platforms.
[13]	Explores blockchain potential beyond cryptocurrency, in the areas of fraud prevention and transparency in crowdfunding.	It develops a prototype on the Rinkeby test network using Ethereum smart contracts to automate funding conditions and include contributors in fund allocation.	It strengthens fraud prevention and transparency, so that crowdfunding opportunities are global and trust is built in the process.	While showcasing shortcomings of traditional crowdfunding including fraud and regulatory gaps, the research also notes a dearth of studies about how to use blockchain for crowdfunding, which points to a call for technical study.

crowdfunding models are fraught with problems, especially a lack of transparency. But often, contributors aren't aware of how their funds are spent, which breeds skepticism and distrust. Additionally, the centralized control over transactions, funds and campaign management adds a single point of failure upon which any platform can be exploited, misused and fun misappropriated even with a chance of hacking. Secondly, high service fees will reduce the actual funds reaching the beneficiaries, so contributions mean less.

Blockchain tech provides an enticing solution to these challenges as it operates alongside immutable and transparent records of all transactions without intermediaries and lower administrative costs. Smart contracts are used in Blockchain based platforms for automatic distribution of fund, approval of milestone and voting of contributors which increases the security, accountability and cost effectiveness of the system. This reduces the room for human error, and releases money as soon as project milestones are completed, once

conditions are set in a smart contract. This level of automation allows contributors to control how fund distribution takes place, and project creators to be held to their stated goals. By utilizing a distributed ledger, all contributions and campaign activities are traceable and publicly verifiable, and users gain trust.

But blockchain on its own does not eliminate the risk of fraudulent campaigns. To overcome this, the proposed framework joins AI powered fraud detection module with blockchain, so that the activities in a campaign can be monitored proactively for detecting suspicious pattern. The module will begin with a rule based system to identify obvious fraud signals, such as unusually large contributions or spikes, and evolve through machine learning to learn the more subtle, complex fraud patterns over time. The framework also includes a blockchain based insurance model that protects contributors from losses in fraudulent or failed campaigns. This model uses smart contracts to manage an insurance pool, funded by small contributions from insured campaigns, and pays contributors in the event of a project failure or fraud detection. Metadata of the campaign is off loaded offchain to the blockchain to provide data transparency and integrity without overloading the blockchain. By integrating blockchain, smart contracts, AI driven fraud detection and decentralized insurance, the proposed framework creates a secure, transparent, accountable crowdfunding solution that tackles the main shortcomings of the existing platforms, and increases user trust.

This framework will be implemented in a phased manner; each component will be tested and refined before advancing to the next phase:

3.1 Development of the Crowdfunding Platform

A data storage layer is implemented by the framework, which utilizes the InterPlanetary File System (IPFS) to ensure that data is transparent and immutable for campaign-related information. This decentralized approach ensures that campaign details like descriptions, images, milestones, and voting outcomes are secure and verifiable public information, allowing users to confirm the accuracy and availability of information. In addition, blockchain networks, like Ethereum or Polygon, make it possible to securely and transparently execute transactions without a third party processing the transaction. All fund transactions on these networks are recorded immutably, without any central authority, making the platform more trustworthy. Smart contract integration on the platform also further promotes efficiency and trust by automating important campaign activities such as campaign creation, milestone tracking and fund release. These smart contracts enable the process to become secure and effective without the involvement of third parties. Project creators and contributors have a user-friendly interface to easily setup campaigns, define milestones, manage funds and track progress of the project as illustrated in Figure 27.1.

3.2 Planned Framework Validation, Testing, and Feedback Process

The proposed framework requires validation to be rigorous in the security, functionality and reliability of all key processes such as campaign management, fund distribution and insurance payouts.

The system architecture, as depicted in Figure 27.1, unites blockchain technology, AI driven fraud detection and decentralized governance in the construction of a secure crowdfunding environment. These technologies will work in combination as a layered defense mechanism, and will be tested on simulated scenarios and real time user inputs.

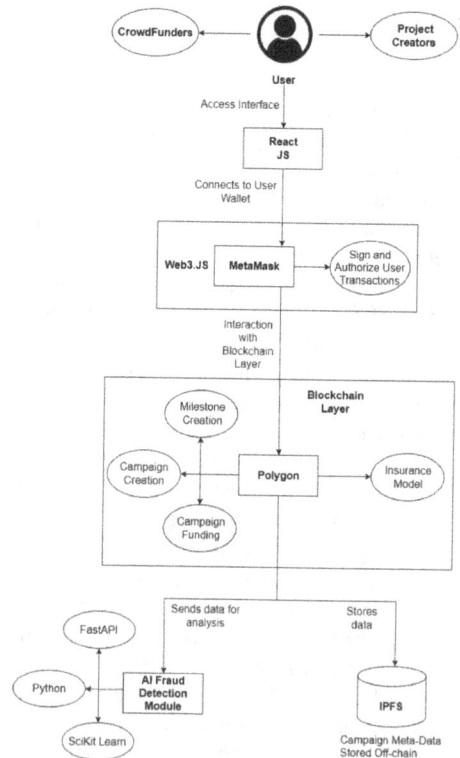

Figure 27.1 Detailed System Architecture for Blockchain-Based Crowdfunding Framework

3.3 AI Fraud Detection and Insurance Mechanism

The fraud detection system starts with rule-based approach to flag suspicious behaviour pattern such as large contributions or sudden donation spike to detect potentially fraudulent activity. As time passes, fraud detection will be powered by a machine learning component, using tools such as Scikit Learn to identify the subtle, evolving fraud patterns of past incidents, so that the system adapts to emerging threats [5].

Further, an insurance mechanism is introduced that protects contributors more, creating a financial safety net with a smart contract managed insurance pool. This pool is transparently managed on the blockchain and each insured campaign contributes a small amount to it. It also means that contributors who want to Projects can choose to have insurance coverage, especially for newer or higher risk projects. This model provides contributors with structured protection against potential losses, which enhances the confidence and trust as illustrated in Table 27.1.

Table 27.2, we compare our proposed decentralized model to traditional crowdfunding systems and discuss how our solution offers improvements in terms of transparency, trust, and automatic allocation of funds. We will test smart contracts that handle contributions and milestones on Ethereum's test network to ensure they're working, and free of errors and in line with acceptable security standards.

The fraud detection module will pass through a two-phase validation process; initial rule-based detections will occur on controlled test cases, and then machine learning driven analysis will be performed. Evaluation of the effectiveness of this module, as illustrated in Table 27.2, is done based on historical campaign data in order to discover potential fraud indicators and patterns. Such tests assure that the module has ability to put a red light for any suspicious activities and minimise false positives.

User feedback will be collected and refined continously to improve interface and overall user experience, giving actionable insights to improve usability. The process of feedback also has a role in future adaptability, as improvements made to the AI module or new blockchain technologies may be added to help scale the platform. Such enhancements will help the platform to cater to user needs and grow to meet the needs of project creators and contributors

Table 27.2 Comparison of Proposed Features Against Existing Limitations

Features	References			Proposed Features
	[1]	[2]	[7]	
Fraud Detection	Lacking automated fraud detection abilities	No built in fraud detection for loans	It uses a rule based trust mechanism for fraud check.	Machine learning based advanced AI driven fraud detection with continuous improvement and accuracy.
User Insurance and Protection	Lacks user protection features, which means that the users are at risk of loss.	No contributor insurance, loss mitigation options or your premium	Lacks insurance for backers	User contributed insurance pool integrated with blockchain; protects against campaign failure and fraud.
Scalability and Data Storage	Partial decentralisation, data off chain	On chain only data storage limits scalability	Lacks large data storage scalability.	Off chain storage of campaign data based on IPFS for better scalability, but without overloading the blockchain

while keeping the platform secure and functional at a high level.

4. Discussion

In this section, the proposed blockchain based crowdfunding framework is compared with the existing models in the literature in terms of fraud detection, user insurance and protection, and scalability and data storage. The proposed framework provides great improvements over traditional approaches since fraud detection is no longer based on a rule based system but an advanced AI driven fraud detection that evolves from a rule-based system to a machine learning model that can be improved continuously.

Besides, it introduces a blockchain insured insurance model for contributors to get a financial safety net in case of project failure or fraud, which will increase user confidence and platform reliability. The framework relies on IPFS off-chain storage to store campaign metadata, and does not place the data integrity burden on the blockchain network itself, making it more scalable and efficient. Combining these features, these seek to overcome the main issues facing existing systems to build a secure, efficient, and transparent crowdfunding environment. In Table 27.2, a detailed comparison of these features with previously developed models is given.

5. Conclusion and Future Work

Through the proposed blockchain based crowdfunding framework, major problems in traditional crowdfunding systems are solved such as security, transparency, and accountability. The platform seeks to achieve security and reduce the need for intermediaries by integrating blockchain technology, smart contracts, an AI driven fraud detection module and a blockchain integrated insurance model. Key functions like fund distribution and milestone tracking are handled by smart contracts, reducing human involvement, and the AI fraud detection module uses a rule based

approach which evolves into machine learning for detecting subtle fraud patterns. By utilizing the insurance model, contributors get financial safety net and smart contracts manage it without any intermediaries, and IPFS based off chain storage improves scalability and cost efficiency.

While these advances have been made, there's still a lot more that can be done. The AI fraud detection module can be improved using advanced techniques such as deep learning to raise detection accuracy. Scalability can be increased while costs are also likely reduced by exploring up and coming blockchain technologies like Avalanche or Solana. This could go a long way towards creating a difference between the campaign management and involving contributors more in their own governance through mechanisms such as DAOs. By simplifying user interaction with blockchain and user interface, user experience of decentralized technologies can be improved and made more accessible. Interoperability with other platforms, such those based on protocol like Polkadot or Cosmos would further expand the platform's versatility. Platform effectiveness, scalability and usability will also have to be validated in real conditions through real world pilot testing. Future work includes the scalability of the insurance mechanism, advanced risk analysis for sustainable coverage, and compliance with the ever changing regulations of decentralized finance and crowdfunding. Legal expertise will be needed for compliance, while maintaining platform decentralization.

In short, this decentralized blockchain framework seeks to bring a whole new level of groundbreaking security, transparency and accountability to the entire crowdfunding space. To overcome the limitations of traditional platforms, we integrate blockchain technology, AI driven fraud detection, decentralized governance and an insurance model. The areas listed above for future research and development when explored, could help cement this framework as a highly scalable, secure and universally trusted crowdfunding platform.

References

[1] Tiganoaia, B., & Alexandru, G. M. (2023). Building a blockchain-based decentralized crowdfunding platform for social and educational causes in the context of sustainable development. *Sustainability*, 15(23), 16205. doi: https://doi.org/10.3390/su152316205.

[2] Asamoah, K. O., et al. (2023). A blockchain-based crowdsourcing loan platform for funding higher education in developing countries. *IEEE Access*, 11, 24162–24174. doi: https://doi.org/10.1109/ACCESS.2023.3252917.

[3] Hassija, V., Chamola, V., & Zeadally, S. (2020). BitFund: A blockchain-based crowdfunding platform for future smart and connected nation. *Sustainable Cities and Society*, 60, 102145. doi: https://doi.org/10.1016/j.scs.2020.102145.

[4] Wan, X., Teng, Z., Li, Q., & Deveci, M. (2023). Blockchain technology empowers the crowdfunding decision-making of marine ranching. *Expert Systems with Applications*, 119685. doi: https://doi.org/10.1016/j.eswa.2023.119685.

[5] Kamali, M., Malek, M. R., Saeedi, S., & Liang, S. (2021). A blockchain-based spatial crowdsourcing system for spatial information collection using a reward distribution. *Sensors*, 21(15), 5146. doi: https://doi.org/10.3390/s21155146.

[6] Syafira, T., Jackson, S., & Tambunan, A. (2024). Fintech integration with crowdfunding and blockchain in industry 4.0 era. *Startupreneur Business Digital (SABDA Journal)*, 3(1), 10–18. doi: https://doi.org/10.33050/sabda.v3i1.433.

[7] Xu, Y., et al. (2023). A decentralized trust management mechanism for crowdfunding. *Information Sciences*, 638, 118969–118969. doi: https://doi.org/10.1016/j.ins.2023.118969.

[8] Bafna, B., Daigavane, V., Shaha, S., Shinde, G., & Shelke, S. (2023). Decentralized transaction system for detection and prevention of fraud in crowdfunding platforms. *Journal of Information and Computational Science*, 13, 133–138. Available: https://www.researchgate.net/publication/376892207_Decentralized_Transaction_System_for_Detection_and_Prevention_of_Fraud_in_Crowdfunding_Platforms

[9] Jadye, S., Chattopadhyay, S., Khodankar, Y., & Patil, N. (2024). Decentralized crowdfunding platform using ethereum blockchain technology. Accessed: Nov. 04, 2024. [Online]. Available: https://www.irjet.net/archives/V8/i4/IRJET-V8I41024.pdf

[10] Sirisawat, S., Chatjuthamard, P., Kiattisin, S., & Treepongkaruna, S. (2022). The future of digital donation crowdfunding. *PLOS ONE*, 17(11), e0275898. doi: https://doi.org/10.1371/journal.pone.0275898.

[11] Yee, W. L. H., & Rahim, N. (2021). Decentralized application for charity organization crowdfunding using smart contract and blockchain. *Applied Information Technology and Computer Science*, 2(2), 236–248. doi: https://doi.org/10.30880/aitcs.2021.02.02.016.

[12] Amin, M. R., & Zuhairi, M. F. (2021). Crowdfunding smart contract: Security and challenges. *Indian Journal of Computer Science and Engineering*, 12(1), 202–209. doi: https://doi.org/10.21817/indjcse/2021/v12i1/211201017.

[13] Saadat, M. N., Abdul Halim, S., Osman, H., Mohammad Nassr, R., & Zuhairi, M. F. (2019). Blockchain based crowdfunding systems. *Indonesian Journal of Electrical Engineering and Computer Science*, 15(1), 409. doi: https://doi.org/10.11591/ijeecs.v15.i1.pp409–413.

28 System-Level Security Framework Using Virtualization and Neural Cryptography for Encryption in Edge Devices

Bharath Devaraju, Kovith Chinthala and Kavitha Agarwal

[1]Department of Computer Engineering and Technology, Chaitanya Bharathi Institute of Technology, Hyderabad, Telangana
[2]Assistant Professor, Department of Computer Engineering and Technology, Chaitanya Bharathi Institute of Technology, Hyderabad, Telangana

Abstract

In the ever-changing landscape of low resource computing the recent developments in Internet of Things (IoT) and allied technologies has paved the way for achieving more throughput locally and more accurate real-time decision making. Despite this development however the age-old problem of securing data against unauthorized access and maintaining substantial levels of integrity is yet to be properly tackled. Aspects like authorization, non-repudiation have been the primary focus of Cryptographic algorithms. Additionally, many current solutions focus on network-level security, neglecting the device-level resource limitations and vulnerability to physical tampering. This paper proposes a Security software that consists of an embedded mini-hypervisor that would abstract each Edge node as a VM and provide isolation with secure interVM communication alongside a Neural cryptography module that generates cryptographic keys with low computational costs to encrypt the data. The framework provides resilient defence mechanisms against common attacks such as MITM, Data breaches, Denial of Service amidst others.

1. Introduction

The nature of the data processing itself has gone distributed toward data sources due to exploded growth in edge computing and Internet of Things, helping with real-time decision-making support. These technologies are in the form of sensors, cameras, and actuators gathering, processing, and transmitting vital data, and hence are integral parts of smart cities, healthcare systems, and industrial automation. Data processing from centralized cloud servers to edge devices shifts real-time response capacity while lowering latency and bandwidth expenses [5]. However, because of billions of linked edge devices, there are serious security issues now.

The key features of the edge devices include low processing power, low energy, and low memory. So, it becomes challenging to use the powerful methodologies used in RSA and AES cryptography for encryption on these types of devices. Moreover, most of the edge devices are also sites of physical access. Hence, it creates a door open for most attacks that can come through the cyberworld: this includes man-in-the-middle and Distributed Denial of Service attacks, and probable physical manipulation/illegal access [2]. There is, therefore a significant need for powerful

DOI: 10.1201/9781003661917-28

but light security measures that could potentially protect private information without putting too much strain on the limited resources that are accessible to edge devices.

Secure boot procedures, trusted execution environments, and encryption methods are some of the major edges of current edge device security strategies. In this regard, Sun et al. introduced a method on lightweight dynamic key generation using neural cryptography in order to deal with scarce resources in the edge devices [1]. VNSFs offer security at the network level, and Canavese et al. introduced this concept. However, these approaches have not focused on the particular problems encountered by specific edge devices. Together, neural cryptography and virtualization can provide an advanced as well as all-round network and device level security.

We present a novel system-level security framework based on two emergent technologies: virtualization and neural cryptography. These have been recently developed to face the unique challenges of security brought about by edge computing environments. Virtualization provides process isolation within secure execution environments of edge devices, thus protecting critical operations from malicious software and unauthorized access [4]. Additional benefit, it uses the use of a neural network, to produce dynamically a nonpredictive but updatable keys cryptographic one with enhanced security but without higher cost computational operations, according to references [3]. It combines these two technologies to provide a robust, lightweight security solution that can be adapted for the dynamic nature of the edge networks. In this framework, virtualization and neural cryptography are well integrated to ensure the confidentiality, integrity, and availability of data even in resource constrained environments.

The rest of this paper is structured as follows:

Section 2: Related Works discusses existing approaches to securing IoT and edge devices, including relevant studies on lightweight cryptography, virtualization, and network security.

Section 3: Proposed System-Level Security Framework presents the design of the proposed framework, detailing how virtualization and neural cryptography can be combined to secure edge devices.

Section 4: Experimental Evaluation and Results evaluates the proposed framework in terms of performance, security, and scalability across different edge computing scenarios.

Section 5: Conclusion and Future Work concludes the contributions of this paper and discusses potential future research directions, such as integrating blockchain-based trust mechanisms and further optimizing neural cryptography models.

2. Related Works

2.1 Lightweight Cryptosystems for IoT

Sun et al. [1] presented a new end-to-end neural cryptosystem for IoT devices that enable authentication, encryption, and distribution of keys. According to their study, the utilization of a neural cryptography technique will be feasible due to the small processing capabilities of IoT devices without any degradation in terms of security problems. In this regard, the key distribution and spreading processes with lightweight CNNs are able to be carried out more efficiently. This research, therefore, comes particularly into our project, showing how neural networks could be applied toward the generation of cryptographic keys, an essential aspect in the encryption mechanisms for the proposed framework.

2.2 Security at Network Level

Canavese et al. [2] introduced a security framework for resource-constrained IoT devices with the help of VNSFs. It contains VPN and IPS systems for network-level security, so it manages security with enhanced efficiency. This technique highlights scalability and modularity toward dynamic security management. It does not consider limitations at the level of particular devices in the IoT due to their limited computing capacity and restricted memory. This calls for including virtualization and light-weight cryptography in our framework for further device and network system protection.

2.3 Neural Cryptography for Dynamic Key Generation

Hagan et al. [3] had designed an edge computing-based adaptive neural cryptography framework focusing on edge computing adaptive key generation. Their work best shows how dynamic cryptographic keys generated through neural network models cannot be determined or reproduced in a session. That is quite important to an IoT network that is being constantly transformed with members joining or leaving it. The results confirm our proposed use of neural cryptography as part of the framework that allows for efficient scalable edge device encryption.

2.4 Virtualization for Process Isolation in Edge Devices

Indeed, the area of embedding virtualization for the improvement of security on the edge devices is promising. Tiburski et al. [4] proposed a lightweight security framework based on embedded virtualization with trust mechanisms. Their system isolates sensitive processes from a compromised application so that the functionalities of the main system will not be damaged. Here, the concept used is that of virtualization, and we have integrated this in our proposed framework for the secure execution environment at the edge device level. We successfully adopted their approach and were able to isolate the critical process. This is accomplished while minimizing the possible damage from attacks and maintaining secure operations in resource-scarce environments.

2.5 Edge Layer Security and Scalability

Errabelly et al. [5] introduced an edge layer security service called EdgeSec, which aims to enhance the security of IoT systems by bringing edge computing closer to cloud architectures. This method uses policies near the origin of the data and, therefore, reduces latency and bandwidth utilization. Our framework uses security at the edge but applies neural cryptography and virtualization for data confidentiality and separation of process. This manuscript will elaborate on the framework applicable to large-scale IoT networks based on the scalability aspects of EdgeSec, which is to be discussed in subsequent sections.

2.6 Blockchain-based Trust Mechanism

Blockchain technology can make decentralized IoT networks more secure, especially in trust establishment and validation of transactions. Khan et al. [7] proposed a hardware-based security framework for IoT edge devices that integrates blockchain technology for distributed trust management. Blockchain technology, while offering strong integrity guarantees, poses severe challenges to resource-constrained IoT devices because of high energy consumption and computation requirements. Our framework presents optimized energy-efficient cryptographic methods, such as neural cryptography, and lightweight blockchain components for future embedded trust management.

2.7 Challenges in Real-Time Edge Computing Security

Blockchain technology has been proposed to enhance the security of decentralized IoT networks, especially in establishing trust and validating transactions. Khan et al. [7] proposed a security framework for IoT edge devices using blockchain for distributed trust management. Although blockchain technology offers strong integrity assurances, it requires high energy usage and computational resources. It thus causes significant overhead issues for IoT devices with limited resources. Our framework will aim at making energy-efficient cryptographic methods like neural cryptography and lightweight blockchain components for the small-sized IoT device for embedded trust management in future research.

2.8 Conclusion of Related Works

As the literature review has indicated, there are numerous strategies to secure IoT and edge computing environments that target different aspects of the security challenge. Neural cryptography is one promising encryption approach that is very flexible with lower computational requirements, while embedded virtualization may provide strong process isolation with secure operation on resource-constrained devices. Our proposed system-level security framework integrates all these advanced technologies to provide a holistic solution for protecting IoT and edge devices while maintaining both data confidentiality and system integrity.

3. Proposed Methodology

The proposed framework shall address the security challenges of the IoT edge devices by encompassing three core elements of lightweight virtualization for a process isolation and protection of functionalities of the device, firewalls with security protocols to limit and manage access to network flow, and neural cryptography used to ensure safe transfer by employing adaptive encryption methods of data transferred among the edge devices. Most edge devices being resource-constrained, there is a guarantee of keeping the integrity, confidentiality, and safety of data transfers among the edge devices in check.

3.1 Virtualization in Edge Devices

In cloud computing, virtualization provides multi-tenancy, resource sharing, and isolation of processes. On the other hand, for devices at the edge having fewer resources,

Table 28.1 Comparative Analysis of Security Approaches for IoT and Edge Devices

Paper	Approach	Strengths	Limitations
Sun et al. (2022) [1]	Neural cryptography for IoT device authentication, encryption, and key distribution	Efficient key management using lightweight CNNs, adaptability to resource constraints in IoT devices	Early-stage implementation, limited deployment in real-world applications
Canavese et al. (2024) [2]	Virtual Network Security Functions (VNSFs) for IoT security	Scalable and modular, enhances network-level security	Does not address device-level resource limitations, no focus on encryption efficiency
Hagan et al. (2020) [3]	Neural cryptography for dynamic key generation	Dynamic, session-specific keys, adapts to changing network environments	Still in early development, lacks widespread deployment
Tiburski et al. (2019) [4]	Embedded virtualization and trust mechanisms for process isolation	Ensures safe interdomain communication, process isolation using virtualization	Lacks robust encryption mechanisms, focuses on virtualization only
Errabelly et al. (2017) [5]	EdgeSec for IoT security at the edge layer	Reduces latency and bandwidth usage, scalability in cloud-edge environments	No built-in cryptography, focuses on edge layer security without addressing encryption
Joar Sikhar (2023) [6]	Edge-driven security framework for IoT	Adaptive to real-time edge computing, addresses the need for lightweight security mechanisms	Lacks focus on cryptography, more emphasis on security policies and frameworks

full virtualization as used in hypervisors is not a feasible approach since it has much overhead. Instead, a micro-hypervisor or container is used to provide process isolation on edge devices so that applications and their respective processes are kept away from each other to achieve better security.

3.1.1 Micro hypervisor Architecture

The proposed system has a micro-hypervisor on every edge node. A micro-hypervisor is an optimized version of the conventional hypervisor, designed to manage VMs with maximum efficiency in terms of resource utilization, thus suitable for the IoT devices that are under capable. Hypervisors create secure and isolated virtual environments where processes may run. For instance, a separate VM can contain the application of encryption and decryption, whereas less sensitive or user-facing applications remain in their respective VMs. This separation ensures that if one application is compromised, the sensitive parts of the system remain safe. All processes run in isolation and thus minimize the overall attack surface, and any compromises are contained within the individual VM, reducing possible damage.

3.1.2 Secure Boot and Chain-of-trust

The framework uses a chain-of-trust model secure boot process to improve the security

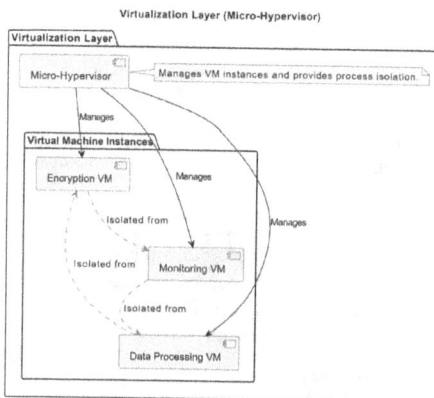

Figure 28.1 Virtualization Architecture for Edge Devices

of the virtualized environment. Secure boot procedure assures only authorized and validated code at the time of the initialization of the device is run. The bootloader authenticates the hypervisor during boot-up at power-up, checking its integrity via cryptographic signatures. Once authenticated, the hypervisor will validate the integrity of virtual machines and their software component prior to running them. This ensures that nothing compromised or malicious would tamper with the system.

3.2 Firewalls and Security Policies in IOT

The system uses firewalls and security protocols to manage data transfer between edge devices and the overall network as well as between devices in the same local network. Such architecture is representative of decentralized, policy-based firewall at the device and network level.

3.2.1 Device-level Firewall

At the edge device, each has a fire wall on board that deals with traffic coming in or out of the device. It works by applying all security policies defined to pass or stop certain types of traffics; for example, reject any incoming traffic from unidentified servers, but allow for those incoming from trusted locations. For outbound traffic to be halted, it must not let data go into unknown places, thus making the harm servers not benefit from that data.

The micro-hypervisor manages the network traffic produced by the individual VMs in tandem with the firewall. VMs can enforce their security policies, which put more constraints on sensitive processes like encryption. The level of detail ensures that a compromised VM cannot easily exfiltrate data or connect to malicious entities.

3.2.2 Centralized Security Policy Management

Managing security policies in a large-scale IoT setup can be complex. To tackle this issue, the framework features centralized management of security policies through a network controller. Administrators have the ability to create and implement security policies across all devices on the network from a unified interface. The network controller transmits these policies to the firewalls on each device, where they are enforced locally.

These policies are adaptable and can be modified in real-time to address emerging threats. For example, if a new vulnerability is identified, the central controller can distribute a rule to block traffic related to that threat, safeguarding all devices without the need for manual action.

3.2.3 Intrusion Detection and Prevention

The structure integrates an IDPS that constantly scans network traffic for anomalies indicating malicious activity. IDPS relies on trained machine learning models running on large datasets of network traffic to ascertain anomalies, such as sudden spikes in data transmission or anomalous access patterns that may point to DDoS attacks or attempts at unauthorized intrusion.

The IDPS is operating at the level of both device and networks for full protection. If the system identifies a probable threat, it can automatically make modifications to the firewall rule set to block the dubious traffic or quarantine affected computers in order to prevent spreading an attack across the entire network.

3.3 Neural Cryptography

This technique of encryption is innovative because it is based on pattern recognition and learning capabilities within neural networks, generating keys to encrypt. It does not rely on the traditional algorithms of cryptography that depend upon mathematical complexity to manage the keys but generates keys dynamically depending on the communication patterns inside the network.

3.3.1 Dynamic key Generation

In the proposed framework, a CNN generates symmetric keys for encryption. It

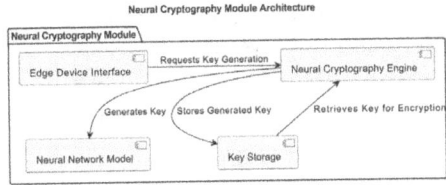

Figure 28.2 Neural Cryptography Key Management

identifies various patterns in edge devices' communication and generates different keys for sessions. This is a form of dynamic key generation so that every session of the communication uses a different key, making it much more challenging for attackers to predict or intercept the data. Neural cryptography offers several advantages:

Adaptability: Neural cryptography adapts to dynamic edge networks, where devices frequently join and leave the network, and where communication patterns constantly change.

Efficiency: Neural networks generate cryptographic keys with minimal computational overhead, making them suitable for resource-constrained edge devices.

Security: The keys generated by neural networks are non-deterministic, making brute force attacks highly challenging.

3.3.2 Secure Key Exchange

Traditional cryptography has issues with key sharing between devices in a safe manner. The proposed framework utilizes neural synchronization to eliminate the need for direct key exchange between devices. Each device generates keys based on shared patterns of communication from their neural networks, ensuring that devices can generate matching keys without ever having to share keys with each other. This concept is called" neural synchronization," which ensures the key is secure and does not become vulnerable during the process of transfer.

Since the keys are actually generated in real time, they can be updated often and without placing undue computational burdens. This would help frequently rotate keys, thereby reinforcing security and minimizing the exposure of keys for long periods.

3.4 Component Integration

The integration of virtualization, firewalls, and neural cryptography provides a comprehensive, multi-layered security framework for edge devices. These components work together to ensure robust security by maintaining data integrity, confidentiality, and availability throughout the system.

3.4.1 Isolation of Process using Virtualization

This three-layer security architecture with virtualization, firewalls, and neural cryptography protects the edge device. It keeps data in integrity, confidential, and available across the whole system through its multi-layered approach.

3.4.2 Network and Data Security

Firewalls and security protocols at both network and device layers provide fine-grained control over the transmission

of data, block unauthorized access, and protect against attacks from the network. Centralized management of security policies will allow real-time updates in response to new threats that arise.

3.4.3 Data Confidentiality with Neural Cryptography

Neural cryptography makes session-specific keys so all communication between edge devices is encrypted. Neural synchronization eliminates the insecurity of the key exchange protocol. The above approaches combined ensure the framework works properly even under difficult conditions while using commonly limited computational and energy resources available on an edge device.

3.5 Methodology Summary

It focuses on system-level security in edge devices combined with lightweight virtualization, firewalls, and neural cryptography, addressing the security issues in IoT settings by providing a layer-by-layer guarantee for preserving data integrity and confidentiality

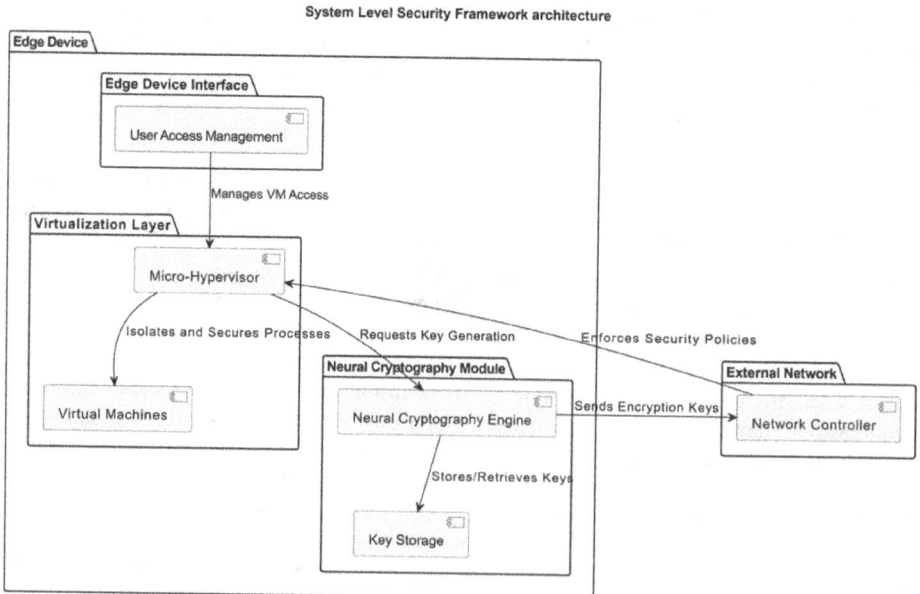

Figure 28.3 System Level Security Framework Architecture

along with secure communication even in resource-limited environments.

4. Conclusions

The rapid growth of IoT devices and edge computing systems has brought about considerable difficulties in securing communication and data processing within resource-limited environments. This paper introduces a novel system-level security framework that combines lightweight virtualization, firewalls, security policies, and neural cryptography to address these challenges proficiently.

The innovations presented by the framework are multiple:

Process Isolation through Virtualization: Critical operations on an edge device are kept distinct through lightweight virtualization methods-precisely through micro hypervisors-so as to prevent security flaws in one single process from compromising the overall system. The secure boot procedure, based on a chain-of-trust model, ensures only approved software runs on the device.

Security Policies and Firewalls: The architecture features distributed, policy-driven firewalls that control traffic incoming to and outgoing from the device and network levels. With the centralization of the management of the security policy, the update of the firewall rules would immediately follow any new emerging threats. This flexibility assures protection against unauthorized access and malicious traffic at the edge devices.

Data Confidentiality with Neural Cryptography: The usage of neural cryptography in the context of recognizing the pattern dynamically and efficiently utilizing the potential of a neural network will be employed in generating dynamic keys to perform encryption within every session. Neural cryptography ensures that within settings with fast-evolving communication patterns, data transmissions remain protected. The usage of neural synchronization eliminates the exchange of keys directly. In this case, this ensures an upgraded security feature within the system.

This, therefore, means that complete integration of these components achieves the fundamental security requirements that IoT and edge devices may need, such as ensuring data confidentiality, integrity, and availability. The findings point to the fact that the developed framework provides robust security in such a way that it still has low computational requirements. This is appropriate for constrained resource devices in dynamic environments.

5. Future Work

While the suggested framework presents a promising approach for securing edge devices, multiple directions for future research can be pursued to boost its performance and scalability:

5.1 Optimizing Neural Cryptography Models

One of the most important benefits of this neural cryptography is its ease and efficiency in generating key pairs. However, considering the growing scale of such networks of IoT and, consequently, the complexity involved in communication patterns, optimized neural cryptographic models are considered necessary for reducing latency along with computational complexities. Future research can go further by exploring more intricate neural network architectures such as RNNs or other attention-based models that potentially improve the efficiency and quality of key generation in more complex large-scale IoT environments.

5.2 Incorporating Blockchain-based Trust Machines

Although this paper primarily centers on securing communication and process isolation, managing trust in decentralized edge environments is an essential concern. The integration of blockchain technology, which offers a distributed ledger to monitor device interactions and transactions, could be added to the proposed framework to strengthen trust and accountability. By utilizing smart contracts and decentralized identity solutions, blockchain could establish a tamper-resistant system for confirming the authenticity and integrity of devices within a distributed IoT ecosystem.

5.3 Improving Intrusion Detection Capabilities

Furthermore, the IDPS defined in the framework utilizes machine learning models, which are implemented for anomaly detection in network traffic. Being effective, there is always room for improvement with real-time response and scalability being the main areas for that. Future research might thus focus on the use of advanced deep learning methods with the aim of making further improvements in anomaly detection capabilities and quicker threat response. Another interesting avenue of exploration in federation learning is to help the edge devices collaborate in developing security models that are even more efficient by avoiding exposure.

References

[1]　Sun, Y., Lo, F. P. W., & Lo, B. (2021). Lightweight internet of things device authentication, encryption, and key distribution using end-to-end neural cryptosystems. *IEEE Internet of Things Journal*, 9(16), 14978–14987. doi:10.1109/JIOT.2021.3067036.

[2]　Canavese, D., Mannella, L., Regano, L., & Basile, C. (2024). Security at the edge for resource-limited IoT devices. *Sensors*, 24(2), 590. doi:10.3390/s24020590.

[3]　Hagan, M., Siddiqui, F., & Sezer, S. (2020).Enhancing security and privacy of next-generation edge computing technologies. *Proceedings of the 17th International Conference on Privacy, Security, and Trust (PST)*, IEEE, 2020. doi:10.1109/PST47121.2019.8949052.

[4]　Tiburski, R. T., Moratelli, C. R., Johann, S. F., Neves, M. V., de Matos, E., Amaral, L. A., & Hessel, F. (2019). Lightweight security architecture based on embedded virtualization and trust mechanisms for IoT edge devices. *IEEE Communications Magazine*, 57(2), 67–73.https://doi.org/10.1109/MCOM.2018.1701047

[5]　Errabelly, R., et al. (2017). EdgeSec: Design of an edge layer security service to enhance IoT security. *IEEE/ACM 1st International Conference on Fog and Edge Computing*, 2017. doi:10.1109/ICFEC.2017.7.

[6]　Jo, H., & Sikhar, R. (2023). Enhancing IoT security through an edge-driven framework. *The Computer Bulletin*, June 2023. doi:10.371416842.

[7]　Khan, M., et al. (2024). A novel trusted hardware-based scalable security framework for IoT edge devices. *Discover Internet of Things*, 4(4). doi:10.1007/s43926024-00056-7.

[8]　B., Bokhari, M. U., & Siddiqui, M. A. (2024). Cruciality of securing user data due to increasing Digital Learner traffic over the Internet using Adversarial Neural Cryptography. *Journal of Electrical Systems*, 20(3), 1585–1603.

[9]　Hanafi, B., & Bokhari, M. U. (2023, May). Enhancement of security in connection establishment for IoT Infrastructure through Adversarial Neural Cryptography using GANs. In *ICIDSSD 2022: Proceedings of the 3rd International Conference on ICT for Digital, Smart, and Sustainable Development, ICIDSSD 2022, 24–25 March 2022, New Delhi, India* (p. 300). European Alliance for Innovation.

[10]　Zhao, W., Mahmoud, Q. H., & Alwidian, S. (2023). Evaluation of GAN-based model for adversarial training. *Sensors*, 23(5), 2697.

[11]　Tang, X., Yin, P., Zhou, Z., & Huang, D. (2023). Adversarial perturbation elimination with GAN based defense in continuous-variable quantum key distribution systems. *Electronics*, 12(11), 2437.

[12]　Khan, N. A., Awang, A., & Karim, S. A. A. (2022). Security in Internet of Things: A review. *IEEE Access*, 10, 104649–104670.

[13]　Kinzel, W., & Kanter, I. (2002). Neural cryptography. *Proc. 9th Int. Conf. Neural Inf. Process.*,3, 1351–1354.

[14]　Yayik, A., & Kutlu, Y. (2014). Neural network-based cryptography. *Neural Netw. World*, 2(14), 177–192.

[15]　Frustaci, M., Pace, P., Aloi, G., & Fortino, G. (2018). Evaluating critical security issues of the IoT world: Present and future challenges. *IEEE IoT-J*, 5(4).

[16]　Golam, M., Lee, J. M., & Kim, D. S. (2020, October). A uav-assisted blockchain based secure device-to-device communication in internet of military things. In *2020 International Conference on Information and Communication Technology Convergence (ICTC)* (pp. 1896–1898). IEEE.

29 Plasmonic Structured CuO Nanofoams for Optical Absorption Enhancements Towards Water-Splitting Applications

Venumbaka Maneesh Reddy[1], D. Nagadevi[2], Saravanan Gengan[3], Selvakumar Duraisamy[1,*], T. Vasist Vikas Reddy[2], U. Harshith Varma[2], S. Srikar Reddy[2] and Marepally Bhanu Chandra[2,*]

[1]Deparment of ECE, PSG Institute of Technology and Applied Research, Coimbatore, T.N.
[2]Deparment of ECE, Chaitanya Bharathi Institute of Technology, Hyderabad, T.S. India
[3]Department of Chemistry, Saveetha School of Engineering, SIMATS, Chennai T.N.

Abstract

Hydrogen energy is one of the highly sort out energy generation pathways leading towards the goal of creating a sustainable and renewable world energy cycle. In this article, we present a detailed study on the plasmonic enhancements of the CuO (Cupric oxide) nanofoam structures grown on thin Cu foil as a substrate, which are utilised as photo-electro catalytic electrodes for water splitting and CO_2 reduction applications. The synthesis route used to grow the nanofoams is the electrochemical method, which is quite simple, cost-effective, and easily scalable. The selectivity of the CuO nano-foam structural dimensions is achieved via tuning the solvent concentration and the electro-potential values. Material characteristics were obtained using various techniques. In addition, FDTD simulations were carried out to study the optical properties. The FDTD studies were modelled based on the synthesised CuO nanofoams to have a good comparison between the results of simulations vs synthesis. The observed optical absorption values show an enhancement of near 2-fold for CuO nanofoams with specific dimensions vs CuO thin film in the visible spectra. Moreover, there is an increase in the effective surface area (catalytic sites) by 6-fold, which is an added advantage when using the material as a photo-catalytic electrode for water splitting applications.

Keywords: Nanofoam, CuO, electrochemical, water splitting, optical

1. Introduction

The ever-rising population and industrialization lead to the world energy deficit. Currently, the world relies on fossil fuels, to meet this demand, which are non-renewable and having adverse effects on the environment like global warming, pollution, polar ice-melting, etc. This created the urgent need to look for alternative energy sources that

Email: bhanuchandram_ece@cbit.ac.in

DOI: 10.1201/9781003661917-29

are renewable and environmentally friendly. Among the various renewable energy sources, solar-based energy production grabs attention due to its abundance and potential to meet the current and future energy demands having a very small environmental footprint. Even though solar energy is abundant but it has a limitation that the sun is not available all the time. This requires energy produced from the sun to be stored which can be achieved using batteries, solar fuels, etc., depending on the application.

Fuels are the primary sources for the production of energy. Among the various fuels, Hydrogen is an attractive and promising green fuel due to its properties like high energy density, high efficiency, etc. [1–2]. Generation of hydrogen can be achieved from different non-renewable and renewable methods like steam methane reforming (SMR), electrolysis, pyrolysis, thermochemical water splitting, photoelectrochemical water splitting, photobiological process, etc. [3]. Fuels generated using solar energy, are called solar fuels, which use water, light, and CO_2 to reduce to fuels that can be stored efficiently and for a long period. This also, reduces CO_2 in the atmosphere and contributes to achieving a renewable energy based sustainable environment.

Photoelectrochemical water-splitting was first demonstrated by Fujishima and Honda in early 1971 by using TiO_2 as an electrode for dissociation of water into H_2 and O_2 [1]. Since then, many materials like semiconductors [4], metals, etc., were reported as catalysts till date to improve the efficiency of H_2 generation [2]. Among various catalysts, Copper (Cu) can produce a wide range of solar fuels [5]. Also, cupric oxide (CuO) exhibits high optical absorption, excellent charge transport properties [6]. To get the advantages of both the materials, in this article, we present CuO nanofoams that were grown on thin Cu foil as the substrate using electrochemical method and also modelled the structures similar to those obtained through synthesis and studied the optical properties obtained from both experimentation and simulation respectively.

2. Materials and Methods

2.1 Synthesis

CuO nanofoams were synthesised on a Cu foil (Sigma Aldrich) by electrodeposition technique using a two-electrode electrochemical cell. Cu foils formed both the Cathodic and Anodic electrodes, which are pretreated which include—mechanical polishing by P280 abrasive sheet followed by etching using 0.5M HCl aqueous solution, washing with DI water and then drying. These electrodes were separated by a distance of 1 cm and the electrolyte solution used is a mixture of 0.4 M $CuSO_4$, 1.5 M H_2SO_4 and 50 mM HCl [5]. CuO was grown on the polished Cu substrate at a constant current of 1.2 amperes and the time as 2 min chosen from the best results previously reported [5]. The deposited substrates were cleaned with distilled water and dried overnight at 70 °C in the air followed by calcination at 350 ºC for 1 hr. The synthesised CuO nanofoam was characterised using UV-Vis spectroscopy, XRD, and SEM techniques to study the optical, structural, and morphological properties respectively.

2.2 Modeling and Simulation

Structures of CuO nanofoams similar to those synthesised were modelled and simulated using FDTD method. Designing steps involved are defining the geometry and material, defining the source and monitors followed by setting the computational domain and boundary conditions. Initially, two Cu substrates with height 74 and 224 microns were created and above them CuO based nano frustum hollow structures of 26 microns height were created (i.e., total thickness of 100 & 250 microns—in line with experimental values) by defining plane wave as a source and choosing mesh size based on the structure of nanofoams. Reflection and transmission monitors were placed just above the surface of nanofrustums and below the substrate to study the absorption and transmission properties.

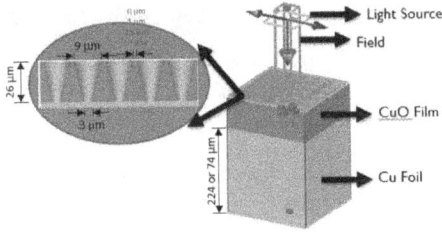

Figure 29.1 CuO Nanofoam Structures on Cu Substrate

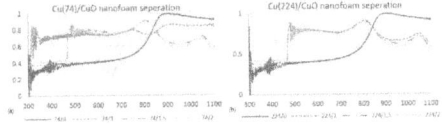

Figure 29.3 Absorption Studies of CuO Nanofoam Structures on (a) 74 Microns Cu Substrate (b) 224 Microns Cu Substrate

Figure 29.1 represents the modelled structure of CuO nano frustums of 26 microns height grown on Cu substrate of 74/224 microns height.

The optical properties of these modelled structures was seen in the wavelength range of 300 nm to 1100 nm by varying the nanofrustum periodicity on both the substrates and compared the results with the experimentally grown, 26 μm height CuO film on Cu substrates.

3. Results

3.1 SEM with EDS

Figure 29.2(a) represents the SEM image which confirms the foam structures with various pore diameters and found that the average diameter of foams is about 9 microns and Figure 29.2(b) confirms the presence of Cu and O with atomic concentrations of 69.5% and 30.5%. XRD results can be referred from the previous reported results [5].

3.2 Simulation

CuO film and CuO nanofoam structures which are grown on 74 microns Cu substrate were simulated and compared as

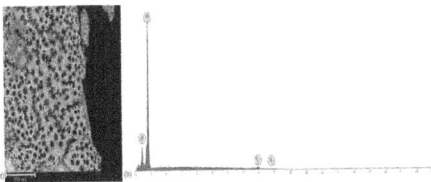

Figure 29.2 CuO Nanofoam Structures on Cu Substrate (a) SEM Image, (b) EDX Graph

shown in Figure 29.3. Here, periodicity of nano-foam structures were varied by 1, 1.5, 2 microns and represented as 74/1, 74/1.5, 74/2 respectively. The CuO film was represented as 74/0. The obtained results shown that nanofrustum based structures enhanced the absorption of visible light by 2 times than the pure CuO film due to increase in the overall surface area and moreover a peak shift towards visible range from near IR.

Similarly, CuO nanofrustums grown on 224 μm Cu foil were also simulated by varying periodicity and compared with the CuO grown on 224 μm, Cu foil and the results obtained were shown in Figure 29.4. A similar 2 fold rise in the visible light absorption due to increase in surface is observed. The observed absorption data was scaled with the ASTM solar spectrum values and found that among the nano-frustum structures the one with periodicity 1 μm shows better performance than the others. The UV-Vis spectra for the experimental CuO NFs is shown in the Figure S1.

4. Conclusion

CuO nanofoams were successfully synthesised using simple and economical, electrochemical method and similar structures were modelled to compare with the physical film structures. The absorption studies revealed that the nanofrustum structures enhanced the optical absorption by two-fold due to enhancement in the surface area by 6 times. A good matching in the experimental and simulated optical properties is observed. Frustum structures were confirmed by SEM analysis and found the average diameter of nanofrustums to be

9 μm. The obtained results show promising outcomes for photocatalytic water splitting and CO_2 reduction applications, which were reported previously [5].

Acknowledgment

This work was supported by the DST-SERB under CRG [CRG/2019/005985]. Dr. M. Bhanu Chandra & Mr. V. Maneesh, are thankful for the support of DST-SERB.

References

[1]   Ma, P., & Wang, D. (2018). The principle of photoelectrochemical water splitting. In Wang, D., Cao, G., eds. *Nanomaterials for energy conversion storage, World Scientific*, pp. 1–69.

[2]   Ambal, S. R., Sivaranjani, K. S., & Gopinath, C. S. (2015). Recent developments in solar H2 generation from water splitting. *J. Chem. Sci.*, 127(1), 33–47.

[3]   Nazir, H., Louis, C., Jose, S., & Prakash, J. (2020). Is the H2 economy realisable in the foreseeable future? Part I: H_2 production methods. *International Journal of Hydrogen Energy*.

[4]   Sivula, K., & Van de Krol, R. (2016). Semiconducting materials for photoelectrochemical energy conversion. *Nat. Rev. Mater.*, 1, 15010.

[5]   Marepally, B.C., Ampelli, C., Genovese, G., & Tavella, F. (2019). Electrocatalytic reduction of CO_2 over dendritic-type Cu- and Fe-based electrodes prepared by electrodeposition. *Journal of CO$_2$ Utilisation*.

[6]   Panah, S. M., Moakhar, R. S., & Tan, H. R. (2015). Nanocrystal engineering of sputter grown CuO photocathode for visible light driven electrochemical water splitting. *ACS Appl. Mater. Interfaces*.

[7]   Phutanon, N., Pisitsak, P., Manuspiya, H., & Ummartyotin, S. (2018). Synthesis of three-dimensional hierarchical CuO flower-like architecture and its photocatalytic activity for rhodamine b degradation. *Journal of Science: Advanced Materials and Devices*, 3, 310–316.

30 Design and Implementation of Automatic Photovoltaic Panel Cleaning Using IOT

M. Perarasi[1], Priscilla Whitin[2], Pavaiyarkarasi R.[1], Chintala Venkatesh[3], Sai Sharan P.[1] and B. Sarala[1]

[1]Department of Electronics and Communication Engineering, R.M.K Engineering College, Chennai
[2]Department of Electrical and Electronics Engineering, Vel Tech Rangarajan Dr. Sagunthala R&D Institute of Science and Technology, Chennai
[3]Department of Electrical and Electronics Engineering, Meenkashi Sundararjan Engineering College, Chennai

Abstract

In this article, smartphone is used in the solar cleaning system. In this system, Node-MCU chip acts as the main component between the smartphone and the cleaning system. To connect this system to the smartphone, the Blynk app acts as the medium of connection between these devices. This system has a camera for monitoring purposes. There will be a water pump motor and it is connected to the relay and then to node-MCU. This camera and motor control will be in the Blynk app as it is connected to the node-MCU server. When we use the Blynk app, we can on the water pump through this same app. When the panel looks clear then the water pump will be set to off. So, with the help of this project, we can able to clean the solar panels at a low cost. And also, this kit will be available at a low cost so that no one can overspend their hard work money. There are some places where solar panels will be used on a large scale like industries etc. for these areas requires a lot of human intervention. With the help of this system, we can get out of this problem. And also, we can able to observe the changes in efficiency before and after cleaning the solar panel.

Keywords: Peizo electric actuator, Blynk app, water sprinkling system

1. Introduction

In this century, the usage of solar panels PV cell panels has rapidly increased in both research and also in domestic areas. For the decarburization purpose in the environment, this type of renewable energy system made a huge contribution. There was an upward increase in solar power said by ministry of government of India. In last six years they told more than 11 times increased from 2.6gw in March 2014 to 3gw in June 2019.

Email: mpi.eee@rmkec.ac.in[1], priscillawhitin@veltech.edu.in[2], rpi.ece@rmkec.ac.in, saisharan2k3@gmail.com[1], bsa.ece@rmkec.ac.in[1]

DOI: 10.1201/9781003661917-30

Photovoltaic power is directly coming from the sunlight. When the sunlight strikes a pv cell, electrons come from the p-n junction. And this electrons travel from the state and creates the electric power. However, there are some regions with high irradiation levels which they suffer from high dust and less water source. Due to this, if we install solar panels in these areas, the dust accumulation on the solar panel will be increased which leads to the output efficiency decrease in the solar panel. And cleaning this accumulated dust on PV panels is a major criterion. There are some areas where solar panels are used on a large scale, here also the same criteria can be seen.

To overcome this, we had invented a cleaning system in which a human can clean the solar panels by operating the cleaning system with help of a mobile phone. There is a reduction in the generation of electric power due to the accumulation of dust particle [1], the potential for reduced solar energy generation and decreased solar energy transmittance to photovoltaics exists due to atmospheric particulate matter (PM). Solar power production is reduced by ~ 17–25% across these regions.

The main goal is to preserve and enhance solar panel efficiency by consistently eliminating accumulated dirt and debris. Through automated cleaning, the system strives to maximize sunlight exposure, consequently boosting energy generation and prolonging the panels' lifespan. Ultimately, it seeks to diminish energy loss caused by soiling, minimize manual upkeep, and maintain a reliable and efficient solar panel operation.

An exploratory concentrate on the impact of residue on power misfortune in sun oriented photovoltaic modules [2], the horizontal and vertical axis moved by the dual motor and crawler motor with the help of cleaning brush. Moreover, the length of the sunlight powered charger cluster can be identified by position changes to keep the SPCR in the ideal working region. Savvy sunlight based photovoltaic board cleaning framework [3] the planned framework can clean dry residue amassed over the board's surface. Besides, by connecting the metal rail tracks to a long sun powered exhibit, the framework is by all accounts implementable for an enormous scope sun-based ranch. Ready to clean the whole sun-based exhibit to and from, driven cleaning rotatory brush.

DC Jordan [4] audits that Corruption rates is expected to know to anticipate power conveyance. The estimation on the corruption rates on the panels individually is done for the past 40 years. The rate is 0.5% each year. J Antonanzas [5] expressed about different strategies for estimating sun-based power. First technique is getting exact irradiance figures. Second technique is foreseeing yield utilizing AI and measurable strategy and a half breed model utilizing first and second strategies. Brabec [6] surveys about execution factual strategy advantage as past information isn't needed and result can be anticipated. This needs transitory and spatial goals which is one of the reasons for mistakes. Lorenz [7] tracked down better approach to eliminate mistakes by the model result measurements. This strategy can be utilized for weather conditions conjecture and further developing goal by interjecting values. Almeida [8] has organized the preparation sets in view of comparable clearness list and experimental appropriation on the irradiance expectation for every day. For 30 days' information is gathered and in light of observational dissemination. Some cleaning ways include compound and different specialists which draw in grimy undesirable particles as well as keeps sun-oriented irradiance from entering PV cells of board. So this decreases execution and subsequently proficiency.

M J Adinoyi [9] expressed that aggregation of residue relies upon different elements like PV module direction, covering, surface harshness and incline. Dust molecule aggregation relies upon size, weight, shape and compound properties of residue.

After doing small research on the available solar cleaning devices on market, we have noticed some cons in them. They are like high cost of the device, higher maintenance cost, lack of efficient shower, limited service and lack of remote monitor and control.

2. Present Technologies and Solar Panel Cleaning Practices

Megaprojects for solar energy production can be created in semi-arid or desert areas that receive a lot of sunlight all year long. However, more possibility also brings greater difficulty. In these barren areas with increased air dust concentrations and infrequent rainfall, panel surface soiling is unavoidable. Cleaning these panels is necessary for the solar cells to produce electricity as efficiently as possible. For cleaning the solar panel, there are some methods available. These methods are in manual mode or in automatic mode. Nearly half of O&M cost in a solar facility is attributable to cleaning charges. IoT revolutionizes photovoltaic setups through the implementation of intelligent monitoring and management systems. These systems utilize sensors to gather immediate information on panel efficiency, energy generation, and surrounding conditions, enabling distant oversight and proactive maintenance. This innovative approach enhances energy efficiency, quickly identifies issues, and empowers informed decision-making, contributing to more sustainable solar energy production.

2.1 Current Technologies

Solar panel washing has traditionally been done with rainwater as a medium. A clean panel is better at capturing solar energy, which produces more electricity. The need for manual cleaning of solar panels is gradually disappearing because to advances in material science and robotics.

To effectively clean a panel, a lot of modern technology is available, including EDS, robots, and cleaning kits. The following subsections explain some of the techniques and tools:

(i) **Manual cleaning:** An operator can manually clean the solar panel using a variety of tools and methods.

There are some methods for manual cleaning such as cleaning equipment kit, it contains brushes, towels, hose connectors, carrying bags and multiple extension poles [10]. For cleaning solar panels, many businesses manufacture manufacturing brushes and hose systems, which are easily accessible on the market. In manual cleaning method it is very tough to clean manually because of cost and demands a lot of labor. It is also increased the O & M cost.

(ii) **Piezoelectric system:** The second method to clean solar panel is piezoelectric system. In this method Piezo electric actuators are employed in the variety of methods. They are optical adjustments, manipulation of biomedical instruments, space expeditions and other electronic equipment because of flexible design, higher torque to volume ratio and accuracy. Solar panel cleaning is done using a piezoelectric method. The solar panel's surface is cleaned by the acoustic piezoelectric system while the compression waves are in their rarefaction cycle by spreading 0.1 to 1 mm of water around it. When a liquid undergoes rarefaction, an ultrasonic cavity is generated, and it is found that the cavity cleans the solar panel by drawing up any dust that may be on its surface. The method is the same for changing the cleaning medium just when it comes to the medium of air. Based on a linear piezoelectric actuator. There is a one phenomenon explained by adding proper field required for the dust particle mounted over the surface. (iii). It is necessary create electric curtain which is a standing wave. This wave is an electromagnetic wave. This wave particular value of magnitude and is oscillating at a particular frequency. During oscillation, the charged dust particles are bound to form phenomena known as charging dust particle. The oscillation frequency is chosen so that dust particles travel to one of the module's edges following the path of an electric field, cleaning the surface as they go. Uncharged particles that fail to form the electric curtain quickly get charged via polarization or electrostatic induction. An Electro-Dynamic Screen (EDS) is a unique invention that operates on the same electric current standing wave principle. Instead, a travelling wave with significant amount of translational energy has high voltage. This uses three phase electric energy source. In order to use this system in places having low humidity and the precipitation does not occur. In order to prevent vapor based bonding present in the dust particles of the module, dry module is required for cleaning. Based on the survey regarding the concentrated solar pants, their application is restricted to areas with relative humidity levels around 60%, and such systems are anticipated to recover the reactance lost owing to soiling with an efficiency of 90%. A translucent self-cleaning nano-1m can be applied to the surface of a solar panel in order to prevent dust from adhering to the panel's surface. Extremely hydrophilic or super hydrophobic materials are used to create the self-cleaning nano-lm. The super-hydrophilicity approach uses rainfall to clear the dust and disperse it throughout the solar module. Because of this, there is ongoing research and this method is not widely used. When a substance is extremely hydrophobic, water droplets swiftly fall off and bring dust particles with them, much like lotus plant leaves do. Numerous studies are being conducted to create super hydrophobic surfaces by creating micro- or nanostructures. Given that solar farms are located in areas with infrequent rain, using these materials on the surface of the solar panel would be controversial. (v) Robotic system: Compared to the other methods already mentioned above, the robotic system is the most in-demand due to its broad range of use in both small and big

PV systems. The robotic system is a virtual one that comprises of actuators, motors, and gears with some movement above the module surfaces. Operator that thoroughly cleans the module, outperforming human hand cleaning. The development of 3D printing and nanotechnology is supporting the formation of extremely complicated robots that can function as effectively as a person would. The cleaning task is now even simpler because to recent advancements in automation, which reduces complexity between the operator and the robot. The output current value and the sun irradiance value are fuzzicated in the fuzzy logic-controlled method outlined in, which leads to a choice about whether to clean or not. The technology field's burgeoning Internet of Things (IoT) discussion could significantly alter the way solar panels are cleaned. The PV system is sensed using sensors. This will transfer the required data to the cloud. This method is used to reduce the cost system. The cost of inspection is reduced by such an IoT-enabled system, and the cleaner team can only send out their robots when necessary. Even site visits can be avoided with the use of IoT, preventing human participation in the cleaning process. The development of a system with the name smart solar photovoltaic panel Cleaning System has provided evidence for this theory. It is designed with the two components. First one performing cleaning tasks and another unit that makes decisions. The autonomous unit analyses the input and output parameters of solar radiation and generated power to the model and execute cleaning operation. Both the cleaning system and supervisor system have access to internet connectivity, the system can be watched with the help of a person stationed anywhere world. An integrated system leveraging IoT technology is designed to effectively identify, oversee, and autonomously clean soiled PV solar panels, amalgamating diverse technologies for this purpose sensor integration, IoT connectivity, data collection and analysis and threshold setting.

2.2 Present Practices

Any PV glass surface can be cleaned in one of two ways: with water and cleaning solutions, or without water with specific brushes. Most cleaning procedures combine one of the aforementioned two procedures with the use of a robotic device. While other PV sites that have access to plenty of water are periodically cleaned with sprinklers to prevent soiling problems. Location of the place, configurations of the array, size of the plant, fluctuations in weather, and other factors all have a role in the choice of cleaning mechanism, ensuring cost of panel cleaning process is less when compared the power generation improvement after the cleaning process.

i. Cleaning System using Robots

Water is the main component of water-based cleaning solutions, a similar method of cleaning using robotic solution for SPV panels, the GB1 from Greenbotic, is wireless and rechargeable. The mechanism which was used to rotate the brushes that are perpendicular to the PV panel axis. It is not cleaning the panel and additionally it removes the sludge's in the panel. For the purpose wireless transmission using robot, hector is widely used. This carries a water solution tank that can independently move around the modules. In market the robotic system and water is used to clean the process. These systems simply differ in terms of how they use

ii. Robotic cleaning methods without water:

Water less cleaning method is also a most effective method to clean the PV panel. In desert areas and arid areas, the water scarce is more and also the dust accumulated is more. In waterless cleaning technology, the dust is removed by rotating or moving a brush that is normally dust resistant across the panel's surface. In dry areas, this technique is effective. As there is no need to store water or pump high-velocity water into the panel, and as a result, there is less complexity involved in the robot cleaning without the use of water. Solar Brush is used for cleaning the system. The Solar Brush drone from Aerial Power is aiming to spread drone cleaning technology throughout the world's desert regions. In a similar vein, Boson solar farm robot that can clean the panel surface without the use of water. With the help of its solar-powered unit, this IoT-connected system cleans the panel at night. This robot also has a gap-spanning feature, which allows it to avoid the issue with previous robots that they can collide with gaps in solar farm panels. Additionally, a mechanism for automatic cleaning has been introduced to this system to increase its dependability. There are numerous more products with waterless methods to effectively clean the panel available in the market.

iii. Automated water sprinkler system:

In areas abundance of water, a system can be installed that conducts routine cleaning operations using only water. Automation for the regular cleaning must be used should regularly clean so that no bird droppings or soiling particles are permitted to remain for an extended period of time. These Systems practically simulate rain in the solar plant by spraying water into the panels directly. The panels' slant causes them to sweep dirt and water to the ground. Through nozzles linked to the solar panels, it automatically washes and rinses the panels. The controller for these systems automatically offers wash and rinse cycles.

On solar panels, dust can occasionally accumulate because of weather changes and winds, which can lower the efficiency of the solar panel. Additionally, manual cleaning can occasionally put workers in danger because the panels can occasionally be positioned in dangerously high places, making it risky to reach the places frequently. It can also be concluded that the cost of the system is more in the places where there are more number of photovoltaic panels present. When the panels increases, the people working to clean that also

Table 30.1 Efficient Decrease in terms of Duration

Days/Month	Efficiency Drop
7 Days	10%
30 Days	20%
60 Days	50%

increases. This increases the wage to be paid to the workers.

There are different possibilities of efficiency drop when the solar panel is not cleaned. Table 30.1 shows the output drop of the panel in terms of duration.

3. Design and Implementation

LDR in the system is used, which recognizes the. Dust on the PV panel means dust on the LDR, and dust on the LDR results in a voltage drop that may be used to measure the amount of dust present on the photovoltaic panel. The node MCU is supplied with the output from the LDR and solar panel.

For no dust circumstances, we have set the LDR output in this system to 3.3 volts, which results in very high light intensity. The voltage for a clean, dust-free solar panel is 3.3 volts. The water sprinkler system is currently stationary. IoT Based system allows us to see the live streaming of the solar panel. This streaming helps us to see whether dust is accumulated heavy or moderate. And also, we can able to see the cleaning process so that after the solar panel looks clean, we can able to turn off the cleaning system. The accumulation of dust is detected by camera on photovoltaic panel, and this camera is connected to the Node-MCU, its output can be seen through the Blynk app. The Fig. 30.1 shows the solar panel before and after cleaning

Blynk simplifies the process of creating IoT applications, offering a versatile platform. Through its application, we remotely control hardware devices like the NodeMCU ESP8266 WiFi module via an intuitive interface on your smartphone.

Using Blynk, we can personalize your mobile interface to seamlessly interact with the NodeMCU ESP8266. This interface allows for options such as controlling lighting through buttons, adjusting temperature via sliders, or monitoring real-time data on displays. Additionally, the platform enables the reception of notifications and alerts triggered by specific events or conditions from the connected hardware. The user who controls the cleaning system can be seen through the Blynk app and can operate the water sprinkling system. If the user presses ON, the Node-MCU activates the relay, and this relay allows a water line to the water pump and this pump sends the water to the sprinkling system with a rated pressure. Humidity and temperature sensors will show their values every time. With the help of the velocity sensor, we can able to know the efficiency changes of the panel. It's shown in Figure 30.2.

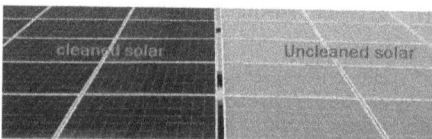

Figure 30.1 Solar Panel View Before and After Cleaning

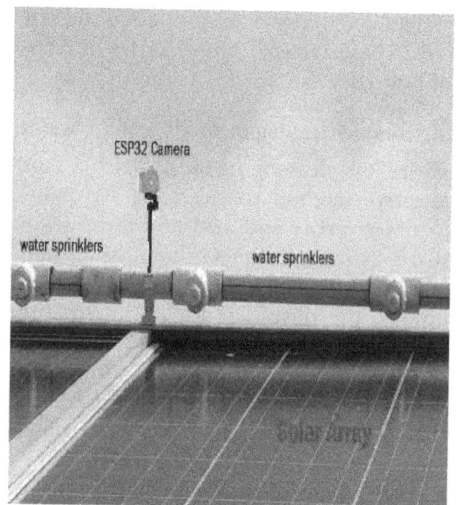

Figure 30.2 Pv Panel Arrangement for Cleaning

A. Blynk APP

It is known as the open-source and free application which is designed for IoT based system connectivity operation. This application allows to connect and interact with the cleaning system with the help of a mobile phone. In this app, we can able to see various data parameters which are shared by different sensors. This app is known to be a flexible application, which means this app can be modified according to the user's need or based on their project requirement. For supporting all types of platforms, Blynk has various libraries. It consists of three sections or columns.

B. Camera Live Stream

This section will allow us to see the live streaming of the solar panel. This streaming helps us to see whether dust is accumulated heavy or moderate. And also, we can able to see the cleaning process so that after the solar panel looks clean, we can able to turn off the cleaning system.

C. Water Sprinkling System

This section is known as the controller part of the system. It looks very simply

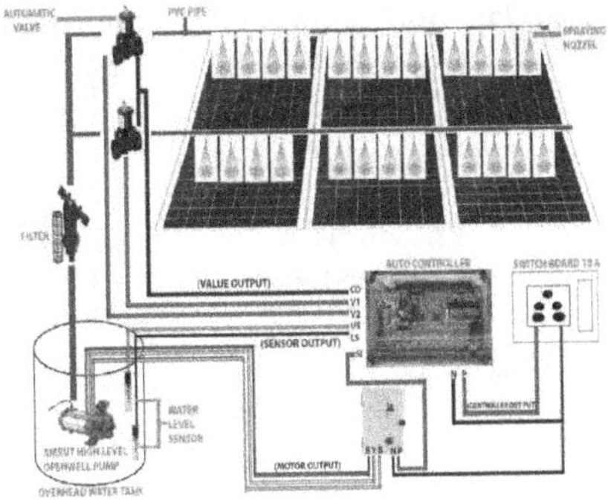

Figure 30.4 Overall Block Diagram

to operate, as it contains just "ON" and "OFF" buttons. Just with one click we can on or off the cleaning system whenever it is required. Fig. 30.3 shows the IOT based with water sprinkling system.

D. Humidity and Temperature sensors

This section is allotted for both humidity and temperature sensors. A humidity sensor is used to know the moisture content present in the air. A temperature sensor is used to detect the atmosphere temperature.

4. Proposed Methodology

This structure is the basic layout of the proposed system. From this, we can assume the various hardware, input, and output connections of this system. The hardware consists of a Solar panel (or) PV cell, Node MCU ESP8266 WIFI Module, Voltage Sensor, Relay Module, Water Pump, sprinkling system, Blynk Application, Humidity Sensor, and Temperature Sensor. A water pump and an operating system make up the system. The device features a microcontroller unit that is designed so that each water spraying segment receives the full water pressure generated by the solar pump separately. As a result, high pressure water is sprayed on each row of solar panels in the solar power plant one by one to clean them all. The device also consists of a timer that enables the system on/off this system automatically at a predefined

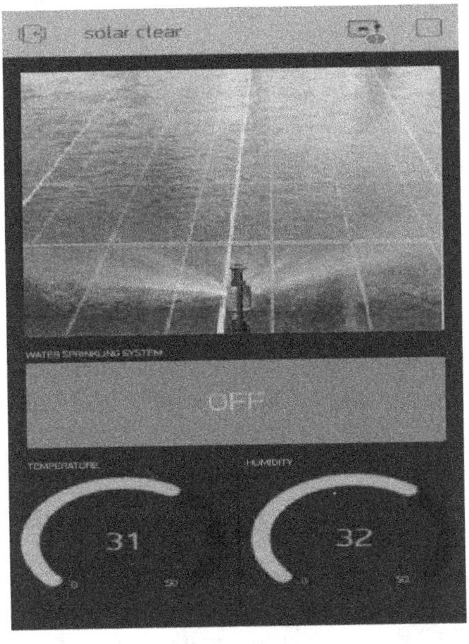

Figure 30.3 IoT based System Results

interval. Fig. 30.4 shows the overall block diagram of proposed methodology.

Overall components of this system are connected to the Node-MCU. And this chip is connected to the Blynk server which is known as the Brain part (or) controller of the system. Camera is used to detect the dust particles, and this camera is connected to the Node-MCU, its output can be seen through the Blynk app. The user who controls the cleaning system can be seen through the Blynk app and can operate the water sprinkling system. If the user presses ON, the NodeMcu activates the relay, and this relay allows the water line to the water pump, and this pump sends the water to the sprinkling system with a rated pressure. Humidity and temperature sensors will show their values every time. When these sensors detect a rise in dust particle levels in the air, they can prompt the water sprinklers to activate, aiding in settling the dust and thereby enhancing the air quality. With the help of the velocity sensor, we can able to know the efficiency changes of the solar panel.

5. Results and Output

The productivity of solar panels has been examined in this work using two distinct ways, and the findings were astounding. The power generation increases by 10%, if the panel is cleaned regularly. Fig. 30.5 shows the experimental setup for the proposed methodology. Table. 30.2 shows the output of before and after cleaning results

The above chart and values show the output. These values are taken in two

Figure 30.5 Experimental Setup

Table 30.2 Output Before and After Cleaning

Date	Before Cleaning	After Cleaning
01-05-2023	15.2	21.9
02-05-2023	17.5	21.6
03-05-2023	18.7	21.7
04-05-2023	17.2	21.4
05-05-2023	16.9	21.7
06-05-2023	17.7	21.6
07-05-2023	16.4	21.5

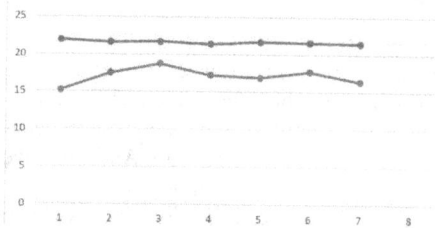

Figure 30.6 Output Before and After Cleaning

conditions. One is before cleaning and after cleaning. Before cleaning it will take readings from the voltage sensor outputs in watts for one day. And after cleaning also the same procedure is done. Fig. 30.6 indicates the response of system before and after cleaning.

After observing the above graph, the power efficiency must be increased in the solar panel after the cleaning. By using water spray and a rubber wiper. After cleaning the solar panel the output decreases. The accuracy of this system is 96%. Most of the dust accumulated because of bird drop, dirt and a speck of dust. There is a another method to clean the PV panel. The name of this method is Dry cleaning.

In this method it uses rechargeable batteries and no other source is need. It reduces the cost of the whole systems and uses light weight components. By using this method this increases the power. Table 30.3 shows the output voltage, current and power before and after cleaning.

Output	Output (Before Cleaning)	Output (After Cleaning)
Voltage 17V DC	12.08V DC	14.06V DC
Power 03 W	1.8W	2.5W
Operating Current 0.18 A	0.12	

6. Conclusion

In this paper, porotype for cleaning solar panels utilizing a Node-MCU, camera module, and Blink services is illustrated. It is employed for cleaning the solar panel when it has gathered dust. In this prototype, a camera module is used for continuously monitoring. The final power of the PV module is analyzed before and after cleaning the solar panel, which clearly shows there is a significant improvement generation of power from the solar panel.

References

[1] Rahkes Anahd atohT, Malika, Gayathri Hari, Mariya Roy, Roopa Ann Mathew3 and Bose Mathew Jos "Solar Rescue Robot" International Research Journal of Engineering and Technology (IRJET), Volume: 04 Issue: 05 | May -2017.

[2] Ashish Saini, Abhishek Nahar, Malika T, Amit Yadav, Arnim, Dhruvash Singh Shekhawat, Mr. Ankit Vijayvargiya "Solar Panel Cleaning System" Imperial Journal of Interdisciplinary Research (IJIR), Vol-3, Issue-5, 2017.

[3] Milan Vaghani, Jayesh Multipara, Keyur Vahani, Jenish Maniya,Deekahea Prof. Rajiv Kumar Gurjwar "Automated Solar Panel Cleaning System using IoT", International Research Journal of Engineering and Technology (IRJET), Volume: 06 Issue: 04 | Apr 2019.

[4] D.C. Jordan and S.R. Kurt z, "Photovoltaic Degradation Rates – An Analytical Review", Progress in Photovoltaics: Research and Applications, vol. 21, no. 1, pp. 12-29, Jan. 2013.

[5] Antonanzas, J., Osorio, N., Escobar, R., Urraca, R., Martinez-de-Pison, F. J., & Antonanzas-Torres, F. (2016). Review of photovoltaic power forecasting. Solar energy, 136, 78-111.

[6] Brabec, M., Pelikán, E., Krc, P ., Eben, K., Maly´, M., Juruš, P, " A coupled model for energy production forecasting from photovoltaic farms", In: ES1002, Workshop March 22nd–23rd, 2011.

[7] Lorenzo, E., Moretón, R., & Luque, I. (2014). Dust effects on PV array performance: in-field observations with non-uniform patterns. Progress in Photovoltaics: Research and Applications, 22(6), 666-670.

[8] Abubakar, A., Almeida, C. F. M., & Gemignani, M. (2021, August). A review of solar photovoltaic system maintenance strategies. In 2021 14th IEEE International conference on industry applications (INDUSCON) (pp. 1400-1407). IEEE.

[9] Adinoyi, M. J., & Said, S. A. (2013). Effect of dust accumulation on the power outputs of solar photovoltaic modules. Renewable energy, 60, 633-636.

[10] M. Mazumder, M. Horenstein, J. Stark, J. N. Hudelson, A. Sayyah, C. Heiling, and J. Yellowhair, "Electrodynamic removal of dust from solar mirrors and its applications in concentrated solar power (CSP) plants," in Proc. IEEE Ind. Appl. Soc. Annu. Meeting, Oct. 2014, pp. 1–7.

31 Exploitation of Spectrum White Spaces by Using Energy Based Primary Transmitter Detection Approach for Cognitive Radio Networks

K. Jeevitha[1], R. Srinath[2], V. Vinoth Kumar[3], C. Shalini[4], B. Sarala[5], J. Jasmine Hephzipah[6], and M. Perarasi[7]

Department of ECE, R.M.K. Engineering College, Department of ECE, SRM Institute of Science & Technology, Kattangulathur, Department of ECE, R&D Institute of Science and Technology, Chennai, Department of ECE, TJS Engineering College, Peruvoyal, Department of ECE R.M.K. Engineering College, Chennai, Department of ECE, R.M.K. Engineering College, Kavaraipettai, Department of ECE, R.M.K. Engineering College Kavaraipettai

Abstract

Advancements in wireless technology and usage of many wireless devices have led to conspicuous concerns on bandwidth, the scarce natural resource. Further development of wireless system plays a pivotal role in utilization of spectrum and also concerned about underutilized ones. With the introduction of efficient and reliable system called cognitive radio the bandwidth shortage problems can be greatly reduced, the reason is it provides dynamic access over the spectrum by sensing the absence of primary used for allocating it to unlicensed users in the case of emergent application. Since the allocation depends on spectrum sensing, a meticulour investigation has been carried out on sensing from which literature states that non co-operative detection techniques performance can be enhanced in large. This paper focuses on energy based detection of primary users, which can be identified by evaluating the detector performance, based on constant rate of detection, alarming and threshold value metrics. An optimal threshold value is derived by ordained detection and alarming probabitlity analysis. Simulative interference reveals that the Monte Carlo based approaches has reduced the computational complexity and moreover modeling points has been increased as with the number of samples.

Keywords: Rate of detection, alarming, threshold value metrics and Monte Carlo simulation

Email: Kja.ece@rmkec.ac.in[1], drsrinathrajagopalan@gmail.com[2], icevinoth@gmail.com[3], drshalinichowdry@gmail.com[4], saralailangovan@gmail.com[5], jjh.ece@rmkec.ac.in[6], mpi.eee@rmkec.ac.in[7]

DOI: 10.1201/9781003661917-31

1. Introduction

Government agencies auctions and allocates the spectrum for use over long duration of time in the name of fixed spectrum allocation and the user guaranteed with a authority is termed as a licensee [24]. This fixed allocation scheme works well for certain application say, telephony and broadcast services, but for emergent applications it demands for high data rates and the survey report of FCC shows that certain portions of spectrum are left underutilized [5, 24].

A viable and sagacious solution to solve this inefficient usage is opportunistic dynamic allocation of spectrum by exploitation of white spaces using cognitive radio, which concerns mainly on two aspects (i) Secondary user communication should not interfere with the primary user, (ii) Effectively identify spectrum gaps to increase system throughput and meet the necessary QoS.

These radios are currently conducting analysis over the radio frequency environment in order to exploit weaknesses, and the main element of this scrutiny is sensing, which determines whether a user is there or not as well as when the principal license holder arrives. In the case of user's absence, it is termed as vacant space. In order to exploit vacant space, the cognitive radio has to sense both temporal and spatial variations, but the underlying problem in licensed band is simultaneous sensing and transmission is quite difficult. Hence Secondary user has to periodically sense the communication opportunity available, which is known as sensing period.

The primary user's performance is entirely dependent on the sensing time and duration. While expanding the sensing duration may boost secondary user throughput may be increased as shown in Fig 31.1, it ignores primary user availability sensing. Minimizing the sensing time achieves higher throughput for secondary user without forgoing primary user consistency over sensing. More over cognitive radio has to model certain uncertainties due to noise. The reason behind modeling of noise uncertainty is

that the radios have to detect the user signal with low SNR else it leads to wastage of transmission opportunity over accessible spectrum or even tend to cause interference. This leads to errors in binary hypothesis test classification [5, 24].

2. Binary Hypothesis Test

The primary user's performance is only influenced by the sensing time and length. While primary user availability sensing is ignored when the sensing length is extended, secondary user throughput may be increased.

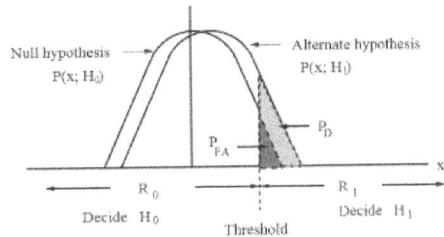

Figure 31.1 Binary Hypothesis Formulation.

Hypothesis test states

i. When the primary user is not active (assuming the channel is vacant), the received signal at the unlicensed or secondary user is expressed as Null Hypothesis H0.

ii. H_0: $x(n) = \omega(n)$
Where $\omega(n) =$ noise

iii. Otherwise, the received signal is expressed as Alternate Hypothesis H1 if the principal user is active (assuming that the channel is occupied).
H_1: $x(n) = s(n) + \omega(n)$
$\omega(n) =$ noise and $s(n) =$ Primary User Signal

But the uncertainties due to noise present and also limited observations has led to some hinder in the above classification, which might lead to either false alarming or missed detection of signals i.e., basically

i. Channel is Unoccupied, but spectrum sensors decide as Occupied which leads to wastage of transmission opportunity. $H_1 | H_0$

ii. Channel is occupied, but spectrum sensors decide as Unoccupied which leads to Interference. $H_0 | H_1$

$$y(n) | x(n) = \{\omega(n)s(n) + \omega(n)\}$$

h-channel gain, which is a complex variable $h = hr + jhi$ constant, and $w(n) = wr(n) + jwi(n)$ is a noise sample with $Var(\omega(n)) = 2\sigma^2\omega$.

3. Performance Metrics

i. Probability of false alarm is the chance that the signal will exist while H0 is true. (P_f).

$$P_f = p_r[\Lambda > \lambda | H_0]$$

Where λ = Threshold value for detection

ii. The missed detection probability, or PMD, is the chance that the signal will be absent when H1 is true.

$$P_{md} = p_r[\Lambda < \lambda | H_1]$$

Missed detection roots unpredicted intrusion to primary users

iii. The Probability Detection (Pd) - likelihood that the signal will be detected while H1 is true.

$$p_d = p_r[\Lambda > \lambda | H_1]$$

$$p_d = 1 - p_{md}$$

Where $p_r[.]$ is the event probability To better exploit spectrum holes, the detection probability should be as high and P_f and P_{MD} would be less than or equal to 0.1. The URL http://www.ieee802.org

4. Threshold Calculation

i. **To determine the likelihood of a false alert, use**

$$p_{fa} = \text{Prob}(\mu > \lambda | H_0)$$

$$= Q\left(\frac{\lambda - \sigma_n^2}{\sigma_n^2 / \sqrt{\mu/2}}\right)$$

Where $Q(.) = Q(x) = \dfrac{1}{2} erfc\left(\dfrac{x}{\sqrt{2}}\right)$

Erfc - Complementary error function (x).

$$\underset{=}{\Delta} \frac{2}{\sqrt{\pi}} \int_x^\infty e^{-t^2} dt$$

The value of threshold needed to detect the primary signals based on predetermined value of false alarming probability λ_{fa}

$$\lambda_{fa} = \sigma_n^2 \left(1 + \frac{Q^{-1}(p_{fa})}{\sqrt{\mu/2}}\right)$$

ii. The likelihood of detection is provided by

$$\lambda_d = \sigma_n^2 (1 + SNR)\left(1 + \frac{Q^{-1}(p_d)}{\sqrt{\mu/2}}\right)$$

The detection probability should be as high as possible and alarming probability values as low as possible for Non interference communication

H_0 represents no signal transmitted and H_1 be signal transmitted $\dfrac{x(t)}{s(n)}$ be deterministic signal transmitted and $\omega(n)$ be AWGN with zero mean and var $\sigma_n^2 = \omega_{n0}$

The PU Signal's mean feature is used to increase the channel sensing's effectiveness. And the SNR is estimated for the channel observation and is signified as $\gamma = \dfrac{\sigma_s^2}{\sigma_n^2}$

Where σ_s^2 and σ_n^2 variance of noise & Signal

iii. **Probability of detection for AWGN channel**

$$\omega(n) = \omega_c(t)Cos\omega_c t + \omega_q(t)Sin\omega_c t$$

$$\int_0^T \omega^2(t) \cdot dt = \frac{1}{2} \int_0^T \left[\omega_i^2(t) + \omega_g^2(t)\right] dt$$

$$\omega_i(t) = \sum_{k=-\infty}^\infty a_{ik} \, Sin \, c(2\omega t - k)$$

Where $Sincx = \dfrac{Sin\,\pi x}{\pi x}$

$$a_{ik} = \omega_i\left(\frac{k}{B_N}\right) \text{ Gaussian random variable}$$

with zero mean a variance.

$$\sigma_k^2 = 2N_0 B_N, \forall k$$

$$\int_{-\infty}^{\infty} \text{Sin}c\left(B_N t - k\right) \text{Sin}c(B_N t - m)dt$$

$$= \begin{cases} \dfrac{1}{B_N}, k = m \\ 0, \quad k \neq m \end{cases}$$

Therefore $n_i(t)$ becomes

$$\int_0^T n_i^2(t)\,dt = \left(\frac{1}{B_N}\right)\sum_{k=-\infty}^{\infty} a_{ik}^2$$

As $n_i(t)$ has degree of freedom $B_N T$ over time duration $(0, T)$, $n_i(t)$ becomes

$$n_i(t) = \sum_{k=1}^{B_N T} a_{ik} \text{Sin}c\left(2B_N t - k\right) \quad 0 < t < T$$

Therefore the $\int_0^T n_i^2(t)\,dt$ over Time duration

T can be expressed as

$$\int_0^T n_i^2(t)\,dt = \left(\frac{1}{B_N}\right)\sum_{i=1}^{TB_N} a_{ik^2}$$

Substituting $\dfrac{a_{ik}}{\sqrt{2B_N N_0}} = b_{ik}, \dfrac{a_q k}{\sqrt{2B_N N_0}} = b_{qk}$

We get $\int_0^T n^2(t)\cdot dt = \left[\sum_{i=1}^{TB_N} b_{ik^2} + \sum_{i=1}^{TB_N} b_{qk^2}\right]\cdot N_0$

Similarly, for transmitting s(n)

$$\int_0^T s^2(t)\cdot dt = \left[\sum_{i=1}^{TB_N} b_{ik^2} + \sum_{i=1}^{TB_N} b_{qk^2}\right]\cdot N_0 = \frac{E_s}{N_0}$$

where $b_{ik} = \dfrac{x_i\left(\dfrac{k}{B_N}\right)}{\sqrt{2B_N N_0}}, b_{qk} = \dfrac{x_i\left(\dfrac{k}{B_N}\right)}{\sqrt{2B_N N_0}}$

$$E_s = \int_0^T s^2(t)$$

Where the output of integrator in equal to

$$\frac{1}{T}\int_0^T y^2(t)\cdot dt$$

Assuming y' as test inference,

$$y' = \frac{1}{N_0}\int_0^T y^2(t)\cdot dt$$

i. If hypothesis H_0 states received signal is of noise the test inference can be of

$$y' = \sum_{k=1}^{B_N T}\left(d_{ik^2} + d_{qk^2}\right)$$

ii. If hypothesis H_1 states received signal is of signal an

$$\int y^2(t)dt = \left[\sum_{i=1}^{TB_N}(b_{ik} + d_{ik})^2 + \sum_{i=1}^{TB_N}(b_{qk} + d_{qk})^2\right]\cdot N_0$$

Here H_0 is Si square distribution and H_1 is non si square distribution

5. Detection Techniques

Certain sensing techniques have been documented in the literature based on the aforementioned notion. Among these methods of detection Since energy detection is less complicated and does not require the previously specified information about the principal user, it is a simpler and more well-known method as opposed to matching

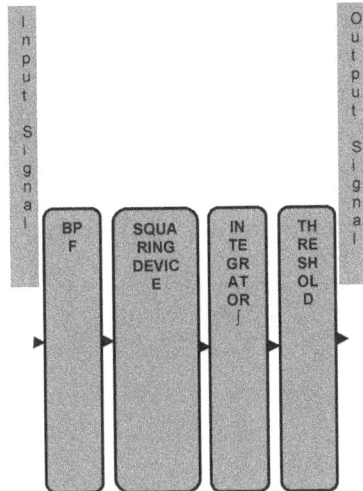

Figure **31.2** System Model for Energy Based Spectrum Sensing [5,7, 30, 41]

filter and feature recognition, which takes a long time to identify user signals [24].

According to the literature, radiometry methods can be used in both the frequency and temporal domains. Zero-Mean Constellation signals, which quantify signal strength received over a certain period and bandwidth, are best detected using this method. Signal power samples are measured in the time domain based on the input signal that is sent to the BPF, while the time domain data can be transformed into a frequency domain signal using FFT. This allows one to determine the channel occupancy status.

To choose the bandwidth array, the contribution sign is permitted to pass through a BPF. After that, output is squared with a squaring device and integrated for a predetermined amount of time. Figure 31.2 illustrates how the integrator's signal output is compared to a predefined threshold value, resulting in test interferences that allow for the inference of the primary user's presence or absence [5, 7, 30, 41].

The transfer function can be expressed as

$$H(f) = \begin{cases} \dfrac{2}{\sqrt{N_{01}}}, & |f - f_c| \le w \\ 0, & |f - f_c| \ge w \end{cases}$$

where W cycles per second is the noise bandwidth. This is normalized to calculate the P_f and P_D using this associated transfer function. V', is then compared to the threshold after the filter's output has been squared and integrated throughout the interval. Thus, the receiver determines that the target signal has been detected if and only if the threshold is crossed.

The noise signal that has been rebuilt is

then $n(t) = \displaystyle\sum_{i=-\infty}^{\infty} n_i \, \text{Sinc} \, (2\,\omega t - i)$

$n_i = n\left(\dfrac{i}{2\omega}\right)$ is the sound of the i-th sample

$V = \displaystyle\int_0^T \dfrac{(n(t))^2}{(\omega(n))^2} \, dt \approx \dfrac{1}{2\omega} \sum_{i=1}^{2T\omega} \omega_i^2$

On simplification it becomes

$$V = \dfrac{1}{\omega N_0} \sum_{i=1}^{2T\omega} |y(i)|^2 \sum_{i=1}^{2T\omega} \overline{y}i^2 \begin{matrix} H_1 \\ \lessgtr \\ H_0 \end{matrix} \lambda$$

Therefore, both cases the test inferences

are $V \sim \begin{cases} H_0 : \chi_{2T\omega}^2 \\ H_1 : \chi_{2T\omega\,(2\gamma)}^2 \end{cases}$

Where
$\chi_{2T\omega}^2$ is Si Square of distribution,
$2T\omega$ is Degree of liberty and
$\chi_{2T\omega\,(2\gamma)}^2$ is distribution with the same DoF that is not Si Square.

6. Design Flow

This flow discusses the initial phase of work started using Matlab-Simulink, where the inferred results shows that allocating the white space to secondary user is very crucial due to less bandwidth availability.

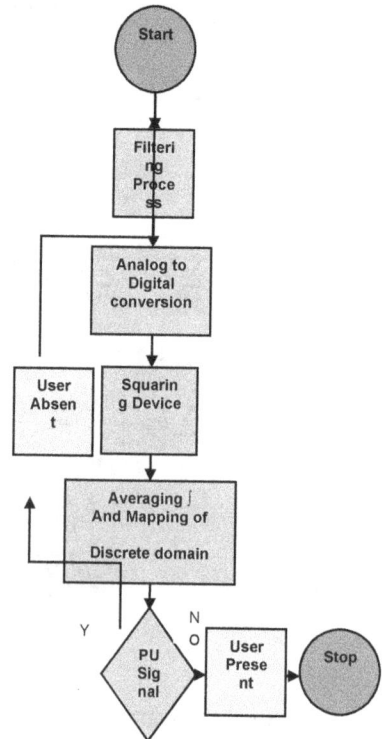

Figure 31.3 Simulation Work Design Flow [17, 19, 24]

To trounce the above issues, the same block has been implemented using system generator tool kit of Xilinx14.2, has successfully achieved the placing of unlicensed user in that particular bandwidth which is free of interference i.e., reduced interference. But a challenge of 1mSec time delay in sensing has occurred as shown in Fig 31.3. To implement it real time with respect to circuit level, the process has been extended and taken to Cadence Virtuoso, with a thought of reducing noise and also achieved it with an added benefit of area optimization.

7. Results and Discussion

Inference from Figure 31.5 shows that Detection is more accurate but collisions occurs, moreover Simulative results reveals that detection probability as well as false alarming rate increases due to presence of AWGN, furthermore the detection levels is of a predetermined value of 1. If the detection instants are reduced by half the harmonics dominates. If the threshold value is less than predetermined, false alarming rate tends to be maximum. The false alarm rate can be decreased by enhancing threshold levels and also by reducing AWGN thereby probability detection gets improved as shown in Fig 31.4. To overcome these collision issues the design has been implemented using System Generator as shown in Figure 31.6.

A secondary user can be readily positioned between the major users without causing any collisions, as Figure 31.7 illustrates, although the likelihood of discovery is generally quite low. Additionally, parametric analysis was performed, as seen in Figures 31.8–31.10 below.

If Threshold points are high error probability will be lesser. For instance, 10 samples have been taken and 5 modeling points, where midpoint peak value will be higher as shown in Figure 31.8.

With a reduced CPU running complexity as shown in Fig. 31.12 and 31.13, the waveform obtained is oscillatory; henceforth so prediction of range is difficult; with peak detection occurs at the earlier stage (0.5 Peak) as shown in Figure 31.9. Figure 31.10 shows that with increased CPU running time the results obtained are as expected, with peak detection at 1.0 and moreover signal is not oscillatory in comparing with previous Figure 31.8. For accurate prediction of User signal we have made circuit level implementation where three bands have been analyzed. Band Pass Filter using RC Network has been designed and AC Analysis had been carried out. Output obtained for Coverage of Bands with centre frequency as 600mV, 10^3 frequency. For Suppression of Negative Cycles, A Squaring Device Using NMOS Current Mirror Has Been Used But Peak Value Detection For

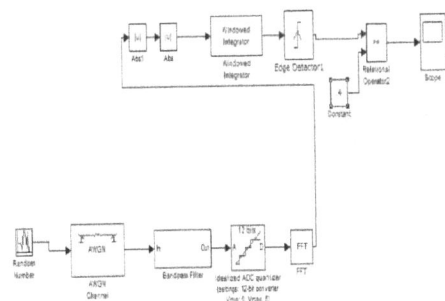

Figure 31.4 Simulink Block Diagram

Figure 31.5 Signal Detection Output

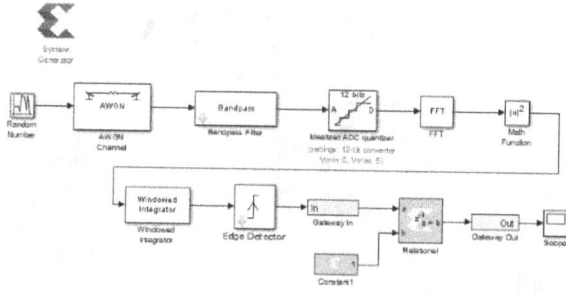

Figure 31.6 System Generator Block Diagram

Figure 31.7 Signal Detection Output

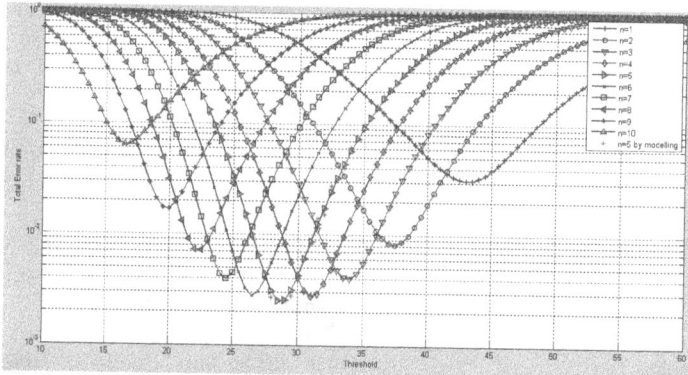

Figure 31.8 Error Rate Vs. Threshold Calculation

Positive Cycle Appears To Be Very Low, So Folded Cascade Design has Been Opted Which Produced Better Output Where Detection Reaches Supply Voltage Of 1.8V. Initially Integrator using RC Network has been considered where User Signals got attenuated and output Range is very low in terms of Pico Volts which leads to misconception of signal whether its User Signals or Noise. Hence Integrator has been designed using Op-Amp, which produced enhanced Output Range of 0–200ms which is in increasing order as shown in Fig 31.11. Then Threshold Device for Comparing System Output with predetermined threshold Reference Signal Range

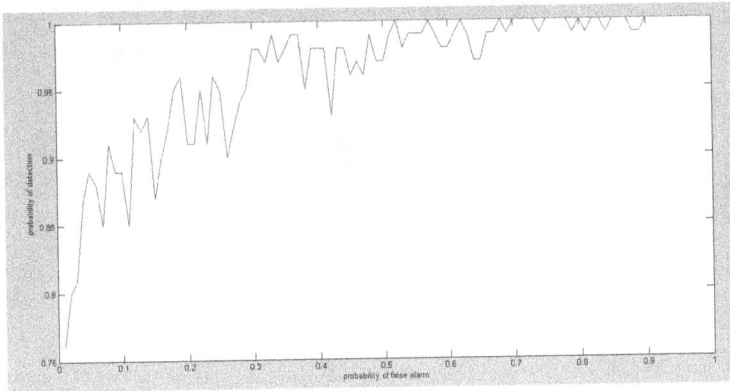

Figure 31.9 P_D Versus P_f.

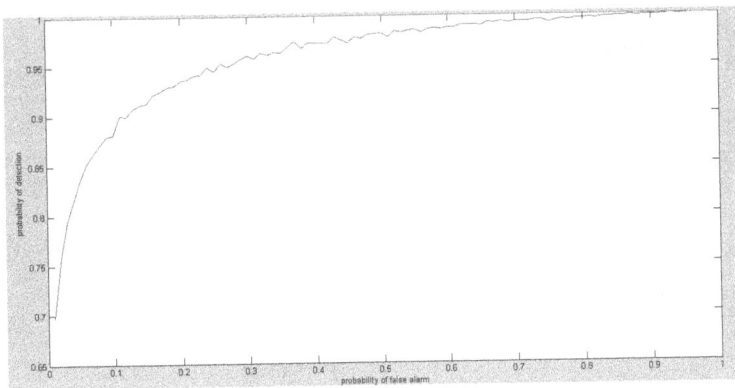

Figure 31.10 P_f Versus$_{PD}$.

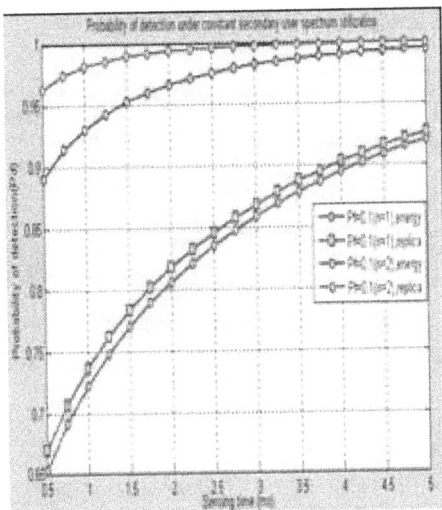

Figure 31.11 Probablity of Detection Under Uncertainity

has been designed with Value - 1, 47.5ms – 87.5ms, 1.8V for User Present Indication. Finally, upon Integrating all Blocks the Output obtained is in range is of 0–12.5ms, 22–127.5 ms for One Cycle of Input shows two number Users present.

8. Conclusion

The theoretical formulation and detection strategies for cognitive radio are analyzed, and it is discovered that energy detection is a straightforward method. Using an energy-based detection system has the advantage of having lower implementation and processing costs. The effectiveness of spectrum sensing techniques is confirmed by numerical and simulative observations. Energy detection works well at low SNR and is not as complicated as other sensing

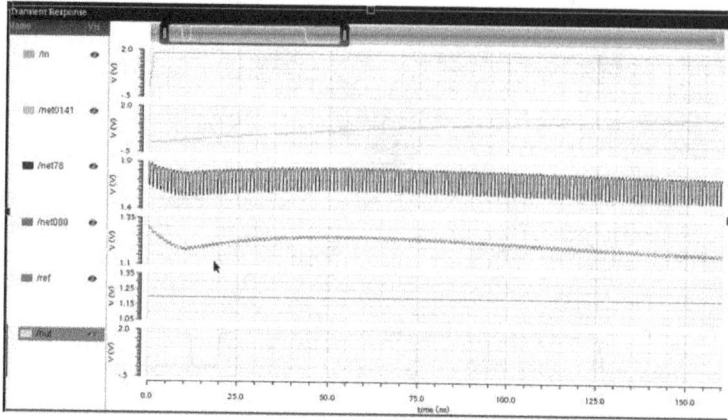

Figure 31.12 Simulated Result

S. NO	POWER	NOISE
BAND PASS FILTER	480.8μW	1.66×10^{-19}
SQUARING DEVICE	56.77μW	3.83×10^{-17}
INTEGRATOR	197.1mW	2.91×10^{-14}
THRESHOLD DEVICE	52.54μW	3.12×10^{-17}
FULL DESIGN	25.32mW	2.45×10^{-14}

Figure 31.13 Simulated Results with Power & Noise

techniques, despite the possibility of missing detection. SNR, false alarm probability, and other contributing factors all affect the likelihood of detection. SNR affects the likelihood of detection; when the former rises, the latter rises on its own.

9. Future Work Enhancements

Cognitive radio is a relatively new field of study where many different spectrum sensing techniques are desired to be investigated, in contrast to the rest of communication theory. Better results can be obtained by using an energy detection method based on the Multiple Input Multiple Output (MIMO) methodology. Using cooperative communication while accounting for antenna diversity and other considerations is another method of spectrum sensing. In the future, the Double Threshold Energy detector can be used to detect energy for cooperative approaches and implement it on hardware. For improved recital, we must apply this concept to cooperative spectrum sensing. Additionally, a comparison performance analysis may be conducted employing both cooperative and non-cooperative sensing. Similar to other hybrid spectrum sensing methods,

References

[1] Bhowmick, A., Roy, S. D., & Kundu, S. (2012). Performance of Spectrum Sensing Scheme Using Double Threshold Energy Detection in the Presence of Sensor Noise. *International Journal of Energy, Information and Communications*, 3(2), 75–84..

[2] Singh, A., Bhatnagar, M. R., & Mallik, R. K. (2011). Cooperative spectrum sensing in multiple antenna based cognitive radio network using an improved energy detector. *IEEE Communications Letters*, 16(1), 64–67.

[3] Jemaa, A. B., Turki, M., & Guibene, W. (2012, June). Enhanced energy detector via algebraic approach for spectrum sensing in cognitive radio networks. In *2012 7th International ICST Conference on Cognitive Radio Oriented Wireless Networks and Communications (CROWNCOM)* (pp. 113–117). IEEE.

[4] Rao, A. M., Karthikeyan, B. R., Mazumdar, D., & Kadambi, G. R. (2010). Energy detection technique for spectrum sensing in cognitive radio. *SASTech-Technical Journal of RUAS*, 9(1), 73–78.

[5]   Jayaprakasam, A., & Sharma, V. (2009, December). Sequential detection based cooperative spectrum sensing algorithms in cognitive radio. In *2009 First UK-India International Workshop on Cognitive Wireless Systems (UKIWCWS)* (pp. 1–6). IEEE.

[6]   Singh, A., & Saxena, V. (2012). Different spectrum sensing techniques used in non cooperative system. *International Journal of Engineering and Innovative Technology*, 1 (2).

[7]   Ebrahimzadeh, A., Najimi, M., Andargoli, S. M. H., & Fallahi, A. (2014). Sensor selection and optimal energy detection threshold for efficient cooperative spectrum sensing. *IEEE Transactions on Vehicular Technology*, 64(4), 1565–1577.

[8]   Avila, J., Thenmozhi, K. (2013). Simulink Based Spectrum Sensing. *International Journal of Engineering and Technology (IJET)*, 5(2), 872–877

[9]   Chaitanya, G. V., Rajalakshmi, P., & Desai, U. B. (2012, July). Real time hardware implementable spectrum sensor for cognitive radio applications. In *2012 International Conference on Signal Processing and Communications (SPCOM)* (pp. 1–5). IEEE.

[10] Wijenayake, C., Madanayake, A., Kota, J., & Bruton, L. (2013). Space-time spectral white spaces in cognitive radio: Theory, algorithms, and circuits. *IEEE Journal on Emerging and Selected Topics in Circuits and Systems*, 3(4), 640–653.

[11] Korumilli, C., Gadde, C., & Hemalatha, I. (2012). Performance analysis of energy detection algorithm in cognitive radio. *International Journal of Engineering Research and Applications*, 2(4), 1004–1009.

[12] Joshi, D. R., Popescu, D. C., & Dobre, O. A. (2011). Gradient-based threshold adaptation for energy detector in cognitive radio systems. *IEEE Communications Letters*, 15(1), 19–21.

[13] Oh, D. C., & Lee, Y. H. (2009). Energy detection based spectrum sensing for sensing error minimization in cognitive radio networks. *International Journal of Communication Networks and Information Security*, 1(1), 1.

[14] Eghbali, Y., Hassani, H., Koohian, A., & Ahmadian-Attari, M. (2014). Improved energy detector for wideband spectrum sensing in cognitive radio networks. *Radioengineering*, 23(1), 430–434.

[15] Digham, F. F., Alouini, M. S., & Simon, M. K. (2007). On the Energy Detection of Unknown Signals Over Fading Channels. *IEEE Transactions on Communications*, 55(1), 21–24.

[16] Fanan, A. M., Riley, N. G., Mehdawi, M., Ammar, M., & Zolfaghari, M. (2014). Survey: A Comparison of Spectrum Sensing Techniques in Cognitive Radio. *International Conference Image Processing, Computers and Industrial Engineering (ICICIE'2014)*, 65–69.

[17] Nautiyal, G., & Kumar, R. (2013). Spectrum sensing in cognitive radio using matlab. *International Journal of Engineering and Advanced Technology (IJEAT)*, 2(5), 529–532.

[18] Gomathi, K., Leela, D., Kanmani Ruby, E. D. (2014). Spectrum sensing methods for Cognitive Radio. *International Journal of Communication and Computer Technologies*, 02(01), 65–68.

[19] Ghosh, G., Das, P., & Chatterjee, S. (2014). Simulation and analysis of cognitive radio system using matlab. *International Journal of Next-Generation Networks*, 6(2), 31–45.

[20] Yu, G., Long, C., Xiang, M., & Xi, W. (2012). A novel energy detection scheme based on dynamic threshold in cognitive radio systems. *Journal of Computational Information Systems*, 8(6), 2245–2252.

[21] Wang, H., Noh, G., Kim, D., Kim, S., & Hong, D. (2010). Advanced sensing techniques of energy detection in cognitive radios. *Journal of Communications and Networks*, 12(1), 19–29.

[22] Harinessha, K., & Roselin Suganthi, J. (2014). Comparison Between Spectrum Sensing In Cognitive Radio Systems. *International Journal of Electronics, Communication and Instrumentation Engineering Research and Development (IJECIERD)*, 4(3), 73–78.

[23] Patil, H., Patil, A. J., & Bhirud, D. S. (2015). Multichannel Cooperative Sensing in Cognitive Radio: A literature Review. *International Journal of Advanced Research in Computer and Communication Engineering*, 4(5), 425–429.

[24] Gadze, J. D., Oyibo, A. M., & Ajobiewe, N. D. (2014). A performance study of energy detection based spectrum sensing for cognitive radio networks. *International Journal of Emerging Technology and Advanced Engineering*, 4(4), 21–29.

[25] Wu, J. Y., Wang, C. H., & Wang, T. Y. (2011). Performance analysis of energy detection based spectrum sensing

with unknown primary signal arrival time. *IEEE Transactions on Communications*, 59(7), 1779–1784. [26]. Kabilesh, T., Vaithianathan, V., & Malarvizhi (2015). Time Domain Analysis of Spectrum sensing in Cognitive Radio by Energy Detection method. *International Conference on Futuristic Trends in Computing and Communication (ICFTCC-2015)*, 11–14, www.internationaljournalssrg.org

[26] Abdulsattar, M. A., & Hussein, Z. A. (2012). Energy detection technique for spectrum sensing in cognitive radio: a survey. *International Journal of Computer Networks & Communications*, 4(5), 223–242.

[27] Mahmood, A., Abdulsattar, K., & Hussein, Z. A. (2012). Energy Detector with Baseband Sampling for Cognitive Radio: Real-Time Implementation. *Wireless Engineering and Technology*, 3, 229–239. Published Online October 2012.http://www.SciRP.org/journal/wet

[28] Subhedar, M., & Birajdar, G. (2011). Spectrum sensing techniques in cognitive radio networks: A survey. *International Journal of Next-Generation Networks*, 3(2), 37–51.

[29] Hossain, M. S., Abdullah, M. I., & Hossain, M. A. (2012). Energy detection performance of spectrum sensing in cognitive radio. *IJ Information Technology and Computer Science*, 11, 11–17. Published Online October 2012 in MECS http://www.mecs-press.org

[30] Mohamad, M. H., & Sani, N. M. (2013). Energy detection technique in cognitive radio system. *International Journal of Engineering and Technology*, 13(5), 69–73.

[31] Murali, K., & Ramesh Babu, N. (2014). Performance Analysis of Non Cooperative Spectrum Sensing Using Search Based Energy Detection. *Journal of Research in Electrical and Electronics Engineering (ISTP-JREEE)*, 3(2), 40–44.

[32] Muthumeenaksh, K., & Radha, S. (2014). Optimal Techniques for Sensing Error Minimization With Improved Energy Detection in Cognitive Radios. *International Journal on Smart Sensing and Intelligent Systems*, 7(4), 2014–2034.

[33] Fatema, N., & Ikram Ilyas, D. M. A. R. (2014). Simulink based Cooperative (Hard Decision Fusion) and Non-Cooperative Spectrum Sensing in Cognitive Radio Using Cyclostationary Detection. *International Journal of Science and Research (IJSR)*, 3(8), 1017–1022.

[34] Beaulieu, N. C., & Chen, Y. (2010). Improved energy detectors for cognitive radios with randomly arriving or departing primary users. *IEEE Signal Processing Letters*, 17(10), 867–870.

[35] Yadav, P., & Mahajan, R. (2014). energy Detection for Spectrum Sensing In Cognitive Radio Using Simulink. *International Journal of Advanced Research in Electrical, Electronics and Instrumentation Engineering*, 3(5), 9367–9372.

[36] Raut, R., & Kulat, K. (2009). Software defined adaptive codec for cognitive radio. *WSEAS Transactions on Communications*, 8, 1243–1252.

[37] Sarala, B., Devi, D. R., & Bhargava, D. S. (2019). Classical energy detection method for spectrum detecting in cognitive radio networks by using robust augmented threshold technique. *Cluster Computing*, 22, 11109–11118.

[38] Sarala, B., Devi, S. R., & Sheela, J. J. J. (2020). Spectrum energy detection in cognitive radio networks based on a novel adaptive threshold energy detection method. *Computer Communications*, 152, 1–7.

[39] Sarala, B., Rukmani Devi, S., Hepzibah, J., Gunasekhar, P., & Sheela, J. J. J. (2021). Simulation and comparison of single and differential ended CG-CS LNA for Cognitive Radio. *International Journal of Wavelets, Multiresolution and Information Processing*, 19(05), 2150013.

[40] Fernandes, R., & Mahamuni, S. (2015). Software Defined Radio Signal Detector Implementation using FPGA. *International Journal of Science and Research (IJSR)*, 4(6), 2476–2479.

[41] Kalamkar, S. S., & Banerjee, A. (2013, February). On the performance of generalized energy detector under noise uncertainty in cognitive radio. In *2013 National Conference on Communications (NCC)* (pp. 1–5). IEEE.

[42] Kalamkar, S. S., Banerjee, A., & Gupta, A. K. (2013, August). SNR wall for generalized energy detection under noise uncertainty in cognitive radio. In *2013 19th Asia-Pacific Conference on Communications (APCC)* (pp. 375–380). IEEE.

[43] Saleem, S., & Shahzad, K. (2012). Performance evaluation of energy detection based spectrum sensing technique for wireless channel. *International Journal of Multidisciplinary Sciences and Engineering*, 3(5), 31–34.

[44] Hanchate, S. M., Nema, S., Pawar, S., & Dethe, V. K. (2014). Implementation of spectrum sensing algorithms in cognitive radio. *International Journal of Advanced Research in Computer Science and Software Engineering*, 4(7), 156–160. [46]. Mahamuni, S., Mishra, V., & Wadhai, V. M. (2012). Efficient energy detection technique in cognitive radio ad-hoc network. *International Journal of Computer Applications*, 46(11), 10–14.

[45] [ED17]. Rasheed, T., Rashdi, A., & Akhtar, A. N. (2014). Spectrum Sensing for Cognitive Radio Users using Constant Threshold Range in Energy Detector. *International Journal of Computer Science Issues (IJCSI)*, 11(5), 125.

[46] Sain, T., & Sharma, K. (2015). Optimization of cooperative spectrum sensing with energy detection in cognitive radio networks using voting rule. *International Journal of Computer Applications*, 127(16), 5–9.

[47] Zhang, W., Mallik, R. K., & Letaief, K. B. (2009). Optimization of cooperative spectrum sensing with energy detection in cognitive radio networks. *IEEE transactions on wireless communications*, 8(12), 5761–5766.

[48] Chen, Y. (2010). Improved energy detector for random signals in Gaussian noise. *IEEE Transactions on Wireless Communications*, 9(2), 558–563.

32   Solar PV Panel Cleaning Techniques and Dust Accumulation Analysis-Review

M. Perarasi[1], Priscilla Whitin[2], Anil Kumar N.[3], Chintala Venkatesh[4], Sai Sharan P.[5] and B. Sarala[5]

[1]Department of Electronics and Communication Engineering, R.M.K. Engineering College, Chennai

[2]Department of Electrical and Electronics Engineering, Vel Tech Rangarajan Dr. Sagunthala R&D Institute of Science and Technology, Chennai

[3]Department of Electronics and Communication Engineering, Mohan Babu University, Tirupathi, Andhra Pradesh

[4]Department of Electrical and Electronics Engineering, Meenakshi Sundararajan Engineering College, Chennai

[5]Department of Electrical and Electronics Engineering, R.M.K. Engineering College, Chennai

Abstract

Solar PV cells are being used to meet with energy requirement of various applications since 1990s. Solar panel efficiency can be reduced by various parameters like bird drop, dust particles etc. These ecological factors reduce the efficiency of the photovoltaic modules and hence the aggregated residues should be removed at regular intervals and it becomes mandatory for improving the efficiency. Since residue removal is carried out by injecting water, the requirement of water increases further leading to water scarcity. In places where water scarcity is very high, the cost of cleaning is exorbitant and unaffordable. Residues are mainly contributed by numerous components like sun-oriented force, the atmospheric conditions, properties of semiconductor material used, soil properties and so on. Therefore, focus of research on solar PV cells is now getting shifted to identifying the most suitable methods for maintaining the PV cell neat and clean. Various researchers have proposed different methods including automated technology for cleaning and an improvement of approximately 15–25% in efficiency is reported. A lot of research has happened in this area leading to different methods of cleaning each having its own advantages and disadvantages. This work is an attempt to study various methods available in literature along with the merits and demerits. This paper is an exhaustive study of residual issues and advancements made in monitoring mechanism for the cleaning arrangement of PV modules. This gives a detailed review of various strategies adopted for cleaning solar PV cells.

Keywords: Solar photovoltaic cells, residue, temperature, humidity

Email: mpi.eee@rmkec.ac.in[1], priscillawhitin@veltech.edu.in[2], aneelstrides@gmail.com, venki7c@gmail.com[4], saisharan2k3@gmail.com[5], bsa.ece@rmkec.ac.in

DOI: 10.1201/9781003661917-32

1. Introduction

Renewable energy resources are generally found in the form of solar, kinetic, thermal and tidal energy. These renewable resources are converted into electricity for human needs. Progress towards sustainable power source sources, like solar, wind, hydropower, biomass and tidal power are driven by numerous components that shift the attention to natural effects (environmental change and an unnatural weather change) leading to immediate impact in fuel cost. Solar power is not only a natural resource but also a very powerful high potential resource for the generation of electrical energy. PV modules are very costly and during the period 2008 to 2011, there had been almost 60% fall in the cost per MW. PV cell is a gadget which converts active electromagnetic radiation into electrical energy by utilising the properties of semi-conductors. Residue is also one of the ecological variables that ought to be considered in enhancing power conversion rate. The most primitive and secure countermeasure for maintaining the surface clean is manual cleaning using a brush and water. Be that as it may, manual cleaning is difficult in the desert conditions is very difficult and unaffordable mainly due to the unavailability of water leading to increased labour charges for water transportation.

Also, cleaning without machines can lead to minor damages on the PV board surface because of brutal brushing. The impacts of residue on sun PV board productivity can be seen from two alternate points of view; impacts of residue dissipating in the environment and impacts of residue stored on the board surface.

2. Solar Cell and its Power Conversion Capacity

The basic parameter for comparing the performance of any element is its power conversion rate. This productivity of a solar cell depends on spectrum, intensity, incidence of sunlight and its temperature. Thus the elements determining efficiency of a solar cell must be carefully measured to estimate the performance of one device to another.

The maximum power output of solar cell is defined in equation 1,

$$P_{max} = V_{oc} I_{SC}\, FF \tag{1}$$

The output efficiency can be calculated in the equation 2,

$$\eta = \frac{V_{oc} I_{SC}\, FF}{P_{in}} \tag{2}$$

Where,
V_{oc} represents open circuit voltage.
I_{sc} represents short circuit voltage.

3. Dust Accumulation in Solar Panel Systems

In the recent times, many researchers have carried out research to identify the most suitable method to improve generation of electricity using PV cells. Desert areas, arid regions are the places where huge amount of power is tapped for electricity generation and also these are the regions where maximum dust accumulation on the panels is possible due to desert sand or sandstorms causing decrease in output power. There are a few difficulties and advantages identified with each and every cleaning system. Each procedure basically concentrates on lessening the attachment security between residue particles and the board. The Figure 32.1 given below gives a glimpse of various categories of PV cleaning methods.

This paper is an attempt to collect details on the work carried out by researchers in the following areas

Figure 32.1 Glimpse of Various Categories of PV Cleaning Methods

- Brush based dry cleaning method
- Dust accumulation study in Desert regions
- Piezoelectric actuator based solar cleaning system
- Dust effect on PV modules in temperature climate zone.
- Electrical characteristics of Photovoltaic cells with respect to shading due to soiling.
- Effect of foreign particles on solar panel
- Greenbotic's robotic cleaning system
- Cleaning system using PLC

3.1 Brush Based Dry Cleaning Method

Motasem Saidan et al. [1], in their study, pointed out that power on-board surface is the principle element that influences the yield of a PV board. Soil and other pollutants also bring down the efficiency of PV cells by approximately 15%. Al Shehri et.al [2] studied brushing the dirt on the glass pane of the module, when a residue particle settles down. This test is conducted, keeping in mind the environmental factors of several locations with a strong potential for photovoltaic module application, such as Saudi Arabia, India, Egypt, and Kuwait. A basic mechanical structure is assembled to evaluate the effect of brush cleaning. Figure 32.2 shows the experimental setup adopted by them for their study.

A specialized soft, black brush made of synthetic material of filament size 0.008 inches

Figure 32.3 Images of Glass Samples

Figure 32.3(a) Clean Glass Sample

Figure 32.3(b) Samples of Dust in Glass

Figure 32.3(c) Dust Particles and Contamination not Completely Expelled

Figure 32.3(d) Brushed and Dusty Glass Sample

is adopted for use in the experiment. They have to take a critical decision as it is essential to maintain the ideal cleaning recurrence to boost cleaning effectiveness and power age, while limiting cleaning cost and other obstructive factors. Images (a), (c) & (d), shown in Figure 32.3 below, are captured using a Scanning Electron Microscope (SEM) and image (b) is taken using an optical microscope. These images reflect the state samples at various stages of the experiment.

Thus, the glass samples once brushed and cleaned with water is restored to its original state of optical clarity thereby suggesting that no functional damage is caused as a result of dry-brushing the surface.

3.2 Novel Method to Clean Dirty Solar Panel

Jiang, Y et al. [3] developed a new model to gauge the frequency of solar panels in desert regions. This is depended on the rate of dust deposition, the impact of the deposited dust density in reducing the PV module power yield. The parameters considered in

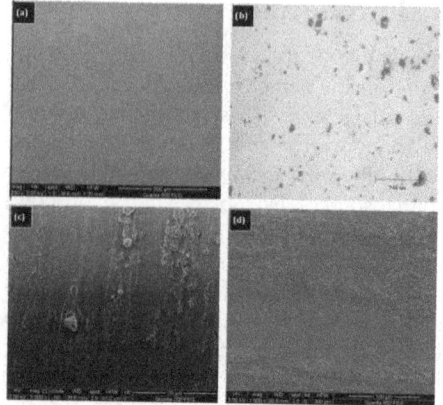

Figure 32.2 Experimental Setup

the development of the model are the installation inclination angles, the concentration of dust in the surrounding atmosphere and the representative average particle diameter. In the view of this model, frequency of cleaning a dusty PV cell is analysed to be approximately once in every 20 days. This analysis is carried out on the basis of 5% of spherical particle concentration at 100 lg/m³. The hypothetical model to gauge spherical particle deposition speed in the slanted surfaces is based on a model which has three layers.

The deposition particle flux in the model can be depicted as shown in equation (3) below:

$$J = -(\varepsilon_p + D)\frac{\partial C}{\partial y} - iv_s C + V_t C \quad (3)$$

where Ep represents the eddy diffusivity of the particle near the surface, D is the particle Brownian diffusivity. In general, mechanical cleaning and protection covering are two major techniques to remove the residues on PV modules. One of these methods is employed to avoid dust contamination on PV module surfaces, following which the PV

This equation (4) given below represents the time taken to clean the dust particles from Solar cells:

$$T = M_d \times A \div (A \times C_d \times V_d) = \frac{M_d}{C_d \times V_d} \quad (4)$$

where, Md represents the particle accumulation density for a specific power loss;
As mentioned above in this experiment, for every 20 lm of dust particle, the width and tilt angle are taken as 0 and 100 lg/m³ respectively. Furthermore, the cleaning process increases the power yield by 5%, and in this manner the aggregated residue thickness is measured as 2 g/m2and the main residue is found to be silicon dioxide (Darwish et al., 2015). Consequently, the kind of dust utilized in their work is a spherical silicon di-oxide particle.

These kinds of particles are found in the actual deposition process of airborne dust particles in desert regions. They have

carried out their work in a desert and with a cleaning time of 20.7 days. It implies that in desert zones, the prescribed frequency of cleaning process is at the rate of once in three weeks for designers to keep up the high yield and conversion efficiency of PV modules.

3.3 Electrostatic Cleaning System

Kawamoto et al. [5] elaborated, a refined cleaning method. The outcomes are very powerful; it has been shown that over 90% of the sand is removed from the outside of the tilted board in the wake of cleaning by utilizing this strategy. This examination is led by utilizing parallel wire electrodes of 0.3mm breadth, inserted in the spread glass plate. Because of the shadow thrown by the electrodes aggravating light assimilation, a pitch of 7mm between the terminals is picked for this experiment. A standing wave is utilized to repulse sand particles from the plate, which when airborne are transported downwards by the power of gravity. The single-stage rectangular voltage important to build up the standing wave is produced by utilizing a microcomputer to control the positive and negative enhancers, exchanged by semiconductor transfers. Six different kinds of sand are collected from various low-latitude desert regions, namely the Namib (A), Japan (B), Eurasia (C), Oceania (D), North America (E) and Africa (F). Photographs showing microscopic level differences between the particles of the associated region are shown in the following Figure 32.4.

Figure 32.4 Six Different Kinds of Sand are Collected from Various Low-latitude Desert Regions

A gadget for use in the fundamental examination of the procedure is made with specifications mentioned below; the substrate glass plate of measurements 100 x 100 x 3 mm. When the 0.3mm width copper wires are spread out properly, a slender glass plate of thickness 0.1 mm is neatly appended using clear, restricting glue to influence the surface smooth just as to stay away from protection breakdown. The cross sectional arrangement of the experiment is shown Figure 32.5.

The equipment is tilted, and the dust particles are reliably dispersed over the spread glass. A solitary stage rectangular voltage is then connected to the parallel anodes. This experiment is directed in a cooled research office (20–25C, 40 – 60 RH) and the dust particles on the glass plate are spurned and transported downwards, as affirmed by inspection of molecular movements using a rapid microscopic camera.

The Figure 32.6 shown the states

The left-hand side of the above figure exhibits the condition of the solar-based board secured with sand that gathered over the long run, and right-hand side demonstrates the after-effects of the cleaning methods used, as it is connected to the other side of the sun PV board for 3 minutes. The above figure clearly exhibits the powerful procedure to remove sand

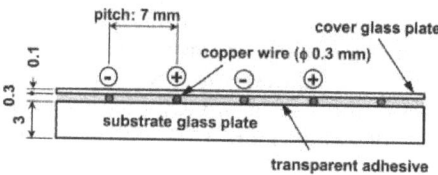

Figure 32.5 Illustration of Cross-Sectional Diagram of the Gadget Used in the Experiment

Figure 32.6 State of the PV Panel

from a photovoltaic board. The power yield diminished to 60% when the plate is secured with residue, and it is recouped to 90% after discontinuous cleaning.

3.4 Automatic Cleaning System

J.B. Jawale et al. [6] investigated the automatic cleaning method using microcontroller and aurdino. The incident light from the sun is hindered by the residue that gets deposited on the front portion of the module. It diminishes the age limit of the photovoltaic system. The yield decreases by half if the frequency of cleaning is more than a month. To clean the residue normally, a programmed cleaning framework has been planned, which detects the residue on the solar board and furthermore cleans the module consequently. This computerized framework is executed utilizing ATMEGA 328 microcontroller which controlled the DC gear engine. This system comprised of a sensor (LDR). While for cleaning the PV modules, an instrument comprised of a sliding brush has been created. In past innovation, PV board is fixed on the rooftop best and it recognizes sun beams just in east west bearing. In any case, in this innovation that we have built up the PV board distinguishes solar beams in east west course as well as in north-south heading. To accomplish this component, the PV board pivots in 180° and the base of entire gathering turns in 360° with the assistance of DC engine. In Heliotex innovation, cleaning of PV board is done physically. Be that as it may, for this innovation, cleaning is finished via programmed framework for example shower system. DMU will enact the shower system through microcontroller by utilizing a clock. The panel gives up 12v DC after cleaning.

3.5 Robot Controlled Cleaning Mechanism

Anderson et al. [8] presented the first robot controlled system for clearing sun PV board at solar parks. The PV Cleaner Robot V1.0 has two trolleys controlled by

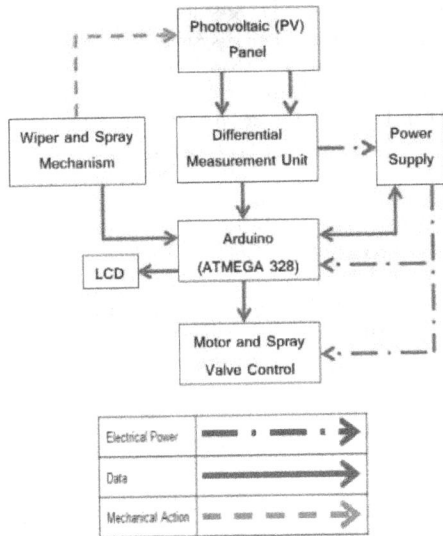

Figure 32.7 Block Diagram of Automatic Solar Panel Cleaning System

generation. They are very costly and hence people are hesitant to switch over to solar powered system; however the cost of solar system can be recovered over a period of time by selling the extra power generated to electrical utility system.

4. Conclusion

This is mainly because of dust birds excrete and accumulation of some other foreign particles on the solar PV panel. The most discussed methods along with the advantages and disadvantages are listed below 1. From the analysis it can be concluded that automatic is preferable. The most preferred one is automatic cleaning mechanism which is controlled by a robot, superior in performance. The authors are working on various alternatives to enhance the performance of automatic cleaning.

motor and one cleaning head. The cleaning process depends on the rotating sprayers for spraying water, along with brushing action by the brushes and scrubber. This helps in cleaning very efficiently at lesser time with less man power.

3.6 Mechanical Removal of Dust

Williams R. Brett [9] and his group have contemplated an ultrasonic method. These brushing technique for cleaning PV panel uses a device similar to the floor brush or like a windscreen-wiper. Since the size of the brush is small with strong adhesive, the cleaning technique is highly inefficient. However, it has the disadvantage that the working is very turbulent because of the abhorrent working state of the PV cell [8]. The proposed method overcomes the above mentioned issues as it is using the blower to remove the residues. The blowing technique controls the wind pressure for improved cleaning. Evacuating the tidies with vibrating and ultrasonic is an appropriate mechanical cleaning technique with fewer disadvantages [9-10].

PV cells are gaining popularity as one of the major elements of electrical energy

References

[1] Saidan, M., Albaali, A. G., Alasis, E., & Kaldellis, J. K. (2016). Experimental study on the effect of dust deposition on solar photovoltaic panels in desert environment. *Renewable Energy*, 92, 499–505.

[2] Al Shehri, A., Parrott, B., Carrasco, P., Al Saiari, H., & Taie, I. (2016). Impact of dust deposition and brush-based dry cleaning on

[3] Jiang, Y., Lu, L., & Lu, H. (2016). A novel model to estimate the cleaning frequency for dirty solar photovoltaic (PV) modules in desert environment. *Solar Energy*, 140, 236–240.

[4] Lu, X., Zhang, Q., & Hu, J. (2013). A linear piezoelectric actuator based solar panel cleaning system. *Energy*, 60, 401–406

[5] Kawamoto, H., & Shibata, T. (2015). Electrostatic cleaning system for removal of sand from solar panels. *Journal of Electrostatics*, 73, 65–70.

[6] J awale, J. B., Karra, V. K., Patil, B. P., Singh, P., Singh, S., & Atre, S. (2016, March). Solar panel cleaning bot for enhancement of efficiency—An innovative approach. In *2016 3rd International Conference on Devices, Circuits and Systems (ICDCS)* (pp. 103–108). IEEE.

[7] Huang, C. H. (2014). Applying Fuzzy Logic Controller to Intelligent Solar Panel

Cleaning System. *Journal of Marine Science and Technology*, 22(6), 716–722.

[8]   Anderson, M., Grandy, A., Hastie, J., Sweezey, A., Ranky, R., Mavroidis, C., & Markopoulos, Y. P. (2010). Robotic device for cleaning photovoltaic panel arrays. In *Mobile Robotics: Solutions and Challenges* (pp. 367–377).

[9]   Williams, R. B., Tanimoto, R., Simonyan, A., & Fuerstenau, S. (2007). Vibration characterization of self-cleaning solar panels with piezoceramic actuation. In *48th AIAA/ASME/ASCE/AHS/ASC Structures, Structural Dynamics, and Materials Conference* (p. 1746).

[10] Al Shehri, A., Parrott, B., Carrasco, P., Al Saiari, H., & Taie, I. (2017). Accelerated testbed for studying the wear, optical and electrical characteristics of dry cleaned PV solar panels. *Solar Energy*, 146, 8–19.

[11] Abderrezek, M., & Fathi, M. (2017). Experimental study of the dust effect on photovoltaic panels' energy yield. *Solar Energy*, 142, 308–320.

[12] Humood, M., Beheshti, A., Meyer, J. L., & Polycarpou, A. A. (2016). Normal impact of sand particles with solar panel glass surfaces. *Tribology International*, 102, 237–248.

[13] Javed, W., Wubulikasimu, Y., Figgis, B., & Guo, B. (2017). Characterization of dust accumulated on photovoltaic panels in Doha, Qatar. *Solar Energy*, 142, 123–135.

[14] Gholami, A., Khazaee, I., Eslami, S., Zandi, M., & Akrami, E. (2018). Experimental investigation of dust deposition effects on photo-voltaic output performance. *Solar Energy*, 159, 346–352.

[15] Tanesab, J., Parlevliet, D., Whale, J., & Urmee, T. (2016). Dust effect and its economic analysis on PV modules deployed in a temperate climate zone. *Energy Procedia*, 100, 65–68.

[16] Kawamoto, H. (2019). Electrostatic cleaning equipment for dust removal from soiled solar panels. *Journal of Electrostatics*, 98, 11–16.

[17] Syafiq, A., Pandey, A. K., Adzman, N. N., & Rahim, N. A. (2018). Advances in approaches and methods for self-cleaning of solar photovoltaic panels. *Solar Energy*, 162, 597–619.

[18] Kawamoto, H., & Shibata, T. (2015). Electrostatic cleaning system for removal of sand from solar panels. *Journal of Electrostatics*, 73, 65–70.

[19] Bousbaa, C., Iferroudjene, N., Bouzid, S., Madjoubi, M., & Bouaouadja, N. (1998). Effects of duration of sand blasting on the properties of window glass. *Glass technology*, 39(1), 24–26.

[20] Marouani N, Bouaouadja Y, Castro AD. Effect of the sandstorms on the solar panels. In: Proceedings of the international symposium on innovative technologies engineering and science (ISITES, Sakarya University); 2013.

[21] Johnson, K. L., & Johnson, K. L. (1987). *Contact mechanics*. Cambridge university press.

[22] Said, S. A., & Walwil, H. M. (2014). Fundamental studies on dust fouling effects on PV module performance. *Solar Energy*, 107, 328–337.

[23] Razykov, T. M., Ferekides, C. S., Morel, D., Stefanakos,E.,Ullal,H.S.,&Upadhyaya,H.M. (2011). Solar photovoltaic electricity: Current status and future prospects. *Solar Energy*, 85(8), 1580–1608.

[24] Bouzid, S., & Bouaouadja, N. (2000). Effect of impact angle on glass surfaces eroded by sand blasting. *Journal of the European Ceramic Society*, 20(4), 481–488.

[25] Landis, G. A., & Jenkins, P. P. (2002, May). Dust mitigation for mars solar arrays. In *Conference Record of the Twenty-Ninth IEEE Photovoltaic Specialists Conference, 2002* (pp. 812–815). IEEE.

[26] Calle, C. I., Buhler, C. R., Johansen, M. R., Hogue, M. D., & Snyder, S. J. (2011). Active dust control and mitigation technology for lunar and Martian exploration. *Acta Astronautica*, 69(11–12), 1082–1088.

[27] Gaier, J. R., & Perez-Davis, M. E. (1991). Effect of particle size of Martian dust on the degradation of photovoltaic cell performance.

[28] Wang, Y. C., & Chang, S. H. (2006). Design and performance of a piezoelectric actuated precise rotary positioner. *Review of Scientific Instruments*, 77(10), 105101.

[29] Kawai, Y., Ono, T., Esashi, M., Meyer, E., & Gerber, C. (2007). Resonator combined with a piezoelectric actuator for chemical

analysis by force microscopy. *Review of scientific instruments*, 78(6), 063709.

[30] Zhao, C. (2011). *Ultrasonic motors: technologies and applications*. Springer Science & Business Media.

[31] Dang, D. H., Friend, J., Oetomo, D., & Yeo, L. (2009). Triple degree-of-freedom piezoelectric ultrasonic micromotor via flexural-axial. *IEEE transactions on ultrasonics, ferroelectrics, and frequency control*, 56(8), 1716–1724.

[32] Vimala, M., Ramadas, G. Perarasi, M., Manokar, A. M, & Sathyamurthy, R. (2023). A review of different types of solar cell materials employed in bifacial solar photovoltaic panel. *Energies*, 16(8), 3605.

[33] Perarasi, M., & Ramadas, G. (2023). Detection of Cracks in Solar Panel Images Using Improved AlexNet Classification Method. *Russian Journal of Nondestructive Testing*,

33 Smart Home Security and Monitoring System

Kavitha P. , Sukhi Y.[1], Mehul Anand G.[1], Arigonda Roshan Kiruthik[1], Amith R. R.[1], Fayaz Ahamed A.[1], Perarasi M.[1], Sarala B.[1] and Jeyashree Y.[2]

[1]R.M.K. Engineering College, Kavaraipettai, Tamil Nadu
[2]SRM Institute of Science and Technology, Kattankulathur, Tamil Nadu

Abstract

This paper proposes an IoT-based home security and automation system designed to transform a house into a Smart Home. The system uses a Wi-Fi enabled ESP-01 microprocessor to accomplish the house via a smartphone with internet access, while the ATmega328P microcontroller handles sensors and actuators. Data is stored on the ThingSpeak global web server. Security features comprise an infrared sensor and reed switch to detect motion and monitor the front door's status. The ATmega328P also controls motorized windows and doors through a mobile device. A buzzer is kept to alert and relays to control appliances. Environmental sensors that notify users of gas leaks and provide real-time data on room conditions. Automatic adjustment of room light brightness based on occupancy levels is also proposed in this system.

Keywords: Sensor, internet of things, home security, smart home

1. Introduction

The Internet of Things is advancing so quickly that a huge number of actual objects are connecting to the internet. Using IoT [1–2] technology, which can be linked with household equipment to function as a single entity. The absence of automation [3] and security in one's house is a serious issue that affects everyone in the modern world. An excellent example of an inexpensive IoT [4] application that greatly benefits people is the home security system. In order to add security features to a smart home, this project shows how to use an Arduino microcontroller to gather input data from sensors [4], process the data, and display the results. Another microcontroller called ESP-01 is intended to link the security system to a Wi-Fi network so that it can notify the homeowner if there are any strange activities occurring in the home. Using signals from the smartphone, one can operate the home's motorized doors and windows. The security system will notify the customer with a buzzer if someone approaches the main door. The largest risk to users is forced entry, which may be avoided by utilizing our system, which can identify forced entry and issue an emergency call to the relevant authority helplines. In the event of an emergency, the buzzer will automatically close the doors to ensure the user's safety

Email: pkt.eee@rmkec.ac.in[1], ysi.eee@rmkec.ac.in[2]

DOI: 10.1201/9781003661917-33

and inform the others within the house. If the user is not home at that time, then a call will be placed to the fire department in the nearby area for instant assistance, and a notification will be sent to him as well. Moreover, the appliances within will be remotely operated using a phone rather than using actual switchboards. With the Arduino IDE, which enables C/C++ programming, both of the microcontrollers used in this project are programmed. An electrical system for automation of home is done using microcontroller general-purpose input/output ports in paper [1] which is a Bluetooth device and is used to transfer the status of the embedded system board to the ARM7 using IoT [7–9]. They created a system authentication so that only authorized users could access household appliances. The ARM9 remote control feature offers support and aid, particularly to the aged and crippled. An inbuilt Raspberry Pi is utilized in [3] to implement control of doors and household appliances through the internet connection. It made use of a wireless camera, PIR sensor [6], and embedded system. The block diagram for the security system is shown in Figure 33.1. This helped the home residents with automation with intelligent systems. In Paper [4], the digital technology with network is implemented in home area to provide connectivity. The overall system is used for development of the architecture. The remote control can also be implemented with a monitoring for better safety. The gateway in the home provides simple interface as well as remote access.

An automation system with security purpose is implemented in homes with the use web-based system in paper [5]. It is made up of ethernet module and a microcontroller to provide a wireless network connection. In order to have communication via TCP/IP which are connected to the same router, it is assigned a static IP address. The authors of paper [6] suggest the uID-CoAP architecture. The purpose of this novel technology is to host internet of things for the typical embedded devices, such as household appliances. When compared to simple sensor nodes, they frequently have to perform a variety of complex duties. It blends the ubiquitous ID (uID) identification with the restricted use of protocol. The system design using the microcontroller is discussed in section 2. The system setup for development of the work flow is discussed in section 3. The block to be setup for the hardware development is discussed in section 4. The security system involved for the developed system is discussed in section 5. The results and the components used are discussed in section 6. Finally, the conclusion is given in the section 7.

2. System Design

The software design and hardware design are the two main components of the system designing approach. PCs can be used to write the program for the microcontroller. There are some simple methods for writing the program. While programming is written and uploaded into the microcontroller, hardware design involves the arrangement of sensors, actuators, and microcontrollers. The proposed system has a microprocessor that is linked to actuator and sensor modules to monitor and manage home appliances. The configuration of various hardware components is displayed in this design part. The details and specifications for the different parts are explained below. The following are the different functional units that the system uses Atmega 328P Microcontroller model which is the

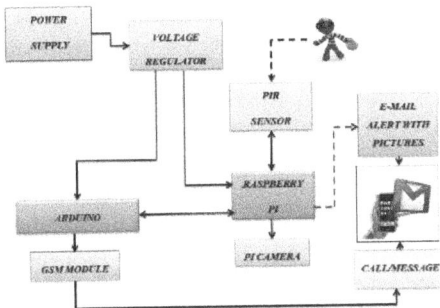

Figure 33.1 Block Diagram of Home Automation

primary CPU. It is in charge of managing every sensor and actuator in the model house. It uses the serial port of the ESP01 to connect with it. ESP01 Microcontroller is the original microcontroller in the well-known ESP8266 series. It is a Wi-Fi SOC with a 32-bit RISC CPU operating at 80 MHz that is integrated with the TCP/IP protocol stack. It is in charge of establishing an internet connection for this model. The Thingspeak server gets all of the model data through it, and in a similar manner, user commands are finally carried out after being received by this microcontroller over the internet and sent to the Atmega328P via the serial port. In the presence of a magnetic field, reed magnetic switch changes connections. The model has motorized doors and windows that can be operated manually or automatically. L293D motor driver integrated circuit (IC) receives signals from the Atmega328P microcontroller and uses them to drive the DC motors in this type of house. The smartphone app obtains information about the state of the model house's sensors and actuators, such as the temperature and status of the main door, by reading the data fields that the ESP01 has written on the Thingspeak server. It publishes the information to fields on when it receives signals from the user. The Thingspeak website hosts this free server. Writes sensor and actuator to the ESP01 information on particular data fields that the program reads. Operating on a local network rather than the internet, this server, which is housed on the ESP01 itself, has the same features as the Thingspeak server. The simple block diagram for the developed system is given in Figure 33.1.

3. Block Diagram

The first step in the implementation process is setting up Thingspeak Cloud. After registering, we must register an application on ThingSpeak in order to utilize its services. We create channels for the application once it has been registered. Block diagram of the system is shown in Figure 33.2. The application downloads and processes the

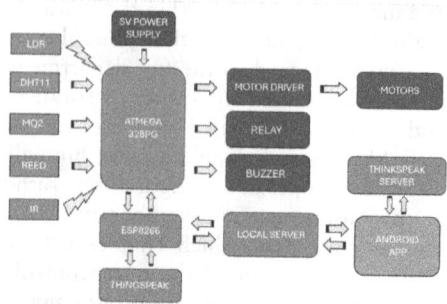

Figure 33.2 Block Diagram

CSV file from the cloud when it needs to display the energy consumption. The information is extracted and processed, and then the energy use is displayed on the screen in kWh. Power can be computed using the formula that follows.

$$Power = \frac{Energy}{Time} \tag{1}$$

$$Energy = Power \times Time \tag{2}$$

The equations (1) and (2) are used to determine the energy in kilowatt hours if we know the power in kilowatts and the time in hours. the block diagram of the proposed system is shown in Figure 33.2. A mobile application connects directly to the cloud inserts data and retrieves feed data at the same time. The Raspberry Pi is used for necessary data downloads and uploads, and the rest of the system is cloud-connected. The Raspberry Pi communicates with the Arduino Client, which controls the appliances and their acknowledgements. In addition to communicating with the Raspberry Pi as a client, Arduino may also function as a stand-alone module that reacts to sensor feedback data to alternating current and 1mW due to DC current. These losses account for 0.34% of the output power. It is noted that large air core inductor has high resistance. On the other hand, the ceramic capacitors have lesser value of equivalent series resistance (ESR) of 200mΩ [3].

The power supply for the implementation of the automated system require a step-down transformer to reduce the voltage level, a full bridge rectifier to get a dc voltage from the available ac voltage, and a regulator to provide constant voltage with LM7805. Serial Communication protocol is used by the ESP01 and the Atmega328P to communicate with one another. The sensors are connected to the Atmega328P's digital input pins as set up in the Arduino IDE. The motors and buzzer actuators are interfaced to pins that are set up as output. The ESP01 establishes a connection with the global Thingspeak server and uses HTTP requests to read and write data. For every two seconds, the data from the sensor is collected and uploaded into the server. For every two seconds, the android application is used to connect the data to the server, updates its data fields with the values from the server, and uploads data to the server. The circuit schematic is displayed as flow chart in Figure 33.3. There is a specific algorithm that drives the whole model. The system starts by measuring each sensor value. Atmega328P transmits this information to the ESP01, which then relays user commands to it. Next, it determines if a gas leak or break-in has occurred. If the case of catastrophe, it follows the emergency protocol; if not, it follows the user instructions.

Additionally, the system continuously monitors the front door for activity and sounds the alarm whenever it senses any it continuously counts the number of individuals inside by keeping an eye on the IR sensors.

4. Overall Security System

The overall security system developed for proper functioning is shown in Figure 33.4. The use of LCD display for the use of public make the system to be simple to use and convenient in all aspects of interfacing. The use of the display unit helps the security system to be more convenient to user with an implementation of feedback in the developed system from the user. The sensitivity of the sensor play an important role in the turning on of the system and in the recent activity taken place in the system for updating the user.

5. Results and Analysis

The overall security system developed using the simulation for proper functioning is shown in Figure 33.5. The use of different interfacing components can produce an increase in the cost but it is essential to ensure long life of the components with good quality components.

Thus, the implementation of the safe and security system is useful to the home resides. The effective use of the security system for the need of the home resides give a good mode of living in the homes.

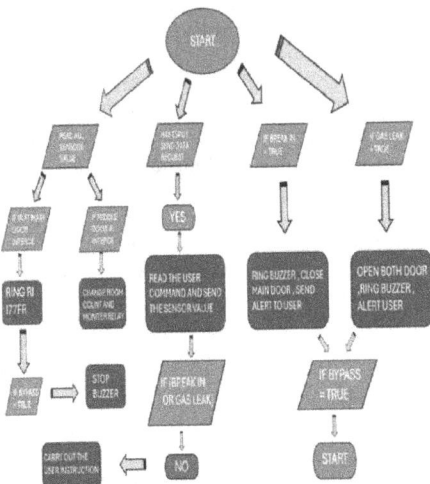

Figure 33.3 Flowchart

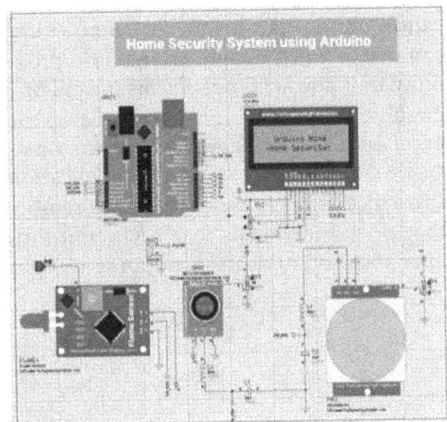

Figure 33.4 Overall Security System

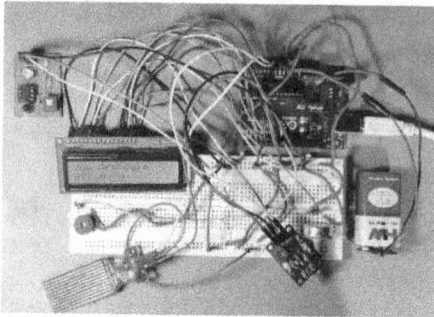

Figure 33.5 Hardware Working Model

S. No.	Sensor	Condition	Action
1	LDR	After 6 PM	Lights on
2	DHT	temperature > 28 °C	Fan is on
3	MQ2	If any gas	Buzzer goes
4	IR	any human/ car enters	Open the door/garage

6. Conclusion

The implementation of the proposed system for the safety is very effective and helpful to the main users. There are two ways to communicate data with it: via a local server and a global server. The sensors and actuators in the home are controlled by the Thingspeak data IoT service, which is similar to the systems that are used in a commercial sense. We have integrated the Internet of Things with a local communication option like LAN. The smart home can detect when people are in close proximity to it and send out alarms in case of a gas leak or break-in. It is proved that in the event of an emergency, phoning an emergency number directly from the application can expedite response times. It has sensors that track individuals in a room and turn appliances on and off in response to their presence or the ambient brightness level. It also features temperature, humidity, and brightness data.

References

[1] Naser, H., Siavash, M., Khajavi, H., Alireza, J., & Jan, H. (2018). An IoT-based Automation System for Older Homes:A Use Case for Lighting System. *11th International Conference on Service Oriented Computing and Applications*, IEEE 2018.

[2] Motlagh, N. H., Bagaa, M., & Taleb, T. (2017). Uav-based iot platform: A crowd surveillance use case. *IEEE Communications Magazine*, 55(2), 128–134.

[3] Kang, Y., & Zhongyi, Z. (2012). Summarize on internet of things and exploration into technical system framework. *IEEE Symposium on Robotics and Applications (ISRA)*.

[4] vGill, K., Yang, S. H., Yao, F., & Lu, X. (2009). A zigbee-based home automation system. *IEEE Transactions on Consumer Electronics*, 55(2), 422–430.

[5] Akyildiz, I. F., Weilian, Y., Sankarasubramaniam, Y., & Cayirci, E. (2002). A survey on sensor networks. *Communications Magazine, IEEE*, 40(8), 102–114.

[6] Sukhi, Y., Jeyashree, Y., Perarasi, M., & Sarojini, B. (2017). Standalone PV-fed LED Street Lighting Using Resonant Converter. *Electric Power Components and Systems*, 45(5), 548–559.

[7] Takayuki, S., Yasue, K., & Futoshi, N. (2014). Abstracting IoT devices using virtual machine for wireless sensor nodes. Internet of Things (WF-IoT). *IEEE World Forum*, Seoul.

[8] Takeshi, Y., Shinsuke K., Noboru, K., & Ken, S. (2013). An Internet of Things (IoT) Architecture for Embedded Appliances IEEE Region 10 Humanitarian Technology Conference.

[9] Junaid, K. A. M., Sukhi, Y., Jeyashree, Y., & Sivakumar, S. (2023). IoT based lifesaving helmet for two-wheeler. *AIP Conf. Proc. International Conference on Inte lligent System (ICIS-2022)*, 2878(1), 020028.

34 Design and Implementation of an IoT-Enabled Smart Energy Meter Using Arduino

Sukhi Y.[1], Kavitha P.[1], Adlin Sharo E. J.[1], Bavithra Jayamohan[1], Kanimozhi M. N.[1], Fayaz Ahamed A.[1], Sarala B.[1], Perarasi M.[1] and Jeyashree Y.[2]

[1]R.M.K. Engineering College, Kavaraipettai, Tamil Nadu
[2]SRM Institute of Science and Technology, Kattankulathur, Tamil Nadu

Abstract

A cutting-edge approach to cost optimization has emerged, offering energy consumers a fresh perspective through rational energy management. Our research enterprise focuses on developing a customized energy consumption meter for residential householders, utilizing Arduino technology. The primary notation of the endeavour is to measure electricity usage and calculate associated costs using Internet of Things (IoT) technology. This study explores the practicality of smart meter gadgets, which automatically track the units consumed, compute the respective expenses, and displays the respective information to user's mobile phone. To ensure effective communication, a GSM module is utilized to transmit real-time notifications to the customer, providing timely updates on energy utilization. Additionally, voltage meter element of the energy management complex correctly measures AC voltage position with the support of the voltage transformer. Alongside monitoring usage, energy meter is associated for issuing alerts to the customer when consumption exceeds predefined values, providing awareness and to provide effective energy usage.

Keywords: Energy meter, arduino, gsm module, internet of things, voltage sensor

1. Introduction

In the modern era, electricity has become an essential requirement for individuals. The increase in electricity usage in recent years and the matter of the electricity theft, where unrecognized individuals steal electricity without being detected, are significant concerns. To tackle these challenges, the implementation of intelligent electric meters has been suggested. Electricity is used for various things with a large portion of environment energy consumption being used in households. Efficient electricity usage is important as it helps individuals, governments, and businesses save costs, while also assisting power plants in avoiding costly upgrades. The main focus of "Smart Energy Meter" project is to offer users real-time data on their daily energy usage and

Email: ysi.eee@rmkec.ac.in[1], pkt.eee@rmkec.ac.in, jeyashry@srmist.edu.in[2]

DOI: 10.1201/9781003661917-34

associated expenses through dedicated website [1–3]. This system also offers predictive insights into the monthly electricity bill [4–6]. However, the time intervals between updates on power usage may not be conducive for consumers to monitor instantaneous changes in their electricity consumption [7]. Furthermore, utility bills often present usage information in a complex manner, making it challenging for consumers to discern alterations in their power usage from the previous billing cycle [8–9]. The Energy meter not only displays the units consumed but also transmits this data to both the consumer and the electrical board, thereby reducing the need for manual intervention. The system components include an Energy meter, Arduino UNO, Wi-Fi module, LDR Sensor, a display unit, and two loads, all working in tandem to streamline the monitoring and management of electricity consumption. Section 2 deals with the block diagram of the node MCU. The schematic representation of system setup is discussed in section 3. The development of system implementation is dealt in section 4. The hardware implantation discussed in section 5. The last section deals with the conclusion and is given in the section 6.

2. Block Diagram

Supervising and tracking the usage of electricity consumption has become a cumbersome task in the present situation. A lot of consumers do not have knowledge of using the electricity amount daily. This is because the consumer does not track the amount of electricity consumed daily. They are using more amount of electricity than they want to use. The block diagram of the developed system for measuring the energy is shown in Figure 34.1. This leads to unexpected billing amount for the electricity consumption in the bill. This led to high energy cost during the billing period.

It is essential for customers to have a clear knowledge of their energy usage and be charged accordingly. To address this issue, an automated system that enables

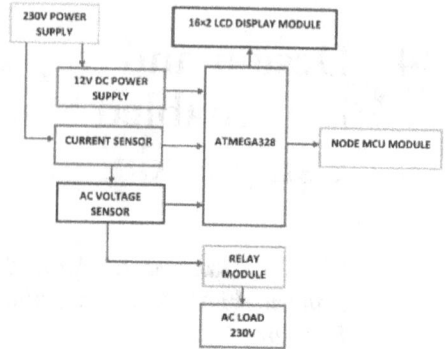

Figure 34.1 Block Diagram

residents to monitor their energy consumption readings is proposed by using the internet. This system utilizes the LCD display, Wi-Fi module and Arduino UNO. Wi-Fi module plays a central role in the Internet of Things operation. Arduino board is connected to a detector that detects the readings of the energy meter. These readings are then transformed and modernized via the Wi-Fi module, allowing users to access them through a web page. Additionally, users will receive SMS notifications when their current consumption reaches a predetermined threshold value. This feature enables users to effortlessly track their energy usage using a user-friendly software application. Overall, our system provides an effective solution for users to monitor their electrical energy meter readings and conveniently check their billing information in real-time.

3. Schematic Representation

The schematic representation of the developed system for the implementation is shown in Figure 34.2. There are three main sections to the flowchart. The first section shows how the ignition works by pressing a button, which causes the DC motors to start up and the LED to light up is activated. The alcohol sensor, which measures alcohol by breathing straight into the sensor, is the main part. A controlling unit, which makes up the second portion, is equipped with the essential features to

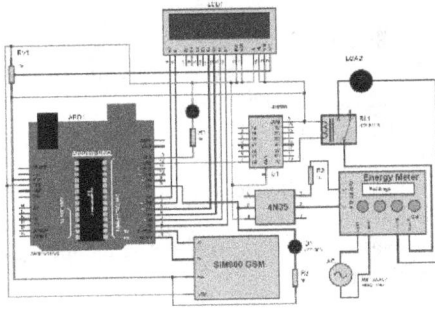

Figure 34.2 Schematic Representation

read the output of the alcohol sensor. The amount of alcohol content is detected first and then the microcontroller will direct the DC motors to run or stop based on the alcohol content, and a LED will visibly indicate the condition of the device. In the third section, the required output is shown on the serial monitor and the GSM module sends a brief message to the authorities.

The Arduino IDE serves as the software requirement for this project, functioning primarily as a text editor that facilitates code compilation and error detection. Once the code has been thoroughly checked for errors, it can then be uploaded to the Arduino board. On the other hand, the hardware requirements encompass the Arduino UNO, along with additional components such as the current sensor, voltage sensor, and jumper wires. The components required are Arduino uno, energy meter, wi-fi module, gsm module, voltage sensor, current sensor, relay.

Energy meters are also known as power meters and vice versa. Power, specifically active power, is a quantification of the energy needed/utilized to carry out productive tasks. For instance, a incandescent lamp rated at 100 watts consumes 100W of active power to generate light energy and heat energy. The Energy, on the other hand, is a measurement of the total work done within a specified time frame. Using the incandescent lamp scenario, if the light bulb remains illuminated for one full hour, it will utilize 100 watts x 3600s = 360000Ws = 0.1kWh of energy. The ESP-12 Lua Nodemcu wi-fi development

board, equipped with the ESP8266 chip, is a comprehensive micro-controller and wi-fi platform that simplifies the creation of projects involving wi-fi and IoT applications.

4. Development of System Implementation

IoT has revolutionized the field of energy by providing a highly desirable framework where customers can have control over his/her energy consumption. By being aware of their usage of energy in real-time, customers can effectively manage their energy administration and pay their bills on time. In cases where the bill payment is not made, the system has the capability to autonomously disconnect the electric power supply from a remote location. The system consists of various units, each with its own modeling and functionality. The components of the system, including the IoT technology, micro-controller, and its architecture, are discussed in detail. The main objective is to reduce energy consumption and monitor the consumed units. It aims to create electrical appliances intelligent, providing comfort to the users while also minimizing power usage in web applications. The proposed system utilizes the ARDUINO UNO controller and IoT technology. In case any modifications are made, the controller sends the data to the server and automatically cuts off the energy supply. The meter used by the customer displays the maximum energy demand, and if this demand is exceeded, it is necessary to disconnect the meter automatically using the embedded system which is available in the meter sensor. The system incorporates a LDR sensor which is placed on the energy meter to sense the blinking pulse of an LED.

The micro-controller then sends this reading via a GSM module to the electricity board. Each customer unit in this system is equipped with a smart energy meter, while the service provider maintains a server. Both the meter and the server are equipped with a GSM module, allowing them to communicate with each other using the existing

GSM infrastructure. To recharge their energy meter, customers simply need to send a hidden PIN number from a scratch card to the server using SMS. The Internet of Things (IoT) has been steadily evolving with each passing day, encompassing Machine-to-Machine (M2M) interactions that involve the integration of electronics, software, sensors, and actuators to enable users to remotely and efficiently monitor and control devices. Within an IoT based framework, both objects and living entities are assigned unique identifiers that allow for seamless data transfer, leading to a more interconnected system. The convergence of micro-electromechanical systems, wireless technologies, and the Internet has resulted in a considerable expansion of the IoT and its extensive application across numerous sectors, including energy, gas, and water. Among others, to automate and streamline routine tasks. As a result of the increase in internet usage, these industries have gone digital and have integrated online payment methods for greater convenience. Manual procedures, such as reading meters, continue to be used in spite of technological improvements. This presents a danger of errors that might result in significant financial losses. An effective way to assess energy use and analyse data for billing and payment purposes is through the use of Automatic Meter Reading (ARM) technology.

ARM technology utilizes wireless connection to show that it is a more affordable choice than traditional wired media, and because Wi-Fi is readily accessible in home settings, it has become a viable option for connectivity. The potential of IoT applications is further demonstrated by the widespread usage of smartphones, with 2.87 percent billion people using them globally in 2017, and consumers of Android holding 79.21% of the smartphone market share in the country at the end of 2017. These figures demonstrate how common Android smartphones are throughout the nation, which makes it a wise decision to deploy IoT systems that are customized to local tastes and usage

habits. As of January 2017, the country's installed generation capacity amounted to 15,351 megawatts (MW), serving 92 percent of the country's urban and 67% of its rural populations. Despite these advancements, manual electricity meter readings remain a challenge for some individuals in Bangladesh, emphasizing the need for innovative solutions like ARM technology to streamline energy monitoring and billing processes for greater efficiency and accuracy. Consequently, their lack of knowledge regarding the amount of electricity used is evident. The primary objective of this study is to create and develop an affordable energy monitoring system that is seamlessly integrated with cloud services and accompanied with mobile application. The developed system involves the utilization of a Wi-Fi based system, which is built upon the IoT concept.

5. Hardware Implementation

An innovative proposal has been put forth for a system incorporating an ARM based with Wi-Fi, developed using Linux software. The survey conducted for the usage of smart meters led into crucial features of the metering process. The prototype of the developed model is shown in Figure 34.3.

In this there are some challenges arises due to the use of big data and the increased use of cloud environments were underscored. Our project also suggested a setup where an Arduino UNO, coupled with an Ethernet shield, could oversee various activities related to electricity flow, current

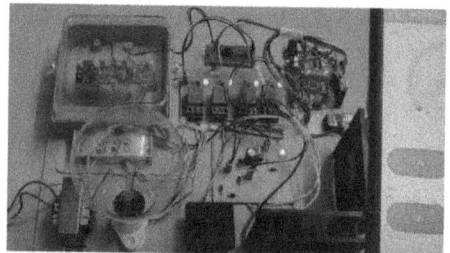

Figure 34.3 Hardware Working Model

usage, and electricity expenses. The primary objective is to tackle payment-related issues and precisely compute the expenses of electricity units. The introduction of a smart energy meter concept utilizing IoT and Arduino was proposed, enabling the transformation of existing energy meters in households into smart meters without necessitating replacements. Users are set to receive notifications concerning their energy consumption. Amount for repaid tally can be prepared using RFID-based recharge tags or supported wallets. Users can access billing information on the web server by providing their details. The hardware setup includes an ARM7 micro-controller, Energy Meter, and appliances controlled by relays. Communication with the server requires Wi-Fi/GPRS, RFID reader, and prepaid tags. The RFID reader validates the authenticity of the tag, verifies the information on it, and ensures that money is available for power subscriptions. Reducing household power usage is the main goal of the Internet of Things (IoT)-based E-meter's development. It does this to save expenses, avoid the need for human intervention, and save human resources. This metre may be used in a flexible manner because it functions in both automatic and manual modes. It can also deliver billing information straight to mobile devices before the due date, all without the need for human intervention. The approach shown here has a number of advantages, including lower costs and increased accuracy and efficiency. It focuses mostly on smart cities that have open-air WiFi hotspots. The proposed system uses IoT to upgrade energy meters, allowing for automatic power readings and optimizing energy usage. The meter data is accessible to both the service provider and customer, providing insights into energy consumption. It utilizes the PIC16F MCU to calculate energy usage and includes a mechanism for remote power disconnection in case of non-payment. This proposed system ensures errorless meter readings and reduces billing mistake.

6. Conclusion

The smart energy meter system, as presented in the paper, was implemented through the use of IoT technology. This system provides multiple advantages such as wireless data transfer, decreased workload, and cost efficiency. It allows for seamless collection of meter readings without the need for human involvement. The integration of an embedded micro-controller and Wi-Fi module improves the dependability of wireless data transmission. Through this system, customers can easily keep track of their energy usage and expenses whenever they wish. Moving forward, there is a possibility for this project to progress towards smart cities by utilizing IoT sensors on an international level.

References

[1] Omar, M., Adolfo, R., Pedro F. R. E., Alexis, A., Alejandro, S., Fernando, L., Ruben, A. R. Z., & Angel, R. (2024). Development of an IoT smart energy meter with power quality features for a smart grid architecture. *Sustainable Computing: Informatics and Systems*, 43, 100990.

[2] Suresh, D. (2024). Multifunctional IOT-based smart energy meter. *Smart Metering*, 2, 39–50.

[3] Alkawsi, G. A., Ali, N., & Baashar, Y. (2020). An Empirical Study of the Acceptance of IoT-Based Smart Meter in Malaysia: The Effect of Electricity-Saving Knowledge and Environmental Awareness. *IEEE Access*, 8, 42794–42804.

[4] Challa, K. R., Sarat, K. S., & Franco, F. Y. (2024). A literature review on an IoT-based intelligent smart energy management systems for PV power generation. *Hybrid Advances*, 5, 100136.

[5] Waheb, A. J., Tan, M. T., Fikri, M., Hamidun, I., Ajwad, H., Kamarudin, C., Wenyan, W., Jamil, S., Abdul, R. A., Alsewari, Mohammed, & Ali, A. H. (2024). Development of LoRaWAN-based IoT system for water quality monitoring in rural areas. *Expert Systems with Applications*, 242, 122862.

[6] Sukhi, Y., Jeyashree, Y., Perarasi, M., & Sarojini, B. (2017). Standalone PV-fed

LED Street Lighting Using Resonant Converter. *Electric Power Components and Systems*, 45(5), 548–559.

[7]   Cabrera, R. S., Cruz, A. P. D. L., & Molina, J. M. M. (2020). Sustainable transit vehicle tracking service, using intelligent transportation system services and emerging communication technologies: A review. *Journal of Traffic and Transportation Engineering (English Edition)*, 7(6), 729–747.

[8]   Mohammed, K., Abdelhafid, M., Kamal, K., Ismail, N., & Ilias, A. (2023). Intelligent driver monitoring system: An Internet of Things-based system for tracking and identifying the driving behavior. *Computer Standards & Interfaces*, 84, 103704.

[9]   Junaid, K. A. M., Sukhi, Y., Jeyashree, Y., & Sivakumar, S. (2023). IoT based lifesaving helmet for two-wheeler. *AIP Conf. Proc. International Conference on Intelligent System: (ICIS-2022)*, 2878(1), 020028.

35 A Framework for Patient Specific Drug Recommendation and Side-Effect Prediction System

Morarjee Kolla[1], Vemula Ishitha Reddy[2], and Rddhi Reddy[2]

[1]Associate Professor, Chaitanya Bharathi Institute of Technology, Hyderabad
[2]BE CSE (4/4), Chaitanya Bharathi Institute of Technology, Hyderabad

Abstract

At present, systems for disease prediction, drug recommendation, and adverse drug reaction (ADR) prediction often operate independently, limiting their ability to provide a cohesive, personalized healthcare solution. These systems lack integration, struggle with real-time data adaptability, and fail to account for individual patient factors. This paper aims to propose a framework that can accurately predict the likelihood of various diseases based on a patient's symptoms and medical history, as well as suggesting suitable medications for the predicted diseases, considering factors such as effectiveness, dosage, and patient-specific attributes, including any potential side effects in order to provide tailor-made recommendations that reduce risk. The system unifies three modules: disease prediction using patient symptoms and medical history, drug recommendation based on efficacy, dosage, and side effect profiles, and a sentiment analysis module to refine recommendations using patient feedback. Techniques like Random Forest and Neural Networks (NNs) are employed, with support from knowledge graphs and collaborative filtering for enhanced personalization. Results demonstrate significant improvements in accuracy, precision, and recall. By integrating disease prediction, drug recommendation, and side effect analysis, the proposed system provides a comprehensive approach that enhances outcomes and minimizes risk.

Keywords: Disease prediction, drug recommendation, sentiment analysis, neural networks, collaborative filtering

1. Introduction

In recent years, multiple data-driven frameworks for decision-making have emerged, incorporating drug recommendation systems within healthcare infrastructures. Traditional methods often rely on rigid guidelines for prescribing drugs, disregarding patients' medical histories, which can lead to delayed or suboptimal outcomes. Adverse drug reactions (ADRs) remain a significant public health challenge, contributing to hospitalizations, treatment failures, and fatalities. This underscores the need for systems that predict both the efficacy and safety of drugs

Email: morarjeek_cse@cbit.ac.in[1], ishithareddy2308@gmail.com[2], rddhireddy@gmail.com[2]

DOI: 10.1201/9781003661917-35

by cross-referencing an individual's medical profile.

This project addresses these issues by developing a data-driven system designed to deliver accurate drug prescriptions while minimizing ADR risks. The system utilizes machine learning techniques to analyze datasets that encompass drug interactions, patient characteristics, and diagnoses. Algorithms such as Collaborative Filtering [1], Transformer-based models [2], and Graph Neural Networks [3] are employed to parse complex relationships between patients, drugs, and side effects. By processing both unstructured data (e.g., patient reviews and medical records) and structured data (e.g., drug compositions and interactions), the system offers a comprehensive perspective on treatment options. Ultimately, this project aims to establish a more individualized healthcare paradigm, delivering drugs tailored to patient needs while minimizing risks.

2. Literature Review

Over the years, as drug treatments have advanced, data-driven approaches for drug and adverse drug reaction (ADR) recommendations have been extensively researched. Early systems, such as rule-based models discussed by Nayak et al. [1], lacked flexibility and failed to account for dynamic patient data, history, or real-time conditions, resulting in generalized treatments. Raj et al. [2] explored machine learning for multi-disease prediction, paving the way for more individualized healthcare solutions. Luo et al. [3] introduced collaborative metric learning, integrating patient similarities into drug recommendation models, while Job et al. [4] applied Transformer-based techniques to improve opinion-based recommendations through sentiment analysis.

Zhang et al. [5] enhanced drug recommendationsusing heterogeneous graph representations, leveraging electronic health record networks to understand complex drug-patient interactions. Ariyachaipong et al. [6] optimized drug selection through the Bee Algorithm, refining treatment

recommendations. In ADR prediction, Uner et al.'s *Deep Side* model [7] combined gene expression and chemical structure data to assess risks in preclinical testing. Saadat et al. [8] and Zheng and Xu [9] employed knowledge graphs to integrate diverse datasets, uncovering intricate drug-disease-patient relationships for more comprehensive recommendations.

Bongini et al. [10] advanced ADR prediction accuracy using deep learning on molecular graphs, while Dongre and Agrawal [11] leveraged social media data to analyze real-time patient feedback, improving drug safety. Despite these advancements, challenges remain, as highlighted by Mian et al. [12], Symeonidis et al. [13], and Li et al. [14]. This literature emphasizes the critical need for integrated, patient-centric frameworks combining genetic, molecular, and real-world data to optimize drug recommendations and minimize adverse effects.

2.1 Machine Learning Classifiers

Machine learning classifiers enhance diagnostic accuracy by efficiently handling large datasets and complex data beyond the scope of traditional models. Decision trees, as highlighted by Nayak et al. [1], are effective for disease prediction, analyzing patient data like symptoms, demographics, and medical history to detect subtle indicators. Techniques like Random Forest, Decision Trees, and SVM are especially useful for high- dimensional data. Raj et al. [2] emphasize that SVM can distinguish between diseases with overlapping symptoms, while Random Forest identifies high-risk factors, assisting healthcare professionals in dynamic environments. These models adapt to new data, making them ideal for healthcare scenarios [5].

2.2 Recommendation Systems

Today's recommendation systems use methods such as collaborative filtering, which groups users with similar profiles, and content-based filtering, which recommends based on item attributes. Luo et al.

[3] demonstrated the effectiveness of collaborative filtering in identifying patient preferences, while Job et al. [4] showed how Transformer-based techniques can improve opinion- based drug recommendations. Knowledge graphs, as described by Saadat et al. [8], enhance these systems by integrating diverse data sources to capture complex drug- patient relationships. Machine learning techniques like those discussed by Ariyachaipong et al. [6] further refine these models, ensuring safe and adaptable drug recommendations.

2.3 Deep Neural Networks

Deep neural networks (DNNs) and Graph Neural Networks (GNNs) are powerful tools for analyzing extensive data, including patient demographics, medication properties, and clinical history. Uner et al.'s *DeepSide* model [7] leverages gene expression and molecular data to predict drug interactions and side effects. Bongini et al. [9] emphasized the advantage of GNNs in detecting non-linear relationships, which are crucial for identifying latent features that may indicate side effects or interactions. By evaluating these relationships, DNNs and GNNs recommend the most suitable medications while assessing associated risks [11].

2.4 Sentiment Analysis

Sentiment analysis models offer unique insights into patient feedback, providing an additional layer for assessing satisfaction, recommending drugs, and detecting adverse drug reactions (ADRs). Saadat et al. [8] highlighted the integration of sentiment analysis with knowledge graphs for enhanced recommendations. Dongre and Agrawal [10] demonstrated the value of analyzing social media data to detect ADRs and improve drug safety. Li et al. [14] explored sentiment-based collaborative filtering to enhance the system's ability to predict outcomes using real-world patient reviews, particularly in cases with limited clinical data.

3. Patient Specific Drug Recommendation and Side Effect Prediction System

The system requires input on the patient's symptoms, including their severity and duration, to accurately predict possible diseases. Additionally, details on the patient's current medications, allergies, or pre- existing conditions help tailor safe and effective drug recommendations. Information for the system is gathered from multiple sources such as clinical trial databases, Electronic Health Records (EHR), patient medical histories, drug interaction databases, documented side effects, and patient reviews on social media platforms. The preprocessing pipeline includes encoding categorical features using one-hot or label encoding, handling missing data via imputation (using tools like pandas and sklearn), and standardizing numerical data with MinMaxScaler or StandardScaler for consistency. Unstructured text data undergoes tokenization (using NLTK or spaCy), followed by stop-word removal and stemming/lemmatization with libraries like NLTK or spaCy to prepare for sentiment analysis as shown in Figure 35.1. For disease prediction, Random Forest (using scikit-learn) creates an ensemble of decision trees on various data subsets, where each tree independently predicts a disease based on patient symptoms. The final prediction is made by majority vote across trees, improving accuracy and reducing overfitting. Recurrent Neural Networks (RNNs), built with TensorFlow or Keras, are used for unstructured data (clinical text), capturing sequential patterns across symptoms and sentences to predict diseases from narrative data. In drug recommendation, Collaborative Filtering (CF), implemented with Matrix Factorization (using Surprise or TensorFlow) or K-Nearest Neighbors (KNN), identifies similar patients and recommends drugs based on shared characteristics. Deep Neural Networks (DNNs), built using TensorFlow or PyTorch, learn

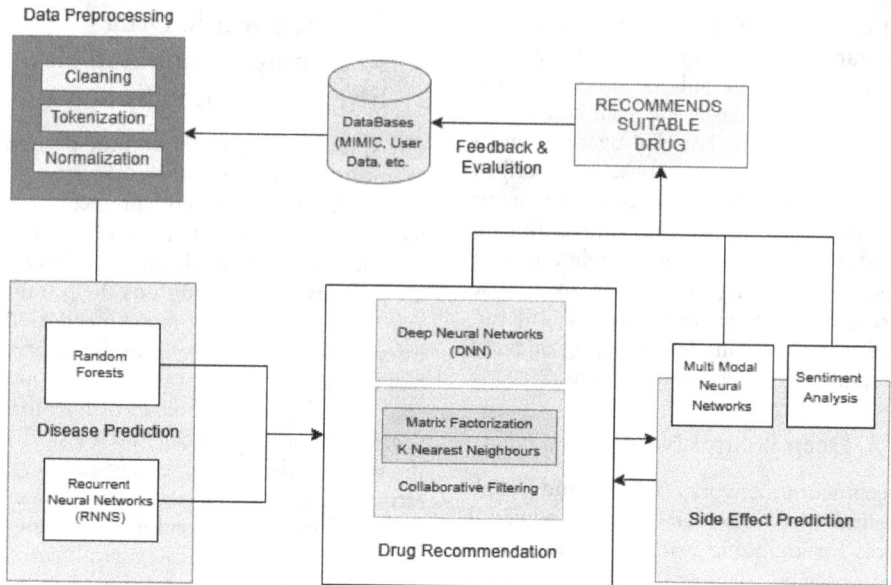

Figure 35.1 Architecture for Patient Specific Drug Recommendation System

complex relationships between patient data, symptoms, disease history, and drug efficacy, providing personalized medication suggestions.

For side-effect analysis, Multimodal Neural Networks, using TensorFlow or PyTorch, integrate structured data (e.g., medical history) and unstructured data (e.g., patient reviews) to identify adverse drug reactions (ADRs). These models cross-reference side-effect databases like SIDER and evaluate textual feedback to identify and assess ADRs. Sentiment Analysis tools such as VADER or TextBlob classify patient feedback as positive, negative, or neutral, allowing the system to detect unreported ADRs and adjust recommendations accordingly. This integration of clinical data and sentiment analysis improves prediction accuracy and enhances patient safety by identifying ADRs not documented in clinical trials. The output of the proposed methodology includes disease predictions, personalized drug recommendations, ADR predictions, sentiment analysis of patient feedback, and alternative drug suggestions to ensure safer and more effective treatment outcomes.

4. Implementation Details and Result Analysis

The proposed paper is implemented using a modular approach, combining multiple techniques to achieve personalized drug recommendations. Preprocessing steps include encoding categorical features, imputing missing data, and normalizing numerical data. For disease prediction, Random Forest and Recurrent Neural Networks (RNNs) are used, leveraging patient symptoms and clinical text data to identify potential conditions. Collaborative Filtering and Deep Neural Networks (DNNs) are applied in the drug recommendation module to analyze patient profiles and drug attributes. For side effect analysis, Multimodal Neural Networks integrate structured clinical data and unstructured patient feedback, while sentiment analysis tools like VADER classify reviews to predict adverse drug reactions.

4.1 Data Sets Used

Various datasets support disease prediction, drug recommendation, and ADR analysis in healthcare systems. MIMIC-III provides real-world clinical data for disease prediction, while the UCI Machine Learning Repository supplies essential patient records for symptom-based predictions [1, 2]. For drug recommendations, DrugBank provides data on drug compounds, interactions, and side effects, supporting tailored suggestions [5]. SIDER offers information on known ADRs, while Drugs.com Reviews adds patient feedback on drug efficacy, enhancing sentiment analysis [8, 10]. Together, these datasets enable a data-driven approach to safer, personalized healthcare.

4.2 Evaluation Metrics

The proposed paper uses several evaluation metrics to ensure accuracy and reliability in healthcare applications. In the Disease Prediction Module, Accuracy measures overall prediction correctness, while Precision evaluates the proportion of correctly diagnosed diseases, reducing false positives. In the Drug Recommendation Module, Mean Average Precision (MAP) assesses the relevance of drug recommendations to patient conditions, and Top-K Accuracy checks if the correct drug ranks within the top-K recommendations. For the Side Effect Analysis Module, True Positive Rate (Sensitivity) measures the ability to identify adverse drug reactions (ADRs), while False Positive Rate gauges incorrect ADR predictions. AUC- ROC evaluates the model's sensitivity and specificity across thresholds. To enhance robustness, RMSE estimates recommendation errors, and Log-Loss compares predicted probabilities to actual outcomes, optimizing drug on patient data.

4.3 Result Analysis

The evaluation results demonstrate the robust performance of the proposed system across multiple healthcare modules as shown in Table 35.1. The Disease Prediction module shows high accuracy (90%) and precision (85%), ensuring reliable disease

detection. recommendations based The Drug Recommendation module achieved a MAP of 0.88 and Top-K accuracy of 0.92, validating its ability to suggest relevant medications based on patient profiles. For Side Effect Prediction, the True Positive Rate (0.89) and AUC-ROC (0.91) indicate strong detection of adverse drug reactions. The Sentiment Analysis module, with an accuracy of 85%, effectively identifies patient feedback, enhancing the system's ability to recommend safer and more effective treatments.

4. Conclusion

In conclusion, the proposed system offers a comprehensive and efficient approach to personalized healthcare by integrating multiple machine learning models for disease prediction, drug recommendation, side effect analysis, and sentiment analysis. The strong performance metrics across all modules demonstrate the system's ability to make accurate predictions, recommend

Table 35.1 Evaluation Metrics and Results of Modules of the Proposed System Across Modules

Module	Evaluation Metric	Result
Disease Prediction	Accuracy	92.33%
	Precision	0.88
Drug Recommendation	Mean Average Precision (MAP)	0.85
	Top K Accuracy	93.81%
Side Effect Prediction	Sensitivity	90.34%
	False Positive Rate	0.07
	AUC - ROC	0.91
Sentiment Analysis	F1-Score	0.83
General Performance	Log Loss	0.15
	RSME	0.35

relevant treatments, and identify potential adverse drug reactions. By leveraging both structured and unstructured data, including clinical records and patient feedback, the system ensures personalized, safe, and effective medication management. This work aligns with current trends in precision medicine and holds significant potential for transforming healthcare practices by improving treatment outcomes and patient safety.

References

[1] Nayak, S. K., Garanayak, M., Swain, S. K., Panda, S. K., & Godavarthi, D. (2023). An Intelligent Disease Prediction and Drug Recommendation Prototype by Using Multiple Approaches of Machine Learning Algorithms. *IEEE Access*, 11, 99304–99318.

[2] Raj, M. K., Malardhas, J. P., & Devapriya, I. (2024). Machine Learning Approach to Predict Multiple Diseases Based on Symptoms. In *2024 10th International Conference on Communication and Signal Processing (ICCSP)*, 1195–1199. Melmaruvathur, India.

[3] Luo, H., Wang, J., Yan, C., Li, M., Wu, F.-X., & Pan, Y. (2021). A Novel Drug Repositioning Approach Based on Collaborative Metric Learning. *IEEE/ACM Transactions on Computational Biology and Bioinformatics*, 18(2), 463–471.

[4] Job, S., Tao, X., Li, Y., Li, L., & Yong, J. (2023). Topic Integrated Opinion-Based Drug Recommendation with Transformers. *IEEE Transactions on Emerging Topics in Computational Intelligence*, 7(6), 1676–1686.

[5] Zhang, H., Yang, X., Bai, L., & Liang, J. (2024). Enhancing Drug Recommendations via Heterogeneous Graph Representation Learning in EHR Networks. *IEEE Transactions on Knowledge and Data Engineering*, 36(7), 3024–3035.

[6] Ariyachaipong, K., Senaluang, P., Songmuang, P., & Kongkachandra, R. (2024). Drug Recommendation Based on Drug Details and Optimization Using Bee Algorithm. In *2024 IEEE International Conference on Cybernetics and Innovations (ICCI)*, 1–6. Chonburi, Thailand.

[7] Uner, O. C., Kuru, H. I., Cinbis, R. G., Tastan, O., & Cicek, A. E. (2023). DeepSide: A Deep Learning Approach for Drug Side Effect Prediction. *IEEE/ACM Transactions on Computational Biology and Bioinformatics*, 20(1), 330–339.

[8] Saadat, H., Shah, B., Halim, Z., & Anwar, S. (2024). Knowledge Graph- Based Convolutional Network Coupled with Sentiment Analysis Towards Enhanced Drug Recommendation. *IEEE/ACM Transactions on Computational Biology and Bioinformatics*, 21(4), 983–994.

[9] Zheng, Y., & Xu, S. (2023). A Knowledge Graph-Based Clustering Approach for Drug Side Effects Prediction. In *2023 IEEE International Conference on Bioinformatics and Biomedicine (BIBM)*, 1764–1769. Istanbul, Turkiye.

[10] Bongini, P., Messori, E., Pancino, N., & Bianchini, M. (2023). A Deep Learning Approach to the Prediction of Drug Side-Effects on Molecular Graphs. *IEEE/ACM Transactions on Computational Biology and Bioinformatics*, 20(6), 3681–3690.

[11] Dongre, S., & Agrawal, J. (2023). Deep-Learning-Based Drug Recommendation and ADR Detection Healthcare Model on Social Media. *IEEE Transactions on Computational Social Systems*, 10(4), 1791–1799.

[12] Mian, S. M., Khan, M. S., Shawez, M., & Kaur, A. (2024). Artificial Intelligence (AI), Machine Learning (ML) & Deep Learning (DL): A Comprehensive Overview on Techniques, Applications and Research Directions. In *2024 2nd International Conference on Sustainable Computing and Smart Systems (ICSCSS)*, 1404–1409. Coimbatore, India.

[13] Symeonidis, P., Chairistanidis, S., & Zanker, M. (2022). Deep Reinforcement Learning for Medicine Recommendation. In *2022 IEEE 22nd International Conference on Bioinformatics and Bioengineering (BIBE)*, 85–90. Taichung, Taiwan.

[14] Li, Y., et al. (2023). A Collaborative Cross-Attention Drug Recommendation Model Based on Patient and Medical Relationship Representations. In *2023 IEEE International Conference on Bioinformatics and Biomedicine (BIBM)*, 2036–2039. Istanbul, Turkiye.

36 Enhancing Electricity Forecasting with Random Forest: Utilizing Correlation Metrics for Advanced Feature Engineering

Balasubbareddy Mallala[1,3], Azka Ihtesham Uddin Ahmed[1],
Sastry V. Pamidi[2], Md Omar Faruque[2], Rajasekhar Reddy M[3],
and V. Lakshmi Chetana[3]

[1]Chaitanya Bharathi Institute of Technology,Gandipet, Hyderabad, India
[2]Centre for Advanced Power Systems, Florida State University, Tallahassee, USA
[3]Amrita School of Computing, Amrita Vishwa Vidyapeetham, Amaravati Campus, Andhra Pradesh, India

Abstract

Electricity forecasting is a very crucial tool for energy management and also ensures the stability of any grid system. This paper discusses the application of Random Forest algorithms to improve electricity demand forecasting capabilities by using feature engineering techniques based on correlation metrics. We carried out a rigorous analysis of historical data about electricity consumption, establishing important features that affect such trends greatly. We then trained the Random Forest model using these engineered features, which yielded very impressive performance metrics results. The study yielded the most favorable outcomes in terms of metrics, including a "Mean Squared Error (MSE) of 0.45", a "Mean Absolute Error (MAE) of 0.17", a "Mean Absolute Percentage Error (MAPE) of 0.0004", and a "R2 Score (R²) of 0.9996". These numbers confirm that our model predicts electricity consumption with high accuracy and reliability. The following results of the application of correlation metrics for feature engineering in the development of forecasting models suggest that such metrics could greatly enhance the forecasts, giving strategic insights to both energy providers and policymakers. This study adds to the developing body of information in electricity forecasting, detailing the potential that machine learning techniques can bring to the optimization of energy management strategies.

Keywords: Electricity bill, machine learning, MSE, MAE, MAPE, R2 score, feature engineering, random forest

1. Introduction

Electricity bills is an important part of modern life, which reflects the cost of energy consumed in households, businesses, and industries. These bills provide detailed info about the amount of electricity used over a billing cycle, the cost of that electricity, and often include various charges and taxes. Understanding electricity bills is crucial for managing energy consumption, budgeting expenses, and contributing to environmental sustainability. They typically keep both the consumer and the energy

DOI: 10.1201/9781003661917-36

provider in check from over consumption and extra generation of electricity.

Electricity bills typically consist of several components: the basic charge, the consumption charge, and various additional fees. The basic charge, or fixed charge, is a flat rate applied regardless of the amount of electricity consumed. The consumption charge, or variable charge, depends on the actual amount of electricity used, measured in kilowatt-hours (kWh). Additional fees can include environmental levies, service fees, and taxes.

The layout and details of electricity bills can differ greatly across different countries, states, or utility providers. For instance, in the United States, utility companies are regulated by state public utility commissions, which dictate the format and information that must be included in the bills. In India, similar regulations are enforced by the state regulatory authorities to ensure transparency and consumer protection.

Energy Management: By examining their electricity bills, consumers can identify patterns in their energy usage and find ways to reduce consumption, thereby lowering costs and reducing their carbon footprint.

Budgeting: Accurate knowledge of electricity costs helps households and businesses budget more effectively, ensuring that they allocate sufficient funds to cover their energy expenses.

Efforts to simplify electricity bills and enhance consumer understanding have been ongoing. Many utility companies now offer online tools and resources to help consumers better understand their bills and manage their energy use more efficiently.

Electricity bills are not only a critical aspect for consumers but also for energy providers, who rely on accurate billing to maintain financial stability, ensure customer satisfaction, and promote efficient energy use. From the energy providers' perspective, the billing process involves complex logistics, advanced technology, and stringent regulatory compliance to accurately capture, calculate, and communicate energy consumption costs.

For energy providers, electricity bills consist of several key components:

Generation Costs: This includes the costs associated with producing electricity, whether through fossil fuels, nuclear power, or renewable energy sources. Providers must balance production costs with market demand and regulatory mandates.

Transmission and Distribution Costs: These are the costs of transporting electricity from power plants to end-users. Infrastructure maintenance, upgrades, and energy loss during transmission are significant factors.

Operational Costs: This encompasses the day-to-day operational expenses of running the utility, including staffing, customer service, and administrative functions.

Regulatory and Compliance Costs: Energy providers must adhere to regulations set by governmental bodies, which often include environmental compliance costs, grid reliability standards, and customer protection laws.

Additional Charges: These can include taxes, surcharges for renewable energy, and fees for energy efficiency programs.

Energy providers face several challenges in the billing process:

Accurate Data Collection: With the introduction of smart meters, service providers are able to gather up-to-the-minute information on how much energy is being used. Yet, incorporating this information into the systems that calculate bills with precision and safety continues to be a hurdle.

Regulatory Compliance: Providers must navigate a complex web of local, state, and federal regulations, ensuring that billing practices meet all legal requirements.

Consumer Engagement: Educating consumers about their energy usage and charges is vital. Providers use detailed bills and online platforms to enhance transparency and help consumers manage their energy consumption.

To address these challenges, energy providers are leveraging technological innovations:

Smart Metering: Smart meters provide detailed, up-to-the-minute information on how much energy is being used, leading to more precise invoicing and improved control over energy demand. They also enable dynamic pricing models, which can help to balance grid load and reduce costs [1].

Renewable Energy Integration: By incorporating a diverse array of renewable energy resources, including solar and wind into the grid, it's possible to distribute and share the load during peak hours. This approach helps in mitigating the production greenhouse gases associated with the utilisation of fossil fuel in conventional energy generation system [5].

Advanced Billing Systems: Modern billing systems use data analytics and machine learning to improve accuracy, detect anomalies, and provide personalized energy-saving recommendations to consumers.

Customer Engagement Tools: Web platforms and smartphone applications enable users to monitor their energy consumption, check their utility statements, and get advice on how to lower their usage. These resources contribute to enhancing customer happiness and retention.

Regulation has a significant impact on how energy companies bill their customers. Regulatory bodies confirm that billing practices are fair, transparent, and that customers are protected from unfair charges. Compliance with these regulations often involves substantial investment in technology and staff training for energy providers.

2. Literature Review

[4] suggested integrated energy management is crucial for electrical energy providers to balance demand and supply. It involves combining diverse energy sources, storage systems, and demand-side control to maximize resource usage, reduce costs, and enhance reliability. It is a strategic need for sustainable energy and the environment. Machine Learning algorithms integrated into energy management can significantly improve load forecasting.

[6] They developed a machine learning model with IoT infrastructure to improve intelligent power quality management

in smart grids. It uses multiple sensors to obtain environmental data on renewable energy sources. The integrated system forecasts electricity production based on real-time environmental conditions, supporting informed decision-making for smart grid stability.

[7] explained how identifying electrical faults is crucial to ensuring a constant electricity supply and taking care of safety concerns. The use of machine learning methods shows great potential in improving systems for identifying and diagnosing faults. The suggested approach involves applying a decision tree classifier and the SMOTE (Synthetic Minority Over-sampling Technique) technique. This study highlights the efficacy of machine learning techniques in fault diagnosis as well as their potential for real-world implementation in power systems and industrial settings.

[3] This research project focuses on the comprehensive research framework for electric vehicle charging systems within smart grids. It employs Genetic Algorithms and machine learning models to reduce waiting periods and increase utilization. The findings highlight the effectiveness of these methods, showing that Artificial Neural Networks achieve almost 99% accuracy. There's a significant potential for Artificial Intelligence to transform the way energy is distributed, balancing the needs of consumers with environmental goals. The architecture of the proposed forecasting system is represented in Fig. 36.1.

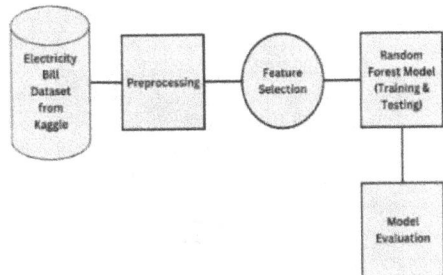

Figure 36.1 Proposed Architecture for Electricity Bill Forecasting system

3. Materials and Methods

Dataset was sourced from the Kaggle platform. It consists of data from various categories that influence the electricity bill, such as fan, refrigerator, Air conditioner, Television, Monitor, Motor pump, Month, Monthly hours, Tariff rate and amount of electricity bill generated for the energy consumed. Additionally, it offers details on the city and the company associated with the data.

3.1 Data Preprocessing

The raw dataset taken from Kaggle had many errors and missing values, which were affecting the actual v/s predicted analysis. So, depending on the size and shape that particular column was either imputed or dropped. Some object entries were there such as data in city and company column, they were converted into our desired data type. After getting a clean dataset of 11 variables with 45,345 samples, it was split into two i.e., train and test data with 80:20 ratio. By taking a note from pearson correlation plot the authors selected "Monthly Hours" as the target variable. The correlation plot with clean dataset is shown in the Figure 36.2.

3.2 Feature Selection

The preliminary analysis indicates that not all the features are correlated. There is a need for a feature selection method to find the features that share highest correlation with our target variable. The authors

decided to go with the correlation coefficient feature selection method. In this author applied a command which dictates the program to select the set of features which have correlation greater than 80% and less than 100% with the target variable. So that the optimised set of features can be taken out and the best results can be achieved. The plot in the Figure 36.3 shows the heatmap obtained after the application of Correlation Coefficient Feature Selection method.

3.3 Random Forest Algorithm

Regression and classification tasks are the main applications for the Random Forest ensemble learning technique. It constructs multiple decision trees during training and outputs each tree's mean prediction (for regression) or class mode (for classification).

Here are some unique aspects of Random Forest:

1. Ensemble Learning: Overfitting, which is typical in single decision trees, is less likely with Random Forest since it combines the predictions of several trees.

2. Bootstrap Aggregating (Bagging): It trains each tree using a bagging technique, which involves sampling subsets of the training data with replacement. This facilitates the creation of varied trees that identify various data patterns.

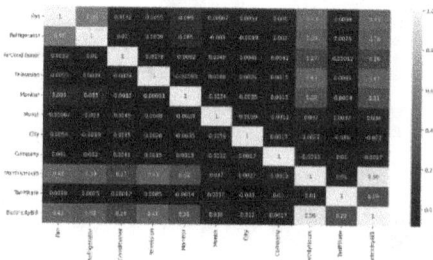

Figure 36.2 Before Feature Selection

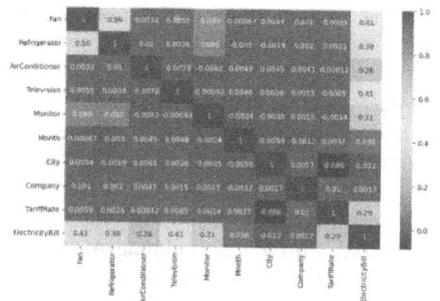

Figure 36.3 Shows Heatmap of Correlation Coefficient Based Feature Selection Method

3. Feature Randomness: Random Forest randomly selects a subset of features while dividing a node in the process of building a tree. By lowering correlation between the trees, this unpredictability contributes to the model's resilience.
4. Variable Importance: Random Forest can provide insights into the importance of different features in making predictions, allowing for better understanding and interpretation of the model.
5. Robustness to Noise: Compared to other algorithms, it is less susceptible to noisy data and outliers, which makes it appropriate for applications in real life where data can be imperfect.
6. Scalability: Random Forest is useful in many fields, including image processing, finance, and healthcare, since it can effectively handle big datasets and high-dimensional feature spaces.

Overall, Random Forest is a powerful and versatile algorithm that balances accuracy and interpretability and allows user to get accurate outputs.

4. Result and Discussion

The system was evaluated by various factors which are Mean Squared Error (MSE), Mean Absolute Error (MAE), Mean Absolute Percentage Error (MAPE) and R-Squared Error (R2). Upon rigorous evaluation of the model the best result was achieved by using Correlation Coefficient Feature Selection method. "The least result that was achieved are MSE: 0.45, MAE: 0.17, MAPE: 0.0004 and R2: 0.9996".

The matrix model visualisation is shown in Fig. 36.4. The Actual V/S Predicted values of electricity consumption (kW) is given in the Table 36.1.

The regression plot based on the actual v/s predicted values is given in the Figure 36.5. From the Figure 36.5 it can be clearly seen that there is very negligible error between actual v/s predicted values and the proposed system gave very accurate results.

To further evaluate the systems performance, its output results are compared

Table 36.1 Shows Actual v/s Predicted Values

	Actual	Predicted_prob
40840	427	426.95
18221	525	524.98
16244	529	529.02
7638	398	397.98
6080	661	660.99
...	...	...
34560	514	514.00
43067	514	514.00
20928	450	450.00
263	689	688.89
27281	288	288.43

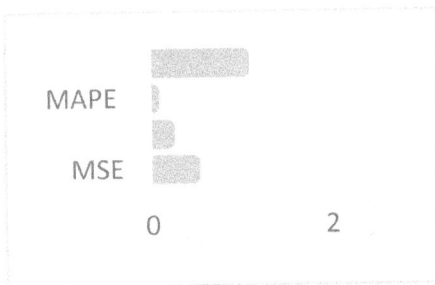

Figure 36.4 Visualizes the Performance Matrix of the Model

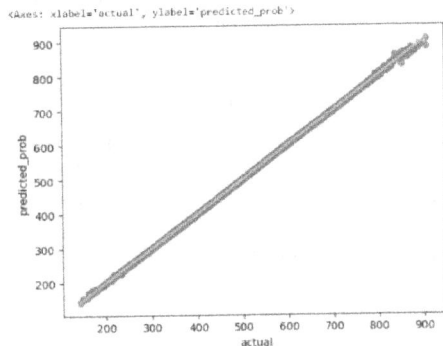

Figure 36.5 Shows Regression Plot based on Actual V/S Predicted Values Table

Table 36.2 Comparative Evaluation of the Prediction System with Existing System

References	Algorithms	Evaluation Parameters
[2]	Linear Regression	MAE = 2.91
	Decision Tree Regression	MAE = 4.39
	Support Vector Regression	MAE = 2.93
	Artificial Neural Network	MAE = 6.70
[1]	KNN	MAPE = 5.15%
	SVM	MAPE = 4.13%
Proposed System	Random Forest Regression	MAE = 0.45 MAPE = 0.0004 R2 Score = 0 .9996

against the outputs of existing systems as provided in the Table 36.2. The proposed system achieved lowest error rates and made predictions with the highest accuracy and precision as indicated in Table 36.1. The random forest algorithm, due to its unique ability to merge predictions from various trees and enhance the systems output, significantly increased the accuracy of the proposed system.

5. Conclusion

The prosed system stood out of the existing prediction systems and gave most accurate results. This paper emphasizes the need of awareness regarding the global climatic changes, the wastage of resources, the causes of pollution and the utilisation of renewable energy sources to tackle the rising concern of the climatic changes. By mandating the use of Renewable Energy Sources the cost of electricity generation can be reduced, the electricity can be provided at a low cost to the consumers and the resources (Fossil fuels) which are depleting at faster pace due to rapid urbanization can be preserved.

Future Scope

This dataset facilitates the execution of data analysis in relation to each city and electricity provider. Furthermore, it possesses the capability to examine the monthly consumption patterns of each city, alongside the electricity bills incurred and the corresponding tariffs.

References

[1]   Bashawyah, D. A., & Qaisar, S. M. (2021). Machine learning based short-term load forecasting for smart meter energy consumption data in London households. In *2021 IEEE 12th International Conference on Electronics and Information Technologies, ELIT 2021 - Proceedings. Institute of Electrical and Electronics Engineers Inc.*, pp. 99–102. https://doi.org/10.1109/ELIT53502.2021.9501104

[2]   Bhandarkar, M., Suryawanshi, A., Deoghare, S., Isaac, N., Hegu, A., & Wankhede, P. (2023). Electricity Billing and Consumption Prediction using Machine Learning. In *2023 7th International Conference On Computing, Communication, Control And Automation, ICCUBEA 2023. Institute of Electrical and Electronics Engineers Inc.* https://doi.org/10.1109/ICCUBEA58933.2023.10392239

[3]   Joshua, K. P., Ranga, J., Prasad, P. V., Mallala, B., Rajendiran, M., & Maranan, R. (2024). Optimized Scheduling of Electric Vehicles Charging in Smart Grid using Deep Learning. In: *Proceedings - 2024 International Conference on Expert Clouds and Applications, ICOECA 2024. Institute of Electrical and Electronics Engineers Inc.*, pp. 408–412. https://doi.org/10.1109/ICOECA62351.2024.00079

[4]   Mallala, B., Khan, P. A., Pattepu, B., & Eega, P. R. (2024). Integrated Energy Management and Load Forecasting Using Machine Learning. In: *2nd International Conference on Sustainable Computing and Smart Systems, ICSCSS 2024 - Proceedings. Institute of Electrical and Electronics Engineers Inc.*, pp. 1004–1009. https://doi.org/10.1109/ICSCSS60660.2024.10625623

[5]   Mallala, B., Uddin Ahmed, A. I., Prasad, P.V., & Kowstubha, P. (2023). Development of Renewable Energy

System for Enhancing Reliability of Power. In: *Procedia Computer Science. Elsevier B.V.*, pp. 1–10. https://doi.org/10.1016/j.procs.2023.12.055

[6]   Rajendran, P., Ranga, J., Prasad, P. V., Mallala, B., Senthilkumar, G., & Natrayan, L. (2024). Development of Intelligent Power Quality Management in Renewable Energy System in Smart Grid using Deep Learning. In: *7th International Conference on Inventive Computation Technologies, ICICT 2024.* Institute of Electrical and Electronics Engineers Inc., pp. 1178–1182. https://doi.org/10.1109/ICICT60155.2024.10544835

[7]   Vivek, B., Teja, B. H., Mallala, B., & Srinitha, G. (2024). Electrical Fault Detection and Localization Using Machine Learning. In: *Proceedings - 2024 International Conference on Expert Clouds and Applications, ICOECA 2024.* Institute of Electrical and Electronics Engineers Inc., pp. 820–825. https://doi.org/10.1109/ICOECA62351.2024.00145

37 Image Processing Smart Glasses for Visually Impaired

K. Kanchana[1], V. Babu[2], and S. Nithyasree[2]

[1]Associate Professor, EEE Department, Saveetha Engineering College
[2]UG Student, EEE Department, Saveetha Engineering College

Abstract

This research is trying to solve the navigation problem for blind and visually impaired people, especially during low light conditions. Current solutions mostly lack accessibility and functionality and rely on other people for assistance. Our smart glasses take the concepts of IoT and AI and merge these ideas through four interconnected models: low-light image enhancement using ESRGAN, audio feedback for object recognition, salient object detection, and a text-to-speech/ tactile graphics generation model. Utilization of ESRGAN would be beneficial to enhance quality in low-light imagery and, hence, could increase the accuracy of object recognition. It uses deep convolutional neural networks for accurate cases of detection with real-time client-server architecture. Early evaluations display competing performance with an accuracy as high as 99.7% on benchmarking datasets. Our approach innovatively aims to empower Blind and Visually Impaired (BVI) individuals, promote independence and safety far better, and enhance their quality of life through the usage of advanced technology.

Keywords: Blind and Visually Impaired (BVI), Smart Glasses, ESRGAN, object recognition, assistive technology, low-light image enhancement

1. Introduction

The problems the BVI user faces with accessing and mobility are very deep-seated and so complex. Those that are cane or guide dog-based, which in themselves compromise the user, only occur in unintelligible lighting conditions. Risks due to a lack of real-time perception of information visually are very high, and thus there is a real need to look for innovative solutions that empower BVI people [1].

In light of all these developments in AI and IoT, all avenues have been opened to enhance mobility solutions for the demographic population. Advanced image processing capabilities provided with smart glasses can bridge the gap between a user and his or her environment by providing critical visual information that can, ultimately, create independence and safety in navigation [2]. This study shall be specifically designed and developed to provide a prototype through smart glasses targeted for BVI individuals and most suitable to use in navigational applications [3–5]. Advanced AI models, for example, ESRGAN, in conjunction with deep

Email: kanchana@saveetha.ac.in[1], 007babubabuv@gmail.com[2], nithyasree20092004@gmail.com[2]

DOI: 10.1201/9781003661917-37

convolutional neural networks, shall be implemented for the enhancement of sharpness and availability of real-time visual information. Smart glasses are of course a technological innovation but also a step closer to more inclusive solutions that take into consideration the needs that individuals with visual impairments have. This will be of great interest to the researcher, as the world ages, there is going to be a straight increase in the visually impaired, thus a need for more effective assistive technologies. The existing mobility solutions do not cater to the BVI-specific needs during challenging conditions, such as low light where visibility is compromised.

Therefore, research into technological interventions may be of extreme utility in enhancing further the quality of life for these people. The present study shall contribute to the body of knowledge regarding the degree to which such technologies, implanted wisely, may enable enormous improvements in the navigation experiences of BVI individuals. This research would not focus merely on optimizing individual mobility by granting power and enhancing autonomy for users but contribute far more importantly to general social welfare. Despite considerable advances in assistive technologies, most solutions developed and currently available fail to adequately support people with BVIs when performing a range of tasks in low light. Most navigation devices still rely on either tactile or auditory feedback modality that fails to meet the need of the user for real-time visual information leading to safe navigation. Concomitantly, BVI persons are becoming increasingly reliant on others to move about, an arrangement that is both very disabling and detracts from full involvement with their environment. Thus, the question of mobility aids reflects an imperative need for innovative solutions tapping into the potentiality of AI and IoT in empowering users.

This system fills the gaps in building a system that processes vision-related information, integrates feedback mechanisms, and optimizes BVI movement. Some of the main scientific questions motivating this work are: how adding ESRGAN impacts BVI residents' visual information in low-light conditions? The results of super-resolution from ESRGAN are promising, and one now understands how BVI users might catch critical visual clues in poor illumination.

Effects on BVI navigation: real-time object recognition with auditory feedback Smart glasses now provide rapid auditory input for barriers and objects of interest to enhance situational awareness and trusted navigation. This question assesses subjective user experience and objective performance measures for overall effectiveness.

This research tackles BVI problems in a holistic manner. IoT and AI will be used to develop a smart glass prototype that helps people move. ESRGAN will be used to build low-light image improvement models to improve photographs taken in poor lighting. The project will also create an item recognition model that gives users real-time feedback to understand their environment. The salient identification of relevant stimuli model assists navigation in busy and complex environments. The effort will culminate in text-to-speech and haptic visuals to provide non-static environmental information. Advanced AI would be used in BVI assistive device applications for the proposed project. The initial set of smart glass prototype testing will show that the ESRGAN model enhances low-light image quality and assesses how accurate and sensitive the object recognition and aural feedback system is to real-time user input. User trials and comments are used to determine how effectively smart glasses improve navigation freedom and safety. This will gather qualitative and quantitative data to show how such technology may affect BVI mobility. This study is chaptered. Chapter 1 was introduction to the research question, context, and significance. Section 2 would be literature of current technologies to uncover gaps in BVI mobility aids and AI/IoT navigation. Methodology for designing and assessing smart glass is provided in Section 3. The results and consequences of the prototype assessment are explained in Section 4. Section 5 conclusion provides key results, field contributions, and future study directions towards how this still has much potential to improve the quality of lives for BVI people. By building a prototype of smart eyewear, our research significantly expands the scope of assistive technology for the blind and visually handicapped. It deals with issues of low light navigation and provides BVI users with real-time input to encourage autonomy and safety across different contexts. This project will advance excellent design in assistive mobility devices for people who are visually impaired and for their freedom and inclusion.

2. Literature Survey

One of the major and challenging tasks in computer vision is object detection. Deep learning-based image-based object detection techniques perform well with clear images but do not favor poor weather conditions such as snow due to degradation in images. Current research aims to incorporate techniques of image restoration to restore degraded images before the detection of the object. Sometimes direct restoration introduces new disturbances which slows down the pace of improvement of the performance of detection. To address this, [6] proposed a combined framework, in which the iterative des now module was coupled with the end-to-end detection module. A good model for object identification, which can be applied in MOT, is YOLOv8.

[7] It can detect all objects in an image since it is a one-pass detector. This is ideal for real-time applications. Frames from video. Therefore, MOT may employ YOLOv8. Each frame re-identifies things using a model from previous frames. The background-object attention transformer (BOA former) is proposed in this paper to improve object recognition by establishing a background-to-object connection of attributes. A pre-designed classification model calculates object prediction [8] box scores. It writes that the scoring method will remove high-scoring item prediction boxes to simplify network calculations.

The [9] features a Wireless Power Transfer MOD and LOD FOD sensor. A sensing device with adaptive excitation and detection will better detect random metal and living objects. This adaptive technique for generating an excitation signal with adjustable frequency components introduces a new advanced slicing-aided inference framework, analogous to Region Proposal Networks [10]. The elimination of fixed-step slicing further enhances the detection accuracy for many item sizes. ESAHI generates patches with all the large object properties from high-resolution object-centric regions. This paper [11] employs long and short focal-length cameras. The short focal length cameras capture wide-angle photos, and the long focal length cameras capture telephoto photos. The DLA-34 feature extraction network will utilize long and short-focal image features for the first time in this paper. Key point detection and regression of 3D box parameters on features have identified both types of pictures.

With attention approaches, UAV infrared imaging can recognize tiny things more accurately [12]. A modified Recursive Feature Pyramid with an improved YOLOX baseline network may enhance vehicle detection in the VisDrone-Vehicle dataset. Thus, for tiny objects, we used a separate focus module and sigmoid activation to link the main and side FPNs. This work proposed adaptive feature fusion technology [13] to optimize feature information usage at any level by replacing simple splicing with a dynamic fusion technique based on relevance in the feature map. The idea behind this approach is to enhance the accuracy and robustness of object identification through smarter and more efficient utilization of feature extraction. New research reviewed and proposed a machine learning and deep learning strategy to enhance these abilities [14].

Multi-layer perceptron networks have enhanced object segmentation and identification [15, 16]. Indexing massive picture databases using object segmentation made data retrieval and management easier [17]. Additionally, quantitative models of object recognition have improved picture classification for surveillance systems [18] and deep learning has inspired real-time object identification [19]. Self-driving automobiles with improved lane identification and algorithms for machine vision enhance navigation and safety in driverless vehicles [20]. The Night vision camera-wired spying robots are another military and surveillance application of such a technology [21]. Advanced studies in segmentation and object detection techniques have allowed multi-object labeling in aerial images to occur with great accuracy [22]. Above study demonstrates how the picture segmentation and object detection can be used to advance any system [23, 24]. Many segmentation methods enhance object detection and feature extraction [25, 26]. Steady, without fanfare improvement approaches to image processing for object recognition from complex settings exist [27, 28]. Feature Analyst and others, though, have shown how segmentation can be applied in aid of object recognition in remote sensing [29].

3. Methodology

The study assesses the effectiveness of smart glasses that have been proposed for BVI individuals using a mixed-method approach. This mixed methodology-based study makes use of both qualitative and quantitative methodologies to provide a comprehensive determination of the effectiveness of the developed prototype. Figure 37.1 depicts the architecture flow of the proposed video processing method. The process begins with object identification and image segmentation of the input video, followed by modification of segmented regions through frame decomposition or masking. The segmented regions are enhanced or downgraded as needed, and the processed frames are reassembled to produce the output video. Figure 37.2 illustrates the proposed architecture, where a user interacts with the system by sending images and receiving text/visual feedback. The process integrates ESRGAN

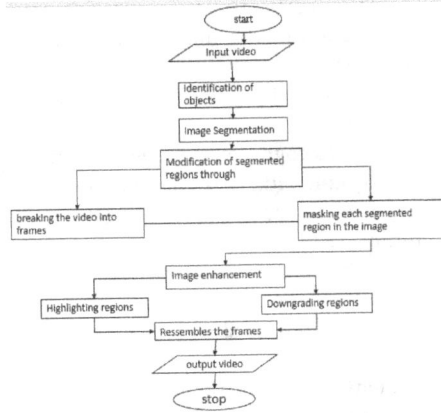

Figure 37.1 Architecture Flow Model

for image enhancement, MobileNet for object recognition, and a salient object detection module to identify key objects, with audio feedback generated by the TTS system.

3.1 Technology Integration

Technology integration Incorporates a lot of state-of-the-art technologies in the following:

ESRGAN Super-Resolution Generative Adversarial Network: ESRGAN is a model that improves images captured in low-light environments but offers clear visual information during navigation. It uses a GAN structure, which is made up of a generator

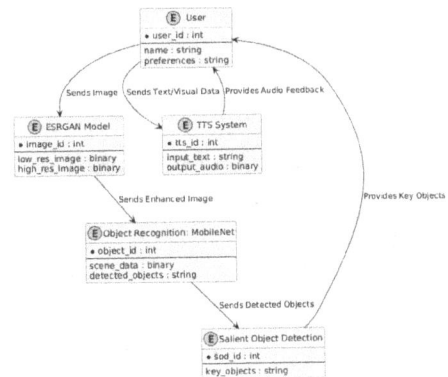

Figure 37.2 Proposed Architecture

and discriminator. The generator improves the resolution of the image while the discriminator checks the quality of the output image, thereby iterating to enhance its output.

Object Recognition Model: The model is deep convolutional neural networks which tries to recognize the objects in the real-time. BVI trained a diverse dataset, usually the one that comes into contact with CNN, that aims to perfectly identify as well as localizing as shown in Figures 37.3 and 37.4.

Salient Object Detection Model: This model identifies significant objects for the user in focus while disregarding uninteresting information and stressing the important stimuli for the user to respond appropriately. It further enhances a user's ability to navigate by pointing out obstacles and relevant scenery within the scene.

The TTS system: generates tactile graphics or provides text-to-speech for

Figure 37.3 Object Detection

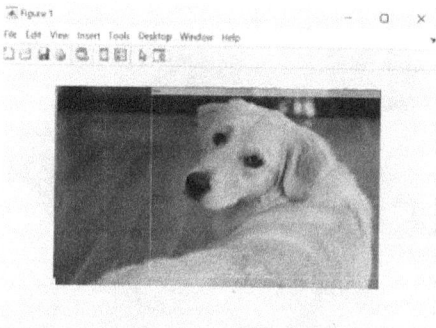

Figure 37.4 Object Identification

received recognized text and important visual information in order to give a better environmental understanding. This will combine the two and present an all-inclusive package for solving the specific issues that the BVI individual is faced with.

3.2 Development Process

Normally, the development process is split into these broad steps:

3.2.1 Hardware Elements

Smart glasses come designed with light materials to ensure long hours of comfortable wear. Some of the more important hardware parts include:

Camera Module: High-definition camera is mounted on the glasses that capture real-time images. The low-light sensitivity of a camera module is one requirement for its effective operation in dim light.

Processing Unit: a compact unit for processing will be used running the AI algorithms on the local machine to reduce latency and having in real-time performance; this unit houses the GPU that will handle computations associated with image processing and object recognition.

Audio Output Device: A bone conduction speaker or a conventional earphone is used to provide audio feedback without mulating the ambient sounds perception of the user.

Power Supply: In this scenario, it has a rechargeable battery that lasts for a long time with lightweight.

3.2.2 Software Development

This would be software development: build and combine numerous AI models and functionalities.

Image processing algorithms: Yes, it is found that ESRGAN is trained on a diversified set of low-light images by exploiting high-resolution image pairs. Consequently, it uses the implementation for both TensorFlow and PyTorch in using libraries for Python. This encourages effective and efficient training and evaluation of the model.

Object detection: It will heavily depend on architectures like pre-trained CNNs, ResNet or MobileNet fine-tuned toward a relevant set of objects from the viewpoint of BVI. It can be run in real time with its software embedded in glasses.

UI development: The developed user-friendly interface is designed to ease interfacing with the smart glasses. Ease of use increases due to the fact that user commands and setting statements can be made with voice commands.

Integration Testing After each of the units is developed integration testing is conducted to see at what level the hardware interfaces with the software In this stage communication by the processing unit, camera, and audio output device are tested.

3.3 Evaluation Techniques

For measuring performance objectively from the smart glasses, several measurement metrics have been studied.

Accuracy of the objects identified with the object recognition model by the sum of correct object recognition during trials as percentages.

Processing Speed: Latency of time between image capture and audio feedback delivery may be measured to ensure real-time performance.

User feedback scores: The responses to the questionnaires are graded based on ratings for usability, comfort, and effects on navigation ability.

3.4 Data Analysis

These are further analyzed through statistical methods to derive meaningful conclusions based on data collected during user trials, surveys, or performance metrics.

Quantitative Analysis: Summary statistics of participant demographics, performance metrics, and self-report measures would summarize the population. Inferential statistics, such as t-tests or ANOVA, would be used to compare performance across task conditions.

Qualitative Analysis: Thematic analysis will be applied on interview transcripts and open-ended survey responses to grasp important themes as well as sentiments of users toward smart glasses.

This chapter describes the methodological approach in developing and evaluating smart glasses for BVI users. Combining cutting-edge technologies and applied through rigorous user trials, this could potentially yield meaningful insight into the possibility of improving mobility and independence for the BVI users. The following chapters summarize the evaluation results with implications for both the field of assistive technologies and the users.

4. Result and Discussion

4.1 General Overview of Evaluation Result

User trials were conducted to assess the smart glasses for BVI users. Such trials provided useful information regarding the effectiveness of the system, usability, and performance in general. The results from the user trials conducted for this study are presented in this chapter and discussed in relation to the implications for enabling enhanced mobility and independence among BVI users.

4.2 Demography Participant

During user trials, the subjects chosen were 30 subjects chosen deliberately with varying levels of visual impairment. Their demographics are summarized below.

It cuts across all ages from 18 to 65 years with an average of 35 years. Sex: Male 15, Female 15.

Visual Impairment Category: S consisted of 10 totally blinded, 10 with severe impairment, and 10 with moderate visual impairment.

Previous use of Augmentative and Alternative Communication: 60% had some kind of previous experience using assistive technology.

Such diversity among the participants made it feasible for other user needs and backgrounds to make an assessment of smart glasses.

4.3 Performance Indicators

It achieved an approximate mean accuracy of about 99.7% in trials. The accuracy is based on how well the model functions under varied environments and lighting condition. The Table 37.1 compares various object detection algorithms based on their accuracy and specific remarks about their performance. The proposed model (ESRGAN + CNN) achieves the highest accuracy (99.7%), excelling in diverse conditions, particularly low-light environments. YOLOv3 provides a faster performance but sacrifices some accuracy, especially for small objects. Faster R-CNN is precise but slower, making it less ideal for real-time applications. SSD offers a balance between speed and accuracy but struggles in complex scenarios, while MobileNet SSD is lightweight and fast but less effective for small or distant objects. InceptionV3 and ResNet-50 demonstrate strong classification and detection capabilities, though ResNet-50 is slower for real-time tasks. VGG-16 is reliable in feature extraction but has the slowest real-time processing speed.

A participant completed the navigation task under the support of smart glasses during the task-based evaluation. Results:

Task Completion Rate: 85% of participants completed their tasks independently.

Average Completion Time: The entire tasks had to be completed in an average of 3.5 minutes, telling how efficiently smart glasses help one navigate solo.

Error Rate: The average error rate, i.e., the percentage of misidentifying obstacles or requiring additional prompts, stood at 10%, reflecting areas of improvement in the

More features: Some users have been asking for GPS navigation, route guidance, alerting parents accident for greater safety.

Comfort and Wears: Most users said that the glasses were comfortable, but some suggested a better weight distribution to ease longer use.

Smart glasses with AI and IoT are possible because the technology improved BVI mobility realistically. A system may enhance the capability for independence and safety mainly because it correctly detects articles and users provide good response. A realistic solution would fill the gap in assistive technology by providing both robust image processing and a user-centric design. BVI users benefit

Table 37.1 Performance Analysis

Algorithm	Accuracy	Remarks
Proposed Model (ESRGAN + CNN)	99.7%	High accuracy for object recognition across various conditions, especially low-light environments.
YOLOv3	95.6%	Faster but slightly less accurate in detecting smaller objects and in low-light conditions.
Faster R-CNN	94.8%	High accuracy, but slower compared to real-time systems. Suitable for precise detection.
SSD (Single Shot Multibox Detector)	92.5%	Balanced between speed and accuracy but underperforms in complex environments.
MobileNet SSD	89.2%	Lightweight, fast, but less accurate for detecting small and distant objects.
InceptionV3	91.7%	Effective for classification, but not optimized for real-time detection.
ResNet-50	93.3%	Strong performance but slower for real-time detection tasks.
VGG-16	90.4%	Good feature extraction but significantly slower in real-time processing.

from smart glasses in low light thanks to ESRGAN's low-light enhancement, fulfilling the market need. The results indicate that smart glasses help BVI people navigate new environments. High item recognition accuracy and positive user feedback make AI and IoT technologies potentially beneficial for BVI people's lives. It empowers individuals and enhances assistive technology by addressing their particular needs.

5. Conclusion

It achieved outstanding object recognition with a 99.7% productivity and improves navigation for the blind and partially-sighted. Low-light enhancement of images and real-time audio feedback were included in the system to ensure it overcomes the biggest challenges that the blind face: the person's great subjective experience with the abilities and methods of self-navigating independently demonstrated high satisfaction and confidence with their independence in navigation skills. But that is part of the progress in assistive technology: it makes possible, for BVI users, life and mobility more independent. With the progress in image processing techniques, such as ESRGAN, complex applications—for instance, mobile navigation in various environments—can be opened.

Further research activities are scaled-up user trials, built to enhance system design, navigation with GPS functionality and constant learning algorithms to personalize experience. Experience would be made more usable and efficient using varied testing environments that include public transport and shopping malls. The society will become progressively more inclusive to help BVI people carry out life's activities safely without any element of fear.

References

[1] Sharma, C., Ghosh, S., Shenoy, K. B. A., & Poornalatha, G. (2024). A Novel Multiclass Object Detection Dataset Enriched With Frequency Data. *IEEE Access*, 12, 85551–85564. doi: 10.1109/ACCESS.2024.3416168.

[2] Jeong, M., Kim, D., & Paik, J. (2024). Practical Abandoned Object Detection in Real-World Scenarios: Enhancements Using Background Matting With Dense ASPP. *IEEE Access*, 12, 60808–60825. doi: 10.1109/ACCESS.2024.3395172.

[3] Choi, K.-H., & Ha, J.-E. (2024). Object Detection Method Using Image and Number of Objects on Image as Label. *IEEE Access*, 12, 121915–121931. doi: 10.1109/ACCESS.2024.3452728.

[4] Qasim, M., Abbas, N., Ali, A., & Al-Ghamdi, B. A. A.-R. (2024). Abandoned Object Detection and Classification Using Deep Embedded Vision. *IEEE Access*, 12, 35539–35551. doi: 10.1109/ACCESS.2024.3369233.

[5] Wang, H., Luo, S., & Wang, Q. (2024). Improved YOLOv8n for Foreign-Object Detection in Power Transmission Lines. *IEEE Access*, 12, 121433–121440. doi: 10.1109/ACCESS.2024.3452782.

[6] Wang, Z., Zhou, G., Ma, J., Xue, T., & Jia, Z. (2024). Beyond the Snowfall: Enhancing Snowy Day Object Detection Through Progressive Restoration and Multi-Feature Fusion. *ICASSP 2024–2024 IEEE International Conference on Acoustics, Speech and Signal Processing (ICASSP)*, Seoul, Korea, Republic of, 2024, pp. 3315–3319. doi: 10.1109/ICASSP48485.2024.10446306.

[7] Jyothi, D. N., Reddy, G. H., Prashanth, B., & Vardhan, N. V. (2024). Collaborative Training of Object Detection and Re-Identification in Multi-Object Tracking Using YOLOv8. *2024 International Conference on Computing and Data Science (ICCDS)*, Chennai, India, pp. 1–6. doi: 10.1109/ICCDS60734.2024.10560451.

[8] Yang, Y., Liang, C., Huang, S., & Wang, X. (2024). BOAformer: Object Detection Network Based on Background-Object Attention Transformer. *2024 6th International Conference on Communications, Information System and Computer Engineering (CISCE)*. Guangzhou, China, pp. 1133–1138, doi: 10.1109/CISCE62493.2024.10653001.

[9] Ye, W., Xu, J., & Parspour, N. (2024). A Hybrid Detector Array for Simultaneous Detection of Living and Metal Object in Wireless Power Transfer Systems. *2024 IEEE Wireless Power Technology*

Conference and Expo (WPTCE), Kyoto, Japan, pp. 618–621, doi: 10.1109/WPTCE59894.2024.10557258.

[10] Qing, T., Xiao, S., Zhang, & Li, P. (2024). Region Proposal Networks (RPN) Enhanced Slicing for Improved Multi-Scale Object Detection. 2024 7th International Conference on Communication Engineering and Technology (ICCET). Tokyo, Japan, pp. 66–70, doi: 10.1109/ICCET62255.2024.00018.

[11] Xiao, H., Li, K., Zhu, Y., & Zhang, J. (2024). 3D Object Detection Based on Long and Short Focal Length Cameras. 2024 5th International Conference on Computer Vision, Image and Deep Learning (CVIDL). Zhuhai, China, pp. 545–548, doi: 10.1109/CVIDL62147.2024.10604155.

[12] Lee, J., Kim, H., Park, C., Jang, J., & Paik, J. (2024). Small Object Detection in Infrared Images Using Attention Mechanism and Sigmoid Function. 2024 IEEE International Conference on Consumer Electronics (ICCE). Las Vegas, NV, USA, pp. 1–3, doi: 10.1109/ICCE59016.2024.10444211.

[13] Liu, Y., Luo, L., Wang, X., & Zheng, Y. (2024). Research on Object Detection Algorithm Based on Twin Attention and Adaptive Feature Fusion. 2024 6th International Conference on Communications, Information System and Computer Engineering (CISCE). Guangzhou, China, pp. 964–968, doi: 10.1109/CISCE62493.2024.10653276.

[14] Manakitsa, N., Maraslidis, G. S., Moysis, L., & Fragulis, G. F. (2024). A review of machine learning and deep learning for object detection, semantic segmentation, and human action recognition in machine and robotic vision. Technologies, 12(2), 15.

[15] Schieber, H., Kleinbeck, C., Theelke, L., Kraft, M., Kreimeier, J., & Roth, D. (2024, January). MR-Sense: A Mixed Reality Environment Search Assistant for Blind and Visually Impaired People. In 2024 IEEE International Conference on Artificial Intelligence and eXtended and Virtual Reality (AIxVR) (pp. 166–175). IEEE.

[16] Naseer, A., Almujally, N. A., Alotaibi, S. S., Alazeb, A., & Park, J. (2024). Efficient Object Segmentation and Recognition Using Multi-Layer Perceptron Networks. CMC-Computers Materials & Continua, 78(1), 1381–1398.

[17] Sikder, J., Islam, M. K., & Jahan, F. (2024). Object segmentation for image indexing in large database. Journal of King Saud University-Computer and Information Sciences, 36(2), 101937.

[18] Wang, J., Hu, F., Abbas, G., Albekairi, M., & Rashid, N. (2024). Enhancing image categorization with the quantized object recognition model in surveillance systems. Expert Systems with Applications, 238, 122240.

[19] Dang, B., Zhao, W., Li, Y., Ma, D., Yu, Q., & Zhu, E. Y. (2024). Real-Time pill identification for the visually impaired using deep learning. arXiv preprint arXiv:2405.05983.

[20] Sujatha, E., Sundar, J. S. J., Raju, D. N., Lakshminarayanan, S., & Suganthi, N. (2024, January). An Intelligent Self-Driving Car's Design and Development, Including Lane Detection Using ROS and Machine Vision Algorithms. In International Conference on Universal Threats in Expert Applications and Solutions (pp. 25–40). Singapore: Springer Nature Singapore.

[21] Anitha, K., Aravind, T., & Praveen Kumar, S. (2024, March). Spying robot for the war field with a wireless night vision camera. In AIP Conference Proceedings (Vol. 2935, No. 1). AIP Publishing.

[22] Helen, R., Gurumoorthy, G., Thennarasu, S. R., & Sakthivel, P. R. (2024, February). Prediction of Osteosarcoma Using Binary Convolutional Neural Network: A Machine Learning Approach. In 2024 Second International Conference on Emerging Trends in Information Technology and Engineering (ICETITE) (pp. 1–7). IEEE.

[23] Fadavi Amiri, M., Hosseinzadeh, M., & Hashemi, S. M. R. (2024). Improving image segmentation using artificial neural networks and evolutionary algorithms. International Journal of Nonlinear Analysis and Applications, 15(3), 125–140.

[24] Naseer, A., & Jalal, A. (2024). Integrating Semantic Segmentation and Object detection for Multi-object labeling in Aerial Images. In ICACS.

[25] Miao, B., Bennamoun, M., Gao, Y., & Mian, A. (2024). Region aware video object segmentation with deep motion modeling. IEEE Transactions on Image Processing.

[26] Derevyanchuk, O. (2024). Use of intelligent fuzzy image segmentation systems in the professional training of future specialists in engineering and pedagogical fields. Professional Pedagogics, 1(28), 103–115.

[27] Naseer, A., & Jalal, A. (2024). Integrating Semantic Segmentation and Object detection for Multi-object labeling in Aerial Images. In ICACS.

[28] Chen, J., Cai, C., Yan, F., & Zhou, B. (2024). Research on insulator image segmentation and defect recognition technology based on U-Net and YOLOv7. *Concurrency and Computation: Practice and Experience*, e8266.

[29] Gao, S., Cheng, Y., Mao, S., Fan, X., & Deng, X. (2024). SSVEP-enhanced threat detection and its impact on image segmentation. *International Journal on Semantic Web and Information Systems (IJSWIS)*, 20(1), 1–20.

38 Integrated Deep Learning Framework for Weapon—Behavior Analysis

*Shanthi P. and Manjula V.**

School of Computer Science and Engineering, Vellore Institute of Technology, Chennai Campus, India

Abstract

In recent days, India has experienced a fussing increase in incidents related to criminal activities, with the National Crime Records Bureau noting a 3.7% rise in illegal arms cases from 2021 to 2022. This paper introduces a novel, integrated proposal for weapon detection and threat evaluation, aiming to meet the imperative demand for improved public safety measures. The proposed comprehensive security framework merges cutting-edge deep learning methods, such as Generative Adversarial Networks (GANs), YOLOv8 to assess the threat by behavior analysis and anomaly detection. The proposed system comprises five essential elements are weapon detection, prevention pipeline, behavior and movement analysis, anomaly detection, and threat assessment. Experiments conducted on a diverse dataset of security camera footage from various Indian urban settings demonstrate the system's efficacy with a weapon detection accuracy of 97.3% and a false positive rate of just 0.5%. The behavior analysis module achieves 89% accuracy in identifying suspicious activities. Furthermore, the integrated approach shows a 28% improvement in overall threat identification compared to traditional single component systems. This research contributes to the field of public safety technology and offers a scalable solution for weapon threat mitigation in complex urban environments.

Keywords: Anomaly detection, convolutional neural networks, generative adversarial networks, long short-term memory, you only look once

1. Introduction

Deep learning frame works such as Generative Adversarial Networks (GANs) and YOLOv8 works greatly increase accuracy and real-time performance in weapon detection. YOLOv8 models can be trained using realistic synthetic data produced by GANs, improving their capacity to identify weapons in a variety of situations. CNN and LSTM networks can be integrated to efficiently capture spatial and temporal features for behavior analysis, respectively. LSTM-CNN models offer a thorough grasp of human behavior by analyzing action sequences to spot questionable behaviors and to safeguard data privacy and defend against hostile attack. DoS, DDoS, MitM, phishing, and malware attacks are examples

Email: shanthi.p2022@vitstudent.ac.in, *manjula.v@vit.ac.in

DOI: 10.1201/9781003661917-38

of traditional cyber-attacks. The main victims of AI-powered attacks are data analysis, synthetic data creation, and data misclassify. In order to produce realistic adversarial samples, GANs are composed of a discriminator and a generator [1]. Three categories are included in the dataset: no violence, cold violence, and hot violence [2].

2. Related Works

Xiaochen Wang et al. [3] a novel hybrid methodology known as the state-based modified artificial bee colony method (SMABCM), which integrates Q- learning to improve the accuracy of target assignment. This approach to address the challenges associated with deviation-based target assignments in weapon systems. Sai Venkateshwar Rao et al. [4] to presents a method for detecting cold steel weapons like knives in surveillance videos using CNN. The main challenge addressed the reflection of light on weapons and make detection difficult using pre- processing procedure called DaCoLT (Darkening and Contrast at Learning and Test stages) to improve detection accuracy under varying light conditions. The R-FCN (ResNet101) model achieved an F1 score of 93.97% in detecting cold steel weapons.

Yadav et al. [5] Detecting weapons in low-light or night time conditions using a modified version of the YOLOv7-DarkVision. The combination of a brightening algorithm and advanced image processing techniques to enhance the detection accuracy and robustness of the model in dark environments and extended efficient layer aggregation network (E-ELAN). Hector Martinez et al. [6] presents a novel surveillance system designed to detect weapons and recognize faces using a distributed computing approach that leverages Edge, Fog, and Cloud computing. Fog layer divided into two levels as first level detects people and weapons, sending only the regions of interest and second level extracts faces from the ROIs and sends them to the cloud layer for identification. YOLOv8 for real-time crowd density evaluation and detection enabling immediate actions to prevent accidents and improve security measures [7].

3. Proposed Systems

To identify and improve possible risks effectively, securing timely responses to security violations. These components can run efficiently by incorporating advanced. Proposing components improves security measures by providing comprehensive situational awareness. In Figure 38.1 workflow as follows video frames are processed consecutively for analysis and in weapon detection utilize yolov8 for identifying weapons and their locations.

In GAN concentrated datasets to improve the detection accuracy such as peculiar and concealed weapons. For behavior analysis to identify the person uses multimodal techniques to cross reference to predict the individuals with the database of authorized weapon transits to track the body movement patterns utilized the LSTM model to detect suspicious activities. After that anomaly detection assess unusual patterns using GAN-based reconstruction errors. To combines the data from weapon detection, behavior analysis, and anomaly detection to evaluate the threat and generates alerts based on the threat level. Using these models' quick response, the system may trigger alerts or preventive actions timely and reduce privacy issues with only real images.

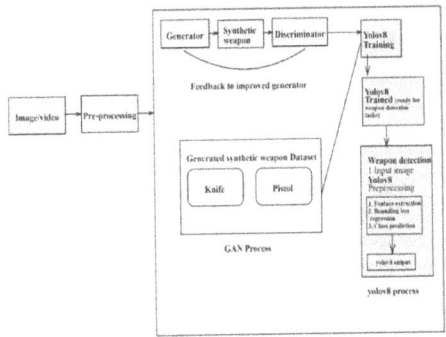

Figure 38.1 GAN and Yolov8 Weapon Detection

3.1 Weapon Detection Module

GAN produces synthetic images to expand the dataset which enhances detection of rare or new types of weapons. Yolov8 detects the real-time object detection model, which is used to locate weapons in video frames with high accuracy. The principal target behind the combination of generative adversarial networks and YOLOv8 as weapons detection module is to improve the overall efficiency, precision and strength of the detection system YOLOv5 object detection algorithm to detect weapons from video frames in real time. The CMS designed to enable timeliness and link a responsive strategy to actual instances where weapons crime may be likely to occur [8].

3.1.1 Generative Adversarial Networks

The GAN driven architecture weapon system also employs the YOLOv8 object detection algorithm to detect weapons from video frames in real time. The synthesis of new DNGB dual n vertical GAN included matrices is a key task performed by the Generator unit of the implemented GAN generator. The Generator, which is often a deep CNN, can be taught to generate images of different weapon types, such as pistol, knives, and against various backgrounds particularly concentrates an invoking difficult images such as weapons that are only partially visible or images of weapons. After generating a synthetic weapon image is juxtaposed with a discriminator that assesses the quality realism of the generated weapon.

3.1.2 YOLOV8

The synthetic data from the GAN is passed to the YOLOv8 training section where the synthesis data is combined with real training data. Weapon detection by YOLOv8 is a complex and efficient method to identify weapons in images or video streams. The backbone of the model is composed by a CSP Darknet that features extract a convolutional layer series from the input image. The produced features are fed into a neck component; however, this is usually PANet, which is used in many applications of one-stage detectors since it creates a feature pyramid to handle objects at varying scales. The multi-scale features processed by the YOLOv8's detection head to produce predictions. After post processing, the technique of non- maximum suppression is employed to minimize redundant detection and include only the most confident predictions. At this stage, the minimum false positives and proper weapon localization in the input image is ensured.

3.2 Prevention Pipeline

The proposed framework with regard to weapon detection and threat assessment forms the core of a prevention pipeline separating authorized from unauthorized weapon carriers as follow below models.

3.2.1 Multi-Model Identification

The system maintains an authenticated database of legal weapon bearers, which holds biometric data besides their credentials and specific weapon details. The databases are continuously updated and encrypted so that data integrity and privacy are maintained. Upon detecting a weapon, the system adopts multi-modal identification such as facial recognition, gait analysis, badge/ID scanning.

3.2.2 Feature Extraction and Fusion

To extract features in the multi-modal system, a unique vector for facial, gait, and badge characteristics is derived facial features derive distinct vectors in ϕf to capture unique facial features, gait features represent movement patterns in ϕg, and badge features encode identification information in ϕb.

Facial features: $\phi f \in RA df$

Gait features: $\phi g \in RA dg$

Badge features: $\phi b \in RA db$

Where df, dg, and db represent the dimensionality of each feature space. Feature fusion is a process of combining facial (φf), gait (φg) features with badge (φb) into a single vector (φ combined) using learned weights (w f, w g, w b). Modalities based on their reliability and relevancy to successfully identify weapon carriers' equation 1.

$$\varphi combined = w\varphi f + wg * \varphi g + wb * \varphi b \quad (1)$$

Where wf, wg and wb are the weights learned for each of the modalities under the constraint: wf + wg + wb = 1.

3.2.3 Matching Algorithm

In contrast, matching approaches include Propensity score matching, Mahalanobis distance matching, and Coarsened Exact Matching Methods; all these methods are used in balancing covariates to have comparable treated and control groups [13]. The algorithm compares the combined feature vector that comes back with authorized database entries by making use of a similarity measure cosine similarity calculated as equation 2.

$$Similarity\ (\varphi combined, \varphi auth) = \varphi combined \cdot \varphi auth / \|\varphi combined\|\|\varphi auth\| \quad (2)$$

Where φ_auth represents the feature vectors stored corresponding to authorized persons. The maximum similarity would surpass a predefined threshold τ; in this case, the person was labeled as an authorized person; otherwise, he or she is considered an unauthorized.

3.3 Behavioral and Movement Analysis System

To combine CNN-LSTM for spatial and temporal analysis the behavioral patterns. This hybrid architecture exploits these two major functionalities. The input sequence that the LSTM takes on the video frames in which the weapons have been spotted and other feature information such as the position, velocity, and pose information of the individuals. Behavioral features such as trajectories movement

and posture/action recognition [10]. Middleware model referred to as HAR-IMB to recognize human activity using wearable sensors. The paper surveys the deep learning models, including CNN, LSTM, and LSTM-CNN with WISDM and UCI-HAR dataset showing promising results and citing scale effectiveness of the model in multiple applications [11].

3.4 Anomaly Detection

In the anomaly detection algorithm with GANs to train normal data and use the generator or the decoder to re-build new input data. Then, you can compute reconstruction error to detect anomalies. Anomalies in network traffic indicative of unusual or suspicious activity over weapon usage, enhancing security measures and threat assessment [9]. A toolkit integrates algorithms for pose estimation and data acquisition ports to automate data processing, which guarantees accurate quantification of human movements and extractions related to gait muscle tone disorder [12]. Let x be an input sample either normal or anomalous and x to be the reconstruction generated from the GAN decoder. The reconstruction error E(x) can be defined as in equation 3:

$$E(x) = \| x - G(zx)\|^2 \quad (3)$$

Where: G(zx) be the reconstructed sample, created by the decoder from the latent vector zx, associated with the input sample x.$\| \bullet \|2$ denotes the squared norm (usually L2 norm). The anomaly score. A(x) is directly proportional to the reconstruction error Anomaly if A(x) > τ where τ is from validation data, usually set to value, optimizing false positives and false negatives.

3.5 Threat Assessment

Threat assessment is the structured process in which one identifies, evaluates, and reduces the possible risks from individuals or situations that might harm others or cause damage to environments.

Algorithm 1 Proposed algorithm
Start
Step 1: Initialize System
Step 2: Monitor CCTV Footage
Step 3: Weapon Detection using YOLO
Step4: Person Identification
Step 5: Behavior Analysis
Step 6: Anomaly Detection
Step 7: Threat Evaluation
Step 8: Trigger Response Based on Threat
Level (Low/ Medium/ High)
Step 9: Log event
Step 10: Continue Monitoring loop back to
capture the next frame and repeat the process.
Step 11: Print (result)
Stop

4. Experimental Analysis

4.1 Dataset Collection

The Weapon Detection dataset model utilized is YOLOv8n (Nano), a lightweight version of the YOLOv8object detection model.
The size of weapon detection dataset such
as total images-5,837, training - 4409, val-
idation-1403 and testing-385. Two class
like weapon detection knives and pistols.
The size of behavior analysis dataset such as
total video clips-1,000, training-536, valida-
tion-324, testing-140 as shown in Table 38.1.
Includes video footage of non-violent activities and various movement patterns and also
captures spatial and temporal variations to
stimulate real-world human behavior.

4.2 Data Pre-Processing

All images are re-sized to 640x640pixels and
the training process uses the albumentations
library for data augmentation. Applies random and median blur with a 1% chance,
blur strength between 3 and 7 Converts
the image to gray scale with a 1% chance
CLAHE (contrast limited adaptive histogram equalization). Combines 4 training

Table 38.1 Dataset Collection

Datasets	Total	Train	Validation	Testing
Weapon	5837	4409	1043	385
Behavior	1000	536	324	140

images into one with a probability of 1.0
(mosaic=1.0). Random horizontal flip: 50%
chance (fliplr =0.5), Random translation:
Up to 10% of the image size (translate=0.1),
Random scaling: Up to 50% of the image
size (scale=0.5). Images are typically normalized to have pixel values between 0 and 1.

4.3 Weapon Detection

Images includes captured in various lighting conditions, orientations and backgrounds. Particularly GANs are used to
generate synthetic images, inserting rare or
unseen weapon types and improves robustness and yolov8 uses multiple prediction
heads at different scales, which improves
its ability to detect both small and large
object s as shown in Figure 38.2.

Figure 38.2 Knife and Pistol Detection

4.4 Behavior Analysis

An effective modeling of human movement
patterns over time, combining CNN and
LSTM is recommended. CNNs excel in spatial feature extraction from frames, learning
relevant patterns and representations of
human actions. The output from CNN layers can be fed into LSTM layers, which are
well-suited for sequential data analysis, capturing temporal dependencies and understanding the dynamics of human movement
across frames as shown in Figure 38.3.

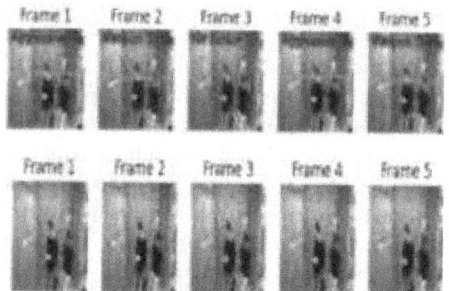

Figure 38.3 Frame and Behavior Analyse

5. Results and Discussion

The weapon detection model, comprising 225 layers and 3,011,238 parameters, demonstrates significant performance improvements during training, with mAP50 rising from 0.3 to 0.782 and mAP50–95 from 0.163 to 0.523. Achieving high precision (0.817) and recall (0.786), it balances accuracy and false positives, with 97.3% accuracy surpassing YOLOv7-DarkVision (93%) and with DaCoLT preprocessing (93.97%) and Hybrid Artificial Bee Colony Method (SMABCM) achieved 91%. Its integration of GANs for synthetic data generation and the CNN - LSTM hybrid for behavior analysis. Its applications include public surveillance, event security, law enforcement, and smart city infrastructure, offering automated threat detection and rapid response.

6. Conclusion

The weapon detection model, leveraging feature pyramid networks with multiple prediction heads, achieved mAP50 of 0.782, mAP50–95 of 0.523, precision of 0.817, recall of 0.786, and 97.3% accuracy in 50 epochs. A behavior analysis system combining CNNs for spatial feature extraction and LSTMs for temporal analysis achieved 89.8% accuracy with precision of 0.717 and recall of 0.726. Future work includes integrating these systems for real-time threat detection, recognizing varied actions, and exploring transfer learning for limited datasets.

References

[1] Yamin, M. M., Ullah, M., Ullah, H., & Katt, B. (2021). Weaponized AI for cyber attacks. *Journal of Information Security and Applications*, 57, 102722..

[2] Nadeem, M.S., Kurugollu, F., Atlam, H.F., & Franqueira, V. N. (2024). Weapon Violence Dataset 2.0: A synthetic dataset for violence detection. *Data in brief*, 54, 110448.

[3] Wang, X., Zhang, Y., & Wang, G. (2024). Target assignment for multiple stages of weapons systems using a deep Q-learning network and a modified artificial bee colony method. *Computers and Electrical Engineering*, 118, 109378.

[4] Rao, A. S. V., Kainth, S., Bhattacharya, A., & Amgoth, T. (2024). An efficient weapon detection system using NSGCU-DCNN classifier in surveillance. *Expert Systems with Applications*, 255, 124800.

[5] Yadav, P., Gupta, N., & Sharma, P. K. (2024). Robust weapon detection in dark environments using Yolov7-DarkVision. *Digital Signal Processing*, 145, 104342.

[6] Martinez, H., Rodriguez-Lozano, F. J., León-García, F., Palomares, J. M., & Olivares, J. (2024). Distributed Fog computing system for weapon detection and face recognition. *Journal of Network and Computer Applications*, 232, 104026.

[7] Sudharson, D., Srinithi, J., Akshara, S., Abhirami, K., Sriharshitha, P., & Priyanka, K. J. P. C. S. (2023). Proactive headcount and suspicious activity detection using YOLOv8. *Procedia Computer Science*, 230, 61–69.

[8] Mukto, M. M., Hasan, M., Al Mahmud, M. M., Haque, I., Ahmed, M. A., Jabid, T., ... & Islam, M. (2024). Design of a real-time crime monitoring system using deep learning techniques. *Intelligent Systems with Applications*, 21, 200311.

[9] Khalaf, L. I., Alhamadani, B., Ismael, O. A., Radhi, A. A., Ahmed, S. R., & Algburi, S. (2024, May). Deep learning-based anomaly detection in network traffic for cyber threat identification. In *Proceedings of the Cognitive Models and Artificial Intelligence Conference* (pp. 303–309).

[10] Paduraru, C., & Paduraru, M. (2022, May). Pedestrian motion in simulation applications using deep learning. In *Proceedings of the 6th International ICSE Workshop on Games and Software Engineering: Engineering Fun, Inspiration, and Motivation* (pp. 1–8).

[11] Song, X., Sun, H., Dong, Y., Pei, Y., & Wang, S. (2024, January). Deep Learning-Based Wearable Human Activity Recognition: Model and Performance Analysis. In *Proceedings of the 2024 8th International Conference on Control Engineering and Artificial Intelligence* (pp. 30–36).

[12] Shen, X., Hui, Z., Zhang, B., Liu, J., & Ma, L. (2023, October). MADS: A Deep Learning-Based Toolkit for Motion Analysis in Movement Disorders. In *Proceedings of the 2023 4th International Symposium on Artificial Intelligence for Medicine Science* (pp. 840–845).

[13] Jung, G., Park, S., Ma, E. Y., Kim, H., & Lee, U. (2024). Tutorial on Matching-based Causal Analysis of Human Behaviors Using Smartphone Sensor Data. *ACM Computing Surveys, 56*(9), 1–33.

39 Optimization of Power Loss Mitigation in Distribution Systems Using Hunter-Prey Optimization Algorithm for Multiple Types of DFACTS

Thirumalai M.[1], Ulagammai M.[2], Suresh T. D.[2], John De Britto C.[2], Hemalatha R.[2], Yuvaraj T.[3] and Farooq Sunar Mahammad[4]

[1]Department of Electronics and Communication Engineering, Saveetha Engineering College, Chennai, India

[2]Department of Electrical and Electronics Engineering, Saveetha Engineering College, Chennai, India

[3]Centre for Smart Energy Systems, Chennai Institute of Technology, Chennai, India

[4]Department of Computer Science & Engineering, Santhiram Engineering College, Nandyal, Andhra Pradesh, India

Abstract

This study tackles critical challenges in modern power systems, including power quality, voltage stability, and loss minimization. It aims to heighten the efficiency of distribution network by optimally deploying DFACTS devices, such as DSTATCOM, DSVC, DSSC, DVR, and DUPQC. These devices are strategically placed to minimize power losses, improve voltage regulation, and strengthen the resilience of the system. The study employs the HPOA to identify the most effective placement of these devices within the Indian 28-bus RDS. The results indicate significant improvements in reducing power losses and enhancing voltage stability, demonstrating the efficacy of the HPOA in optimizing the operation of modern distribution networks and ensuring reliable grid performance.

Keywords: DSTATCOM, DSVC, DSSC, DVR, DUPQC, HPOA

1. Introduction

The modern distribution system is evolving rapidly due to advancements in grid technology and the growing complexity of active networks. These advancements aim to enhance efficiency, reliability, and resilience. However, they also bring challenges such as power quality issues, voltage instability, and inefficient power delivery [1, 2]. Conventional systems often suffer from high losses, poor voltage profiles, and limited reactive power control, reducing efficiency and increasing costs. Advanced power flow management is thus crucial to meet dynamic, localized demands [3, 4].

Email: thirumalai3788@gmail.com[1], ulagammaimeyyappan@gmail.com[2], tdsureshresearch@gmail.com[3], yjohnde@gmail.com[2], hemalathar@saveetha.ac.in[2], yuvaraj4252@gmail.com[3], farooq.cse@srecnandyal.edu.in[4]

DOI: 10.1201/9781003661917-39

DFACTS strategies provide effective solutions by improving power quality, reducing losses, and enhancing stability. Unlike traditional FACTS devices used in transmission systems, DFACTS devices— such as the Distribution Static Compensator (DSTATCOM), Distribution Static VAR Compensator (DSVC), Distribution Static Series Compensator (DSSC), Dynamic Voltage Restorer (DVR), and Distributed Unified Power Quality Conditioner (DUPQC)—are tailored for distribution networks, addressing localized issues [5].

Each DFACTS device serves specific functions. DSTATCOM and DSVC improve voltage stability by compensating reactive power. DSSC optimizes power flow by controlling line impedance. DVR addresses voltage sags and swells, enhancing stability, while DUPQC combines DSTATCOM and DVR functions for comprehensive quality improvement. Power loss reduction is critical as it directly impacts efficiency and costs. Losses arise from reactive power flow, unbalanced loads, and voltage fluctuations, which DFACTS devices mitigate effectively [6–8].

Strategic deployment of DFACTS devices enhances voltage profiles, reduces losses, and strengthens resilience, ensuring efficient power delivery even under varying loads and renewable energy integration.

2. Literature Review

The deployment of DFACTS devices in distribution networks has been extensively researched, with many studies showcasing their effectiveness in enhancing power quality, voltage stability, and reducing power losses. This review examines recent contributions to the field, focusing on optimization-based approaches for the strategic allocation of DSTATCOM, DSVC, DSSC, DVR, and DUPQC devices.

DSTATCOM, a shunt-connected device, provides reactive power support, improving voltage stability and power factor. Studies demonstrate its ability to minimize power losses through reactive power compensation. For instance, [9] showed that optimally placing DSTATCOM using the harmony search algorithm significantly reduced losses and improved voltage profiles in RDS. Similarly, [10] used the bat algorithm to optimize DSTATCOM size and placement, achieving enhanced voltage regulation and loss reduction under variable loads.

DSVC, another shunt-connected device, uses passive components for reactive power and voltage regulation, making it cost-effective. Research [11] optimized DSVC allocation using the adaptive firefly algorithm, achieving significant power loss reductions. In addition, [12] demonstrated DSVC's role in mitigating losses in a smart distribution system with solar and EVCS integration. DSSC, a series-connected device, controls line impedance to optimize power flow and reduce losses. For instance, [13] applied the bat algorithm to DSSC placement in radial networks, achieving notable reductions in power losses, particularly in high-impedance networks.

DVR, also series-connected, compensates for voltage sags and swells, indirectly reducing losses by stabilizing voltage profiles and limiting reactive power demand. Optimization techniques for DVR placement in renewable-integrated systems have demonstrated improved voltage stability and loss reduction [14]. The DUPQC, combining DSTATCOM and DVR functions, provides shunt and series compensation, making it highly effective for comprehensive power quality improvement and loss reduction. Research [15] has shown that optimized DUPQC placement significantly enhances power quality and reduces losses.

Strategically deploying DFACTS devices is essential to reduce losses, improve efficiency, and enhance grid resilience, particularly as power quality challenges and demands grow [16]. Advanced optimization techniques provide a framework for their effective allocation, offering valuable guidance for utilities seeking modernization.

3. Mathematical Modeling of DFACTSs in RDS

By injecting or absorbing reactive power, DFACTS devices improve voltage stability

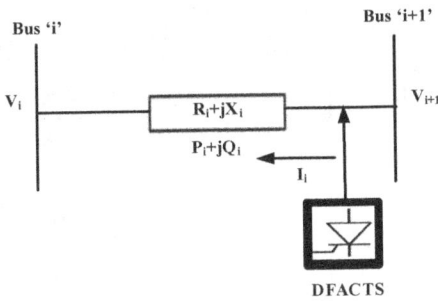

Figure 39.1 Single Bus Diagram of RDS with DFACTS

and reduce losses. Figure 39.1 illustrates an abbreviated one-line representation of the RDS with incorporated DFACTS.

3.1 DSTATCOM Modeling

The reactive power injected by the DSTATCOM is given by:

$$Q_{DSTATCOM} = V_i \cdot I_{inj} \qquad (1)$$

3.2 DSVC Modeling

The reactive power delivered by DSVC (Q_{DSVC}) is given by:

$$Q_{DSVC} = \frac{V_i^2}{X_{DSVC}} \qquad (2)$$

3.3. DSSC Modeling

The voltage injected by the DSSC is given by:

$$Q_{DSSC} = I_i \cdot X_{DSSC} \qquad (3)$$

3.4. DVR Modeling

The voltage supplied by the DVR is expressed as:

$$V_{DVR} = |V_{ref} - V_i| \qquad (4)$$

3.5. DUPQC Modeling

The total injected power by the DUPQC can be expressed as:

$$Q_{DUPQC} = Q_{shunt} + Q_{series} \qquad (5)$$

3.6. Objective Function

The primary aim of this study is to mitigate the total power loss in the RDS by optimally placing and sizing DFACTS devices.

Objective Function =

$$\text{Minimize}\left[P_{loss} = \sum_{i=1}^{N}\left(I_i^2 \cdot R_i \right) \right] \qquad (6)$$

Where, P_{loss} represents the overall active power dissipation in the network, I_i denotes the current passing through line i, R_i signifies the resistance of line i, and N refers to the total count of branches or conductors in the RDS.

4. Results and Analysis

4.1 Indian 28-Bus RDS

To verify the effectiveness of the proposed approach, an 11 kV Indian 28-bus RDS is examined, with the radial network configuration and feeder layout depicted in Figure 39.2. The results in Table 39.1 confirm the substantial influence of DFACTS devices on enhancing system performance.

In Case-I (Without DFACTS), the baseline system experiences 68.81 kW active power loss, 46.04 kVAr reactive power loss, a minimum voltage of 0.9123 p.u., and a VSI_min of 0.6927 p.u., highlighting the need for voltage support. In Case-II (Single DFACTS), placing DSTATCOM at Bus 7 (580 kVAr) reduces active power loss by 46.36%, improving voltage to 0.9472 p.u. DUPQC achieves the best

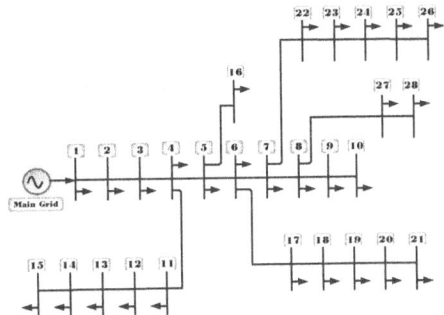

Figure 39.2 Radial Structure of Indian 28-Bus RDS

Table 39.1 Performance of Indian 28-Bus System Under Various DFACTSs

Case	Parameters	DFACTS				
		DSTATCOM	DSVC	DSSC	DVR	DUPQC
Case-I (Without DFACTS)	P_{LOSS} (kW)	68.81	68.81	68.81	68.81	68.81
	Q_{LOSS} (kVAr)	46.04	46.04	46.04	46.04	46.04
	V_{min} (p.u)	0.9123	0.9123	0.9123	0.9123	0.9123
	VSI_{min} (p.u)	0.6927	0.6927	0.6927	0.6927	0.6927
Case-II (With single DFACTS)	Position and capacity (kVAr)	580 (7)	580 (7)	580 (7)	580 (7)	580 (7)
	P_{LOSS} (kW)	36.89	37.30	37.85	38.16	36.52
	P_{LOSS} reduction (%)	46.36%	45.77%	45.01%	44.54%	46.92%
	Q_{LOSS} (kVAr)	24.22	24.60	25.11	25.46	23.88
	V_{min} (p.u)	0.9472	0.9468	0.9461	0.9455	0.9483
	VSI_{min} (p.u)	0.7999	0.7987	0.7969	0.7955	0.8014
Case-III (With two DFACTSs)	Position and capacity (kVAr)	480 (7), 230 (12)	480 (7), 230 (12)	480 (7), 230 (12)	480 (7), 230 (12)	480 (7), 230 (12)
	P_{LOSS} (kW)	33.75	34.05	34.51	34.89	33.26
	P_{LOSS} reduction (%)	50.94%	50.50%	49.85%	49.32%	51.66%
	Q_{LOSS} (kVAr)	22.28	22.56	23.08	23.46	21.88
	V_{min} (p.u)	0.9476	0.9473	0.9468	0.9464	0.9487
	VSI_{min} (p.u)	0.8035	0.8028	0.8012	0.7999	0.8052

performance, with a 46.92% loss reduction and a voltage of 0.9483 p.u. In Case-III (Two DFACTSs), devices at Buses 7 and 12 enhance performance further. DUPQC reduces power loss by 51.66% and raises VSI_min to 0.8052 p.u., showcasing superior stability. Figures 39.3 and 39.4 illustrate real power loss reductions with single and dual DFACTS placements.

For instance, at Bus 2, DSTATCOM reduces losses from 13.5043 kW to 6.7802 kW (approximately 50%), and at Bus 3, it cuts losses from 18.5285 kW to 9.1321 kW (50.7%). Similarly, DUPQC effectively reduces Bus 11's loss from 1.6509 kW to 0.8581 kW (48%). Figure 39.4 highlights even greater improvements with dual DFACTS configurations. For example, at Bus 2, a combination of DSTATCOM and DSVC lowers power loss from 13.5043 kW to 6.3269 kW (53.1%), while at Bus 5, DSTATCOM-DSSC reduces the loss from 9.251 kW to 4.2751 kW

(53.8%). High-loss buses, such as Bus 22, experience remarkable benefits; DVR and DUPQC together reduce losses from 6.531 kW to 0.5826 kW (91%). Overall, dual DFACTS configurations outperform single devices by an additional 5–10%,

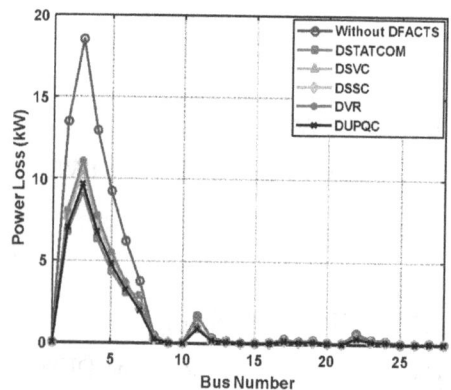

Figure 39.3 Power Loss Profile of the System with Single DFACTS

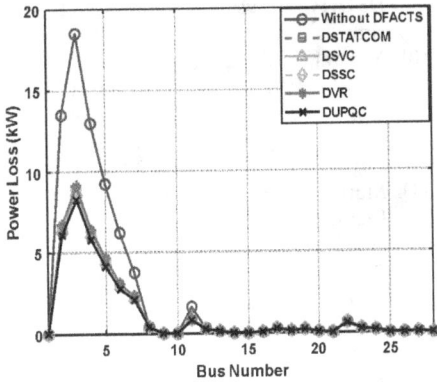

Figure 39.4 Power Loss Profile of the System with Two DFACTS

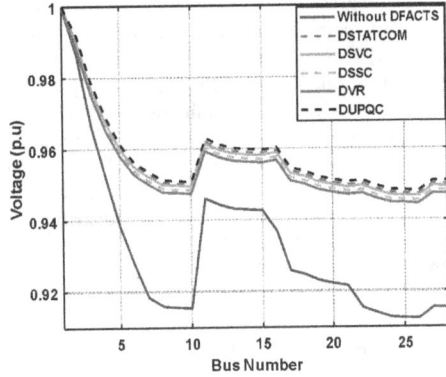

Figure 39.5 Voltage Profile of the System with Single DFACTS

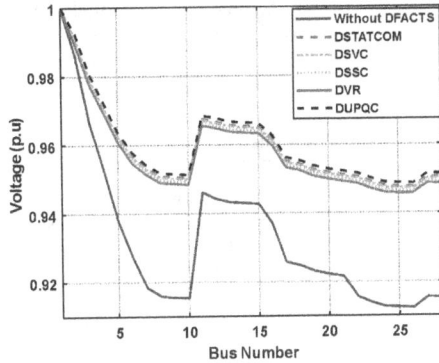

Figure 39.6 Voltage Profile of the System with Two DFACTS

underscoring their superior capacity for power loss reduction and system performance enhancement.

The voltage profile analysis across Cases I, II, and III highlights significant stability improvements with DFACTS configurations. In Case II, single devices (e.g., DSTATCOM, DSVC, DSSC, DVR, and DUPQC) enhance voltages, with DUPQC providing the highest gains. Bus 2's voltage rises from 0.9862 to 0.9917 with a single DUPQC and to 0.9925 in Case III with dual DFACTS. Low-voltage buses like Bus 11 improve from 0.9461 to 0.962 (DSTATCOM) and 0.9686 (DUPQC), a 2.2% increase. Dual setups achieve a 2.5% average boost, significantly strengthening voltage stability, particularly in low-voltage areas (Figures 39.5 and 39.6).

5. Conclusion

This research assessed the impact of DFACTS devices—DSTATCOM, DSVC, DSSC, DVR, and DUPQC—on power quality, voltage regulation, and loss reduction in the IEEE 33-bus RDS using the HPOA for optimal placement. A single DFACTS unit reduced power losses from 68.81 kW to 36.89 kW (DSTATCOM) and 36.57 kW (DUPQC), achieving a 46.92% reduction. Dual configurations further lowered losses to 33.26 kW (DUPQC), marking a 51.66% reduction. Voltage stability improved,

with DUPQC attaining a minimum VSI of 0.8052 p.u. These findings emphasize DFACTS devices' effectiveness in enhancing grid performance and supporting distributed generation integration.

References

[1] Yuvaraj, T., Devabalaji, K. R., Anish Kumar, J., Thanikanti, S. B., & Nwulu, N. (2024). A comprehensive review and analysis of the allocation of electric vehicle charging stations in distribution networks. *IEEE Access*.

[2] Thirumalai, M., Hariharan, R., Yuvaraj, T., & Prabaharan, N. (2024). Optimizing distribution system resilience in extreme weather using prosumer-centric microgrids with

integrated distributed energy resources and battery electric vehicles. *Sustainability*, 16(6), 2379.

[3] Yuvaraj, T., Suresh, T. D., Meyyappan, U., Aljafari, B., & Thanikanti, S. B. (2023). Optimizing the allocation of renewable DGs, DSTATCOM, and BESS to mitigate the impact of electric vehicle charging stations on radial distribution systems. *Heliyon*, 9(12).

[4] Devabalaji,K.R.,Imran,A.M.,Yuvaraj,T.,& Ravi, K. J. E. P. (2015). Power loss minimization in radial distribution system. *Energy Procedia*, 79, 917–923.

[5] Gupta, A. R., & Kumar, A. (2022). Deployment of distributed generation with D-FACTS in distribution system: a comprehensive analytical review. *IETE Journal of Research*, 68(2), 1195–1212.

[6] Chawda, G. S., Shaik, A. G., Mahela, O. P., Padmanaban, S., & Holm-Nielsen, J. B. (2020). Comprehensive review of distributed FACTS control algorithms for power quality enhancement in utility grid with renewable energy penetration. *IEEE Access*, 8, 107614–107634.

[7] Yuvaraj, T., Devabalaji, K. R., Srinivasan, S., Prabaharan, N., Hariharan, R., Alhelou, H. H., & Ashokkumar, B. (2021). Comparative analysis of various compensating devices in energy trading radial distribution system for voltage regulation and loss mitigation using blockchain technology and bat algorithm. *Energy Reports*, 7, 8312–8321.

[8] Yuvaraj, T., Arun, S., Suresh, T. D., & Thirumalai, M. (2024). Minimizing the impact of electric vehicle charging station with distributed generation and distribution static synchronous compensator using PSR index and spotted hyena optimizer algorithm on the radial distribution system. *e-Prime-Advances in Electrical Engineering, Electronics and Energy*, 8, 100587.

[9] Yuvaraj, T., Devabalaji, K. R., & Ravi, K. (2015). Optimal placement and sizing of

DSTATCOM using harmony search algorithm. *Energy Procedia*, 79, 759–765.

[10] Yuvaraj, T., Ravi, K., & Devabalaji, K. R. (2017). DSTATCOM allocation in distribution networks considering load variations using bat algorithm. *Ain Shams Engineering Journal*, 8(3), 391–403.

[11] Muthukumar, P., Ramesh, M. V., Babu, P. V., Rohinikumar, P., & Satyanarayana, S. V. (2023). Optimal integration of multiple D-SVCs for voltage stability enhancement in radial electrical distribution system using adaptive firefly algorithm. *International Journal of Intelligent Engineering and Systems*, 16(3), 378–387.

[12] Suresh, M. P., Yuvaraj, T., Thanikanti, S. B., & Nwulu, N. (2024). Optimizing smart microgrid performance: Integrating solar generation and static VAR compensator for EV charging impact, emphasizing SCOPE index. *Energy Reports*, 11, 3224–3244.

[13] Thirumalai, M., & Yuvaraj, T. (2021). Application of Bat Optimization Algorithm for Power System Loss Reduction Using DSSSC. In *Smart Computing Techniques and Applications: Proceedings of the Fourth International Conference on Smart Computing and Informatics*, 1, pp. 87–94. Springer Singapore.

[14] Soomro, A. H., Larik, A. S., Mahar, M. A., Sahito, A. A., Soomro, A. M., & Kaloi, G. S. (2021). Dynamic voltage restorer—A comprehensive review. *Energy Reports*, 7, 6786–6805.

[15] Iswariya, M., & Yuvaraj, T. (2022). Voltage profile improvement in distribution networks using DSTATCOM and UPQC by reducing power loss. In *2022 14th International Conference on Mathematics, Actuarial Science, Computer Science and Statistics (MACS)*, pp. 1–5. IEEE.

[16] Naruei, I., Keynia, F., & Sabbagh Molahosseini, A. (2022). Hunter–prey optimization: Algorithm and applications. *Soft Computing*, 26(3), 1279–1314.

40 Optimal Placement of Renewable Based DG and DSTATCOM in Distribution Systems to Alleviate the Effects of EVCS

Suresh T. D.[1], Thirumalai M.[2], Ulagammai M.[1], Hemalatha R.[1], John De Britto C.[1], Yuvaraj T.[3], and Farooq Sunar Mahammad[4]

[1]Department of Electrical and Electronics Engineering, Saveetha Engineering College, Chennai, India
[2]Department of Electronics and Communication Engineering, Saveetha Engineering College, Chennai, India
[3]Centre for Smart Energy Systems, Chennai Institute of Technology, Chennai, India
[4]Department of Computer Science and Engineering, Santhiram Engineering College, Nandyal, Andhra Pradesh, India

Abstract

The global rise in EV usage increases electricity demand, challenging distribution system efficiency and reliability. Proper siting of EVCS is crucial to mitigate these impacts. This study employs the HPOA and other optimization methods to identify the best locations and capacities for RDGs, DSTATCOMs, and EVCSs in a RDS, focusing on reducing power losses and improving voltage stability, with the IEEE 34-bus system used as a test case. The findings highlight substantial enhancements in power loss mitigation, voltage levels across buses, and overall system reliability.

Keywords: Electrical vehicle, hunter prey optimization algorithm, PV distributed generations, DSTATCOM, EVCS, radial distribution system

1. Introduction

EVs play a crucial role in the transition to renewable energy, providing significant decreases in carbon emissions and fuel costs due to their high energy efficiency [1]. However, the integration of EVs and their charging infrastructure into power grids presents challenges, such as increased energy losses, voltage fluctuations, and grid congestion [2]. One potential solution is the incorporation of RDG sources. Although RDGs can reduce reliance on conventional grid power, their financial and operational performance has often fallen short of expectations [3]. Moreover, RDGs introduce new management complexities, requiring additional grid flexibility [4]. To address these challenges, microgrids and battery energy storage systems (BESS) have been developed to improve grid stability and flexibility. Proper sizing and placement of BESS are crucial to avoid infrastructure overload and reduce costs [5].

Additionally, shunt capacitors in RDS provide vital reactive power support, helping

Email: tdsureshresearch@gmail.com[1], thirumalai3788@gmail.com[2], ulagammaimeyyappan@gmail.com[3], hemalathar@saveetha.ac.in[4], yjohnde@gmail.com[5], yuvaraj4252@gmail.com[6], farooq.cse@srecnandyal.edu.in[7]

DOI: 10.1201/9781003661917-40

minimize power losses, regulate voltage, and enhance overall power quality [6, 7]. However, improper allocation and sizing of RDGs, EVCSs, BESS, and DSTATCOMs can exacerbate system strain and increase costs. Therefore, optimizing the placement and sizing of these components is essential for improving system efficiency and facilitating the integration of clean energy.

While existing research has focused on the allocation of RDGs, while existing research has focused on the apportionment of RDGs, capacitors, and BESS in RDS, studies specifically addressing EVCS placement remain limited. Some studies have explored the integration of EVCSs with RDGs, capacitors, and BESS, examining their simultaneous installation and optimization. Methods like probabilistic fuzzy constraint programming, PSO, and combined optimization strategies have been employed to regulate the optimum locations for EVCSs and DGs, aiming to optimize the various objectives [8–15]. However, a unified approach that optimizes RDGs, capacitors, and EVCSs together offers greater benefits, including reduced power losses, improved voltage levels, and enhanced system stability. The Fig. 40.1 indicates the Schematic diagram of proposed approach

2. Problem Statement

This study uses the DLF method to assess power losses and voltage variations across the network. The DLF approach is chosen for its ability to effectively handle the unique features of RDS, offering a more accurate power flow analysis [16]. The power loss is calculated using the following equation:

$$P_{Loss}(i, i+1) = \left(\frac{P_{i,i+1}^2 + Q_{i,i+1}^2}{|V_i|^2} \right) R_{i,i+1} \quad (1)$$

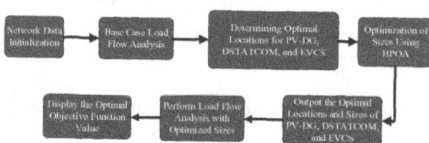

Figure 40.1 Schematic Diagram of Proposed Approach

The following expression is used to aggregate the losses from all individual branches:

$$P_{Loss}^T = \sum_{i=1}^{nb} P_{Loss}(i, i+1) \quad (2)$$

The voltage stability index (VSI), defined in equation (3) [3], is used to measure the stability of the RDS. For the system to remain stable, the VSI value must be non-negative (t ≥ 0). Lower VSI values at certain nodes suggest a higher likelihood that compensators will be required to stabilize voltage levels.

$$VSI(i+1) = |V_{i+1}|^4 - 4\big[P_{i,i+1} * X_{i,i+1} -$$
$$Q_{i,i+1} * R_{i, i+1}\big]^2 - 4\big[P_{i,i+1} * R_{i,i+1}$$
$$+ Q_{i,i+1} * X_{i,i+1}\big]|V_{i, i+1}|^2 \quad (3)$$

2.1 Research Goal

The main goal of this optimization is to minimize energy losses within the system and improve the voltage stability of the distribution network. The objective function is formulated as:

$$\text{Minimize}\,(F) = \text{Min}\left[\left(P_{Loss}^T\right) + \left(\frac{1}{VSI}\right)\right] \quad (4)$$

2.2 Hunter Prey Optimization Algorithm

The following outlines the main steps of the modified HPOA, incorporating the specific variables and objectives related to the RDS [16].

Step 1: Initialization
- *Population Initialization:* Randomly initialize hunters, with each hunter representing a configuration of EVCS, PV-DG, and DSTATCOM locations and sizes.
- *Parameter Setting:* Set parameters such as the maximum number of iterations T_{max}, population size P, and exploration factor α.

Step 2: Exploration Phase
Hunters explore the search space by moving randomly:

$$X_j^{t+1} = X_j^t + \alpha \cdot r\,\text{and}\cdot\left(X_{best}^t - X_j^t\right) \quad (5)$$

This phase allows the algorithm to cover a wide area to find potential optimal solutions.

Step 3: Exploitation Phase

Hunters refine their positions towards the best solution:

$$X_j^{t+1} = X_j^t + \beta \cdot \left(X_{best}^t - X_j^t \right) + \gamma \cdot \mathrm{randn} \cdot \left(X_j^t - X_{worst}^t \right) \tag{6}$$

This phase focuses on fine-tuning solutions in promising regions.

Step 4: Escape Mechanism

To avoid local optima, some hunters randomly escape to new regions:

$$X_j^{t+1} = X_j^t + \delta \cdot \mathrm{rand} \cdot \left(X_{upper} - X_{lower} \right) \tag{7}$$

This ensures diversity and prevents premature convergence.

Step 5: Fitness Evaluation

Each hunter's fitness is evaluated based on total power loss:

$$f(X_j) = \left[\left(P_{Loss}^T \right) + \left(\frac{1}{VSI} \right) \right] \tag{8}$$

Additionally, voltage stability is assessed to ensure optimal system performance.

Step 6: Termination Criteria

The algorithm concludes when either the maximum iteration count is reached or there is no improvement in the best solution. The optimal configuration, denoted as X_{best}, determines the most efficient allotment of EVCS, PV-DGs, and DSTATCOM devices.

3. Results and Discussion

This study evaluates the show of the planned optimization methodology on the IEEE 34-bus RDS, using system data from reference [17, 18]. The system, shown in Figure 40.2 as a single-line diagram. The approach determines the ideal placement of EVCSs at buses 9 and 12. The optimal sizes for these EVCSs are determined using the HPOA. The simulation results are presented in Table 40.1, comparing the performance

Table 40.1 Simulation Results for 34-Bus RDS Under Various Algorithms

Scenario	Parameters	HPOA	PSO
Base Scenario	P_{LOSS} in kW	221.28	221.28
	VSI_{min} in p.u	0.7875	0.7875
With EVCS	Location and capacity in kW	966 (8)	966 (8)
	P_{LOSS} in kW	296.02	296.02
	VSI_{min} in p.u	0.7673	0.7673
With EVCS & DSTATCOM	Location and capacity in kW (EVCS)	966 (8)	966 (8)
	Location and capacity in kVAr (DSTATCOM)	1250 (27)	1245 (12)
	P_{LOSS} in kW	254.65	268.27
	VSI_{min} in p.u	0.7839	0.7758
With EVCS & PV-DG	Location and capacity in kW (EVCS)	966 (8)	966 (8)
	Location and capacity in kW (PV-DG)	2000 (27)	2075 (12)
	P_{LOSS} in kW	161.08	175.68
	VSI_{min} in p.u	0.8558	0.8104
With EVCS, DSTATCOM & PV-DG	Location and capacity in kW (EVCS)	966 (8)	966 (8)
	Location and capacity in kVAr (DSTATCOM)	1250 (27)	1245 (12)
	Location and capacity in kW (PV-DG)	2000 (27)	2075 (12)
	P_{LOSS} in kW	121.62	148.78
	VSI_{min} in p.u	0.8650	0.8192

of HPOA with PSO across various scenarios for the 34-bus RDS.

In the base case, with no additional devices, both HPOA and PSO result in a power loss of 221.28 kW and a minimum VSI of 0.7875 p.u., indicating stable system operation. Introducing EVCS at buses 9 and 12 leads to an increase in power loss to 296.02 kW for both algorithms, and the VSI drops slightly to 0.7673 p.u., indicating a reduction in voltage stability due to the additional load. Integrating DSTATCOM alongside EVCS reduces power loss, with HPOA yielding 254.65 kW, outperforming PSO's 268.27 kW. The VSI improves to 0.7839 p.u. with HPOA, better than PSO's 0.7758 p.u., reflecting improved voltage stabilization.

Further integration of PV-DG alongside EVCS reduces power losses even more. HPOA achieves a power loss of 161.08 kW, compared to PSO's 175.68 kW, demonstrating its superior optimization of both EVCS and PV-DG placement. The minimum VSI improves significantly to 0.8558 p.u. for HPOA, surpassing PSO's 0.8104 p.u. The results confirm that HPOA consistently delivers better performance, achieving lower power losses and improved voltage stability. Figures 40.2 and 40.3 illustrate the comparison of system loss profiles under both HPOA and PSO, with HPOA achieving superior results in high-loss scenarios.

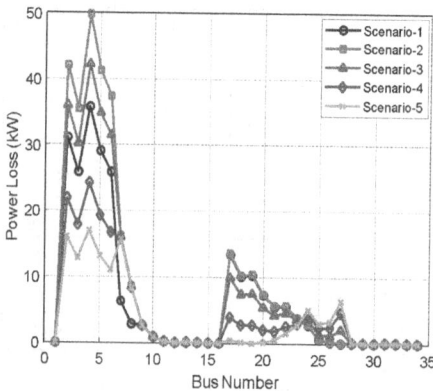

Figure 40.2 Power Loss Profile Comparison Across Different Scenarios (HPOA)

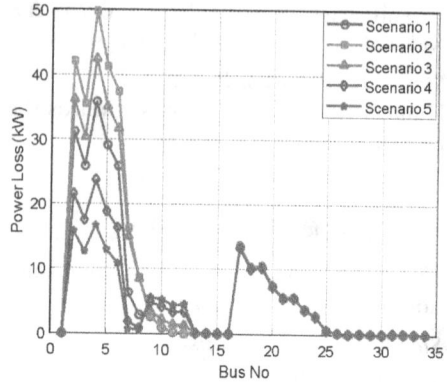

Figure 40.3 Power Loss Profile Comparison Across Different Scenarios (PSO)

5. Conclusion

The growth of EV adoption increases the demand for EVCSs. This paper presents an innovative approach for determining the best positions and capacities of EVCSs, PV-DGs, and DSTATCOMs to reduce energy losses. The study utilizes the HPOA and contrasts its performance with the conventional PSO. The results show that HPOA outperforms PSO in all scenarios. In cases with EVCS and DSTATCOM, HPOA reduces power losses to 254.65 kW, compared to PSO's 268.27 kW. With PV-DG integration, HPOA further reduces losses to 161.08 kW, while PSO results in 175.68 kW. In the most optimized scenario, involving EVCS, DSTATCOM, and PV-DG, HPOA minimizes power losses to 121.62 kW, significantly improving VSI compared to PSO.

References

[1]   Kim, Y., Kim, H., & Suh, K. (2021). Environmental performance of electric vehicles on regional effective factors using system dynamics. *Journal of Cleaner Production*, 320, 128892.

[2]   Yuvaraj, T., Devabalaji, K. R., Kumar, J. A., Thanikanti, S. B., & Nwulu, N. (2024). A Comprehensive Review and Analysis of the Allocation of Electric Vehicle Charging Stations in Distribution Networks. *IEEE Access*.

[3] Yuvaraj, T., Suresh, T. D., Meyyappan, U., Aljafari, B., & Thanikanti, S. B. (2023). Optimizing the allocation of renewable DGs, DSTATCOM, and BESS to mitigate the impact of electric vehicle charging stations on radial distribution systems. *Heliyon*, 9(12).

[4] Thirumalai, M., Hariharan, R., Yuvaraj, T., & Prabaharan, N. (2024). Optimizing Distribution System Resilience in Extreme Weather Using Prosumer-Centric Microgrids with Integrated Distributed Energy Resources and Battery Electric Vehicles. *Sustainability*, 16(6), 2379.

[5] Mirzapour, F., et al. (2019). A new prediction model of battery and wind-solar output in hybrid power system. *Journal of Ambient Intelligence and Humanized Computing*, 10, 77–87.

[6] Devabalaji, K. R., Yuvaraj, T., & Ravi, K. (2018). An efficient method for solving the optimal sitting and sizing problem of capacitor banks based on cuckoo search algorithm. *Ain Shams Engineering Journal*, 9(4), 589–597.

[7] Yuvaraj, T., Arun, S., Suresh, T. D., & Thirumalai, M. (2024). Minimizing the Impact of Electric Vehicle Charging Station with Distributed Generation and Distribution Static Synchronous Compensator Using PSR Index and Spotted Hyena Optimizer Algorithm on the Radial Distribution System. *e-Prime-Advances in Electrical Engineering, Electronics and Energy*, 100587.

[8] Huiling, T., et al. (2020). An optimization framework for collaborative control of power loss and voltage in distribution systems with DGs and EVs using stochastic fuzzy chance constrained programming. *IEEE Access*, 49013–49027.

[9] Suresh, M. P., Yuvaraj, T., Thanikanti, S. B., & Nwulu, N. (2024). Optimizing smart microgrid performance: Integrating solar generation and static VAR compensator for EV charging impact, emphasizing SCOPE index. *Energy Reports*, 11, 3224–3244.

[10] Luo, L., et al. (2019). Joint planning of distributed generation and electric vehicle charging stations considering real-time charging navigation. *Applied Energy*, 242, 1274–1284.

[11] Sriabisha, R., & Yuvaraj, T. (2023, March). Optimum placement of Electric Vehicle Charging Station using Particle Swarm Optimization Algorithm. In *2023 9th International Conference on Electrical Energy Systems (ICEES)* (pp. 283–288). IEEE.

[12] Yuvaraj, T., Suresh, T. D., Ananthi Christy, A., Babu, T. S., & Nastasi, B. (2023). Modelling and Allocation of Hydrogen-Fuel-Cell-Based Distributed Generation to Mitigate Electric Vehicle Charging Station Impact and Reliability Analysis on Electrical Distribution Systems. *Energies*, 16(19), 6869.

[13] Shabbar, R., Kasasbeh, A., & Ahmed, M. M. (2021). Charging station allocation for electric vehicle network using stochastic modeling and grey wolf optimization. *Sustainability*, 13(6), 3314.

[14] Yuvaraj, T., Devabalaji, K. R., Thanikanti, S. B., Aljafari, B., & Nwulu, N. (2023). Minimizing the electric vehicle charging stations impact in the distribution networks by simultaneous allocation of DG and DSTATCOM with considering uncertainty in load. *Energy Reports*, 10, 1796–1817.

[15] Yuvaraj, T., Devabalaji, K. R., Thanikanti, S. B., Pamshetti, V. B., & Nwulu, N. I. (2023). Integration of electric vehicle charging stations and DSTATCOM in practical indian distribution systems using bald eagle search algorithm. *IEEE Access*, 11, 55149–55168.

[16] Teng, J. H. (2003). A direct approach for distribution system load flow solutions. *IEEE Transactions on Power Delivery*, 18(3), 882–887.

[17] Naruei, I., Keynia, F., & Sabbagh Molahosseini, A. (2022). Hunter–prey optimization: Algorithm and applications. *Soft Computing*, 26(3), 1279–1314.

[18] Das, D., Nagi, H. S., & Kothari, D. P. (1994). Novel method for solving radial distribution networks. *IEE Proceedings-Generation, Transmission and Distribution*, 141(4), 291–298.

41 Intelligent Residential Energy Control with Demand Response Incorporating Renewable Power Sources and Electric Vehicle Integration

Ulagammai M.[1], Hemalatha R.[1], John De Britto C.[1], Suresh T. D.[1], Thirumalai M.[2] and Yuvaraj T.[3]

[1]Department of Electrical and Electronics Engineering, Saveetha Engineering College, Chennai, India
[2]Department of Electronics and Communication Engineering, Saveetha Engineering College, Chennai, India
[3]Centre for Smart Energy Systems, Chennai Institute of Technology, Chennai, India

Abstract

This study examines the effect of DR-based P2P energy trading between prosumers and consumers within smart homes. Prosumers, who produce power via renewable sources such as solar energy, wind turbines, and electric vehicles, have the opportunity to sell surplus energy to lower their costs. A two-step scheduling approach is introduced to reduce energy expenditures for consumers. During the initial phase, household appliances are scheduled using RTP, optimized through the HBOA. The second phase involves P2P transactions, facilitated by DR. Results show that this approach significantly lowers electricity costs and reduces grid demand during peak times.

Keywords: Honey badger optimization algorithm, prosumer, consumer, demand response, P2P energy transactions

1. Introduction

The integration of intelligent communication and data management tools in metering systems has enabled smart home consumers to implement demand response (DR) strategies [1]. DR, a strategy for managing energy demand, motivates consumers to modify their usage behaviors to lower electricity expenses. The residential sector, being the second-largest electricity consumer, offers considerable potential for energy management [2]. Research has focused on scheduling time-shiftable appliances such as dishwashers, dryers, washing machines, and electric vehicles (EVs) based on real-time pricing (RTP) to reduce costs [3]. Some home energy management systems (HEMS) use RTP to minimize electricity expenses without considering distributed generation (DGs), often relying on load shifting and shedding, which may impact user comfort [4]. Other strategies, including two-stage stochastic models and

Email: ulagammaimeyyappan@gmail.com[1], hemalathar@saveetha.ac.in[2], yjohnde@gmail.com[1], tdsureshresearch@gmail.com[1], thirumalai3788@gmail.com[2], yuvaraj4252@gmail.com[3]

DOI: 10.1201/9781003661917-41

differential evolution, reduce costs and emissions through demand and DG scheduling [5]. However, many studies fail to incorporate the impact of DGs and thermal loads, limiting their effectiveness [6].

The role of DGs, particularly photovoltaic (PV) systems, in enhancing DR benefits has been explored. DR strategies using Time-of-Use (ToU) tariffs can reduce costs and peak demand by incorporating PV systems and battery energy storage systems (BESS) [7]. However, studies often overlook the impact of thermal loads on comfort and restrict appliance modeling [8]. Increased PV penetration can provide significant cost reductions, especially with RTP-based DR strategies and DG and battery integration [9].

P2P energy exchange enables prosumers to sell excess DG power to consumers, improving economic benefits and alleviating peak demand [10]. Despite these advancements, there is a need to improve DG and battery utilization for greater financial benefits from DR and P2P trading [11–13].

Figure 41.1 arrangement provides consumers the freedom to choose the lowest-cost energy provider, enabling them to minimize their electricity bills.

2. Formulation of the Problem

This model aims to minimize electricity costs in smart homes by optimizing power usage from the grid, PV, WT, battery storage, and EVs. The objective functions for prosumers and consumers consider both energy costs and revenues from energy sales [14, 15].

Figure 41.1 Proposed layout for P2P energy trading

2.1 Objective Function for Prosumer

The cost minimization objective for a prosumer who can both consume and generate power (e.g., through PV, wind, etc.) is defined as:

$$\text{Minimize Cost} =$$

$$\sum_{t=1}^{24} \begin{cases} \left(P_g(t) \cdot C_g(t) \cdot U(t)\right) + \\ \left(P_{PV}(t) \cdot C_{PV}\right) + \left(P_{WT}(t) \cdot C_{WT}\right) + \\ \left(P_{b,d}(t) \cdot C_b\right) + \left(P_{ev,d}(t) \cdot C_{ev}\right) - \\ \left(P_{sell}(t) \cdot C_{sell} \cdot X(t)\right) - \\ \sum_{c=1}^{n} \left(P_c(c,t) \cdot C_c(t) \cdot Y(c,t)\right) \end{cases} \quad (1)$$

Where, $P_g(t)$ is the power from the grid; $C_g(t)$ is the grid power cost; $P_{PV}(t)$ and $P_{WT}(t)$ are the solar PV and WT power; C_{PV} and C_{WT} are the PV and wind maintenance costs; $P_{b,d}(t)$ and $P_{ev,d}(t)$ are the battery and EV discharges; C_b and C_{ev} are the battery and EV costs; $P_{sell}(t)$ is the power sold to the grid; C_{sell} is the sale price; $P_c(c, t)$ is the power sent to consumers; $C_c(t)$ is the charge to consumers; $U(t)$, $X(t)$ and $Y(c, t)$ are the binary variables for grid, selling, and power transfer status.

2.2 Objective Function for Consumer

The ultimate aim for a consumer who can only consume power (and buy from neighboring prosumers) is:

$$\text{Minimize Cost} =$$

$$\sum_{t=1}^{24} g \begin{cases} \left(P_g(t) \cdot C_g(t) \cdot U(t)\right) \\ + \sum_{p=1}^{m} \left(P_p(p, t) \cdot C_p(t) \cdot Y(p,t)\right) \end{cases}$$

$$(2)$$

Where, $P_p(p, t)$ is power from prosumers; $C_p(t)$ is the prosumer cost.

3. Simulation Results and Discussion

This section explores how DR based P2P energy trading helps reduce electricity costs for consumers and prosumers. The aim is to optimize household appliance schedules, considering solar, wind, storage, and EVs

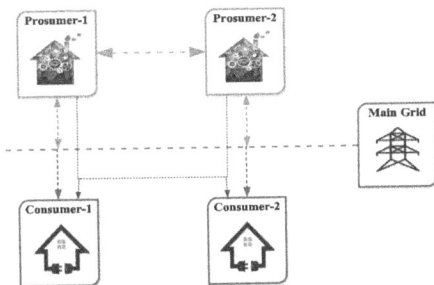

to lower expenses. The process begins with scheduling appliances for both groups, with prosumers equipped with PV, WT, storage, and EVs, using data from [15]. The appliance scheduling of the proposed work is shown in Table 41.1

Daily energy usage and expenses for prosumers and consumers are presented in Table 41.2. similarly, Energy costs are given in Fig. 41.2. In an energy management system for smart homes, the interaction between prosumers and consumers is crucial for optimizing energy consumption and reducing costs. Energy exchanges, along with their associated costs, can be modeled through objective functions to better understand the relationships between consumers, prosumers, and the grid.

Prosumer-1 purchases 35 kW from the grid at INR 3.00 per kW, costing INR 105.00. They also use 10 kW from their EV for V2H applications, reducing grid dependence. Prosumer-1 supplies 5 kW to other consumers, earning INR 25.00, and provides 10 kW to Consumer 1, generating INR 30.00. Their energy-related transactions involve grid purchases, EV usage, and power supply to consumers.

Prosumer-2 purchases 30 kW from the grid at a cost of INR 90.00. Like Prosumer 1, they use 12 kW from their EV for V2H applications. They supply 7 kW to other consumers, earning INR 35.00, and provide 15 kW to Consumer 1, generating INR 45.00. Similar to Prosumer 1, they balance energy usage with grid purchases and consumer energy exchanges, maintaining a dynamic energy system.

Consumer-1 buys 40 kW from the grid at INR 120.00 and receives 15 kW from Prosumer-2 at INR 45.00. They do not use EV power (V2H), so their total expenditure is the sum of the grid purchase cost and the cost of power received from Prosumer-2. Consumer-1 does not sell power to others, limiting their energy-related costs.

Consumer-2 purchases 20 kW from the grid at INR 60.00 and receives 10 kW from Prosumer-1 at INR 30.00. Like Consumer-1, they do not use EV power (V2H) and do not sell power, so their costs are only from grid purchases and power received from Prosumer-1.

These transactions illustrate the complex interaction between prosumers, consumers, and the grid in a smart home setting. Prosumers act as energy suppliers, balancing grid purchases and EV usage, while consumers rely on grid purchases or energy received from prosumers. By

Table 41.1 Appliance scheduling for the proposed work

Sl. No	Appliances	Power Rating (kW)	Operation Duration (hours)	Start Time (hours)	End Time (hours)	Prosumer-1	Prosumer-2	Consumer-1	Consumer-2
1	Clothes Washer	1.5	2	8	11	✓	✓	✓	
2	Clothes Dryer	2.2	1	14	17	✓		✓	✓
3	Refrigerator	0.2	24	0	24	✓	✓	✓	✓
4	Electric Oven	2.0	2	16	18	✓	✓		✓
5	Electric Kettle	1.0	1	7	9		✓	✓	
6	Air Conditioner (AC)	1.25	12	12	24	✓			✓
7	Water Heater	1.8	3	6	9	✓	✓		
8	Television	0.25	4	18	22		✓	✓	✓
9	Desktop Computer	0.35	5	10	15	✓			✓
10	Vacuum Cleaner	1.7	1	9	20		✓	✓	

Table 41.2 Daily energy usage and expenses for prosumers and consumers

	Grid-Purchased Power (kW)	Grid Power Cost (INR)	V2H power (kW)	Power Supplied by Prosumers (kW)	Cost for Power Supplied by Prosumers (INR)	Power Distributed to Consumers (kW)	Cost for Power Delivered to Consumers (INR)
Prosumer-1	35	105	10	5	25	10	30
Prosumer-2	30	90	12	7	35	15	45
Consumer-1	40	120	-	15	45	-	-
Consumer-1	20	60	-	10	30	-	-

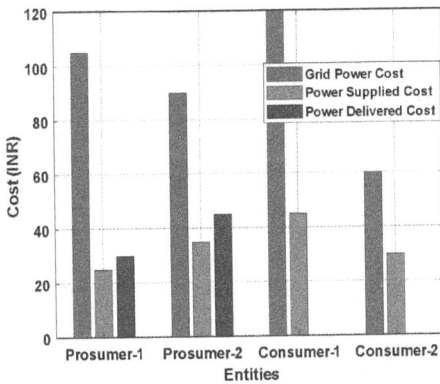

Figure 41.2 Energy costs representation

Table 41.3 Power and cost of P2P energy trading

	To Consumer-1		To Consumer-2	
	Power (kW)	Cost (₹)	Power (kW)	Cost (₹)
From Prosumer-1	27.9	55.98	21.5	44.26
From Prosumer-2	24.7	50.3	14.7	30.3

optimizing energy flows and exchanges, all parties can minimize costs, ensuring the economic feasibility and efficiency of the smart home energy network.

Table 41.3 illustrates P2P energy exchanges between prosumers and consumers. Prosumer 1 shared 27.9 kW with Consumer 1 and 21.5 kW with Consumer 2, while Prosumer 2 shared 24.7 kW with Consumer 1 and 14.7 kW with Consumer 2. Consumer 1 paid ₹55.98 for energy from Prosumer 1 and ₹50.3 from Prosumer 2, while Consumer 2 paid ₹44.26 and ₹30.3,

respectively. These figures demonstrate that P2P trading offers a cost-effective alternative to grid electricity, reducing grid dependency, transmission losses, and providing financial and environmental benefits. Further optimization could enhance its potential in decentralized energy systems.

4. Conclusion

This study demonstrated the benefits to minimize electricity costs. By integrating RDG technologies like PV, wind WT, and EVs, prosumers sold surplus energy, reducing grid dependency. Prosumer-1 purchased 35 kW from the grid for ₹105 and supplied 5 kW to consumers, earning ₹25. Prosumer-2 bought 30 kW for ₹90 and supplied 7 kW, earning ₹35. P2P trading details revealed prosumer-1 supplied 27.9 kW to consumer-1 (₹55.98) and 21.5 kW to consumer-2 (₹44.26). Results showed that P2P trading reduced costs, optimized renewable use, and eased grid demand.

References

[1] Avancini, D. B., Rodrigues, J. J. P. C., Martins, S. G. B., Rabêlo, R. A. L., Al-Muhtadi, J., & Solic, P. (2019). Energy meters evolution in smart grids: A review. *Journal of Cleaner Production*, 217, 702–715.

[2] Garg, A., Maheshwari, J., & Mahapatra, D. (2009). Energy conservation and commercialization in Gujarat. *Technical Report*.

[3] Sharma, H., & Mishra, S. (2020). Techno-economic analysis of solar grid-based virtual power plant in Indian power sector: A case study. *International Transactions on Electrical Energy Systems*, 30(1), e12177.

[4] Jamil, M., & Mittal, S. (2020). Hourly load shifting approach for demand side

management in smart grid using grasshopper optimisation algorithm. *IET Generation, Transmission & Distribution*, 14(5), 808–815.

[5] He12, S., Liebman, A., Rendl12, A., Wallace12, M., & Wilson, C. (2014). Modelling RTP-based residential load scheduling for demand response in smart grids.

[6] Zhao, R., & Zheng, W. (2024). Efficient operation of combined residential and commercial energy hubs incorporating load management and two-point approximation for uncertainty modeling. *Computers and Electrical Engineering*, 116, 109197.

[7] Philipo, G. H., Chande Jande, Y. A., & Kivevele, T. (2020). Demand-side management of solar microgrid operation: Effect of time-of-use pricing and incentives. *Journal of Renewable Energy*, 2020(1), 6956214.

[8] Ali, S., Khan, I., Jan, S., & Hafeez, G. (2021). An optimization based power usage scheduling strategy using photovoltaic-battery system for demand-side management in smart grid. *Energies*, 14(8), 2201.

[9] Xu, D., Zhong, F., Bai, Z., Wu, Z., Yang, X., & Gao, M. (2023). Real-time multi-energy demand response for high-renewable buildings. *Energy and Buildings*, 281, 112764.

[10] Soto, E. A., Bosman, L. B., Wollega, E., & Leon-Salas, W. D. (2021). Peer-to-peer energy trading: A review of the literature. *Applied energy*, 283, 116268.

[11] Li, Z., & Ma, T. (2020). Peer-to-peer electricity trading in grid-connected residential communities with household distributed photovoltaic. *Applied Energy*, 278, 115670.

[12] Tushar, W., Saha, T. K., Yuen, C., Azim, M. I., Morstyn, T., Poor, H. V., ... & Bean, R. (2020). A coalition formation game framework for peer-to-peer energy trading. *Applied Energy*, 261, 114436.

[13] Hoque, M. M., Khorasany, M., Azim, M. I., Razzaghi, R., & Jalili, M. (2024). A framework for prosumer-centric peer-to-peer energy trading using network-secure export–import limits. *Applied Energy*, 361, 122906.

[14] Shah, M. I. A., Wahid, A., Barrett, E., & Mason, K. (2024). Peer-to-peer energy trading in dairy farms using multi-agent systems. *Computers and Electrical Engineering*, 118, 109437.

[15] Kanakadhurga, D., & Prabaharan, N. (2021). Demand response-based peer-to-peer energy trading among the prosumers and consumers. *Energy Reports*, 7, 7825–7834.

[16] Hashim, F. A., Houssein, E. H., Hussain, K., Mabrouk, M. S., & Al-Atabany, W. (2022). Honey Badger Algorithm: New metaheuristic algorithm for solving optimization problems. *Mathematics and Computers in Simulation*, 192, 84–110.

42 Strengthening Resilience in Distribution Networks through Microgrid Development and Integration of Distributed Energy Resources Using the Group Instruction Optimization Algorithm

John De Britto C.[1,2], *Suresh T. D.*[1], *Hemalatha R.*[1], *Thirumalai M.*[2], *Ulagammai M.*[1], *and Yuvaraj T.*[3]

[1]Department of Electrical and Electronics Engineering, Saveetha Engineering College, Chennai, India
[2]Department of Electronics and Communication Engineering, Saveetha Engineering College, Chennai, India
[3]Centre for Smart Energy Systems, Chennai Institute of Technology, Chennai, India

Abstract

Severe weather events have exposed vulnerabilities in traditional power grids, emphasizing the need for greater grid resilience. Microgrids, capable of operating independently or alongside the main grid, offer a solution to enhance grid stability. By integrating DERs such as solar PV, energy storage, and BEVs, microgrids reduce reliance on the central grid, ensuring a stable and sustainable energy supply. This study aims to optimize distribution system resilience by minimizing END with DER integration, using GTOA. Applied to a 6-bus radial distribution system, the approach identifies optimal DER placements and sizes, improving grid resilience and energy efficiency during extreme weather events.

Keywords: Microgrid, resilience, distributed energy resources, group teaching optimization algorithm, radial distribution system

1. Introduction

As urbanization and technological growth drive demand, the interconnectedness of critical infrastructures—such as energy, healthcare, finance, transportation, and telecommunications—has become essential. These networks support public services and economic productivity, but maintaining reliable power delivery is a significant challenge. The energy grid faces threats from cyber-attacks, physical sabotage, natural disasters, and geomagnetic disturbances, all of which require resilient infrastructure [1, 2]. Severe weather events, like monsoons and cyclones, further stress the

Email: yjohnde@gmail.com[1], tdsureshresearch@gmail.com[1], hemalathar@saveetha.ac.in[1], thirumalai3788@gmail.com[2], ulagammaimeyyappan@gmail.com[1], yuvaraj4252@gmail.com[3]

DOI: 10.1201/9781003661917-42

need for robust systems that can quickly recover and minimize disruptions. Global resilience focuses on a system's ability to recover swiftly while minimizing negative impacts [3, 4].

The integration of electric vehicles (EVs) offers a promising solution for a more sustainable and resilient energy grid. EVs, using vehicle-to-grid (V2G) technology, can provide portable power during emergencies, supporting grid stability [5, 6]. BESS and BEVs help stabilize the grid and provide backup power, ensuring availability during critical times [7–10].

Recent research has focused on improving power distribution network resilience through DER integration, with methods to optimize MG operations and economic performance. Incorporating renewable energy and energy storage has proven effective in enhancing economic efficiency [11]. Networked MGs offer localized power generation and energy-sharing capabilities, further improving resilience [13, 14]. However, challenges remain in defining resilience metrics and quantifying the effects of natural disasters on energy consumers. This study explores new approaches to strengthen power distribution resilience using DERs, networked MGs, and new resilience indices for enhanced grid stability, validated on Indian distribution systems [16–17].

2.   Problem Formulation

2.1 Formulation of Resilience Metrics

(i) *Total Number of Household-Hours during Outage*

$$H_H = \sum_{i=1}^{N} \left(P_i * D_i \right) \qquad (1)$$

Where, H_H is the total household-hours during outage.

(ii) *Total Number of Household-Energy Not Supplied*

$$E_{ns} = \sum_{i=1}^{N} \left(P_i \cdot E_i \right) \qquad (2)$$

Where, E_{ns} is the total household-energy not supplied (in kWh) and E_i is the energy consumption per household iii during normal operation (in kWh)

(iii) *Total Number of Households Affected During Outage*

$$N_A = \sum_{i=1}^{N} A_i \qquad (3)$$

In this context, N_A denotes the overall count of impacted households, and A_i serves as an indicator variable, representing whether a particular household iii experienced an outage (assigned a value of 1 if affected and 0 if unaffected) [18].

(iv) *Average Number of Households Affected During Outage*

$$\underline{N_A} = \frac{1}{T} \sum_{t=1}^{T} N_A(t) \qquad (4)$$

Where, N_A is the average number of households affected, T is the total number of time intervals (outage periods) and $N_A(t)$ is the number of affected households at time t.

(v) *Total Revenue Loss from the Utilities During Outage*

$$R_L = \sum_{i=1}^{N} \left(E_i \cdot C_i \right) \qquad (5)$$

Where, R_L is the total revenue loss from the utility (in USD), and C_i is the cost of electricity per unit for household i (in USD/kWh) [19].

2.2 Proposed Resilience Index

The resilience index (R) can be formulated as a weighted metric combining the outage impact parameters (household-hours and energy loss), normalized by the total number of households and average outage duration.

$$R = \frac{1}{1 + \left(\dfrac{H_H + E_{ns}}{N \cdot T_{avg}} \right)} \qquad (6)$$

Where, T_{avg} is the average outage duration.

2.3. Formulation of the Objective Function

The objective function is typically designed to minimize the impact of outages and enhance resilience by reducing energy not supplied. The objective can be written as:

Objective Function = Maximum (R) (7)

3. Results and Discussion

This study assesses the effectiveness of a microgrid approach on a modified IEEE 6-bus RDS [20], assuming all DERs remain operational after a severe event. Multiple scenarios are evaluated for resilience, with each compared to a baseline where no DERs are present. Repairs on lines 2 to 6 are conducted simultaneously over 5 hours, from 1 p.m. to 6 p.m., after a storm causes moderate damage. A microgrid is set up across these branches, with a BEV in V2G mode to support during the disaster. Table 42.2 provides key parameters for the case study, including household details, load requirements, DER capacities (PV-DG, BESS, BEV), and resilience indicators. Simulation data for 6-bus RDS is given in Table 42.1. The following scenarios were analyzed:

i. Faulted system without DER
ii. Faulted system with DER

3.1 Scenario-I (Fault without DER)

In Scenario-I, the system experiences a complete outage across all six buses, with no DERs available to support the grid (Figure 42.1). The outage lasts 5 hours,

Table 42.2 DERs data for the simulation of 6-bus RDS

Parameters	Scenario-I	Scenario-II
PV-DG Size (kW) & Location	–	50
BESS Size (kW) & Location	–	50
BEV Size (kW) & Location	–	50
Total Number of Household-Hours during Outage (h)	1800 (6 buses * 5 hours)	900 (6 buses * 3 hours)
Total Number of Household-Energy Not Supplied (kWh)	5500	1000
Total Households Impacted by Outage	6	3
Average Households Impacted per Outage	6	3
Utility Revenue Loss During Outage ($/kWh)	550	100
Resilience Index	0.3	0.7

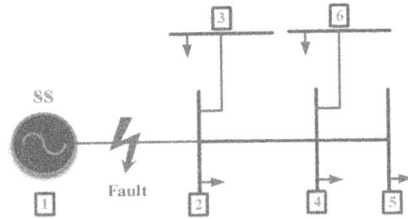

Figure 42.1 Faulted 6-bus system without DER

affecting all 553 households, resulting in a total of 1800 household-hours. The energy not supplied is 5500 kWh, reflecting the total demand during the outage. This significant energy loss occurs due to the absence of DERs to assist the grid. The resilience index is calculated at 0.30, indicating low resilience due to the extended outage and substantial energy loss.

3.1.1 Scenario-II (Fault with DER)

In Scenario-II, the system utilizes DERs, including PV generators, BESS, and BEVs,

Table 42.1 Simulation data for 6-bus RDS

Bus Number	Households	Power Demand (kW)
1	0	0
2	133	200
3	100	150
4	120	180
5	107	160
6	93	140

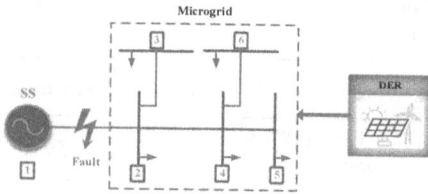

Figure 42.2 Faulted 6-bus system with DER

to support faulted buses (Figure 42.2). This reduces the outage duration to 3 hours, resulting in 900 household-hours. The DERs supply power, cutting the energy not supplied to 1000 kWh. This reduction in energy loss is attributed to the DERs meeting household energy demands during the outage. Consequently, the resilience index rises to 0.70, indicating a notable improvement in system resilience. The system shows enhanced ability to maintain power during faults, evidenced by reduced outage duration and ENS.

The comparison between Scenario-I and Scenario-II illustrates the substantial improvement in system resilience when DERs are integrated into the grid. In Scenario-I, the lack of DER support leads to a high number of household-hours during the outage and significant energy loss, resulting in a low resilience index. However, in Scenario-II, the deployment of DERs reduces the outage duration, decreases the total ENS, and increases the resilience index. The numerical results emphasize the importance of integrating RES and storage systems to enhance the reliability and stability of RDS during faults.

Figure 42.3 presents a comparison of key resilience parameters across two fault scenarios in the distribution system. The results highlight significant improvements in system performance with the integration of DERs. Specifically, the total household-hours during outages were reduced by 50%, and the total household-energy not supplied decreased by 81.82%. Additionally, the impacted households and average households affected during outages decreased by 50%. Utility revenue loss

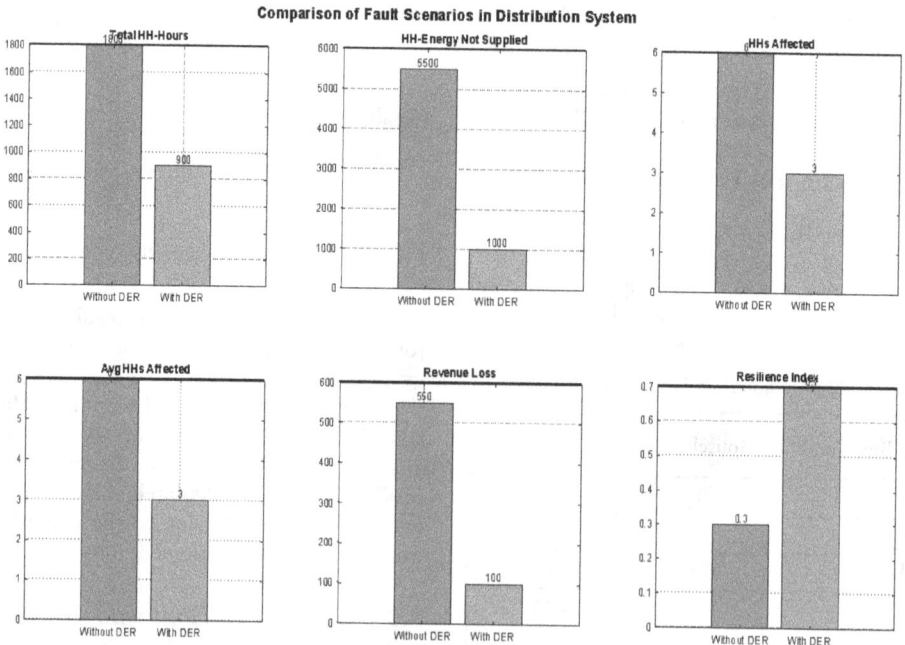

Figure 42.3 Comparison of various resilience parameters

dropped by 81.82%, while the resilience index improved by 133.33%. These results highlight how DERs significantly boost system resilience, reducing outage duration, energy losses, and financial impacts. The sharp rise in the resilience index reflects the system's enhanced ability to endure and recover from disruptions.

4. Conclusion

This study demonstrated the effectiveness of DERs in enhancing resilience within a 6-bus RDS during extreme weather. By optimizing DER placement and sizing with the GTOA, significant improvements in outage metrics were achieved. The integration of DERs reduced total household-hours during outages by 50%, from 1800 to 900 hours, and decreased ENS by 81.82%, from 5500 kWh to 1000 kWh. The resilience index rose from 0.3 to 0.7, indicating enhanced system robustness against disruptions. These results underscore DERs' potential to reduce grid dependency, minimize outage impacts, and improve overall system resilience, offering a sustainable model for future power grid modernization.

References

[1] Mall, R. K., Attri, S. D., & Kumar, S. (2011). Extreme weather events and climate change policy in India. *Journal of South Asia Disaster Studies*, 4(2), 37–56.

[2] Yuvaraj, T., Devabalaji, K. R., Kumar, J. A., Thanikanti, S. B., & Nwulu, N. (2024). A Comprehensive Review and Analysis of the Allocation of Electric Vehicle Charging Stations in Distribution Networks. *IEEE Access*.

[3] Xu, L., Guo, Q., Sheng, Y., Muyeen, S. M., & Sun, H. (2021). On the resilience of modern power systems: A comprehensive review from the cyber-physical perspective. *Renewable and Sustainable Energy Reviews*, 152, 111642.

[4] Thirumalai, M., Hariharan, R., Yuvaraj, T., & Prabaharan, N. (2024). Optimizing Distribution System Resilience in Extreme Weather Using Prosumer-Centric Microgrids with Integrated Distributed Energy Resources and Battery Electric Vehicles. *Sustainability*, 16(6), 2379.

[5] Yuvaraj, T., Devabalaji, K. R., Suresh, T. D., Prabaharan, N., Ueda, S., & Senjyu, T. (2023). Enhancing Indian Practical Distribution System Resilience Through Microgrid Formation and Integration of Distributed Energy Resources Considering Battery Electric Vehicle. *IEEE Access*, 11, 133521–133539.

[6] Yuvaraj, T., Devabalaji, K. R., Thanikanti, S. B., Pamshetti, V. B., & Nwulu, N. I. (2023). Integration of electric vehicle charging stations and DSTATCOM in practical indian distribution systems using bald eagle search algorithm. *IEEE Access*, 11, 55149–55168.

[7] Yuvaraj, T., Suresh, T. D., Ananthi Christy, A., Babu, T. S., & Nastasi, B. (2023). Modelling and Allocation of Hydrogen-Fuel-Cell-Based Distributed Generation to Mitigate Electric Vehicle Charging Station Impact and Reliability Analysis on Electrical Distribution Systems. *Energies*, 16(19), 6869.

[8] Suresh, M. P., Yuvaraj, T., Thanikanti, S. B., & Nwulu, N. (2024). Optimizing smart microgrid performance: Integrating solar generation and static VAR compensator for EV charging impact, emphasizing SCOPE index. *Energy Reports*, 11, 3224–3244.

[9] Yuvaraj, T., Arun, S., Suresh, T. D., & Thirumalai, M. (2024). Minimizing the Impact of Electric Vehicle Charging Station with Distributed Generation and Distribution Static Synchronous Compensator Using PSR Index and Spotted Hyena Optimizer Algorithm on the Radial Distribution System. *e-Prime-Advances in Electrical Engineering, Electronics and Energy*, 100587.

[10] Yuvaraj, T., Suresh, T. D., Meyyappan, U., Aljafari, B., & Thanikanti, S. B. (2023). Optimizing the allocation of renewable DGs, DSTATCOM, and BESS to mitigate the impact of electric vehicle charging stations on radial distribution systems. *Heliyon*, 9(12).

[11] Mousavizadeha S., Ghanizadeh Bolandib, T., Haghifama, M. R., Moghimic, M., & Luc, J. (2020). Resilience analysis of electric distribution networks: a new approach based on modularity concept. *Int J Electric Power Energy Syst*, 117.

[12] Rosales-Asensio, E., Martín, M. S., Borge-Diez, D., Blanes-Peiro, J. J., &

Colmenar-Santos, A. (2019). Microgrids with energy storage systems as a means to increase power resilience: an application to office buildings. *Energy*, 172, 1005–1015.

[13] Barnes, A., Nagarajan, H., Yamangil, E., Bent, R., & Backhaus, S. (2019). Resilient design of largescale distribution feeders with networked microgrids. *Electr Power Syst Res*, 171, 150–157.

[14] Zhua, J., Yuana, Y., & Wang, W. (2019). An exact microgrid formation model for load restoration in resilient distribution system. *Int J Electr Power Energy Syst*, 116.

[15] Hussain, A., Bui, V. H., & Kim, H. M. (2019). Microgrids as a resilience resource and strategies used by microgrids for enhancing resilience. *Appl Energy*, 240, 56–72.

[16] Gilasi, Y., Hosseini, S. H., & Ranjbar, H. (2022). Resiliency-oriented optimal siting and sizing of distributed energy resources in distribution systems. *Electric Power Systems Research*, 208, 107875.

[17] Yuvaraj, T., Devabalaji, K. R., Thanikanti, S. B., Aljafari, B., & Nwulu, N. (2023). Minimizing the electric vehicle charging stations impact in the distribution networks by simultaneous allocation of DG and DSTAT-COM with considering uncertainty in load. *Energy Reports*, 10, 1796–1817.

[18] Galvan, E., Mandal, P., & Sang, Y. (2020). Networked microgrids with roof-top solar PV and battery energy storage to improve distribution grids resilience to natural disasters. International *Journal of Electrical Power & Energy Systems*, 123, 106239.

[19] Zhang, Y., & Jin, Z. (2020). Group teaching optimization algorithm: A novel metaheuristic method for solving global optimization problems. *Expert Systems with Applications*, 148, 113246.

[20] Al-Anbarri, K. A. (2024). A New Fast Decoupled-like Algorithm for Radial Distribution Power Flow Problem by Using Rectangular Coordinates. *IEEE Access*.

43 Reliability Analysis of Distribution Systems with Solar-Based DG and Capacitor Integration Using the Water Cycle Optimization Algorithm

Hemalatha R.[1], John De Britto C.[1], Yuvaraj T.[2], Ulagammai M.[1], Suresh T. D.[1], and Thirumalai M.[3]

[1]Department of Electrical and Electronics Engineering, Saveetha Engineering College, Chennai, India
[2]Centre for Smart Energy Systems, Chennai Institute of Technology, Chennai, India
[3]Department of Electronics and Communication Engineering, Saveetha Engineering College, Chennai, India

Abstract

Ensuring high reliability in RDS is critical for maintaining consistent electricity delivery and enabling rapid recovery from disruptions. This study explores the role of SDG and capacitors in improving the reliability of RDS. A goal function is established to boost system reliability by minimizing essential reliability metrics, including SAIDI, SAIFI, and CAIDI. To reach this aim, the WCOA is applied and contrasted with the GA. Validation is conducted on the IEEE 12-bus test network, and findings indicate that WCOA considerably improves the system's reliability. This technique provides a promising approach to optimize RDS performance, addressing the increasing energy demands and challenges faced by modern power systems.

Keywords: Reliability, solar DG, capacitor, radial distribution system, water cycle optimization algorithm

1. Introduction

The global shift toward renewable energy highlights the importance of integrating solar-based distributed generation (SDG) and capacitors into radial distribution systems (RDS) to support sustainable energy distribution. While SDG units and capacitors offer several advantages, such as improving power quality, they also pose challenges to RDS operations [1, 2]. Research has focused on evaluating the impact of SDG and capacitor installations on power loss, voltage stability, and reliability in RDS [3–5]. The growing demand to reduce greenhouse gas emissions and increase energy self-sufficiency is driving the rapid adoption of these technologies, causing significant changes in grid operations [6].

Email: hemalathar@saveetha.ac.in[1], yjohnde@gmail.com[1], yuvaraj4252@gmail.com[2], ulagammaimeyyappan@gmail.com[1], tdsureshresearch@gmail.com[1], thirumalai3788@gmail.com[3]

DOI: 10.1201/9781003661917-43

The integration of SDG and capacitors can increase power losses within the RDS due to fluctuating solar generation and dynamic capacitor behavior. Managing these losses is critical for designing efficient, sustainable distribution systems. Furthermore, fluctuations in solar power generation can lead to voltage instability, particularly during peak loads. Consistent voltage levels are essential for maintaining reliable power supply to residential and commercial users [7, 8]. Renewable energy variability, especially in regions with significant solar adoption, can put additional stress on infrastructure like transformers and conductors, potentially leading to higher equipment failure rates. Therefore, ensuring the reliability of RDS is crucial for delivering uninterrupted electricity. Effective management of these challenges is necessary for the smooth integration of SDG and capacitors, requiring innovative solutions to enhance distribution, voltage stability, and overall RDS resilience [9].

Reliability studies are vital for understanding the impact of SDG and capacitors on the overall reliability of RDS, assessing system failures, downtime, and the grid's capacity to accommodate additional generation and reactive power. Research often uses probabilistic models and reliability indices to improve RDS resilience with SDG and capacitors.

A new approach for determining the optimal placement of electric vehicle charging stations (EVCSs) and battery swapping stations (EVBSSs) in RDS has been explored, examining their effects on critical network parameters. However, prior research often assumed perfect integration of renewable DGs and storage systems, which may not reflect real-world conditions [10]. Hybrid methodologies, like combining golden jackal optimization (GJO) with random forest algorithms (RFA), have been used for EV and station allocation, though practical complexities are often overlooked [11]. Studies on hydrogen fuel cell-based DGs (HFC-DG) to mitigate EVCS impacts have made idealized assumptions, disregarding practical challenges like equipment failure

and regulatory restrictions [12]. AI-based methods, such as hybrid grey wolf optimization (GWO) and particle swarm optimization (PSO), have been used for optimal EVCS and DG placement but have limitations, including reliance on conventional DGs and stochastic EV load models [13]. Research on plug-in EV (PEV) charging and discharging has shown improvements in energy not supplied (ENS) and SAIDI but lacked integration of transportation and distribution systems [14]. Studies on the IEEE 33-bus system revealed that locating fast-charging stations at robust buses was more sustainable than at weaker buses, which created operational challenges [15].

This research aims to evaluate the impact of EVCSs on RDS reliability, using the Water Cycle Optimization Algorithm (WCOA) to optimize placement and minimize their effects. Results will be compared with the Genetic Algorithm (GA) to demonstrate WCOA's superior capabilities in optimizing EVCS placement. The findings will inform the design, operation, and upgrading of RDS infrastructure to meet the growing demand for EVCS while ensuring a stable and reliable power supply.

2. Problem Formulation

2.1 Reliability matrices formulation

The reliability of RDS is crucial for ensuring a stable power supply and minimizing disruptions. Key performance indicators like SAIDI, SAIFI, and CAIDI assess system reliability: SAIDI measures average outage duration, SAIFI indicates the frequency of interruptions, and CAIDI reflects the average outage time for affected customers [9]. Improving these metrics enhances system resilience, reduces outage durations, and boosts customer satisfaction. Methods such as network reconfiguration, renewable energy integration, and DG are commonly used to enhance reliability and minimize outages. These reliability metrics are mathematically defined in equations (1) to (3).

$$SAIDI = \frac{\sum_{i=1}^{n} \text{Duration of Interruptions at bus } i \,(\text{in hours})}{\text{Total number of customers}} \quad (1)$$

$$SAIFI = \frac{\sum_{i=1}^{n} \text{Number of Interruptions at bus } i}{\text{Total number of customers}} \quad (2)$$

$$CAIDI = \frac{\sum_{i=1}^{n} \text{Duration of Interruptions at bus } i \,(\text{in hours})}{\text{Number of Interruptions at bus } i} \quad (3)$$

2.2 Objective Function

The objective function aimed at enhancing the reliability of the RDS can be defined by minimizing the combined weighted total of the three indices. This can be represented as:

$$\text{Objective Function} = \text{minimize} \left(\omega_1 \cdot SAIDI + \omega_2 \cdot SAIFI + \omega_3 \cdot CAIDI \right) \quad (4)$$

Where ω_1, ω_2 and ω_3 are the weight factors representing the relative importance of each index. The goal is to minimize the overall impact of outages by adjusting system parameters like SDG and capacitor placement, and other reliability-enhancing strategies [16].

3. Results and Discussion

The analysis in this study focuses on the IEEE 12-bus RDS, with system data obtained from reference [17]. Figure 43.1 shows the system's single-line diagram, which features 12 buses and 11 branches,

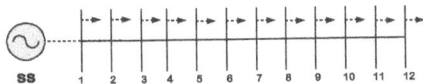

Figure 43.1 Radial structure of IEEE 12-bus RDS

representing a large-scale distribution network. Based on the proposed methodology, the optimal placement of SDG and capacitors is identified at nodes 7 using the WCOA and at node 9 using the GA. The sizes of these components are optimized using the WCOA. Table 43.1 provides a comparison of the optimal sizes of

Table 43.1 Optimal location and capacity of SDG and capacitor

Cases	Optimal location and capacity (kVAr/kW)	
	WCOA	BA
Base Case	–	–
With capacitor	1100 (7)	1120 (9)
With SDG	1600 (7)	1640 (9)
With capacitor and SDG	1100 (7) 1600 (7)	1120 (9) 1640 (9)

SDG and capacitors determined by both WCOA and GA at the identified locations. Moreover, Table 43.2 displays the simulation results for different reliability indices related to the 12-bus RDS.

4.1 Numerical discussion

(i) Base Case: In the Base Case (common for both WCOA and GA), the system exhibits a SAIFI value of 0.204, a SAIDI value of 0.30, and a CAIDI of 1.47. These values represent the baseline reliability of the 12-bus RDS without any additional optimization or interventions.

(ii) With capacitor: In Case-II, there is a noticeable improvement in reliability, albeit modest. For WCOA, SAIFI reduces to 0.168, reflecting a 17.6% improvement in the frequency of interruptions compared to the base case. SAIDI also shows a reduction to 0.252 (16% improvement), signifying a reduction in total interruption duration. However, the CAIDI value slightly increases to 1.50, indicating that the duration of each interruption has marginally increased. This suggests that capacitors, while improving the frequency and duration of interruptions, may have a limited impact on the overall recovery time from interruptions.

(iii) With SDG: In Case-III, the results show more significant improvements. The SAIFI value further decreases to 0.144 in WCOA (a 29.4% reduction from the base case), showing a substantial reduction in the number of interruptions. Similarly,

Table 43.2 Simulation results of various reliability indices for 12-bus RDS

Cases	SAIFI		SAIDI		CAIDI	
	WCOA	GA	WCOA	GA	WCOA	GA
Base Case	0.204	0.204	0.30	0.30	1.47	1.47
With capacitor	0.168	0.18	0.252	0.264	1.50	1.47
With SDG	0.144	0.168	0.216	0.24	1.50	1.43
With capacitor and SDG	0.132	0.156	0.18	0.216	1.36	1.38

SAIDI reduces to 0.216, demonstrating a 28% decrease in the total interruption duration. CAIDI remains stable at 1.50 in WCOA, reflecting no significant change in the average restoration time per interruption. This indicates that the integration of SDGs leads to a more reliable system with fewer interruptions and reduced overall outage time.

(iv) With Capacitor and SDG: Finally, in Case-IV, WCOA shows the most substantial improvements. SAIFI reaches its lowest value of 0.132, marking a 35.3% reduction from the base case, indicating a significant decrease in the frequency of interruptions. SAIDI also drops to 0.18 (40% improvement), reflecting a substantial reduction in the total duration of outages. Interestingly, CAIDI decreases to 1.36, the lowest among all cases, suggesting that both the capacitors and SDGs together not only reduce the number of interruptions but also improve the restoration time, making the system more resilient to faults.

4.2 Comparative analysis

In comparison, GA's results generally show similar trends but with less significant improvements across all cases. For instance, the reduction in SAIFI and SAIDI is more gradual in GA, and the CAIDI values either remain constant or show smaller reductions, demonstrating that WCOA, as a proposed algorithm, yields better performance in enhancing system reliability. Figure 43.2 shows the comparison of reliability indices of IEEE 12-bus RDS. a) SAIFI compares the frequency of interruptions in the system, with lower values indicating better reliability. The results

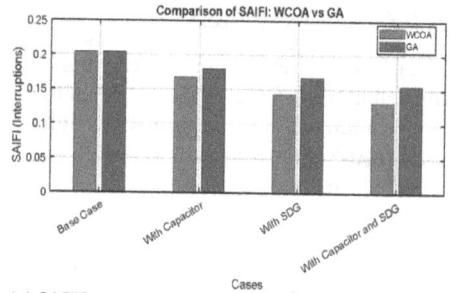

(a) SAIFI

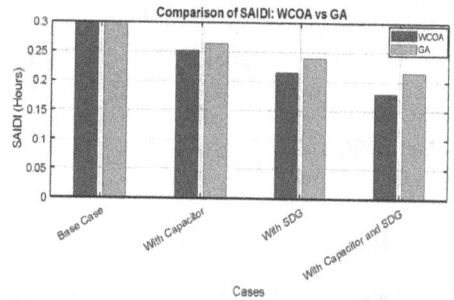

(b) SAIDI

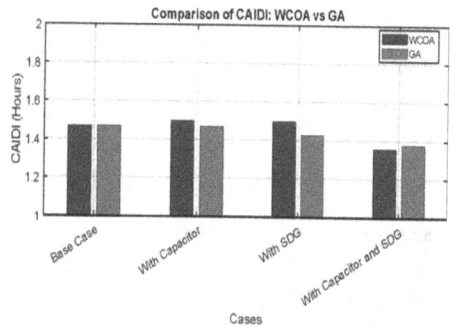

(c) CAIDI

Figure 43.2 Comparison of reliability index of IEEE 12-bus RDS

show that WCOA significantly reduces SAIFI compared to GA, particularly when capacitors and SDGs are applied. b) SAIDI reflects the average duration of interruptions. WCOA again outperforms GA in minimizing SAIDI, suggesting improved system recovery times with WCOA. c) CAIDI, which calculates the average period of disruptions for customers who experience outages, shows that WCOA leads to a greater reduction in CAIDI than GA, indicating faster restoration of service. Overall, WCOA demonstrates superior performance in improving reliability by reducing SAIFI, SAIDI, and CAIDI in the IEEE 12-bus RDS, particularly with the integration of capacitors and SDGs.

4. Conclusion

In this study, the influence of SDG and capacitors on the reliability of RDS was assessed by minimizing reliability indices such as SAIDI, SAIFI, and CAIDI. The WCOA was applied and compared to the GA using the IEEE 12-bus test system. Results displayed that WCOA outperformed GA in enhancing reliability. For the base case, WCOA reduced SAIFI from 0.204 to 0.144, while GA only reduced it to 0.168. Similarly, WCOA lowered SAIDI from 0.30 to 0.216, outperforming GA's reduction to 0.24. The combination of SDG and capacitors further improved the system, with WCOA achieving the lowest values in all reliability indices. These findings confirmed WCOA's effectiveness in optimizing RDS performance, providing better reliability compared to GA.

References

[1]   Ahmad, F., Iqbal, A., Ashraf, I., & Marzband, M. (2022). Optimal location of electric vehicle charging station and its impact on distribution network: A review. *Energy Reports*, 8, 2314–2333.

[2]   Yuvaraj, T., Devabalaji, K. R., Kumar, J. A., Thanikanti, S. B., & Nwulu, N. (2024). A Comprehensive Review and Analysis of the Allocation of Electric Vehicle Charging Stations in Distribution Networks. *IEEE Access*.

[3]   Deb, S., Tammi, K., Kalita, K., & Mahanta, P. (2018). Impact of electric vehicle charging station load on distribution network. *Energies*, 11(1), 178.

[4]   Yuvaraj, T., Arun, S., Suresh, T. D., & Thirumalai, M. (2024). Minimizing the Impact of Electric Vehicle Charging Station with Distributed Generation and Distribution Static Synchronous Compensator Using PSR Index and Spotted Hyena Optimizer Algorithm on the Radial Distribution System. *e-Prime-Advances in Electrical Engineering, Electronics and Energy*, 100587.

[5]   Yuvaraj, T., Devabalaji, K. R., Thanikanti, S. B., Aljafari, B., & Nwulu, N. (2023). Minimizing the electric vehicle charging stations impact in the distribution networks by simultaneous allocation of DG and DSTATCOM with considering uncertainty in load. *Energy Reports*, 10, 1796–1817.

[6]   Yuvaraj, T., Devabalaji, K. R., Thanikanti, S. B., Pamshetti, V. B., & Nwulu, N. I. (2023). Integration of electric vehicle charging stations and DSTATCOM in practical indian distribution systems using bald eagle search algorithm. *IEEE Access*, 11, 55149–55168.

[7]   Hariri, A. M., Hejazi, M. A., & Hashemi-Dezaki, H. (2019). Reliability optimization of smart grid based on optimal allocation of protective devices, distributed energy resources, and electric vehicle/plug-in hybrid electric vehicle charging stations. *Journal of Power Sources*, 436, 226824.

[8]   Yuvaraj, T., Suresh, T. D., Meyyappan, U., Aljafari, B., & Thanikanti, S. B. (2023). Optimizing the allocation of renewable DGs, DSTATCOM, and BESS to mitigate the impact of electric vehicle charging stations on radial distribution systems. *Heliyon*, 9(12).

[9]   Bilal, M., Rizwan, M., Alsaidan, I., & Almasoudi, F. M. (2021). AI-based approach for optimal placement of EVCS and DG with reliability analysis. *IEEE Access*, 9, 154204–154224.

[10] Zhang, H., Hu, Z., Xu, Z., & Song, Y. (2015). An integrated planning framework for different types of PEV charging facilities in urban area. *IEEE Transactions on Smart Grid*, 7(5), 2273–2284.

[11] Yuvaraj, T., Suresh, T. D., Ananthi Christy, A., Babu, T. S., and Nastasi, B. (2023). Modelling and Allocation of Hydrogen-Fuel-Cell-Based Distributed Generation to Mitigate Electric Vehicle Charging Station Impact

and Reliability Analysis on Electrical Distribution Systems. *Energies*, 16(19), 6869.

[12] Amini, M. H., Moghaddam, M. P., & Karabasoglu, O. (2017). Simultaneous allocation of electric vehicles' parking lots and distributed renewable resources in smart power distribution networks. *Sustainable Cities and Society*, 28, 332–342.

[13] Phonrattanasak, P., & Leeprechanon, N. (2012). Optimal location of fast charging station on residentialdistribution grid. *International Journal of Innovation, Management and Technology*, 3(6), 675.

[14] Sadeghi-Barzani, P., Rajabi-Ghahnavieh, A., & Kazemi-Karegar, H. (2014). Optimal fast charging station placing and sizing. *Applied Energy*, 125, 289–299.

[15] Shukla, A., Verma, K., & Kumar, R. (2019). Multi-objective synergistic planning of EV fast-charging stations in the distribution system coupled with the transportation network. *IET Generation, Transmission & Distribution*, 13(15), 3421–3432.

[16] Eskandar, H., Sadollah, A., Bahreininejad, A., & Hamdi, M. (2012). Water cycle algorithm–A novel metaheuristic optimization method for solving constrained engineering optimization problems. *Computers & Structures*, 110, 151–166.

[17] Latreche, Y., Bouchekara, H. R. E. H., Naidu, K., Mokhlis, H., & Dahalan, W. M. (2020). Comprehensive review of radial distribution test systems. *TechRxiv*, 1, 1–65.

44 Blackout Restoration at Short Period of Time using PPSR and PPPSR Algorithms with PHEV Systems in IEEE 39 Bus System

K. Vinitha[1], G. Arul Freeda Vinodhini[2], R. Hariharan[2,] and T. Yuvaraj[3]*

[1]Reseach Scholar, Saveetha School of Engineering, Saveetha Institute of Medical and Technical Sciences, Chennai, India
[2]Associate Professor, Saveetha School of Engineering, Saveetha Institute of Medical and Technical Sciences, Chennai, India
[3]Associate Professor, Center of Smart Energy Systems, Chennai Institute of Technology, Chennai, India

Abstract

This paper proposes a comparative analysis of algorithms for Parallel power system Restoration (PPSR), Novel Parallel Sub-island Power system restoration (PPPSR), and PPPSR with PHEV (Plugin Hybrid Electric Vehicle) System for Blackout restoration for a short period in IEEE 39 and IEEE 57 Bus system. PPSR (Parallel Power system restoration) and PPSR (Parallel Sub-island Parallel Power system restoration Algorithm are simulated by LabVIEW Graphical Coding for restoring the whole power system in a short period IEEE 39 and IEEE 57 Bus system. Based on the obtained results Novel PPPSR has about (160 Minutes), PPPSR with PHEV system has about (140 Minutes) and PPSR algorithm has about (240 Minutes) total restoration time in IEEE 39 bus system. Novel PPPSR and PPPSR with the PHEV system give less restoration time given the PPSR algorithm for the IEEE 39 Bus system, In IEEE 57 Bus tested system also provide better result in PPPSR System than the PPSR system.

Keywords: Novel PPPSR algorithm, PPSR algorithm, power restoration, blackout restoration

1. Introduction

The modern power system has complex networks due to increasing the number of residential areas and many industries erection. The huge demand can be met by an intelligent power system. For example in India, Five years installed capacity of power shows drastically change from 2012 to 2017. In 2012, the installed capacity in India was 199,877 MW. In 2017 the installed capacity in India is 326,841 MW. So the improvement percentage between 5years is 10.31% as the system networks increased. Electrical Power Generation was

Email: vinibabu89@gmail.com[1], arulfreedavinodhini@saveetha.com[2], harinov22@gmail.com[2], yuvarajt@gmail.com[3]

DOI: 10.1201/9781003661917-44

done by conventional methods and Non-Conventional methods to meet the load demand. Conventional methods such as thermal, hydro, and nuclear can adopt some amount of load demand and other power can be adopted from wind, solar and non conventional methods. Wind and solar are renewable energy able to significantly adopt the amount of power demand but it is not consistent, integrally unstable, asymmetrical, and indiscriminate. It causes partial failure, then produces severe problems, and this chain reaction causes a severe blackout in the system. Al, so blackouts can happen due to Natural disasters, malfunctions in the power system equipment, and an overload of the power system [Liu, Yutian 2016].

2. Literature Review

In July 2012, two massive blackouts in India affected 620 million people, 9% of the world's population. The failure began with a tripped circuit breaker on the Bina-Gwalior line, causing a cascade through the northern, eastern, and northeastern grids, worsened by a 32 GW power shortage. Poor transmission management was identified as a key cause [2].

A blackout in South Australia by a typhoon in 2016, it causes affect 1.6 million people [3]. Florida got affected due to hurricane Irma in September 2017 over 3.8 million customers were affected. Due to a lightning attack on a transmission line in London in 2019, 1.1 million users were affected. The Philippines in 2020 due to the typhoon disaster over 31 million people got affected [4].

Power system blackout restoration is a complex, multi-level process involving equipment availability, load balance, restoration time, and operation status [Garciano et al. 2021]. It occurs in three stages: generation capability optimization, transmission path optimization, and load restoration. Generation begins with Black-Start units, facing challenges in starting Non-Black-Start units. Transmission path optimization identifies the shortest path and synchronizes islands. Load restoration ensures

constraints are met while reconnecting the system. The PPSR algorithm reduces restoration time by restoring islands in parallel before synchronization, though it is less feasible for complex networks [6].

This paper introduces the Parallel Sub-Island Power System Restoration (PPPSR) algorithm, which enhances the PPSR method by subdividing islands into sub-islands for faster restoration. Tested on the IEEE 39-bus system, PPPSR showed reduced restoration times while maintaining stability and meeting system constraints. Both algorithms were implemented using LabVIEW.

3. Proposed Methodology

3.1 Research Design

This paper presents the PPPSR algorithm, an improved version of PPSR, enabling faster micro-level restoration while ensuring stability. It optimizes generation, enhances transmission paths through micro-islands, and accelerates load restoration. Each sub-island requires a Black-Start and Non-Black-Start unit, with balanced generation and load within frequency and voltage constraints.

The objective function is expressed as the Minimum time to take an unserved load in the network for a short period. Objective Function considers Capacity of load, Number of loads, and percentage of load being restored. Problem constraints of the function are Generation power and Demand power is equal in each sub-island. It is similar on each island. Frequency and voltage levels should be at the constraint level during the restoration process.

System Restoration depends upon the number of load pickups in the network in a short period. Restoration time may vary depending upon the capacity of load and the number of loads.

Restoration action times are Restart BS unit (P1) in 15 Minutes, Energize a bus from BS Unit in 5 Minutes, Tie line Synchronization in 25 Minutes, NBS unit activation in 15 minutes, and Synchronized subsystems in 20 Minutes and Pick upload in 20 Minutes.

3.2 PPPSR NOVEL Algorithm with PHEV SYSTEM

In PPPSR Algorithm Implements the PHEV system as a Black start unit. Restoration process finding Black start unit and improvise generation capability at initial its challenging task. In the Proposed module Plugin Hybrid vehicle is considered as the Black start unit provide crank power to Non-Black Start unit at a quick manner. PHEV Station is a charging unit for Electric vehicles. Nowadays many countries come forward to use electric vehicles for electrical safety. In the Proposed module Plugin Hybrid vehicle considered the Black start unit to the art Non-Black Start unit at a quick factor. This factor improves the restoration time and improves restoration efficiency.

Island 1 restores the system by 130 Minutes. In Island 1 each sub island as sub-island 1 takes 90 Minutes, Sub- island 2 takes 80 Minutes and Sub-island 3 takes 80 Minutes. After restoring each sub island. It synchronized together as island 1. Total time the restore the system is 130 Minutes. Parallel Island 2 and Island 3 synchronized the system. So System restoration time gets reduced.

Island 2 takes time to restore the system 95 Minutes, and Island 2 Sub island 1 and 2 takes 75 minutes. After restoring each island. It synchronized together as island 2. Total time the restore the system is 95 Minutes.

Island 3 takes time to restore the system 140 Minutes, and Island 2 Sub Island 1 and 2 takes 100 and 120 minutes. After restoring each island. It synchronized together as island 2. Total time the restore the system is 140 Minutes.

After restoring all islands all three islands synchronized together as a system. All island and sub-island restore process run parallel. So the maximum time taken island is 140 Minutes. So three synchronized times is 40 Minutes.

The total time takes to restore the system is 180 Minutes.

3.3 Novel PPPSR with PHEV System

The novel PPPSR Algorithm takes a restoration time of 180 Minutes. Here modified the PPPSR power system restoration algorithm by the PHEV system helps to start up the generator in an instant during the blackout period. Identifying the Black start unit is a complicated process. Here PHEV Charging station act as Crank up power Non- Black Start unit. So the restoration time gets reduced. It tested in IEEE 39 bus system which is compared with previous results.

Island 1 is sectionalized into three sub-island. On each sub-island three NBS Unit Generating Power plants, NBS units as G1, G8, and G9 get to crank up power by PHEV Charging station. After energizing the generating unit, it picks up the load in each sub-island. After recovering the sub-island it synchronized together in island 1. Island 1 takes time to restore the system as 100 Minutes.

Island 2, NBS units as G6, and G4 get to crank up power by the PHEV Charging station. After energizing the generating unit, it picks up the load in each sub-island. After recovering the sub-island it synchronized together in island 2. Island 2 takes time to restore system as 65 Minutes.

Island 3, NBS units as G2, and G3 get to crank up power by the PHEV Charging station. After energizing the generating unit, it picks up the load in each sub-island. After recovering the sub-island it synchronized together in island 3. Island 3 takes time to restore the system as 90 Minutes.

After restoring all the islands maximum time taken from the parallel process is 100 Minutes. After synchronizing all islands to complete the system, it takes 140 Minutes. Compared to previous methods, the Novel PPPSR PHEV charging station reached less restoration time.

3.4 Compared study with the conventional method (PPSR System)

PPSR algorithm Parallel Power system restoration. It is an existing algorithm, the same algorithm tested in IEEE 39 us system. The results are shown in the table. PPSR algorithm split the system into islands all the islands are recovered the island by a parallel manner, then synchronize with the system. The total time taken to restore the system is 245 minutes. It is compared with the proposed methods as the PPPPSR algorithm and PPPSR algorithm with the PHEV system take less restoration time than the existing system.

In all three methods, the same island parameters are considered. In island 1 takes time to restore the system in PPSR algorithm is 205 Minutes, PPPSR algorithm is 130 Minutes and PPPSR algorithm with PHEV system is 100 Minutes. So improvised time in conventional to proposed methods is 36.5%.

In island 2 takes time to restore the system in PPSR algorithm is 140 Minutes, PPPSR algorithm is 95 Minutes and PPPSR algorithm with PHEV system is 60 Minutes. So improvised time in conventional to proposed methods is 32.14%.

In island 3 takes time to restore the system in PPSR algorithm is 200 Minutes, PPPSR algorithm is 140 Minutes and PPPSR algorithm with PHEV system is 90 Minutes. So improvised time in conventional to proposed methods is 30%.

The maximum time chosen PPSR algorithm for Island 1 is 205 Minutes, PPPSR algorithm for Island 3 is 140 Minutes and PPPSR with PHEV System is Island 1 100 Minutes. Synchronization of three islands takes 40 Minutes. So the total time for the PPSR system is 245 Minutes, PPPSR method takes 180 minutes, and PPPSR with the PHEV system is 140 Minutes. So improvised time in conventional to the proposed algorithm is 26.5%. The out put of the Fuzzy logic is shown in Table 44.1.

3.5 Conclusion and Future Enhancement

Due to natural disasters, overload, and equipment malfunction, blackout happens in power systems. power system operators are responsible for recovering the system as much as earlier to recover the system. But the system is very complex due to the large network. here the proposed method provides an effective strategy to recover the system as earliest. PPPSR algorithm results are compared with the Conventional PPSR algorithm, it provides an improvised restoration time as 26.5%.

With the proposed system possible to increase sub-island which helps to reduce the restoration time. But sub-island depends on one NBS unit and balanced load demand. If it is satisfied the subilsand process in the algorithm.

The proposed algorithm also finds a Black start unit to provide startup power

Table 44.1 Fuzzy logic output

| S. N | INPUT VARIABLES | | | | | OUTPUT VARIABLES | | | |
	Soil moisture	Humidity	Temperature	Weather Data	Pest	Water Schedule	Fertilizer	Schedule of workers	Invoked Rule
1	0.823	0.06	0.197	0.11	0.11	0.50	0.50	0.192	17
2	0.823	0.864	0.197	0.11	0.11	0.35	0.65	0.200	24,25
3	0.823	0.86	0.897	0.11	0.11	0.19	0.50	0.199	29
4	0.823	0.86	0.897	0.51	0.11	0.20	0.49	0.204	31
5	0.823	0.86	0.897	0.51	0.71	0.19	0.52	0.199	32,62,64
6	0.123	0.86	0.897	0.51	0.71	0.79	0.79	0.49	16
7	0.123	0.36	0.897	0.51	0.71	0.49	0.79	0.644	8,56
8	0.123	0.36	0.297	0.51	0.71	0.79	0.79	0.5346	4,76

Table 44.2 Data Analysis in Fuzzy logic variables

	Soil Moisture	Humidity	Worker Schedule	Invoked Rule
Mean	0.523	0.584	0.368	18.66
Std	0.349	0.308	0.229	10.78
Min	0.123	0.064	0.199	1
25%	0.123	0.363	0.199	16.25
50%	0.723	0.614	0.202	17.5
75%	0.823	0.864	0.525	26.2
Max	0.823	0.864	0.799	31

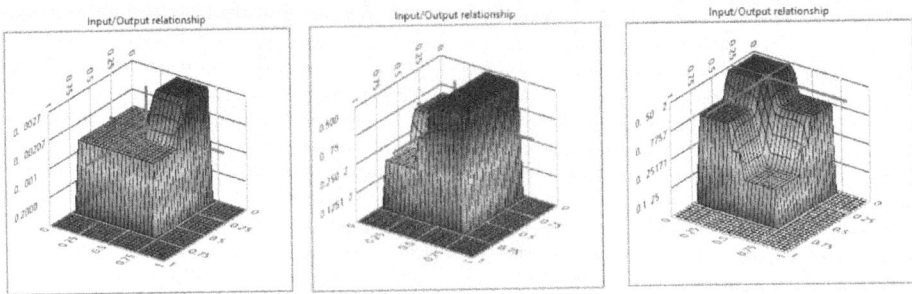

Figure 44.1 Fuzzy sample output analysis

nonblack start unit. So the proposed system was modified with PHEV system. Here this proposed system PHEV Charging station consider a Black start unit. It helps to start the Non-Black start unit. This modified algorithm also tested in IEEE 39 bus system and compared with the proposed algorithm. PPPSR with PHEV system provides an improvised restoration time as 22%. PPPSR with PHEV system motives the E Vehicle Charging station, it helps the Environment Eco System. So the total time for the PPSR system is 245 Minutes, PPPSR method takes 180 minutes, and PPPSR with the PHEV system is 140 Minutes. So improvised time in conventional to the proposed algorithm is 26.5%.

3.6 Key Observations

The dataset presents an advanced agricultural system leveraging environmental indicators and Fuzzy Logic to optimize irrigation, fertilization, and scheduling. Using Virtual Instrumentation, it enhances resource efficiency and productivity

from land preparation to harvest. Table 44.2 indicates data analysis in Fuzzy logic variables.

4. Conclusion

In conclusion, the dataset highlights an advanced agricultural system that optimizes resources and productivity using environmental indicators and refined decision-making. The "Invoked Rule" column reflects the system's adaptability, while Fuzzy Logic integration via Virtual Instrumentation ensures precision by managing uncertainties. This innovative approach enhances efficiency, sustainability, and productivity in agricultural practices.

References

[1] Liu, Y., Fan, R., & Terzija, V. (2016). Power system restoration: a literature review from 2006 to 2016. *Journal of Modern Power Systems and Clean Energy*, 4(3), 332–341.

[2] Lai, L. L., Zhang, H. T., Mishra, S., Ramasu-bramanian, D., Lai, C. S., & Xu, F. Y. (2012, November). Lessons learned from July 2012 Indian blackout. In *9th IET International Conference on Advances in Power System Control, Operation and Management (APSCOM 2012)* (pp. 1–6). IET.

[3] Yan, R., Saha, T. K., Bai, F., & Gu, H. (2018). The anatomy of the 2016 South Australia blackout: A catastrophic event in a high renewable network. *IEEE Transactions on Power Systems*, 33(5), 5374–5388.

[4] Mitsova, D., Esnard, A. M., Sapat, A., & Lai, B. S. (2018). Socioeconomic vulnerability and electric power restoration timelines in Florida: the case of Hurricane Irma. *Natural Hazards*, 94, 689–709..

[5] Garciano, L. E., Tanhueco, R. M., Torres, A., & Iguchi, H. (2021). Assessing Vulnerabilities and Costs of Power Outages to Extreme Floods in Surigao City, Philippines. *GEOMATE Journal*, 20(82), 7–14..

[6] Chopra, S., Qiu, F., & Shim, S. (2023). Parallel power system restoration. *INFORMS Journal on Computing*, 35(1), 233–247.

45 Optimizing mmWave Network Performance by Utilizing Proximal V2I Connected Vehicles during Blockages

Vinodh Kumar Minchula[1,], Venkata Aniruddh Kalyan Tallur[2], and Sharathchand Kodam[2]*

Corresponding Author
[1]Principal Investigator of ANRF-SURE Project, Associate Professor, Department of ECE, CBIT, Hyderabad, India
[2]Student, Department of ECE, CBIT, Hyderabad, India

Abstract

The Wave communication is very crucial for future advancements of vehicular networks as it provides high data rates & very low latency. Nevertheless, there are few challenges associated with mmWave network such an encountering blockage frequently due to high mobility and penetration losses. The primary focus of this paper is to analyse and enhance the performance of mmWave communication in vehicular network, with specific focus on highway scenarios, by addressing the limitations associated with frequent LOS blockages. The LOS blockages are prominently due to high mobility and large dimensions of BVs, leading to significant signal attenuation and interruptions. An evaluation framework is established, considering a highway scenario, where vehicles share data with the network thorough one of the ways- either using V2I link or V2V & V2I link. The parameters used for the analysis of the network are blockage probability, average blockage duration. The expressions for these parameters are derived using steady state solution of continuous time- Markov chain models of blockage events that V2V and V2I links experience. With these results, it is established that vehicular relays (proximal V2I connected vehicles) are quite beneficial for improving the performance of the network by providing alternate communication paths when LOS links are blocked. By enabling V2V links, vehicular relays are reducing blockage probability and average blockage duration ensuring connectivity even in high density traffic scenarios. Relays offer frequent and reliable connections, while maintaining acceptable SINR levels. The paper also illustrates that relays which are located far away from a vehicle produce negligible benefits because the chances those links being blocked are very high, as compared to closer relays. The analytical model proposed provide a more precise and fast evaluation of the network.

Keywords: Line-of-Sight (LOS), Blocking Vehicle (BV), Connecting Vehicle (CV), Standard Vehicle (SV), Vehicle-to-Infrastructure (V2I), Vehicle-to-Vehicle (V2V)

E-mail: vinodhkumarm_ece@cbit.ac.in[1], aniruddhkalyan.talluri@gmail.com[2], sharathchandkodam@gmail.com[2]

DOI: 10.1201/9781003661917-45

1. Introduction

Rapid deployment of connecting vehicles (CVs) leads to a great increase in demand of Resource-Intensive applications such as real-time video streaming, Advanced Driver Assistance Systems (ADAS), Remote Vehicle diagnostics etc. [12]. Such applications need extremely low latency rates (1–10ms) and large data rates (0.1–1 Gbps) [10, 18], raising challenges for vehicular networks. Hence, to meet such demands, fifth-generation (5G) networks using millimeter-wave (mmWave) frequencies are crucial. The low latency and high data rate characteristics of 5G mmWave network were found to be beneficial for critical applications, enhancing safety, autonomous vehicles etc. [3, 7, 9].

However, mmWave communication suffers from high losses of penetration, frequent outages due to blockages, especially in scenarios such as highways with high-mobility [5, 11, 18]. Vehicles such as trucks or buses, with immense size can block LOS links, resulting in significant attenuation of signal strength (above 20 dB), and reduced SINR [10]. This paper is focused on overcoming blockages by utilizing vehicular relay. Especially, in highway scenarios, where connecting vehicles (CVs) transmit data with Roadside units (RSUs) via LOS V2I links. When such links are blocked, V2V links are established by CVs with relay vehicles to forward data, to ensure connectivity [1, 2]. The primary goal of this paper is to provide an analytical model for evaluating the performance of vehicular relay network in a more accelerated and precise way, based on mainly three parameters, namely blockage probability, average blockage duration and SINR distribution. Distribution of SINR a Connecting vehicle (CV) experiences is calculated by incorporating the probabilities of V2I and relay connections. The equations of blockage duration in vehicular relay network are derived and its distributions obtained from simulations in different deployment scenarios are examined and discussed on how blockage duration depends on various metrics such as number of Roadside Units (RSU), Vehicle speeds or CV probability [1, 14].

The primary conclusions of this paper are: With high density traffic, the increased frequency of blockages leads to the rise of blockage probability. However, the chances of finding relays around the vehicle with links of less blockage duration increases significantly, which in turn reduce the average blockage duration. Communication range leads to increase in no. of available candidate relay nodes to a CV [8]. However, it produces a negligible benefit, as V2V link blockage probability escalates as distance between CV and relay node increases. The SINR of Vehicle-to-Infrastructure (V2I) link is more than that connected using a vehicular relay (V2V & V2I link), hence as blockage probability of V2I link increases, there will be decline in SINR of CV. High density of RSU's decreases the blockage probability and average blockage duration. In addition, shorter blockages are removed by deploying tall RSU's, however longer blockages will lead to large average blockage duration.

The core contributions of this paper include the development of continuous time Markov chain model to represent blockages experienced by various links, where a CV communicates with RSU using vehicular relays. The proposed model and its steady state solution are utilized to formulate the equations for performance metrics such as Blockage Probability (P_b) and Average Blockage Duration (T_b). The influence of various highway, vehicle and deployment parameters is examined on these given metrics-namely blockage probability, average blockage duration.

2. System Model

2.1 Modelling Highway Scenario

A highway scenario has been modelled with N_L lanes, where width of each lane is W_L. The given scenario considers three different classes of vehicles, namely Connecting Vehicles (CV), Blocking Vehicles (BV) and Standard Vehicles (SV). Connecting Vehicles (CVs) are the primary entities in our considered vehicular network. They ensure seamless communication is facilitated with network infrastructure (RSUs) and other CVs for efficient data transmission and maintain connectivity. When CVs direct V2I link is blocked due to Blocking Vehicles (BV) in the LOS path, it uses proximal V2I connected CVs, forming a V2V connection to ensure the data is forwarded to RSU through relay mechanism. Standard Vehicles (SVs) are regular, non-communicating vehicles which are used in the model to contribute traffic dynamics, realistically simulating highway traffic scenario. They do not cause blockages in the system. SVs share the same dimensions (height and length) and mobility patterns as CV. It's presence neither support nor obstruct communications links. They indirectly affect the traffic conditions and the probability of CVs finding relays nearby.

Height of CVs = Height of SVs = H_C. The length of CV/SVs are firmly fixed and are exponentially distributed with mean $1/\mu C$. L ~ exponential (μC), where E[L] = $1/\mu C$, representing mean length and μC is rate parameter. Speed of CV/SVs = V_C. The Blocking Vehicles (BV) representing vehicles like buses or trucks, have larger dimensions than vehicles (CVs and SVs). The primary purpose of using BVs in this model is to block direct LOS path between CVs and RSUs, leading to increased blockage probability and signal attenuation. They cause temporary and frequent interruptions creating a need for relay mechanisms. Height of a BV = H_B, Fixed speed of BVs = V_B, Length is exponentially distributed with mean $1/\mu B$. To simplify the analysis, it is assumed fixed heights and exponentially distributed lengths for all the vehicles. Moreover, it's assumed $H_B > H_C$ and $1/\mu B > 1/\mu C$, meaning Blocking vehicles (BVs) are larger than CVs and SVs. Each vehicle type - BV, CV and SV appears with probabilities of P_B, P_C, and P_S respectively.

Primary goal of this paper is to evaluate the coverage performance of CVs, where CVs communicate with RSU's positioned on along the right-most lane. Traffic intensity is modelled using exponential distribution E[d], which represents the distance between two consecutive vehicles, which depends on traffic density. Higher value of E[d] indicates lower traffic density, similarly, Lower value of E[d] implies high traffic density.

2.2 Communication Channel Model

In this paper, very narrow and directional beams are considered to eliminate the challenges faced by mmWave frequencies such as high pathloss and signal attenuations [8, 13, 15]. In this paper, an intermittent blockage model is adopted, where connections obstructed by surrounding vehicles are considered to be temporarily unavailable and disrupted [4, 6]. Both the links—V2I and V2V undergo Rayleigh's fading, characterized by a mean value of $1/\mu$, where h ~ exp (μ), and h denotes the component of small-scale fading. The Received SINR for any LOS link is given by the equation:

$$\text{SINR}(\text{Received}) = \frac{P_T h d^{-\alpha}}{N_0} \qquad (1)$$

where,

$d^{-\alpha}$ is standard power attenuation for path loss of LOS link
P_T is the transmitting power
N_0 is the noise power
d is the Euclidean distance between T_x and R_x [5].

2.3 Model for V2I Link Communication

In this paper, lanes of a highway scenario are indexed from 1 to NL, where right most lane is the slowest lane. The CV moves through the center of ith lane, where i ranges from 1 to N_L, while BVs travel through the center of jth lane, where j ranges from 1 to i-1. To exchange data with the network, a connecting vehicle, for example, CV1 in i^{th} lane ($1 \leq i \leq$ NL), attempts to establish a Line-of-Sight (LOS) V2I connection with the Roadside Unit. If the link is established successfully, the CV communicates with RSU through it and is said to be in coverage and connected. There is a chance that V2I link may be obstructed by Blocking Vehicles (BV) positioned in j^{th} lane, where j ranges from 1 to i-1. When a BV obstructs the LOS path between CV and Roadside Unit, the direct V2I connection to the RSU is blocked, making RSU out-of-reach for the CV. In a situation where all the V2I links are disrupted, CV1 can communicate with the RSU by forming a V2V link with a neighboring CV, such as CV2. The CV2 must have a direct V2I link with at least one RSU, to forward the data between CV1 and network, acting as a relay. The Received SINR at a given distance can determine if transmission of packet has been done successfully or not. To simplify the calculations, a fixed LOS communication range R is assumed. Data transmission over a LOS V2V or V2I link is considered successful only if the distance between Rx and Tx is less than R and Line-of-Sight (LOS) is not blocked by any BV. Total coverage range of a connecting vehicle, denoted by R_C, represents the road segment where RSU's have LOS Link distance less than R to the CV traveling in lane i. R_C is the coverage range, d_R is inter-RSU distance. N_R denotes the number of RSU's that lie within the CV's LOS coverage range, is given by:

$$N_R = \frac{R_C}{d_R} \qquad (2)$$

All the RSU's are deployed at same height h_R and we are interested in those circumstances where their maximum height is restricted due to challenges posed by unavailability of resources, expenses and highway regulations etc. Thus, increasing the height of RSU's can't completely resolve the problem of blockages caused by BVs. A BV has the potential to obstruct the LOS path between the CV in i^{th} lane and Roadside Unit, placed on lane j where $j = 1, 2, .., i\text{-}1$, if the height of the BV is larger than the threshold value. The threshold value is called critical height of the lane j, denoted as $h_B^{(i,j)}$, where this parameter depends on the height of RSU antennas, CVs, and BVs.

$$h_B^{(i,j)} \triangleq h_C(h_R - h_C) \cdot \frac{i-j}{i-\frac{1}{2}} \qquad (3)$$

where, h_C and h_R are the height of CVs and RSU units, respectively. The total coverage range of a CV in lane i is given by:

$$R_C = 2\sqrt{R^2 - |h_R - h_C|^2 - \left[\left(i - \frac{1}{2}\right)W_L\right]^2} \qquad (4)$$

This equation shows that when height of the RSU's is increased, number of potential blockages decreases. Moreover, it reduces the coverage area R_C.

2.4 Model for V2V Link Communication

A CV may establish a V2V connection with another CV if all the V2I links have been interrupted, if they are within LOS communication range R and there are no BVs in the path of Line-of-Sight (LOS) between them. A BV can block a potential V2V link between two CVs which are positioned on i^{th} and j^{th} lanes (where $i, j = 1, 2, .., N_L$) by moving either between the lanes (i and j) or along them.

2.5 Blocking Vehicle (BV) Placement Model

To model the placement of Blocking Vehicle (BV), a simplified two-state continuous time Markov chain model has been established and used. The model characterizes the probability distribution of all vehicle types to assess if a given vehicle has the potential to obstruct a Line-of-Sight (LOS) Vehicle-to-Infrastructure (V2I) connection. In the given deployment setup, all the CVs & SVs are considered to be of identical height, making BVs the only vehicles capable of blocking the path of LOS V2I connection. The rates of transition for this model are derived from inter-BV distance (between consecutive BVs), and the length distribution of BVs. The rates are combined with speed of the vehicles to calculate Markov's chain transitions. For a Connecting Vehicle (CV) located in lane i, the arrival rate of Blocking Vehicle (BV) in lane j, where j = 1, 2, . . . , i-1 is given by the equation:

$$\lambda_B^{(i,j)} = \begin{cases} \left[\dfrac{1-pB}{pB}(1/\mu_C + E[d]) + E[d] \right]^{-1}, \\ h_B > h_B^{(i,j)} \quad 0, \ h_B \le h_B^{(i,j)} \end{cases} \tag{5}$$

h_B is BV height, E[d] is mean-inter vehicle distance, P_B is the probability that a BV will be encountered. The arrival rate equation has been derived based on length of CV, inter-BV distances, to complete the model of BV placement.

2.6 Modeling Blockages to A Vehicle-to-Vehicle (V2V) Link

A two-state continuous time Markov chain model has been employed, to analyse how blockages are affecting the V2V LOS links between two CVs. The system represents the status of V2V connections between the CV in the ith lane and possible relay vehicle in the jth lane. The arrival rate of blockages for V2V link is represented by $\lambda_V^{(i,j)}$, while the departure rate is denoted by $\mu_V^{(i,j)}$. The V2V link blockage probability, denoted by $P_{V2V}^{(i,j)}$, refers to the probability of V2V link between CV on lane i and vehicular relay on lane j getting blocked. The blockage probability of V2V link is given by the steady-state solution of Markov chain:

$$P_{V2V}^{(i,j)} = \frac{\lambda_V^{(i,j)}}{\left(\lambda_V^{(i,j)} + \mu_V^{(i,j)} \right)} \tag{6}$$

In this example using Markov chain, when an LOS V2V link between a reference CV on lane i and relay CV on lane j is blocked by a BV on lane k (*where k= i, j*), the reference CV attempts to communicate with another CV which is located within the LOS communication range. If successful, the system will remain in state L, if not, it transitions into blocked state B. Again, when LOS is restored, it returns back to state L. In this paper, simulations are performed for required combinations of i and j, to derive the transition rates $\lambda_V^{(i,j)}$ and $\mu_V^{(i,j)}$ respectively. A simple simulation environment has been built and used where vehicles can overtake each other without the need of lane switching.

3. Analysis of Communication Coverage

In this section, two-step continuous time Markov chain is used to model the blockages which affect the LOS path of Vehicle-to-Infrastructure (V2I) and Vehicle-to-Vehicle (V2V) communication link. Three parameters have been derived- namely Blockage probability, Average blockage duration, and SINR (Singal-to-Interference-plus-Noise) distribution, to evaluate the quality of communication link in the vehicular networks [6].

3.1 V2I Link—Markov Chain Model

Let the state space of V2I Markov Chain be denoted as S, comprising of $2^{(i-1)N_R}$ states. Each state of the chain is represented as a binary vector, with $(i-1)N_R$ bits. Where, a binary digit shows whether that respective lane is blocked or not. For example, if a binary vector is represented as 011, it indicates that two lanes are blocked whilst the other one remains unblocked. For a connecting vehicle (CV) located in lane i, the relative speed of the projection of RSU in lane j is given by:

$$V_R^{(i,j)} = V_c \frac{j - \dfrac{1}{2}}{i - \dfrac{1}{2}} \tag{7}$$

For a CV in lane i, the LOS coverage duration in lane j is distributed exponentially with the rate $\lambda_B^{(i,j)} \left| V_B - V_R^{(i,j)} \right|$, where $i = 2, \dots, N_L$ and $j = 1, 2, \dots, i\text{-}1$, with $\lambda_B^{(i,j)}$ as defined in (4). Similarly, the blockage duration caused by a BV which is located in lane j, is distributed exponentially with rate $\mu_B \left| V_B - V_R^{(i,j)} \right|$, where $i = 2, \dots, N_L$ and $j = 1, 2, \dots, i\text{-}1$, with $\dfrac{1}{\mu_B}$ indicating the average length of the Blocking Vehicle. Two distinct state sets are defined: S_B for the blocked state of CV, and S_C for its coverage state. Steady-state probability of state S is denoted by π_S. To calculate the V2I link blockage probability, we compute the steady-state probability vector using the $\pi = \begin{bmatrix} \pi_1, \cdots, \pi_{|S|} \end{bmatrix}$ given equation:

$$\pi Q = 0_{1 \times |S|} \tag{8}$$

Where, Q is the Markov chain Infinitesimal generator matrix. For a CV in lane i, the blockage probability $P_{V2V}^{(i)}$ is expressed as:

$$P_{V2I}^{(i)} = \sum_{s \in S_B} \pi_s \tag{9}$$

The average blockage duration, is denoted as $\underline{T}_{V2I}^{(i)}$ which is derived based on total transition rate from unblocked states to blocked states, is given as

$$r_B = \sum_{s \in S_C} \sum_{s' \in S_B} r_{ss'} \cdot \pi_s \tag{10}$$

The average blockage duration $\underline{T}_{V2I}^{(i)}$. is given as:

$$\underline{T}_{V2I}^{(i)} = \frac{E[N_s]}{r_B} \tag{11}$$

where $E[N_s]$ describes the number of blocked lanes expected.

$$E[N_s] = \sum_{s \in S_B} \pi_s N_s \tag{12}$$

The average coverage $\underline{T}_C^{(i)}$ duration is expressed as:

$$\underline{T}_C^{(i)} = \underline{T}_{V2I}^{(i)} \frac{1 - P_{V2I}^{(i)}}{P_{V2I}^{(i)}} \tag{13}$$

3.2 Vehicular Relay—Markov Chain Model

Let the state space for the vehicular relay's be denoted as **S'**, where each state describes the CVs coverage status by vehicular relays in each lane. The blockage arrival and departure rates for each lane j is represented as follows:

$$\lambda_j' = \lambda_V^{(i,j)} \left(1 - P_{V2I}^{(j)}\right) + \frac{1}{\underline{T}_C^{(i)}} \tag{14}$$

$$\mu_j' = \mu_j^{(i,j)} \left(1 - P_{V2I}^{(j)}\right) \cdot P_{V2V}^{(i,j)} + \left(1 - P_{V2V}^{(i,j)}\right) \cdot \frac{P_{V2I}^{(j)}}{\underline{T}_{V2I}^{(j)}} \tag{15}$$

The Blockage Probability of a Connecting Vehicle in i^{th} lane is obtained through solving the steady-state probability vector $\pi' = \begin{bmatrix} \pi', \cdots, \pi'_{|S'|} \end{bmatrix}$ by setting $\pi'Q' = 0_{1 \times |S|}$ as,

$$P_V^{(i)} = \pi_1' \tag{16}$$

When all j^{th}-lanes are blocked, the average blockage duration is given as:

$$\underline{T}_V^{(i)} = \left(\sum_{j=1}^{N_j} \mu_j'\right) \tag{17}$$

3.3 Overall Blockage Probability

For a CV placed in lane i, the overall blockage probability $P_B^{(i)}$ is the multiply of V2I and V2V blockage probabilities

$$P_B^{(i)} = P_{V2I}^{(i)} P_V^{(i)} \tag{18}$$

3.4 Average Blockage Duration

The blockage durations of V2V and V2I links are considered to follow an exponential distribution, characterized by rates $\frac{1}{\underline{T}_{V2I}^{(i)}}$ and $\frac{1}{\underline{T}_V^{(i)}}$ respectively. Using this assumption, the Average Blockage Duration of a Connecting Vehicle located in i^{th} lane, represented as $\underline{T}_B^{(i)}$, can be expressed as follows:

$$\underline{T}_B^{(i)} = \left(\frac{1}{\underline{T}_{V2I}^{(i)}} + \frac{1}{\underline{T}_V^{(i)}}\right)^{-1} \tag{19}$$

3.5 Signal-to-Interference-Plus-Noise-Ratio (SINR) Distribution in V2x Scenarios

The SINR distribution for V2I links is denoted as, is expressed as:

$$f_{SI}(s) = A\mu \left(\begin{array}{c} \Phi\left(0, \dfrac{\mu s N_o}{R^\alpha}\right) - \\ \Phi\left(0, \dfrac{\mu s N_o}{\left((i - 1/2)WL\right)^\alpha}\right) \end{array} \right) \tag{20}$$

The SINR distribution for V2V links is given as:

$$f_{SV}(s) = A\mu \left(\Phi \left(0, \frac{\mu s N_o}{R^\alpha} \right) - \Phi \left(0, \frac{\mu s N_o}{W^\alpha L} \right) \right)$$

(21)

The Cumulative Distribution Function (CDF) for SINR, FS(s) is expressed as:

$$F_S(s) = A\mu \left(\frac{R^\alpha}{\mu N_0} \left(1 - e^{-\frac{\mu s N_o}{R^\alpha}} \right) + s \cdot \Phi \left(0, \frac{\mu s N_o}{R^\alpha} \right) \right)$$

(22)

4. Results

Initially, vehicles are randomly generated by the simulator on the considered highway scenario according to the predefined highway and vehicle properties. The generated vehicles travel at their specified speeds to cover some distance. After every time step (delta), the locations of the vehicles are updated and the LOS link state of the connecting vehicle (CV) with all the RSU's are checked.

If CV finds any LOS V2I link with any of the RSU's, the link state is said to be connected. Similarly, each time interval is processed sequentially. In cases where all the V2I connections are obstructed, potential V2V connections are calculated to find out all the possible relay paths for a CV to communicate efficiently with RSU.

If any of the vehicular relays has an established LOS V2I link with any RSU, the system will be in connected state. Otherwise, relay path to RSU can't be found and the system will be set to blocked state.

At every time step, the connected and blocked intervals are tracked and stored to calculate blockage and coverage durations. MATLAB 2024 software is used to implement the discrete-event simulator for the described highway scenario. The system is reset after every 10^8 steps, or approximately after 10^8 milliseconds. Vehicle positions are reset every 10^4 ms.

The values of Blockage Probability and the Average Blockage Duration are recorded with varying set of available RSU's, for four different scenarios which are described below:

Case A: $R = 800m$, $p_C = 0$, $V_B = 60 km/h$
Case B: $R = 800m$, $p_C = 0.5$, $V_B = 60 km/h$
Case C: $R = 800m$, $p_C = 0.5$, $V_B = 100 km/h$
Case D: $R = 400m$, $p_C = 0.5$, $V_B = 100 km/h$

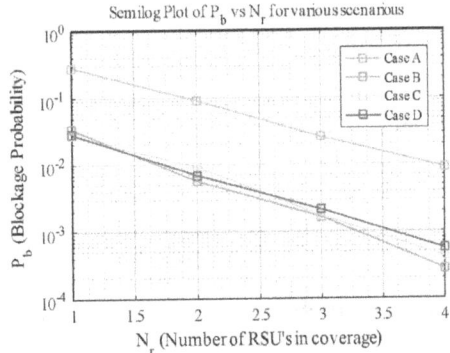

Figure 45.1 P_b as a function of N_R at $h_R = 2m$

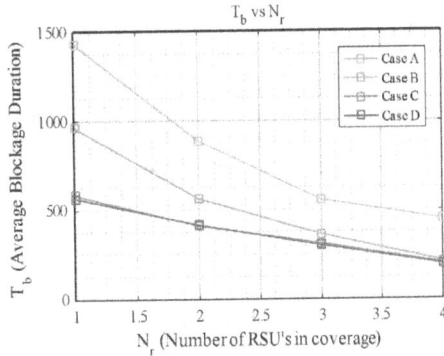

Figure 45.2 T_b as function of N_R at $h_R = 2m$

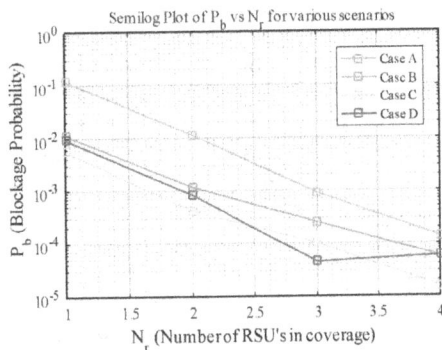

Figure 45.3 P_b as function of N_R at $h_R = 6m$

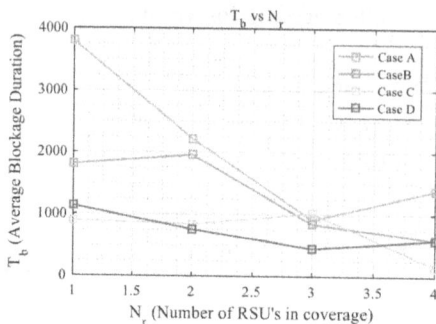

Figure 45.4 T_b as function of N_R at h_R = 6m

It is observed that from Figures 45.1–45.4, as the speed of the Blocking vehicles is increased from 60km/h to 100km/h, blockage duration reduces, while the blockage probability remains unaffected. From Figures 45.1 and 45.3, it is observed that speed of vehicles does not affect blockage probability, while faster vehicles will result in blockages with short duration. Consequently, it reduces average blockage duration.

When the vehicular relays are included (with probability of CV P_C = 0.5), performance in both metrics- blockage duration and blockage probability improved as compared to the scenario where only V2I links are used (P_C = 0). The difference between the scenarios P_C = 0 and P_C = 0.5 indicates that huge improvement can be achieved in terms of performance of the system, by using vehicular relays.

While N_R is kept constant, increasing the LOS communication range will produce negligible effect on the performance of the system. It won't help reduce blockage probability nor the average blockage duration. This occurs because the relay vehicles selected by CV are usually at shorter distances, as relays placed far away are more likely to be obstructed by blocking vehicles (BV).

In Figures 45.5 and 45.6, the performance of the system is based on traffic density. Larger values of E[d] implies lower traffic density and vice-versa. For E[d] = 30m, 50m, 70m, the value of p_C is varied

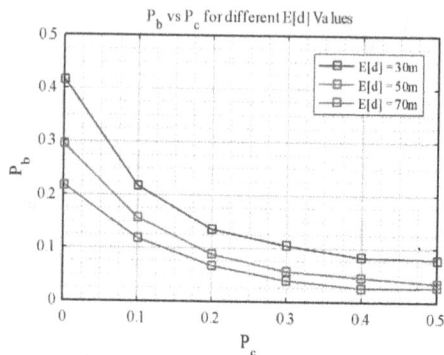

Figure 45.5 Blockage Probability vs CV Probability for various traffic scenarios

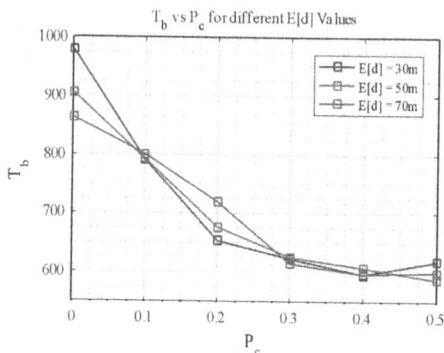

Figure 45.6 Average Blockage Duration vs CV Probability for various traffic scenarios

from 0 (no relays) to 0.5, to evaluate the metrics blockage probability and average blockage duration, as shown in Figures 45.5 and 45.6 respectively.

As traffic density increases (when E[d] - inter vehicular distance is low) the blockage probability increases, however lower average blockage duration is observed. In highly dense traffic scenario, even though no. of blockages increases, more no. of relay vehicles are available which provide frequent connections, thus shortening the blockage duration. Similarly, as traffic density increases, SINR decreases.

5. Conclusion

As mmWave communication is a promising technology to support future vehicle applications, in this paper, constraints of mmWave communication for vehicular networks have been investigated. The mmWave communication between Connecting Vehicles and Roadside units is analyzed across a V2I link or a V2V link using a relay, emphasizing parameters such as average blockage duration, blockage probability or SINR distribution. Using Markov chain model for blockages in V2V and V2I links, analytical expressions for primary performance metrics have been derived. Results indicate that simulation outputs closely lineup with theoretical values for various range of parameters. The influence of system parameters on its performance has been assessed. It is observed that keeping the RSU units fixed, and increasing the LOS communication range doesn't improve the coverage performance of the system.

It is proved that V2V links and relays, enhances the system performance as in heavy traffic conditions, average blockage duration is reduced by using relays, since it provides intermittent and frequent connection opportunities for CV.

6. Future Scope

While this paper focuses on LOS communication, in future work, analysis on NLOS communication for both V2V and V2I links would be a valuable addition. Moreover, machine learning can be implemented for 5G V2X communication, for predicting blockages based on various factors such as vehicle movement, traffic patterns and environmental conditions etc.

Acknowledgment

The above work has been carried out under the project entitled "Machine Learning based Intelligent Indian vehicle network with Relay assisted mm-Wave BS system for 5G NR-V2X communications" sponsored by SURE, DST, New Delhi, India, vide sanction letter. No. SUR/2022/004990, dated: 25th October 2023.

References

[1] Tunc, C., & Panwar, S. S. (2021). Mitigating the Impact of Blockages in Millimeter-Wave Vehicular Networks Through Vehicular Relays. In *IEEE Open Journal of Intelligent Transportation Systems*, 2, 225–239. doi: 10.1109/OJITS.2021.3100856.

[2] Kodam, S., Minchula, V. K., Reddy, A. S., Chalamalasetty, K., & Tammireddi, R. S. (2024, June). Optimizing Relay based V2V Communication in Non-Line-of-Sight Scenarios: Pathloss Modelling and Strategies. In 2024 15th International Conference on Computing Communication and Networking Technologies (ICCCNT) (pp. 1–7). IEEE. doi: 10.1109/ICCCNT61001.2024.10725657.

[3] Elkashlan, M., Duong, T. Q., & Chen, H. H. (2014). Millimeter-wave communications for 5G: fundamentals: Part I [Guest Editorial]. IEEE Communications Magazine, 52(9), 52–54. doi: 10.1109/MCOM.2014.6894452.

[4] Tammireddi, R. S., Minchula, V. K., Reddy, A. S., Chalamalasetty, K., & Kodam, S. (2024, June). Performance Evaluation of IEEE 802.11 ax Using Validated Propagation Loss Models for V2I Applications. In 2024 15th International Conference on Computing Communication and Networking Technologies (ICCCNT) (pp. 1–5). IEEE. doi: 10.1109/ICCCNT61001.2024.10724061.

[5] Al-Shammari, B. K., Hburi, I., Idan, H. R., & Khazaal, H. F. (2021, June). An overview of mmWave communications for 5G. In 2021 International Conference on Communication & Information Technology (ICICT) (pp. 133–139). IEEE. doi: 10.1109/ICICT52195.2021.9568459.

[6] Zugno, T., Drago, M., Giordani, M., Polese, M., & Zorzi, M. (2020, December). NR V2X communications at millimeter waves: An end-to-end performance evaluation. In GLOBECOM 2020–2020 IEEE Global Communications Conference (pp. 1–6). IEEE. doi: 10.1109/GLOBECOM42002.2020.9348259.

[7] Vinodh Kumar, M., Sasibhushana Rao, G. (2021). Performance analysis of Multiple Antenna Systems with New Capacity Improvement Algorithm for MIMO based 4G/5G Systems. In: Hussain Al-Rizzo, ed. Antenna Systems. IntechOpen publisher.

[8] Yamamoto, A., Ogawa, K., Horimatsu, T., Kato, A., & Fujise, M. (2008). Pathloss prediction models for intervehicle communication at 60 GHz. *IEEE Trans. Veh. Technol.*, 57(1), 65–78.

[9] Minchula, V. K., & Rao, G. S. (2018). SVD-based IWFA for next generation wireless MIMO communication. ICT Express, 4(3), 171–174.

[10] Boban, M., Dupleich, D., Iqbal, N., Luo, J., Schneider, C., Müller, R., ... & Thomä, R. S. (2019). Multi-band vehicle-to-vehicle channel characterization in the presence of vehicle blockage. IEEE Access, 7, 9724–9735.

[11] Minchula, V. K., Rao, G. S., & Vijay, C. (2018). Multi-Antenna system Performance under ICSIT and ICSIR channel conditions. Springer Advances in Intelligent Systems and Computing, 695, 505–511.

[12] Amendola, D., et al. (2019). Deliverable D2.1 5G CARMEN use cases and requirements. 5GCARMEN, Rep. D2.1. Accessed: Jan. 31, 2021.

[13] Minchula, V. K., & Rao, G. S. (2018). Investigation of Optimal Cyclic Prefix length for 4G fading channel. Springer Advances in Intelligent Systems and Computing, 695, 429–435.

[14] Dong, K., Mizmizi, M., Tagliaferri, D., & Spagnolini, U. (2022). Vehicular Blockage Modelling and Performance Analysis for mmWave V2V Communications. ICC 2022 - IEEE International Conference on Communications. Seoul, Korea, Republic of, pp. 3604–3609

[15] Minchula, V. K., & Rao, G. S. (2018). SAC channel effects on MIMO wireless system Capacity. Springer Microelectronics, Electromagnetics and Telecommunications, 521, 759–765.

[16] Cassillas, A. M., Kose, A., Lee, H., Foh, C. H., & Yen Leow, C. (2023). Contextual Multi-Armed Bandit based Beam Allocation in mmWave V2X Communication under Blockage. 2023 IEEE 97th Vehicular Technology Conference (VTC2023-Spring), Florence, Italy.

[17] Minchula, V. K., & Rao, G. S. (2018). Carrier frequency Offset impact on LTE-OFDM systems. Springer Microelectronics, Electromagnetics and Telecommunications, 521, 401–408.

[18] Shayea, I., Ergen, M., Azmi, M. H., Çolak, S. A., Nordin, R., & Daradkeh, Y. I. (2020). Key challenges, drivers and solutions for mobility management in 5G networks: A survey. IEEE Access, 8, 172534–172552.

[19] Minchula, V. K., Rao, G. S. (2018). Novel Schemes for minimizing PAPR in LTE-OFDM System. *IEEE International Conference on Recent Innovations in Electrical, Electronics & Communication Engineering, School of Electrical Engineering.* KIIT- Bhubaneswar.

46 Loss Reduction and Bus Voltage Improvement in Distribution Network under Unbalanced Conditions using Cat Swarm Optimization Algorithm

Chava Hari Babu[1] and R. Hariharan[2,]*

[1]Research Scholar, Department of Electrical and Electronics Engineering
Saveetha School of Engineering, Saveetha Institute of Medical and Technical Sciences, Chennai, India
[2]Associate Professor, Department of Electrical and Electronics Engineering
Saveetha School of Engineering, Saveetha Institute of Medical and Technical Sciences, Chennai, India

Abstract

This paper emphasize the network reconfiguration of electrical power networks used in unbalanced systems with the aim to reduce the power losses and improve the voltage profile at each bus. Distribution systems compute power flows in each bus using a forward and backward method. The optimal string is discovered through randomized information exchange across strings via crossover and mutation. Network reconfiguration is a challenging optimization issue involving operational limitations and binary choice variables. Studies have suggested that heuristic approaches perform most quickly with acceptable degree of approximation, we introduce Cat Swarm Optimization for reconfiguration of distribution network for unbalanced system, a new swarm intelligence technique (CSO). In order to optimize cat swarms, we first split the two main behaviours of cats into two smaller models: tracking mode and seeking mode. Better performance can be presented by CSO by combining these two modes in a user-defined ratio. The solution process established by Cat Swarm Optimization Algorithm (CSO) is used for 33 bus radial distribution system. According to research, there is a way to satisfy all of the constraints and appear to give less power loss by changing the pattern of opening or closing.

Keywords: Radial distribution system, feeder reconfiguration, unbalanced system, cat swarm optimization

1. Introduction

A distribution system's losses might range from 5 to 13 percent of the total power produced. Therefore, reducing active power loss in distribution systems is crucial to enhancing the overall effectiveness of the electrical power network. There are numerous strategies to lower the losses, including network reconfiguration, load control, capacitor placement, distributed generation, and more. The distributed generating

Email: haribabu.jntuk@gmail.com[1], hariharan@saveetha.com[2]

DOI: 10.1201/9781003661917-46

technology is receiving increased attention as a result of the widespread concern over the energy problem and global warming. Tie and sectionalizing switches are common in power distribution systems, and their states influence the network's topological topology. The arrangement of these feeders can be altered to facilitate the manual or automatic switching of loads between them.

Reconfiguring the network can help decrease power losses when operating conditions change, as long as protective devices are kept in appropriate coordination and technical operational restrictions are respected. It is highly improbable that a single network architecture will be ideal at all times across a lengthy time horizon due to the high degree of uncertainty surrounding future network conditions. As a result, it is required to periodically modify the distribution network

2. Literature Review

A power-flow strategy for a 3-φ radial distribution network was proposed by Thukaram et al. (1999). They put the recommended method to the test on a number of distribution systems with high R/X ratios and voltage levels [1]. A load-flow solution for imbalanced radial distribution networks was presented by Teng and Chang in 2002. The distribution equipment model might be simply implemented using their suggested method [2]. After comparing the suggested method's outcomes with those of other top approaches now in use, they concluded that the suggested strategy was appropriate for large-scale distribution setups.

Subrahmanyam and Radhakrishna (2009) approach, which relied on the fewest possible equations, resulted in an effective configuration in terms of power losses during network reconfiguration [3]. In order to obtain an effective network design that reduced losses and improved voltage profiles, Rao et al. (2011) proposed an approach for network reconfiguration utilizing a harmony search algorithm

(HSA). The technique eliminated the derivative information by avoiding gradient search and relying solely on random search [4]. The outcomes showed great promise for networks with a balanced composition. After network reconfiguration, Swarnkar et al. (2011) described an effective technique for obtaining a new topological structure of radial distribution systems using Adapted Ant Colony Optimization (AACO) [5], which decreased the system's real power losses.

According to the literature review above, the network reorganization approach necessitates rigorous numerical computation and involves a decision-making process. For the nature of the problem, a heuristic-based method has been presented in the literature. As a conclusion, it appears that a simple CSO Algorithm based reconfiguration strategy for unbalanced systems is required. In this study, for the loss reduction reconfiguration of unbalanced distribution networks, a CSO Algorithm is suggested The problem is under system constraints which are i) load point limits, ii) radial configuration format, iii) no load point distortion and iv) feeder limitations. This is how the remaining portion of the paper is organized: Segment II gives the analysis of Problem formulation. Segment III includes a new Cat Swarm Optimization (CSO). Segment I describes application of CSO for Unbalanced distribution system of network reconfiguration. Segment V exhibits simulation results and segment VI gives conclusions.

3. Problem Formulation

In distribution system, sectionalizing switches and a sequence of tie switches connects it to neighboring feeders. The difficulty of determining where to sectionalize switches in order to minimize distribution losses as much as possible while satisfying the following requirements.

In the 3-φ distribution system, the operation function is specified as follows:

$$\text{Min } T = \sum_{k=1}^{Nl} P_k \tag{1}$$

3.1 Forward- Backward Sweep Load Flow Method

The forward/backward sweep technique is used in numerous power flow systems. Power summation methods and current summation methods are two types of summing procedures.

3.2 Backward Sweep

Backward sweeps are employed to adjust branch currents by incorporating the voltage values from the previous iteration at each node. Starting from the farthest branch, the backward sweep proceeds toward the initial point of the forward path.

The single line diagram of the three-phase system is shown in Fig. 46.1. The parent branch current feeds the sub-laterals attached to the parent branch as well as the load at the qth node. Equation can be used to calculate this current (2).

$$I_k^{abc^j} = IL_q^{abc^j} + \sum_{m \in M} I_m^{abc^j} +$$
$$\sum_{m \in M} \left(Y_{sh_m}^{abc^j} \right) * V_{q_m}^{abc^{j-1}} \qquad (2)$$

The capacitor current must be added if a bank of capacitors is positioned.

3.3 Forward Sweep

The values of the voltages at every node in the network, starting with the source node, are ascertained using the forward sweep method. The voltages at the other nodes are calculated as

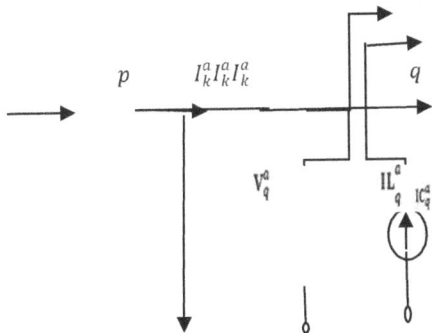

Figure 46.1 Single line diagram of three phase system

$$V_q^{abc^j} = V_p^{abc^j} + Ze_k^{abc} \left(Y_{sh}^{abc} V_p^{abc^j} - I_k^{abc^j} \right) \qquad (3)$$

Where

$V_q^{abc^j}, V_p^{abc^j}$ In the kth iteration, are the voltage vectors.

$$Ze_k^{abc} = \left[ze_k^{aa}\ ze_k^{ab}\ ze_k^{ac}\ ze_k^{ba}\ ze_k^{bb} \right.$$
$$\left. ze_k^{bc}\ ze_k^{ca}\ ze_k^{cb}\ ze_k^{cc} \right] \qquad (4)$$

These calculations will continue until the voltage on each bus falls within the required range. They are utilised to calculate the power losses in each branch. As a result, the losses are represented in the network as:

$$S_k^{abc} = \left(V_p^{abc} - V_q^{abc} \right) \left(I_k^{abc} \right)^* \qquad (5)$$

4. Cat Swarm Optimization (CSO) Algorithm

Any optimization algorithm will require some form of representation for the solution set. For example, in CSO, ants are used as agents, with the paths they create representing the solution sets, while in GA, chromosomes are employed to represent the solution sets. In our proposed algorithm, we address optimization challenges by using cats and their behavioral model, with cats serving as the representation of the solution sets.

4.1 Application of CSO for Network Reconfiguration in Unbalanced Distribution System

The application of CSO for the existing problem is explained .Through studying their actions, we may conclude that cats spend the majority of their waking time sleeping. They move slowly and with caution when they are sleeping, and occasionally they even remain in their original positions. In order to apply this behavior to CSO, we represent it somehow through the searching mode.

The six steps that make up the CSO process are as follows:

Step1: In the process, create N cats.

Step2: The cats should be randomly distributed throughout the M-dimensional solution space.

Step 3: Enter the positions of each cat, which stands for the requirements of our aim, to examine its fitness value. Then, note which cat is the best. Remember that the best cat (xbest) is the best response thus far, so we just need to note where it is.

Step 4: Sort the cats based on their flags; if they are in the searching mode, move them to it; Transfer them to the tracing mode if not. The steps in the process are shown above.

Step5: As directed by MR, pick a fresh set of cats

Step 6: Verify that the prerequisites for terminating the program have been fulfilled; if not, move on to steps three through five.

5. Simulation Results

The suggested method's performance is evaluated on three phase unbalanced radial distribution system test systems in order to minimise power loss due to network reconfiguration.

An IEEE 33-bus radial distribution system with 37 branches is used to illustrate the recommended approach. The open switches in the network's loops L1 through L5 are 33, 34, 35, 36, and 37, as shown in Figure 46.2.

Figures 46.3, 46.4 and shows the power loss at each bus and Tables 46.1 and 46.2 shows the summary of test results of a 33 node distribution network under unbalanced conditions before and after network reconfiguration. In phases a, b, and c, the

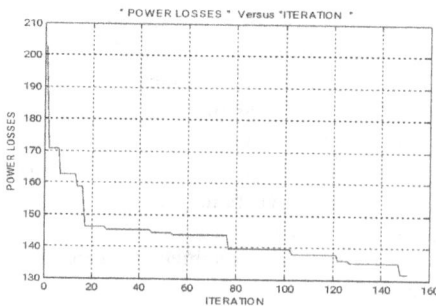

Figure 46.2 Active power loss (phase a) vs Generation number

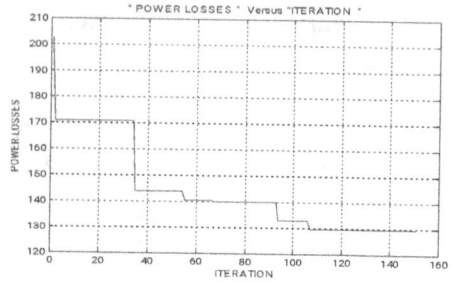

Figure 46.3 Active power loss (phase b)

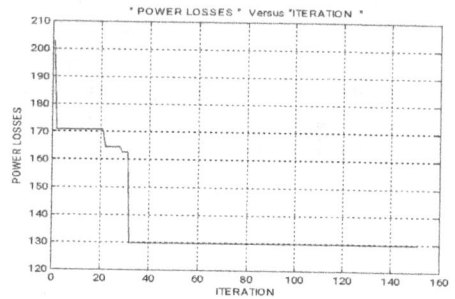

Figure 46.4 Active power loss (phase c)

Table 46.1 Test Results Before Reconfiguration

Summary	Results at Different Phases		
	Phase a	Phase b	Phase c
Switches that are open	33 35	34 37	36
Voltage at Bus (Minimum)	0.9131	0.9132	0.9142
Power loss	235.12	227.58	259.75

Table 46.2 Test Results After Reconfiguration

Summary	Results at Different Phases		
	Phase a	Phase b	Phase c
Switches that are open	7 9	14 25	32
Voltage at Bus (Minimum)	0.9413	0.9418	0.9333
Power loss	131.81	132.95	129.66
Loss reduction (%)	43.93	41.58	50.08

active power loss is reduced and the minimum voltage is enhanced from 0.9131, 0.9132, and 0.9142 to 0.9143, 0.9418, and 0.9333 as shown in Tables 46.1 and 46.2.

6. Conclusion

This paper provides an illustration of the CSO Algorithm-based optimisation technique, to find the voltages at each node and power flows in each branch in appropriate neighborhood foundation for unbalanced distribution system. The amount of CPU time required for switching operations is decreased as a result of the elimination of needless switching activities. Through the use of unbalanced radial distribution test systems, the effectiveness of the suggested method is demonstrated. The 3-phase power flow, bus electricity restrictions, cutting-edge feeder distribution, radial arrangement format, and no load factor distortion are among the main features. This is particularly helpful because the suggested strategy is observed faster than methods with less switching operations in this paper. The proposed technique the switching options were determined to be less than those determined by existing methods.

References

[1] Goswami, S. K., & Basu, S. K. (1992). New algorithm for the reconfiguration of distribution feeders for loss minimization. *IEEE Trans. Power Deliv.*, 7(3), 1484–1491.

[2] Wagner, T. P., Chikhani, A. Y., & Hackam, R. (1991). Feeder reconfiguration for loss reduction. *IEEE Trans. Power Deliv.*, 6(4), 1922–1933.

[3] Hong, Y. Y., & Ho, S. Y. (2005). Determination of network configuration considering multi-objective in distribution systems using genetic algorithms. *IEEE Trans. Power Syst.*, 20(2), 1062–1069.

[4] Chiou, J. P., Chung, C. F., & Su, C. T. (2005). Variable scaling hybrid differential evolution for solving network reconfiguration of distribution systems. *IEEE Trans. Power Syst.*, 20(2), 668–674.

[5] Mendoza, J., Lopez, R., Morales, D., Lopes, E., Dessante, P., & Moraga, R. (2006). Minimal loss reconfiguration using genetic algorithms with restricted population and addressed operators: real application. *IEEE Trans. Power Syst.*, 21(2), 948–954.

47 Hardware in the Loop (HIL) Implemented for an Investigation to Unity Input Power Factor in Presence of Permanent Magnet Synchronous Motor Drive

R. Hariharan[1] and M. Deva Darshanam[2,]*

[1]Associate Professor, Department of Electrical and Electronic Engineering, Saveetha School of Engineering, Saveetha Institute of Medical and Technical Sciences, Chennai, India
[2]Research Scholar, Department of Electrical and Electronic Engineering, Saveetha School of Engineering, Saveetha Institute of Medical and Technical Sciences, Chennai, India

Abstract

This paper presents a voltage control with Direct Current Control (DCC) method designed to maintain unity input power factor, using a vector-controlled Permanent Magnet Motor (PMM) as the load. A three-phase rectifier with a Proportional-Integral (PI) controller is employed at the input side, and the PMSM drive is operated through an inverter controlled by Space Vector Pulse Width Modulation (SVPWM) with an additional PI controller. The proposed method consists of two stages: first, by varying the rectifier load, the system's ability to maintain unity power factor and stabilize the rectifier's DC output voltage is evaluated. In the second stage, the DC voltage is converted into AC using the SVPWM-based inverter to power the PSM. The execution of the PMSM drive is analyzed under variable test conditions, including constant and variable loads, to observe input power factor variations. The detailed parameter design of the rectifier, DCC controller, and SVPWM-based inverter is modeled in SIMULINK/MATLAB, and a (hardware-in-the-loop) OPAL RT-4510 setup is tested on a prototype. Experimental results from both transient and steady-state conditions demonstrate the successfulness and excellent performance of the proposed method.

Keywords: Rectifier, Direct current control (DCC), Hardware In Loop (HIL), Permanent magnet motor (PSM)

1. Introduction

The most common power conversions is from AC to DC, known as rectification, while DC to AC conversion is referred to as inversion. To achieve variable frequencies, cycloconverters are used. During conversion, switches must operate instantaneously for short durations to meet the demands of the applied load. However, variable loads

Email: harinov22@gmail.com[1], devadarshanam12@gmail.com[2*],
Corresponding Author: devadarshanam12@gmail.com[2*]

DOI: 10.1201/9781003661917-47

and switching actions can impact the power factor at the input supply. To achieve a unity power factor, researchers have developed several techniques are Sinusoidal Pulse Width Modulation (SPWM), Hysteresis Loop Control, Space Vector Pulse Width Modulation (SVPWM), Fuzzy Logic Control, Neural Network (ANN) and Direct Current Control (DCC).Among these, Direct Current Control (DCC) is particularly popular one.

The operation rectifier is involving current control and sensing the output current, transforming it into d-q axis voltages based on the reference voltages. To handle this transformation, the DCC method utilizes two loops—an outer loop and two inner loops—each equipped with proportional-integral (PI) controllers.

The outward PI controller generates the reference current for the d-axis I_d while the interior loop PI controllers determine the d-axis V_d and q-axis V_q voltages and got gating pulses of rectifier bridge.

A control methodology is essential, and the PI controller is one of the most commonly used approaches. The load used is a Permanent Magnet Motor (PMM) drive with PI controller and controlled using the vector control method [1].

The input to the PMM motor is AC, which undergoes a two-stage conversion process. In this AC to DC and DC to AC. During the AC to DC conversion stage, the Direct Current Control (DCC) method is employed using a 3-leg, six-switch configuration. In this reference DC voltage with the actual voltage and transformed into I_d and I_q currents using PI controllers. These currents are subsequently converted into V_d and V_q voltages and generated gate with pulses for the rectifier.

The DC is converted to AC with inverters in second stage and control PMM using space vector control. In this paper, various test conditions are conducted on the PMM drive, including constant speed, variable speed, and variable load scenarios. Under these conditions, the effect of the input power factor is analyzed and analyzed the PMM execution under current harmonics and torque behavior [2].

2. Direct Current Control

2.1 Three Phase PWM Rectifier model

It converts three phase AC voltage to DC voltage with different switching times and distinct loads. That is shown Figure 47.1. So, the mathematical modeling of rectifier is, the phase voltages are u_a, u_b and u_c, phase currents are i_a, i_b and i_c, the load current is i_L and DC output voltage is V_{dc}.

The rectifier modeling dynamic are equation in parks transformation are

$$L\frac{di_d}{dt} = u_d - i_d R_1 + \omega Lq - u_{rd} \quad (1)$$

$$L\frac{di_q}{dt} = u_q - i_q R_1 + \omega Li_d - u_{rq} \quad (2)$$

$$C\frac{dV_{dc}}{dt} = -\frac{V_{dc}}{R_L} + \frac{3}{2}\left(S_d i_d + S_q i_q\right) \quad (3)$$

$u_{rd} = s_a v_{dc}$, $u_{rq} = s_q v_{dc}$, $u_{rd} = u_{rq}$, Synchronous-form rectifier voltages. s_d, s_q Switching functions in the synchronous rotating frame.Angular frequency. The Proportional-Integral (PI) controller for current regulation and the controlled variables, u_{rd} and u_q are computed using the following equations:

$$u_{rd} = -\left(K_{ip} + \frac{Kil}{S}\right)\left(i_d^* - i_d\right) + \omega L_q + u_d \quad (4)$$

$$u_{rq} = -\left(K_{ip} + \frac{Kil}{S}\right)\left(i_q^* - i_q\right) + \omega L_d + u_q \quad (5)$$

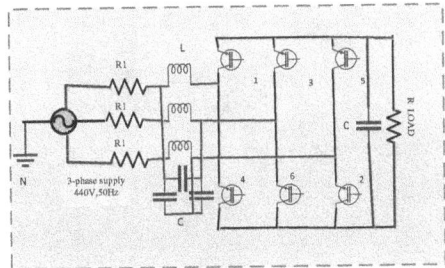

Figure 47.1 PWM rectifier

Source: Author's compilation

2.2 Design of Direct current control loop

The current controller regulates the current vectors I_d and I_q. These vectors are derived from the voltage components V_d and V_q. The output DC voltage is compared with a reference DC output voltage and error is treated through a PI controller. PI controller produced the direct-axis current reference I_d^*. This process is illustrated in Figure 47.2. Here I_d^* can be expressed

$$I_d^* = K_{ip}\left(v_{dc}^* - v_{dc}\right) + \int K_i I\left(V_{dc}^* - V_{dc}\right) dt \quad (6)$$

The actual current I_d is obtained by from I_{abc} phase currents with phase angle using transformation with Phase-Locked Loop (PLL).

The actual I_d compared with the reference direct-axis current I_d^* and generated the direct axis modulating voltage Urd.For operating the rectifier, the quadrature-axis current is $I_q = 0$ and current loop equation can be defined [3].

$$I_d^* - I_d = U_{rd} \text{ and } I_q = U_{rq}$$

The generated currents are generated the pulses using SPWM control.

3. Priciple of Svpwm

In Space Vector Pulse Width Modulation, voltage vectors represented in reference circle. The inverter is shown in Figure 47.3,

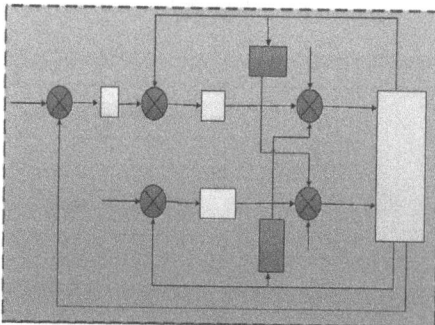

Figure 47.2 d-q dual closed-loop Direct current controller of rectifier

Source: Author's compilation

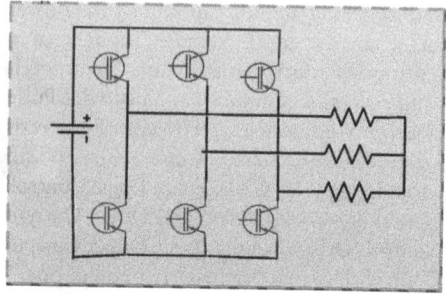

Figure 47.3 Three-phase inverter diagram

Source: Author's compilation

each phase is represented by a switch, denoted as A [4]. The switching functions for the three phases are represented as $A_A(t)$, $A_B(t)$ and $A_C(t)$ respectively.

The space vector of output voltage are expressed as $V(A_A, A_B, A_C) = 2V_{dc}$ $(A_A + \alpha A_B + \alpha^2 A_C)/3$, $\alpha = e^{j120.}$ and V_{dc} is bus voltage. In this context, the upper-stage arm being ON is denoted as "1," and the OFF state is denoted as "0.". The variable T refers to the conduction times within the same sector [8]. The active voltage space vectors, comprising W_4, W_6, W_0 and W_7 can be determined accordingly by

$$W_{out} = T_4 W_4 / T + T_6 W / T$$

4. Modelling of PMM

The d-q model of PMM with linear transformation are

$$\left|V_q\, V_d\right| = \left[R_q + L_q P\,\omega_r\ \ L_d - \omega_r\, L_q\, R_d + L_d P\right]$$

$$= \left|I_q\, I_d\right| + \left[P\omega_r\,\psi_f\ 0\right] \quad (7)$$

$$T_e = \frac{3P}{2}\left[\psi_f I_q + \left(L_d - L_q\right) I_d I_q\right] \quad (8)$$

$$T_e = T_L + J_m\,\rho\omega_r + B_m\,\omega_r \quad (9)$$

These models are used by researchers and consider id = 0, Ψ is control from flux then resultant equations [5]

$$T_e = \frac{3P}{2}\left[\psi_f I_q\right] \quad (10)$$

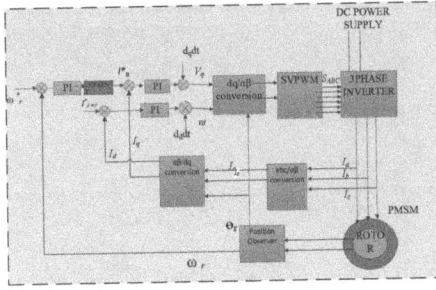

Figure 47.4 Diagram of closed-loop control of PMM motor

Source: Author's compilation

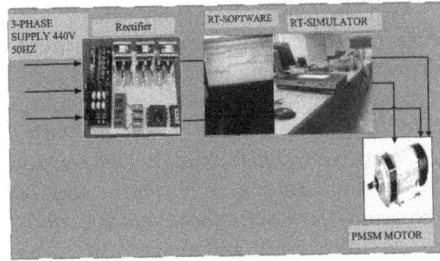

Figure 47.5 Real time implementation hardware setup

Source: Author's compilation

$$T_e = \frac{3P}{2} \left| \psi_f \, I_q + \left(L_d - L_q \right) I_d \, I_q \right| \qquad (11)$$

$$V_d = -P\omega_r L_q I_q \qquad (12)$$

$$PI_q = \frac{1}{L_q} \left[V_q - R_s \, I_r - \rho \, \omega_r \psi_f \right] \qquad (13)$$

This limitation is illustrated in Figure 47.4.

5. OPAL-RT 4510 Hardware Implementation

The OPAL-RT 4510 serves as the central simulation unit, controlling the rectifier and PMSM motor with high-performance processors and optional FPGA modules for precise, high-speed dynamic simulations. It supports analog/digital I/O channels and communication protocols like Ethernet, RS232, CAN, and Modbus for interfacing with external systems. A host computer connects via Ethernet to act as the user interface for model upload, control, and monitoring.

The Device Under Test (DUT) interacts with the simulator through I/O channels or protocols, receiving simulated inputs and providing outputs under real-world conditions. Signal conditioning ensures compatibility, adjusting voltage/current levels, providing isolation, and filtering noise. Measurement tools such as oscilloscopes and DAQs validate system performance in real-time or through offline analysis. Simulation step size is optimized

for real-time execution, with physical I/O ports mapped to model variables. The setup can include specialized components like grid simulators or custom firmware for applications such as renewable energy systems, enabling real-time validation and optimization of control systems, power electronics, and embedded hardware. Figure 47.5 an explanation of the key elements involved in OPAL-RT hardware implementation.

6. Hardware Implemented Results and Discussions

The proposed PI controllers demonstrates significant practical relevance for industrial applications involving PMM drives by ensuring efficient energy conversion and precise control through a PWM rectifier model employing Direct Current Control and SVPWM. The model consists of The AC Source Operates at 50 Hz with a phase-to-phase voltage of 380 V Resistance = 0.01 Ω, Inductance = 5 mH, Output Capacitance = 4700 μF. The DC Voltage Configured to 450 V in steady-state operation. The Rectifier and PMSM PI controllers gain value are $Kp=1.4$, $Ki=10.4$. For the quadrant—axis current: $Kp=3$, $Ki=1$. For the direct—axis current: $Kp=3$, $Ki=1$. These are validated experimentally, maintains a unity power factor on the AC input side, enhancing power quality while complying with industrial standards. In this, the Rectifier Reliable maintenance of a 450 V DC output at a high switching frequency of 10 kHz ensures efficiency and minimal harmonic distortion, reducing operational costs and extending equipment lifespan. The motor reference speed is set to 300 rad/s, with a step time of 0.0001 s and at T=0.5 s, and Motor load torque of 5 Nm is applied. Experimental results show that the motor successfully starts and reaches the reference speed, with minimal rotor fluctuations. Whose major parameters motor are Power is110W to 550W, inertia is 0.8 kgm² Rated load torque is 8 Nm, motor voltage range is 75 V to 390 Rated stator current 9 A, motor stator resistance is 2.775 Ω, motor stator inductance is 8.5 MH, Rated motor peed is 1500 rpm, Frequency varying range is 15 Hz to 50 Hz.

Case (i) Variable Torque or Load

A small load was applied to the rectifier using Direct Current control and to the PMM via the inverter using Space Vector Control. When a load was introduced at T=5 ms, the rectifier exhibited transient dynamics in input voltage and currents, as shown in Figure 47.6(d), but quickly stabilized while maintaining unity power factor. Similarly, transient dynamics were observed in the DC output voltage and current, which also stabilized shortly after. In another case, a load change at T=5 ms with a 5 A increase caused the motor speed to drop from 300 RPM but recover rapidly to the reference speed. The motor exhibited transient currents that adapted to the new load, as depicted in Figure 47.6(b), with rotor speed achieving reference speed and unity power factor, shown in Figures 47.6(c) and 47.6(d). These results highlight that the rectifier and inverter operate independently, with the rectifier consistently maintaining a unity power factor. The industrial applications like robotics, manufacturing, and precision control, where PMSM drives are preferred for their high efficiency and excellent torque characteristics. This result provides a robust and scalable framework for improving industrial PMSM drive performance.

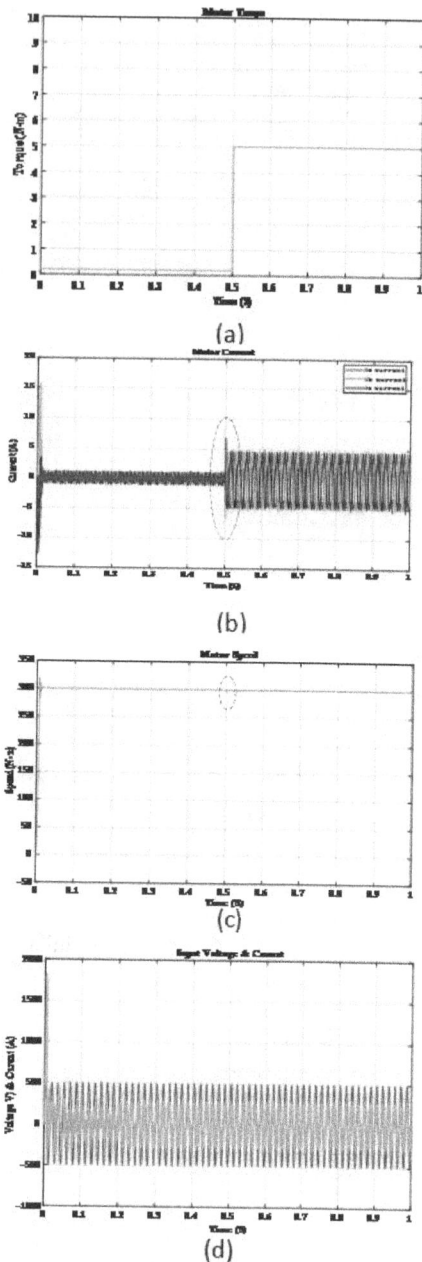

(a)

(b)

(c)

(d)

Figure 47.6 Experimental Performance of DCC method at Variable load (a) PMM Variable torque (b) Three phase currents of PMM (c) Rotor speed of PMM (d). Unity power factor voltage & current at Input side of Rectifier

Source: Author's compilation

7. Conclusion

This study presents advancements in modeling rectifiers, inverters, SVPWM, and PMSMs, enabling analytical and independent controller design with unity power factor under varying loads. A robust rectifier controller using IDA and GSSA addresses load variation challenges, while SVPWM improves DC bus utilization and system stability. Experimental results demonstrate efficient operation, stable steady-state performance, and effective control of PMMs and induction motors.

Conflict of Interest

Authors confirm that no conflicts of work.

References

[1] Wang Jiuhe, Yin Hongren, Zhang Jinlong, and Li Huade, "Study on Power Decoupling Control of Three Phase Voltage Source PWM Rectifiers", Power Electronics and Motion Control Conference, 2006.

[2] Z. Yang, and L. Wu, "A new Passivity-Based Control Method and Simulation for DC/DC Converter", Proceedings of the 5th World Congress on Intelligent Control and Automation, Hangzhou, P. R. China, June 15-19, 2004, pp.5582-5585.

[3] Rastogi, M., Naik, R., and Mohan, N.: 'A comparative evaluation of harmonic reduction techniques in three-phase utility interface of power electronic loads', IEEE Trans. Ind. Appl., 1994, 30, (5), pp. 1149–1155.

[4] Kolar, J.W., and Zach, F.C.: 'A novel three-phase utility interface minimizing line current harmonics of high-power telecommunications rectifier modules', IEEE Trans. Ind. Electron., 1999, 44, (4), pp. 456–467.

[5] R. F. Post, T. K. Fowler, and S. F. Post, "A high-efficiency electromechanical battery," Proceedings of the IEEE, vol. 81, no. 3, pp. 462-474, Mar. 1993.

[6] J. R. Hull, "Flywheels on a roll," IEEE Spectrum, vol. 4, no. 7, pp. 20-25, 1997.

[7] B. Vladimir and K. Vikram, "A new mathematical model and control of a three-phase AC–DC voltage source converter," IEEE Transactions on Power Electronics, vol. 12, no. 1, pp. 116-123, 1997.

[8] R. Wu, S. B. Dewan, and G. R. Slemon, "A PWM AC to DC converter with fixed switching frequency," IEEE Trans. on Industry Applications, vol. 26, no. 5, pp. 880-885, 1990.

48 Testing of Hardware in the Loop to maintain Unity Input Power Factor Based on Artificial Intelligent Controller with Power Electronic Drive

R. Hariharan1 and M. Deva Darshanam[2,*]

[1]Associate Professor, Department of Electrical and Electronic Engineering, Saveetha School of Engineering, Saveetha Institute of Medical and Technical Sciences, Chennai, India
[2]Research Scholar, Department of Electrical and Electronic Engineering, Saveetha School of Engineering, Saveetha Institute of Medical and Technical Sciences, Chennai, India

Abstract

In this paper, the fuzzy logic controller (Artificial Intelligent controllers-AIC) is proposed to control the rectifier and SVPWM and Artificial Intelligent controllers are proposed to inverter with Permanent Magnet motor (PMM) is a load. This work aims to evaluate AI-based controllers to maintain the input power factor with speed regulation accuracy, dynamic response, robustness, and computational complexity in these systems. In this to maintain unity power factor and reactive power control at input side of three phase rectifier used two stages operation. In first stage by varying load on rectifier, check the viability of unity power factor at input side and DC voltage at output side of rectifier. In second stage, first stage DC output voltage is connected input to three phases Inverter with PMSM as with SVPWM control. The PMSM drive is operated at different load conditions such as constant load, variable load and no load to observe variations of input power factor at input side. The detailed parameters design of Fuzzy logic controller, rectifier, inverter, PMM and SVPWM in SIMULINK/MATLAB environment and also test the prototype hardware in loop (HIL). The experimental results in different test conditions show the excellent performance of the proposed method even in both transient and steady states.

Keywords: Rectifier, FLC, direct current control (DCC), permanent magnet motor (PMM), space vector pulse width modulation (SVPWM), Hardware In Loop (HIL)

1. Introduction

In the power converters used driver circuits and switching devices those are Diode, thyrister, MOSFET, IGBT etc. By connecting these switches control the voltage, current, power and power factor [1]. The converter having the resistance and energy storing devices due to this continuous, currents flow in the circuit.

The DC voltage is converted to AC using an inverter, and this AC voltage is

Email: harinov22@gmail.com[1] devadarshanam12@gmail.com[2*],
Corresponding Author:[2*] devadarshanam12@gmail.com

DOI: 10.1201/9781003661917-48

then connected to the Permanent Magnet Synchronous Motor (PMSM). During this process, switches are operated for short durations to allow continuous variation of the applied load. As the switching duration and load vary, the power factor varying and affected. To get a unity power factor at the rectifier input, various techniques are being researched, including: (1) Sinusoidal Pulse Width Modulation (SPWM) and (2) Hysteresis Loop Control (3) SVPWM (4) Artificial Intelligent controllers (5) Artificial neural network (ANN) (6) Direct current control (DCC). Among them Fuzzy logic controller is use for power factor correction at Input side [2].

In this paper two stages conversion, the first stage is rectifying voltage with fuzzy logic controller, second stage is inverting the voltage with AI controller and PMSM as a drive. In first stage of operation the Artificial Intelligent controllers senses the Direct output current and further transformed to d-q axis voltage with given reference voltage. These voltages transform to with Artificial Intelligent controllers and produce the currents. These currents are generating the gate pulses to Rectifier [3].

The PMM drive is operated with constant torque angle control with vector control method, to obtain effective rotor speed, less stator winding current THD and reduced ripples. The gating pluses to inverter fed from Artificial Intelligent controllers with sinusoidal pulse width modulation (spwm) method. That Vα, Vβ as reference vectors are transformed to parks transformation to clarks transformation. The 3 stator currents transformed I_d, I_q with reference speed controller, that I_q current and generate id current, $I_d = 0$ taken as references which are further transformed to V_d, V_q using Artificial Intelligent controller [4].

2. Direct Current Control

2.1 Three Phase PWM Rectifier model

It converts 3-phase AC voltage to DC voltage with varying switching times and different load conditions. Therefore, the

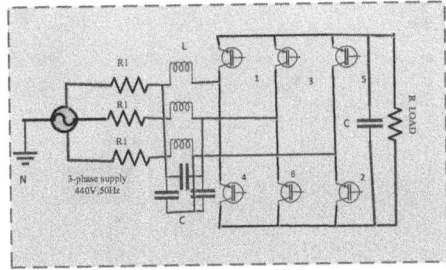

Figure 48.1 PWM rectifier

Source: Author's compilation

mathematical modeling of the rectifier is necessary. So, the mathematical modeling of rectifier, the phase voltages are u_a, u_b, u_c, currents are i_a, i_b and i_c, the rectifier load current is i_L and DC output voltage is V_{dc}. The three phase rectifiers shown in Figure 48.1.

The rectifier modeling dynamic are equation in parks transformation are

$$L\frac{di_d}{dt} = u_d - i_d R_1 + \omega Lq - u_{rd} \qquad (1)$$

$$L\frac{di_q}{dt} = u_q - i_q R_1 + \omega Li_d - u_{rq} \qquad (2)$$

$$C\frac{dV_{dc}}{dt} = -\frac{V_{dc}}{R_L} + \frac{3}{2}\left(S_d i_d + S_q i_q\right) \qquad (3)$$

$u_{rd} = s_d v_{dc}$, $u_{rq} = s_q v_{dc}$, $u_{rd} = u_{rq}$, are synchronous form rectifier voltages s_d, s_q are Switching function in synchronous rotating form. ω Angular frequency. u_d, u_q are voltage source and i_d, i_q are parks current transformation, The AI controller treated to current regulation and and u_{rd}, u_{rq} are controlled from these equations [5].

$$u_{rd} = -K_{fd} fd\left(i_d^* - i_d\right) + \omega L_q + u_d \qquad (4)$$

$$u_{rq} = -K_{fd}\left(i_q^* - i_q\right) + \omega L_d + u_q \qquad (5)$$

2.2 Design of Direct current control loop

The current controller is regulate the current vectors I_d and I_q. The I_d and I_q vectors are generate from the V_d and V_q. The output actual DC voltage is compared to a

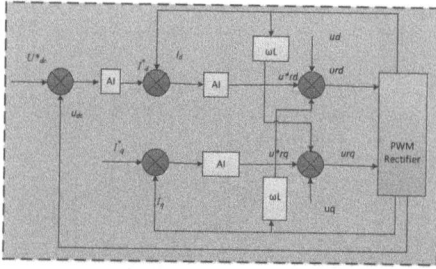

Figure 48.2 d-q dual closed-loop Direct current controller of rectifier

Source: Author's compilation

reference DC voltage and error processed through AI and generated the direct axis current I_d^*, The defuzzified output of of Fuzzy controller is k_{fd} .That is shown Figure 48.2. Here I_d^* can be expressed

$$I_d^* = K_{fd}\left(v_{dc}^* - v_{dc}\right) \qquad (6)$$

The modified signal is then compared with a triangular carrier signal to generate the SPWM (Sinusoidal Pulse Width Modulation) switching pulses [6].

3. Proposed Method of Artificial Intelligent Controller

A FL controller for a rectifier and PMSM combines intelligent control strategies for power conversion and motor operation. In the rectifier, the FLC regulates the DC output voltage and improves power factor by reducing harmonics and ensuring stable operation. For the PMSM, the FLC controls speed, torque, or position by dynamically adjusting the torque- producing current I_q while maintaining efficiency through unity power factor $I_d = 0$. This integrated FLC approach provides robust, adaptive, and efficient control for applications demanding precise motor performance. While control the speed of PMM, the following steps or Algorithms are implemented for Fuzzy logic controller [7].

Step 1: Define the control objective for PMSM motor speed control, specifying the desired set point and the need to maintain it under varying load conditions

Step 2: identify input variables such as speed error and changein speed error, for the Increase or decrease sped

Step 3: Design membership functions for speed error, changein speed and output variables as speed, using shapes Gaussian, and divide the ranges into linguistic terms like Negative Big, Negative medium, Negative Small, Zero, Positive Small, Positive medium and Positive Big

Step 4: Develop a fuzzy rule base with "if-then" rules, for example: "If speed error is Positive Big and change in speed error is Positive Small, then Decrease Speed," and "If speed error is Zero, then Maintain Speed

Step 5: Fuzzify the inputs by mapping crisp speed error and changein speed error values to their respective fuzzy sets using the defined membership functions

Step 6: Applied the inference mechanism to evaluate fuzzy rules using algorithms Mamdani inference to compute fuzzy outputs or for crisp outputs

Step 7: Aggregate the outputs of all rules into a combined fuzzy output, determining the overall control action using methods max-min product composition

Step 8: Defuzzify the aggregated fuzzy output to obtain a crisp control signal using techniques such as Centroid of Area.

Step 9: Train and optimize the controller by manually tuning membership functions and rules.

From this, Validate the controller by testing it under various operating scenarios, including changes in load and reference speed, and measure metrics like speed tracking accuracy, stability, and response time

In this 7 membership input speed variables are used and got total 49 rules. These rules are shown Table 48.1. The rules are same rectifier as well as PMM motor.

The Importance of FLC in current scenario is high efficiency, robustness, and adaptability are essential in various industrial and commercial applications. Below are the key reasons why FLC is particularly

Table 48.1 PMSM Speed Fuzzy Rules

ΔT(n)		(ω_r) Speed Error						
		NB	NM	NS	ZE	PS	PM	PB
Δ(ω_r) Change in Speed Error	NB	NB	NB	NB	NS	NS	NS	ZE
	NM	NB	NB	NM	NM	ZE	NS	NS
	NS	NB	NM	NS	ZE	PS	PM	PM
	ZE	NB	NM	NS	ZE	PS	PM	PB
	PS	NS	NS	ZE	ZE	PS	PM	PB
	PM	NM	ZE	PS	PS	PM	PB	PB
	PB	PS	PS	PM	PM	PB	PB	PB

Source: Author's compilation

important for PMM drives today. Those are nonlinear handling, improved performance in dynamic conditions, efficiency and power factor control, reduced need of mathematical models, cost effective, high stability and reliability in the industries.

Table 48.1. Shows the PMSM Speed Fuzzy Rules which is considered for the process.

4. Priciple Operation of SVPWM

In Space Vector Pulse Width Modulation, voltage vectors represented in reference circle. The inverter is shown in Figure 48.3, each phase is represented by a switch, denoted as A. The switching functions for the three phases are represented as $A_A(t)$, $A_B(t)$ and $A_C(t)$ respectively. The space vector of output voltage are expressed as

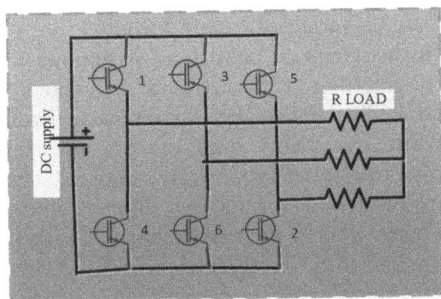

Figure 48.3 Three-phase inverter diagram

Source: Author's compilation

$V(A_A, A_B, A_C) = 2V_{dc} (A_A + \alpha A_B + \alpha^2 A_C)/3$, $\alpha = e^{j120.}$ and V_{dc} is bus voltage. In this context, the upper-stage arm being ON is denoted as "1," and the OFF state is denoted as "0.". The variable T refers to the conduction times within the same sector. The active voltage space vectors, comprising $\mathbf{W}_4$, $\mathbf{W}_6$, $\mathbf{W}_0$ and $\mathbf{W}_7$ can be determined accordingly by

$$\mathbf{W}_{out} = T_4\mathbf{W}_4/T + T_6\mathbf{W}/T$$

5. Modelling of PMM

The d-q model of PMM with linear transformation are

$$\left[V_q V_d \right] = \left[R_q + L_q P \omega_r L_d - \omega_r L_q R_d + L_d P \right]$$
$$= \left[I_q I_d \right] + \left[P \omega_r \psi_f 0 \right] \quad (10)$$

$$T_e = \frac{3P}{2} \left[\psi_f I_q + (L_d - L_q) I_d I_q \right] \quad (7)$$

$$T_e = T_L + J_m \rho \omega_r + B_m \omega_r \quad (8)$$

These models are used by researchers and consider id = 0, Ψ is control from flux then resultant equations

$$T_e = \frac{3P}{2} \left[\psi_f I_q \right] \quad (9)$$

$$T_e = \frac{3P}{2} \left[\psi_f I_q + (L_d - L_q) I_d I_q \right] \quad (10)$$

$$V_d = -P \omega_r L_q I_q \quad (11)$$

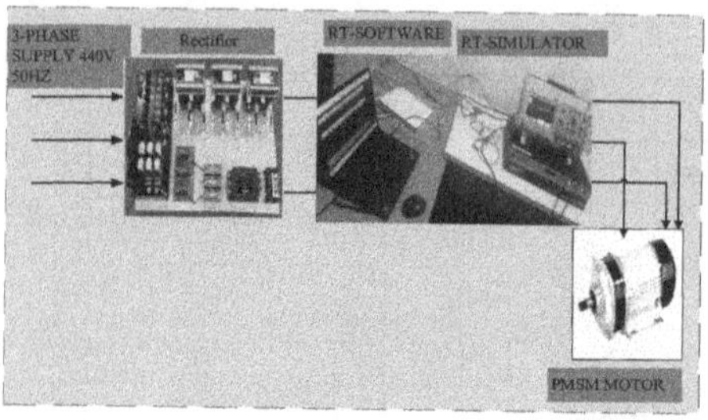

Figure 48.4 Block diagram of closed-loop control of PMSM motor

Source: Author's compilation

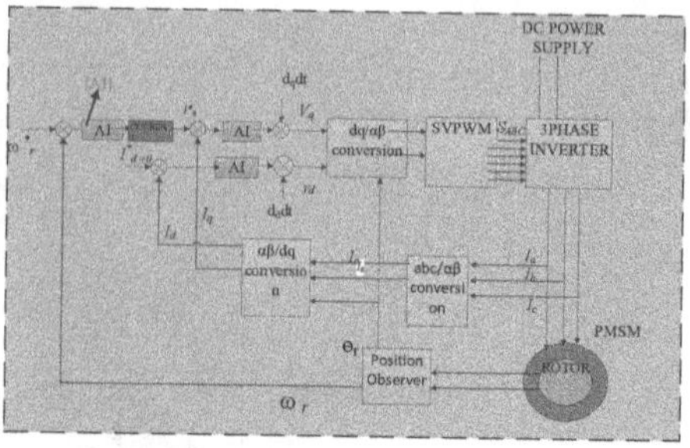

Figure 48.5 Hardware setup of real time implementation

Source: Author's compilation

$$PI_q = \frac{1}{L_q}\left[V_q - R_s I_r - \rho\omega_r \psi_f\right] \qquad (12)$$

From equation results, the motor may not reach its rated speed under rated conditions and show in Figure 48.4.

6. OPAL-RT Hardware Testing:

In this HIL setup, OPAL-RT simulates rectifier and PMSM using a Simulink-based model that includes d-q axis stator voltage equations, electromagnetic torque, and rotor dynamics. Unity power factor control is achieved by setting the d-axis current reference to zero. The simulator also models the inverter and load for a complete system representation.

The physical controller (DSP or FPGA) executes motor control algorithms, including a speed AI controller, using feedback signals senses with sensors like rotor position, speed, and stator currents Digital IO cards, Analog cards from OPAL-RT. The controller outputs SVPWM signals to the simulator, which calculates the applied voltages to the motor model. OPAL-RT ensures real-time execution with deterministic synchronization, enabling accurate system performance. That is show in Figure 48.5.

7. Hardware implemented Results and Discussion

The research model uses a PWM rectifier and an eight-vector inverter implemented in MATLAB/Simulink with a hardware-based closed-loop controller. Experimental results showed a steady 500V DC output, unity power factor, and strong dynamic response with 380V, 50Hz AC input, 0.01Ω resistance, 5mH inductance, 4700μF capacitance, and 10kHz switching frequency. Speed regulator and q-axis current outputs align with model characteristics. During testing, the motor ran for 0.4 seconds, starting unloaded and transitioning to a 3Nm load at 0.2 seconds. It reached reference speed with minimal rotor fluctuations, validating performance and agreement with parameters. Whose major parameters motor are Power range is 110W to 560W, motor Moment of inertia is 0.8 kgm²·Rated load torque is 7 NM, Voltage range is 75 V to 390 Rated current 10 A, motor Stator resistance is 2.775 Ω, motor Stator inductance is 8.4 MH, motor rated speed is 1500 rpm, Frequency varying range is 10 Hz to 50 Hz.

Case (i)-Variable Load or Torque
In this setup, a small load is applied to the rectifier with artificial intelligence controllers using direct current control, while the PMM load on the inverter operates under space vector control. The experimental results are illustrated in Figure 48.6. In the first scenario, a load is applied to the rectifier at t = 0.2 seconds, affecting the DC output voltage and input power factor, as shown in Figures 48.6(d) and 48.6(e), with the eclipse symbol

marking torque changes at t = 0.5 seconds, causing a slight decrease in motor speed from 700 RPM and an increase in motor current. The input power factor remains stable, as depicted in Figure 48.6(d), with transients highlighted by the eclipse symbol in Figures 48.6(c), 48.6(d), and 48.6(e). From these results, the Fuzzy logic controller is more fast and most suitable for dynamic and steady state conditions in the high-power Industries.

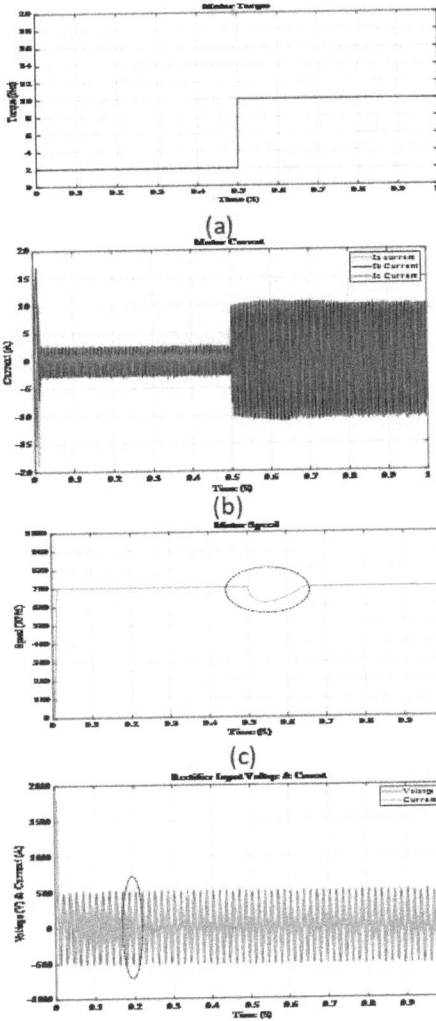

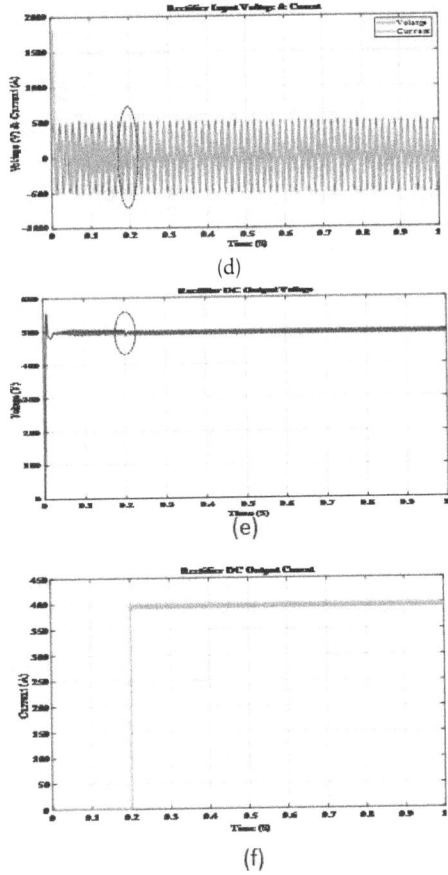

Figure 48.6 Experimental performance of DCC method at variable load

(a) PMM Variable torque (b) Three phase currents of PMM (c) Rotor speed of PMM (d) Unity power factor voltage & current at Input side of Rectifier (e) Rectifier Output DC Voltage (f) Rectifier Output DC current

Source: Author's compilation

8. Conclusion

This paper presented the design and modeling of a rectifier, inverter, Artificial Intelligence (AI) controllers, SVPWM, and PMSM, incorporating Fuzzy Logic Controllers (FLC) for improved adaptability under varying loads. Experimental results highlight challenges in maintaining stable DC voltage regulation due to load fluctuations. The proposed IDA and GSSA approaches, enhanced with FLC, transform the control problem into a regulation problem, achieving better dynamic performance and stability. PMSM, controlled through speed control, demonstrated reliable operation with closed-loop speed, current, and voltage vector methods. The system exhibited good steady-state characteristics, but further validation of the FLC approach in real-world applications is recommended.

Conflict of Interest

The authors declared no conflict of work.

References

[1] Jiuhe, W., Hongren, Y., Jinlong, Z., & Huade, L. (2006, August). Study on power decoupling control of three phase voltage source PWM rectifiers. In *2006 CES/IEEE 5th International Power Electronics and Motion Control Conference* (Vol. 1, pp. 1–5). IEEE.

[2] Yang, Z., & Wu, L. (2004, June). A new passivity-based control method and simulation for DC/DC converter. In *Fifth World Congress on Intelligent Control and Automation (IEEE Cat. No. 04EX788)* (Vol. 6, pp. 5582–5585). IEEE.

[3] Langella, R., Testa, A., & Alii, E. (2014). IEEE recommended practice and requirements for harmonic control in electric power systems. In *IEEE recommended practice*. IEEE.

[4] Rastogi, M., Naik, R., & Mohan, N. (1994). A comparative evaluation of harmonic reduction techniques in three-phase utility interface of power electronic loads. *IEEE Trans. Ind. Appl.*, 30(5), 1149–1155.

[5] Kolar, J. W., & Zach, F. C. (1999). A novel three-phase utility interface minimizing line current harmonics of high-power telecommunications rectifier modules. *IEEE Trans. Ind. Electron.*, 44(4), 456–467.

[6] Post, R. F., Fowler, T. K., & Post, S. F. (1993). A high-efficiency electromechanical battery. *Proceedings of the IEEE*, 81(3), 462–474.

[7] Hull, J. R. (1997). Flywheels on a roll. *IEEE Spectrum*, 4(7), 20–25.

[8] Vladimir, B., & Vikram, K. (1997). A new mathematical model and control of a three-phase AC–DC voltage source converter. *IEEE Transactions on Power Electronics*, 12(1), 116–123.

[9] Wu, R., Dewan, S. B., & Slemon, G. R. (1990). A PWM AC to DC converter with fixed switching frequency. *IEEE Trans. on Industry Applications*, 26(5), 880–885.

49 Improvisation of Electric Vehicle Battery Charging Efficiency Using Supervisory Power Management Control Algorithm Compared With Cell Balancing Optimization Algorithm by Reducing the Power Loss

R. Hariharan[1] and Rambabu Yerra[2]

[1]Associate Professor, Department of Electrical and Electronics Engineering Saveetha School of Engineering, Saveetha Institute of Medical and Technical Sciences, Chennai, India

[2]Research Scholar, Department of Electrical and Electronics Engineering Saveetha School of Engineering, Saveetha Institute of Medical and Technical Sciences, Chennai, India

Abstract

Aim: The aim of this work is to run the Electric Vehicle (EV) efficiently using Novel Supervisory power management control algorithm for maximum efficiency and to run maximum distance. **Materials and Methods:** A total number of 16 samples are collected from matlab simulation, The Proposed Method is a novel Supervisory power management control algorithm and the second Method is normal cell balancing where each method consists of 8 samples. By implementing the novel Supervisory power management control algorithm, the overall efficiency of the vehicle is improved and the power losses are reduced.

Keywords: Plug-In hybrid electric vehicle, HEV, EV, Battery, Novel Supervisory power management control algorithm, Cell balancing algorithm

1. Introduction

This work reveals a novel Supervisory power management control algorithm in electric vehicles (EV, HEV, PHEV) to increase the traveling distance and increase efficiency by reducing the losses. This research is done to implement EV's to increase their efficiency and decrease the pollution levels coming from gasoline vehicles [1]. By using those fuels vehicles emit harmful gasses like Carbon Dioxide (CO_2), Carbon Monoxide (CO), Hydrocarbon (HC), Nitrogen Oxide (NOx), and other toxins are emitted (Li et al. 2009). These toxins are harmful to human beings, animals, etc. To decrease this, a pollution battery is used as a fuel [14] to run the vehicle. The application of

Email: harinov22@gmail.com[1], y.rambabu.mtech@gmail.com[2]

DOI: 10.1201/9781003661917-49

this METHOD is used in normal roads & off-roads [17] and runs the vehicle in 2-wheel drive, 4-wheel drive, and all-wheel drive withgreater speed and torque [7, 8, 18].

More than 400 research articles were published in PubMed, Elsevier springer, and IEEE Xplore in recent 5 years. This author et al. [6]. Enhanced the Cell balancing algorithm and implemented it for IEEE 14-bus system. This author [8] has proposed a power management control algorithm and loss reduction. This authors have proposed the Supervisory power management control algorithm and loss reduction. This author et al., (Mandel 2012, Taylor 2014) have proposed the Supervisory power management control algorithm for better efficiency and instantaneous response. The author (Marco et al. 2020) had proposed an Intelligent algorithm for battery control that would implement the charging efficiency of the battery.

2. Challenges Posed by Electric Vehicle Charging Stations

(EVCS) On Distribution Systems Electric vehicle charging stations, or EVCS, pose serious problems for distribution networks. Increased heat stress on equipment, transformer overload, and voltage fluctuations can all be caused by high power demands during peak charging periods. Inadequate infrastructure and an uneven distribution of charging stations can lead to imbalanced loads and localized congestion. Furthermore, EVCS integration with renewable energy sources is made more difficult by the intermittent nature of renewable energy sources. The grid's dependability is also threatened by the security threats connected systems present. In order to lower operating costs, demand management techniques and grid upgrades are required. Moreover, grid integration and coordination activities are made more difficult by the lack of standards in EVCS.

EVCS (Electric Vehicle Charging Stations) present significant challenges for distribution systems. High power demands

during peak charging times may result in voltage fluctuations, overloading transformers, and increased thermal stress on equipment. There may be localized congestion and imbalanced loads as a result of inadequate infrastructure and uneven distribution of charging stations. Additionally, the intermittent nature of renewable energy sources complicates the integration of EVCS with renewable energy sources. The security risks associated with interconnected systems also pose a threat to the reliability of the grid. It is necessary to upgrade the grid and develop demand management strategies in order to reduce operational costs. Furthermore, the lack of standardization in EVCS further complicates grid integration and coordination efforts. In the previous research, the researchers have not considered solar power in simultaneous battery backup. Hence the novel supervisory power management control algorithm has been proposed to suppress the losses. Consideration of batteries to travel more distance. The aim of this research is to introduce SPMC for better efficiency.

3. Materials and Methods

Using the G Power program, this study examines two compensation strategies using eight samples out of a total of sixteen tests [16] (Belyaev, Izotova, and Kashin 2018). Gpower is 0.80, effect size is 0.5, and the statistical test is the difference between two independent means. This research was carried out using the MATLAB software.

4. Cell Balancing Algorithm

Cell balancing algorithms are a key technology in the electrical vehicles. This algorithm will monitor and control the voltage of the cell and balance the State of Charge of each cell. A cell balancing algorithm is used to check and control the charge present in each cell present in the battery. Here charge is controlled in each cell of the battery. In this system charging and discharging of the battery is controlled in each cell, so that it can easily find out the damaged cell present in the battery. The cell balancing algorithm is also one of the MPPT algorithms mostly used in EV operations [4]. In this algorithm, battery control will be present. So that the Current management, Voltage management, Soc management will be there and by controlling these electrical motors are driven and controlled. The flow chart of the Cell balancing algorithm is shown in Figure 49.1. (Tulpule et al. 2009). The Group 1 sample preparation of the Cell Balancing algorithm (CB)has been developed by collecting the output efficiency value (79.5%) by changing the Input Current (A) to the EV. Figure 49.1 shows the flow chart of Cell Balancing algorithm (CB).

5. Comprehensive Description of the Test System, Scenarios, and Parameters Evaluation

An EV battery pack, Level 2 charging infrastructure, and a real-time simulation environment such as MATLAB/Simulink are all part of the test system. Supervisory Power Management Control (SPMC) versus Cell Balancing Optimization (CBO) is used to evaluate charging efficiency improvements. The scenarios involve normal charging, adaptive charging with SPMC, and cell balancing with CBO under dynamic grid conditions.

A system called supervisory power management control makes sure that electricity capacity and demand are always in balance. Additionally, it makes sure that the power plant's capacity is not exceeded by the load from major consumers.

6. Novel Supervisory Power Management Control Algorithm

This controller will control many individual controls and control loops in a single time. The sample preparation of group 2 was done by collecting the output speed for the Novel Supervisory power management control algorithm; those output values are obtained by varying the accelerations. A

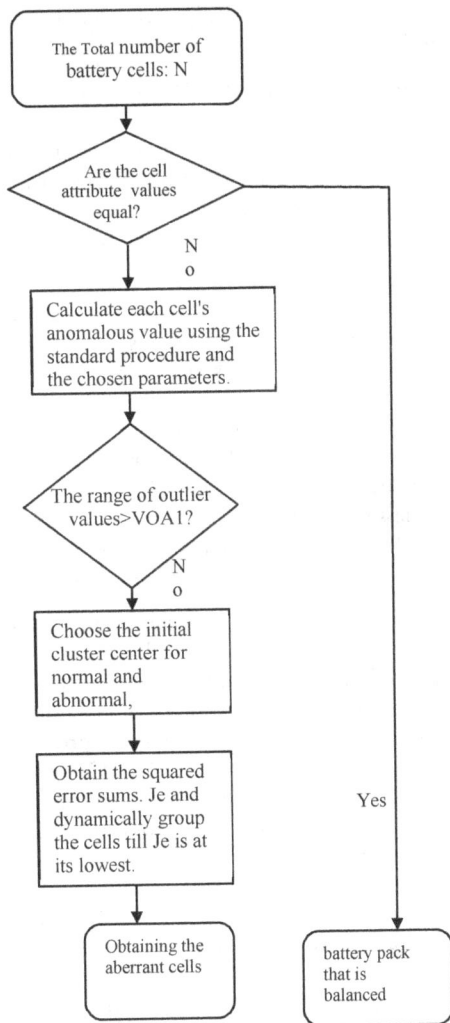

Figure 49.1 Cell balancing algorithm

novel Supervisory power management control algorithm is used in EVs, HEV's, PHEVs, to run the vehicles by shifting the transmission as they require. So that the speed of the vehicle is controlled and fuel is used efficiently [7]. Charging the battery and discharging the battery are also controlled here. Supervisory power management control algorithm/method is widely used in the HEV, PHEV, and many types of vehicles for better operation and working. By using this system the vehicle will run & charge the battery efficiently. The flow chart of the supervisory power management control algorithm is shown in Fig 49.2.

Battery electric vehicles (BEVs)

These completely electric cars have no exhaust emissions and are powered only by a battery pack. They must be charged from an external power source, such as a public charging station or a home charger.

Hybrid electric vehicles (HEVs)

These cars have a tiny battery, an electric motor, and an internal combustion engine. While the motor needs electricity from the batteries, the engine uses fuel. HEVs have the option of using their electric motor and engine separately or simultaneously. Although hybrid electric vehicles (HEVs) use less fuel than internal combustion engine (ICE) vehicles, they still need gasoline and oil and produce emissions from their tailpipes.

Plug-in hybrid electric vehicles (PHEVs)

These vehicles can be charged using regenerative braking or charging equipment and run on a combination of gas and battery power. Although plug-in chargers provide the power, PHEVs are more like EVs even though they can run on both gas and batteries. Charge efficiency, power loss, SOC uniformity, voltage stability, thermal behavior, and grid impact are some of the parameters that are analyzed. Efficiency gains, power loss reductions, consistency of SOCs, thermal performance, and algorithm response time are among the metrics that have been measured. To identify an efficient, balanced, and reliable charging algorithm, the objective is to identify the optimal algorithm.

Figure 49.2 shows the flow chart of Supervisory power management control algorithm (SPMC). The Group 2 sample preparation of Supervisory power management control algorithm (SPMC) has been developed by collecting the output efficiency value (83.3%) by changing the Input Current (A) to the EV.

7. Results

Table 49.1 represents simulation results from the Cell Balancing algorithm (CB) of obtaining output efficiency value (79.5%).

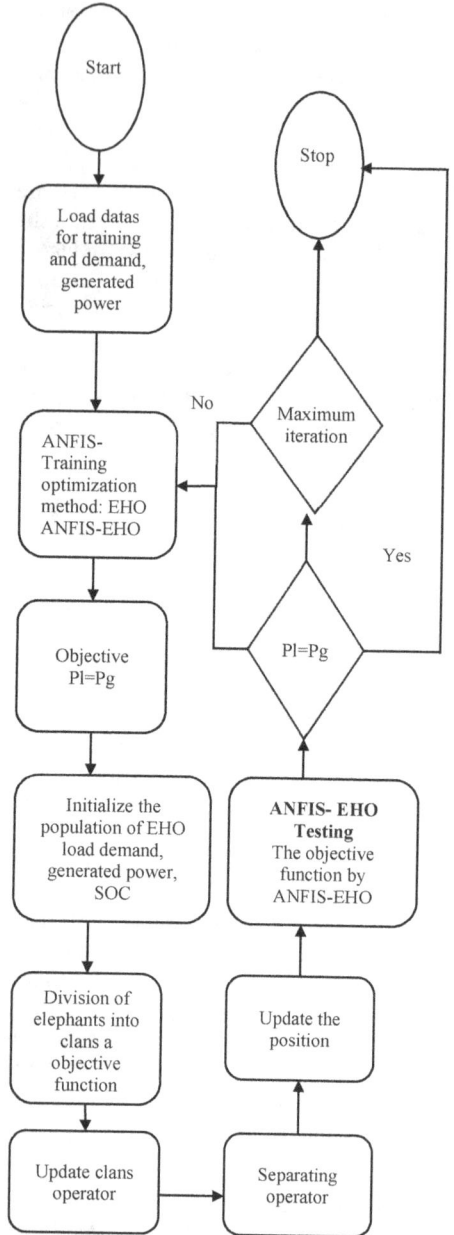

Figure 49.2 Supervisory power management control algorithm

and for the Novel Supervisory Power Management Control Algorithm (SPMC) of obtaining the output efficiency value (83.3%) by changing the Input Current (A). Tables 49.2 and 49.3 show that the

Table 49.1 Efficiency analysis using Cell balancing (CB) algorithm and Supervisory Power Management Control (SPMC) algorithm for various input current

Input Current (A)	Output Efficiency For CB algorithm (%)	Output Efficiency For SPMC algorithm (%)
0	0	0
4	8	14
6	15	25
8	26	36
11	34	50
13	49	59
15	61	63
18	77	80

output of group statistics independent samples T-test have a significance value of 0.908 (p>0.05), t = −0.118 & Mean Difference = −0.375.

Figure 49.3 shows the Battery graph of vehicles working in Supervisory power management control algorithms without solar panels. And it runs very less distance with the charging of the battery. Figure 49.4 shows the Battery graph of vehicles working in the Supervisory power management control algorithm with solar panels, by using this system the vehicle can be moved more distance without stopping and taking charge. So that the vehicle can run at greater speed and more distance while moving the vehicle. The Continuous battery will charge when the sun is present. Figure 49.5 shows the Graph for discharging of cells while using a Cell balancing algorithm for individual cells in batteries in EV's. Figure 49.6 shows the Vehicle graph with different speeds and their Drive torque, Electrical power of Motor, Generator, and Battery with different Accelerations.

Figure 49.7 shows Comparison of CB method and SPMC in terms of mean accuracy. The mean accuracy of SPMC is better than CB and the standard deviation of SPMC is slightly better than CB. From Sample Independent T-Test the Mean difference= -0.375, Standard Deviation = 3.18303, Standard Error Mean = 2.37359, X-Axis: CB vs SPMC. Y-Axis: Mean accuracy detection.

Table 49.2 The mean, standard deviation, and standard error of efficiency of SPMC and CB were statistically analyzed in groups. The efficiency difference is statistically significant. The CB algorithm is the least efficient (79.5%), while the SPMC method is the most efficient (83.3%)

Group	No of Samples	Mean	Std. Deviation	Std. Error Mean
Cell balancing algorithm	8	79.5000	5.99851	2.12079
Supervisory Power Management Control algorithm	8	83.3050	6.71353	2.37359

Table 49.3 T tests for independent samples for both sample groups were statistically analyzed. Statistical analysis of independent sample T tests for both sample groups. Significance value is .908 (2-tailed and p>0.05) with 95% confidence value is calculated

		Levene's Test for Equality Variances				T-test for Equality of means			95% Confidence Interval of the Difference	
		F	Sig	t	diff	Sig (2-tailed)	Mean Difference	Std Error Difference	Lower	Upper
Efficiency	Equal variances assumed	.126	.728	−.118	14	.908	−.37	3.18	−7.20	6.45
	Equal variances not assumed			−.118	13.826	.908	−.37	3.18	−7.20	6.45

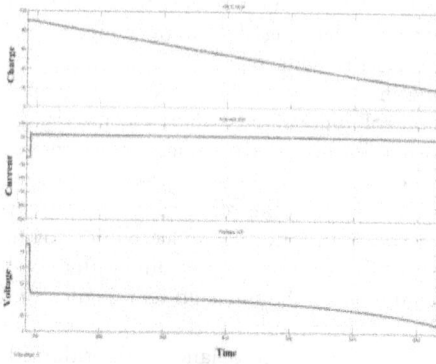

Figure 49.3 Battery graph of Vehicle working in Supervisory power management control algorithm without solar panel

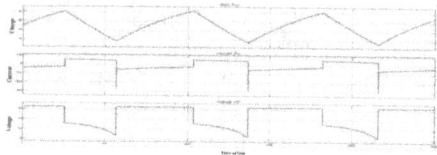

Figure 49.4 Battery graph of Vehicle working in Supervisory power management control algorithm with solar panel

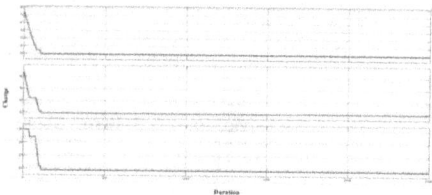

Figure 49.5 Graph for Cell balancing algorithm for individual cell in a battery

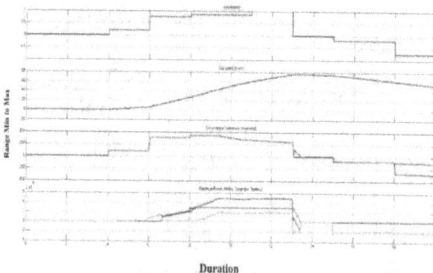

Figure 49.6 Vehicle graph with different speed and there Drive torque, Electrical power of Motor, Generator, Battery with different accelerations

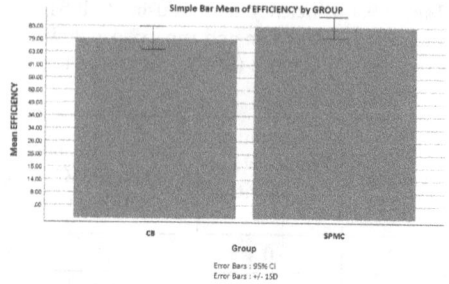

Figure 49.7 Comparison of cell balancing (CB) and Supervisory Power Management Control (SPMC) algorithm in terms of Mean output power. SPMC is better than CB. X-Axis: CB Vs SPMC algorithm and Y-Axis: Mean efficiency with ± 1 SD

8. Discussion

The results of the work reveal that the SPMC algorithm-based compensation provides more promising results to run the vehicle and get high efficiency then CB algorithm. Automatic measurement of traveling distance and battery backup are the most important considerations because of the failure to achieve conclusions based on previous compensation methods. The SPMC method has the highest efficiency (83.3%) where CB method has the lowest efficiency (79.5%). SPMC algorithm got the best efficiency then CB.

Supervisory power management control and Cell balancing algorithms are implemented and its output is analyzed. Supervisory power management control algorithm is suitable and best compared to cell balancing. Based on the simulation results, the comparative analysis of Supervisory power management control and Cell balancing algorithms have been carried out. And the vehicle runs with a Supervisory power management control algorithm runs more distance [13]. And charges continuously while the vehicle is running with the help of regenerative braking and solar power. This method [16] is opposing the Supervisory Power Management Control Algorithm. The Supervisory Power Management Control Algorithm is better than the Cell balancing algorithm, because the power losses in the battery are high in CB. Without a solar panel, the vehicle can run less distance, because the car battery can't charge hence less backup will be achieved [5]. The battery can't charge if the sun is absent [11]. Moreover, Electrical Vehicles (EVs) deliver more torque than gasoline vehicles fuel cost is less for EVs [1, 9, 10].

Although the results of the study showed better performance with limited attributes using the SPMC compensation, there are certain limitations. SPMC devices have no revolutionary parts, for the implementation of battery backup, additional equipment is needed. The size of the device is heavy. This algorithm is not suitable for all types of vehicles. SPMC with the plug-in hybrid electric vehicle (EV, HEV, PHEV) will get high efficiency and this vehicle travels more distance. With advanced preprocessing techniques the vehicle will travel more distance with a single charge and losses are also less [11]. The speed of the vehicle is increased as the vehicle will travel very smoothly without humming and noise. Pollution is decreased while electric vehicles (EV, HEV, PHEV) are using and the fuel cost is also decreased [16].

9. Conclusion

Novel SPMC method is compared with the CB method. Based on simulation results, The Novel SPMC method has the highest efficiency of (83.3%) compared with CB method (79.5%).It is observed that the simultaneous Supervisory power management control algorithm performed significantly better than the existing cell balancing algorithm by reducing the power losses.

Declaration Conflicts of Interest

No conflict of interest in this manuscript.

Funding

No funding support provided by any organization.

References

[1] Aljanad, A., Mohamed, A., Khatib, T., Ayob, A., & Shareef, H. (2019). A novel charging and discharging algorithm of plug-in hybrid electric vehicles considering vehicle-to-grid and photovoltaic generation. *World Electric Vehicle Journal*, 10(4), 61.

[2] Shen, J., & Khaligh, A. (2021). An Energy Management Strategy for an EV with Two Propulsion Machines and a Hybrid Energy Storage System. Accessed October 3, 2021. https://doi.org/10.1109/itec.2015.7165791.

[3] Babu, A., & Ashok, S. (2012, December). Algorithm for selection of motor and vehicle architecture for a plug-in hybrid electric vehicle. In 2012 *Annual IEEE India Conference (INDICON)* (pp. 875–878). IEEE. https://doi.org/10.1109/indcon.2012.6420740.

[4] Badin, F., Scordia, J., Trigui, R., Vinot, E., & Jeanneret, B. (2006). Hybrid Electric Vehicles Energy Consumption Decrease according to Drive Train Architecture, Energy Management and Vehicle Use. *IET Hybrid Vehicle Conference 2006*. https://doi.org/10.1049/cp:20060610.

[5] Butler, K. L., M. Ehsani, and P. Kamath. 1999. "A Matlab-Based Modeling and Simulation Package for Electric and Hybrid Electric Vehicle Design." *IEEE Transactions on Vehicular Technology*. https://doi.org/10.1109/25.806769.

[6] Cheng, R., & Dong, Z. (2015). Modeling and Simulation of Plug-In Hybrid Electric Powertrain System for Different Vehicular Application. *2015 IEEE Vehicle Power and Propulsion Conference* (VPPC). https://doi.org/10.1109/vppc.2015.7352976.

[7] Dathu, K. P. M. Y. V., Dathu, K. P. M. Y., & Hariharan, R. (2021). Modelling of Plug-in Hybrid Electric Vehicle (PHEV) with Multi Source. *Materials Today: Proceedings*. https://doi.org/10.1016/j.matpr.2020.10.215.

[8] Denis, N., Dubois, M. R., Trovão, J. P. F., & Desrochers, A. (2018). Power Split Strategy Optimization of a Plug-in Parallel Hybrid Electric Vehicle. *IEEE Transactions on Vehicular Technology*. https://doi.org/10.1109/tvt.2017.2756049.

[9] Ehsani, M., Gao, Y., Longo, S., & Ebrahimi, K. (2018). *Modern Electric, Hybrid Electric, and Fuel Cell Vehicles*. CRC Press.

[10] ElMenshawy, M., ElMenshawy, M., Massoud, A., & Gastli, A. (2016). Solar Car Efficient Power Converters' Design. *2016 IEEE Symposium on Computer Applications & Industrial Electronics (ISCAIE)*. https://doi.org/10.1109/iscaie.2016.7575059.

[11] Ferahtia, S., Djerioui, A., Zeghlache, S., & Houari, A. (2020). A Hybrid Power System Based on Fuel Cell, Photovoltaic Source and

Supercapacitor. *SN Applied Sciences.* https://doi.org/10.1007/s42452-020-2709-0.

[12] Hesse, H. C., Schimpe, M., Kucevic, D., & Jossen, A. (2017). Lithium-Ion Battery Storage for the Grid—A Review of Stationary Battery Storage System Design Tailored for Applications in Modern Power Grids. *Energies.* https://doi.org/10.3390/en10122107.

[13] Islam, F. R., & Pota, H. R. (2012). Impact of Dynamic PHEV Load on Photovoltaic System. *International Journal of Electrical and Computer Engineering (IJECE).* https://doi.org/10.11591/ijece.v2i5.1239.

[14] Liu, X., Li, L., Hori, Y., Akiba, T., & Shirato, R. (2005). Optimal Traction Control for EV Utilizing Fast Torque Response of Electric Motor. *31st Annual Conference of IEEE Industrial Electronics Society, 2005. IECON 2005.* https://doi.org/10.1109/iecon.2005.1569319.

[15] Li, X., Lopes, L. A. C., & Williamson, S. S. (2009). On the Suitability of Plug-in Hybrid Electric Vehicle (PHEV) Charging Infrastructures Based on Wind and Solar Energy. *2009 IEEE Power & Energy Society General Meeting.* https://doi.org/10.1109/pes.2009.5275171.

[16] Marco, J., Dinh, Q. T., & Longo, S. (Eds.). (2020). *Energy Storage and Management for Electric Vehicles.* MDPI.

[17] Rizzoni, G., Guzzella, L., & Baumann, B. M. (1999). Unified Modeling of Hybrid Electric Vehicle Drivetrains. *IEEE/ASME Transactions on Mechatronics.* https://doi.org/10.1109/3516.789683.

[18] Senanayake, T., Iijima, R., Isobe, T., & Tadano, H. (2016). Z-Source with Rectangular Wave Modulation Inverter for Hybrid/Electric Vehicles. 2016 Applications (EPE'16 ECCE Europe). https://doi.org/10.1109/epe.2016.7695682.

50 India's National Green Hydrogen Mission: Barriers and Climatic Impact

Rajeev Kumar[1], Sonali Goel[1], Renu Sharma[1,*] and Belqasem Aljafari[2]

[1]Department of Electrical Engineering, ITER, Siksha 'O' Anusandhan Deemed to be University, Bhubaneswar, India
[2]Department of Electrical Engineering, College of Engineering, Najran University, Najran, Saudi Arabia

Abstract

India's energy demand is expected to increase dramatically, surpassing that of any other nation because of its size and economic potential. This increased demand must primarily be satisfied through renewable sources. In order to mitigate the effects of global warming, India has committed to achieving net zero carbon emissions by 2070 and obtaining half of its power from renewable sources by 2030. Developing a hydrogen-based economy offers India a sustainable energy transformation and carbon reduction pathway. By implementing blue and green hydrogen technologies, India may diversify its energy portfolio and reduce dependence on carbon-emitting fuels. India needs a comprehensive approach to meet its environmental targets by 2050, including infrastructure development, supportive policies, and technological advancements. Effective and cost-efficient hydrogen storage continues to pose a problem, necessitating substantial investment in infrastructure for hydrogen production, transportation, and refuelling stations. Enhancing public knowledge and acceptance of hydrogen as a secure and practical energy source must be improved.

Keywords: Renewable sources, net zero carbon emission, climate change, blue hydrogen, green hydrogen

1. Introduction

The power sector in India is transforming due to the country's ambitious clean energy goals. As the population expands and rural areas access electricity, there's a growing need for power in homes, industries, and communities. The shift to renewable energy sources will reduce pollution and enable rural areas to become energy-independent. The year 2022 is expected to bring substantial growth to India's renewable energy industry, with anticipated investments of $15 billion, fuelled by government programs focusing on green hydrogen and future electric vehicles. It is anticipated that improvements in battery technology would reduce solar energy costs by 66% when compared to present rates. In India, replacing coal with renewable energy may yield yearly savings of Rs. 54,000 crore. Furthermore, it is anticipated that 15 GW of solar wind hybrid systems will be

Email: mrrjvkmr@gmail.com[1], sonali19881@gmail.com[1], renusharma@soa.ac.in[1*], bhaljafari@nu.edu.sa[2]

DOI: 10.1201/9781003661917-50

implemented between 2020 to 2025. The Central Electricity Authority (CEA) estimates that renewable energy's contribution to India's power generation mix will rise from 18% to 44% by 2029–30, while the share of thermal power may decline from 78% to 52%. Furthermore, according to CEA, India's electricity demand is expected to reach 817 GW by 2030. In the universe, hydrogen reigns as the most plentiful element, predominantly existing in nature as a component of water molecules, bonded with oxygen. To a smaller extent, it is also found combined with carbon in organic materials and fossil fuels, including methane [8, 9]. Among energy carriers, hydrogen is unique because of its high energy density, surpassing natural gas by 2.2 times, gasoline by 2.75 times, and oil by 3 times in terms of energy content per kilogram. However, its status as the lightest gas poses challenges for storage. Hydrogen is also considered hazardous due to its high risk of leakage and increased potential for deflagration and detonation [7]. More than 95% of global hydrogen is derived from fossil fuels, primarily natural gas, by a process known as steam reforming. This method involves heating a mixture of gas and steam to extremely high temperatures, between 700° C to 1000° C, in the presence of metal catalysts. However, this technique is not environmentally friendly, producing 10 kg of CO_2 for every 1 kg of H2 generated [2]. India came in fourth place in 2023 for installed wind, solar, and renewable energy capacity. India's capacity for renewable energy was 125.15 GW in FY23. By 2030, India wants to produce five million tons of renewable hydrogen with the country's electrolyzer production capacity reaching 8 GW per year in 2025. India's green hydrogen industry might be worth $8 billion by 2030, and increasing hydrogen production would require at least 50 gigawatts (GW) of electrolyzers. By March 2024, the combined capacity of renewable energy sources—such as biomass, waste to energy, and waste to power was 143.64 GW. Non-fossil energy sources accounted for 42.25% of India's total installed power

capacity in February 2024. ICRA's analysis suggests that India's renewable energy capacity will expand to approximately 170 GW by March 2025, increasing from 135 GW in December 2023. By 2030, the country aims to achieve 450 GW of energy from renewable sources, with solar energy contributing 280 GW, accounting for over 60% of the total. In FY23, non-hydro renewable energy saw production of 4.2 GW in the first quarter, with 2.6 GW added in FY22. According to CEEW-CEF, in FY23, India's total added electrical power generation capacity reached 416 GW, with renewable energy contributing 125 GW (30%) and hydro sources adding 47 GW (11%). In 2023–24, India achieved a significant milestone by increasing its renewable energy capacity by 18.48 GW, reflecting a 21% growth compared to the previous year. Northern India presents a significant possible capacity of 363 GW, supported by renewable energy policies.

Before green hydrogen can become as affordable as grey or blue hydrogen, the International Renewable Energy Agency (IRENA) projects that 150 billion US dollars in government financing will be required globally over the next ten years to cut the cost of manufacturing "Green Hydrogen." According to the Secretary of the UNECE Group of Experts on Gas, initiatives for the production and application of green hydrogen remain more political than commercial. The Department of Energy USA's Hydrogen Program Plan states that by 2040, only 28% of the hydrogen planned for use in the US will come from renewable energy sources, while 41% will come from coal and natural gas. Kenya's strategy for hydrogen production includes targets for electrolyzer capacity of 0.1 GW and 0.25 GW by 2027 and 2032, respectively, with corresponding renewable energy capacity goals of 0.15 GW and 0.45 GW for hydrogen generation. Turkey aims to achieve electrolyzer capacities of 2, 5, and 70 GW by 2030, 2035, and 2053, respectively. Japan's revised hydrogen plan includes goals for electrolyzer capacity of 15 GW by 2030 as well as consumption objectives of 3 Mt by 2030, 12 Mt by 2040, and 20 Mt by 2050 [6]. The hydrogen as a fuel is shown in Figure 50.1, and the hydrogen production cost in India is shown in Figure 50.2.

Aiming to establish India as an international centre for the production and export of green hydrogen, the Indian government established the National Green Hydrogen Mission in January 2023. The objective of this mission is to reduce carbon emissions and reduce the dependence on fossil fuels. By 2030, the primary objective of the mission is to generate 5 million tons of renewable hydrogen annually. National hydrogen strategies and definitions of green hydrogen by different countries are given in Tables 50.1 and 50.2, respectively.

2. Methodology

The methodology section delineates the approach employed for this study on hydrogen energy systems. A comprehensive literature review was undertaken, analyzing peer-reviewed articles, industry reports, and government publications to acquire relevant information on hydrogen production, storage, distribution, and utilization. The collected data was subjected to critical evaluation to identify key barriers, assess climate impacts, and formulate conclusions regarding the future prospects of hydrogen energy systems. The manuscript's structure is as follows:

The introduction is presented in Section 1, followed by the methodology in Section 2. Section 3 addresses the barriers to using hydrogen energy systems. The consequences

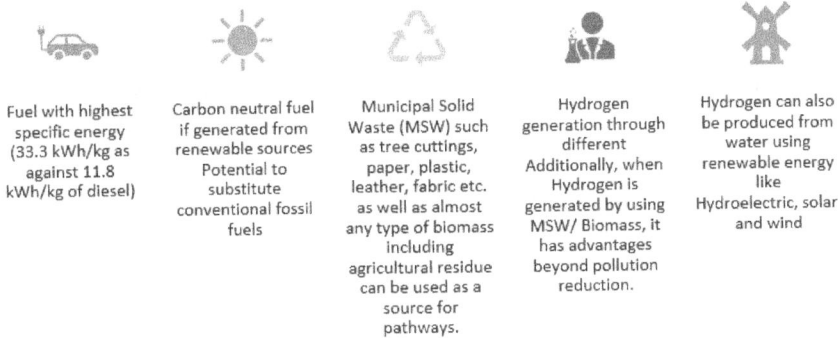

| Fuel with highest specific energy (33.3 kWh/kg as against 11.8 kWh/kg of diesel) | Carbon neutral fuel if generated from renewable sources Potential to substitute conventional fossil fuels | Municipal Solid Waste (MSW) such as tree cuttings, paper, plastic, leather, fabric etc. as well as almost any type of biomass including agricultural residue can be used as a source for pathways. | Hydrogen generation through different Additionally, when Hydrogen is generated by using MSW/ Biomass, it has advantages beyond pollution reduction. | Hydrogen can also be produced from water using renewable energy like Hydroelectric, solar and wind |

Figure 50.1 Hydrogen as a fuel

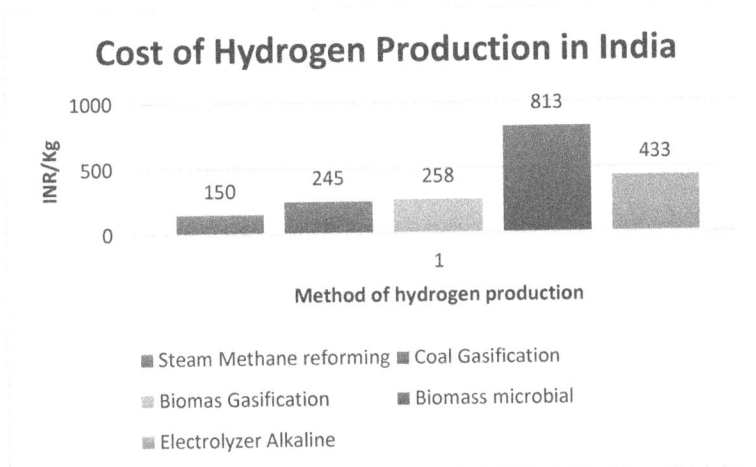

Cost of Hydrogen Production in India

INR/Kg

1000

813

500

433

150   245   258

0

1

Method of hydrogen production

■ Steam Methane reforming ■ Coal Gasification
■ Biomas Gasification ■ Biomass microbial
■ Electrolyzer Alkaline

Figure 50.2 Cost of Hydrogen Production in India

Table 50.1 National Hydrogen Strategy

Timeline	National strategic planning for green hydrogen
2018	Japan
2019–20	Rep of Korea, New Zealand, Australia
2020–21	Netherlands, Norway, Germany, Portugal, EU, Spain, Italy, Chile, France, Canada, Morocco
2021–22	Hungary, Paraguay, Slovakia, Sweden, UK, Czech Rep, Luxembourg, South Africa, Poland, Colombia, Denmark
2022–23	Croatia, China, Austria, Uruguay, Costa Rica, Namibia, Netherlands, Trinidad and Tobago, Singapore, Oman, Belgium, Turkiya, India
2023–24	Finland, Panama, Bulgaria, USA, Japan, Germany, Ireland, Estonia, Ecuador, UAE, Malaysia, Argentina, Kenya, Sri Lanka, New Zealand, France, Indonesia
2024–25	Algeria, Vietnam, Oman, Tunisia

Table 50.2 Green Hydrogen strategy and goal in different countries

Country	Definition of hydrogen	Strategy	Goal	Ref
European countries	Hydrogen is produced via electrolysis using renewable energy sources	1. European nations concentrate on using excess renewable energy (solar and wind) to produce green hydrogen. 2. Investment in storage options, refueling stations, and pipelines. 3. Members of the EU work together to share infrastructure and technology through initiatives such as the EU Hydrogen Strategy.	1. Aim to generate 10 million tons of hydrogen from renewable sources by 2030. 2. Reduce carbon emissions in transportation, heating, and heavy industrial sectors. 3. Establish a hydrogen marketplace by 2030 and transition to a comprehensive hydrogen-based economy by 2050.	[5]
US	Hydrogen is produced with a minimal carbon footprint, primarily through renewable energy	Promoting the development of clean hydrogen and outlining US government initiatives to combat climate change, including a net-zero emissions economy by 2050 and a carbon-free grid by 2035.	By 2030, 2040, and 2050, the target is to produce and use 10 million metric tonnes (MMT), 20 MMT, and 50 MMT of clean hydrogen, respectively, at an increasing annual rate.	[13]
Canada	Hydrogen from water electrolysis powered by renewable energy	1. The hydrogen initiative has the potential to generate more than 350,000 well-compensated positions and yield revenues surpassing CAD 50 billion by 2050. 2. A CAD 1.5 billion Clean Fuels Fund is in place to bolster the development of new clean fuel production facilities, with hydrogen included. 3. Additionally, a CAD 8 billion Net-zero Accelerator program supports efforts to reduce carbon emissions in industrial sectors.	The objective is to achieve carbon neutrality and establish Canada as a global leader in the clean renewable fuels industry by 2050.	[1] [12]

Country	Definition of hydrogen	Strategy	Goal	Ref
Germany	Hydrogen derived from renewable sources like solar and wind through electrolysis	1. The goal is to expand the country's local hydrogen market and support its hydrogen strategy by importing hydrogen and its derivatives. 2. The aim is to achieve 95 to 130 terawatt hours (TWh) by 2030, with further increases anticipated in subsequent years. 3. Plans include enhancing the nation's capacity for hydrogen importation through an expanded fleet of ships and pipelines and establishing partnerships with other countries to promote increased regional and global collaboration on hydrogen initiatives.	1. The objective is to enhance nationwide energy efficiency and electricity consumption while incorporating hydrogen-based energy into areas where electrification is not feasible by 2023. 2. Additionally, the plan aims to expand the country's green hydrogen production capacity from 5GW to 10GW and develop its hydrogen infrastructure.	[3]
Australia	Renewable energy sources, particularly solar and wind power, are used to manufacture hydrogen.	1. For establishing a clean, inventive, safe, and competitive hydrogen sector that boosts the economy and communities of Australia, facilitating net zero transition and establishing Australia as a global leader in hydrogen. 2. Producing a minimum of 15 million tons of hydrogen per year by 2050, with the possibility of reaching 30 million tons per year. 3. Programs that encourage extremely cheap renewable energy, which accounts for the majority of the expenditures associated with producing hydrogen. 4. To promote the green hydrogen market, including energy storage, industrial processes, and transportation.	1. Boosting green hydrogen in various applications, such as power production, transportation (fuel cells), and material and chemical feedstock production. 2. To be among the top exporters of green hydrogen worldwide, focusing on Asian markets.	[11]
China	Hydrogen is produced using renewable energy, particularly in urban areas.	1. By 2025, the annual generation of green hydrogen from renewable feedstock resources is expected to reach 100,000–200,000,000 tons. 2. Clean hydrogen's application in various fields, such as industry, energy storage, and electricity production.	By 2025, 50,000 hydrogen fuel-cell vehicles should be on the road, along with several hydrogen-filling facilities.	[4]

Country	Definition of hydrogen	Strategy	Goal	Ref
India	Hydrogen is mostly created by electro-lyzing water using renew-able energy sources.	1. Establishing India as a world leader in producing and supplying green hydrogen. 2. Creating markets for green hydrogen and its derivatives for export. 3. A reduction in dependency on fossil fuel imports. 4. Creating opportunities for economic development and employment.	To develop a minimum of 5 MMT (million metric tons) of annual capacity for producing green hydro-gen, along with a 125 GW increase in the nation's renewable energy capacity.	[10]

of hydrogen on climate are provided in Section 4, and the conclusions and future scope are discussed in Section 5.

3. Barriers to Hydrogen Energy Systems

Several substantial obstacles hinder the widespread adoption and implementation of hydrogen energy systems. Hydrogen's production, storage, and distribution infrastructure are significant reasons that pose an important obstacle. The current predominant method of hydrogen production relies on fossil fuels, mainly through steam methane reforming, which is not environmentally sustainable. Although eco-friendly hydrogen production techniques utilizing renewable energy sources and electrolysis are emerging, they remain costly and technologically intricate. The lack of a comprehensive network for hydrogen distribution and refuelling stations further hinders the expansion of hydrogen-powered vehicles and other applications. Another significant hurdle involves the technical difficulties related to the transportation and storage of hydrogen. Due to its low volumetric energy density, hydrogen requires high-pressure containers or cryogenic storage methods, increasing the system's complexity and cost. Safety issues also remain a concern because of hydrogen's flammability and potential for leakage. Furthermore, the overall energy efficiency of hydrogen systems, especially in transportation, often falls short compared to direct electrification alternatives. These factors, coupled with regulatory

uncertainties and the need for industry-wide standardization, present considerable challenges to the broad implementation of hydrogen energy systems across various sectors. Key issues and challenges of green hydrogen are presented in Figure 50.3.

Many challenges prevent the widespread adoption of green hydrogen in India, even though it has committed to creating a green hydrogen environment. These are the cost of production, infrastructure development, technology, and policy.

4. Consequences of Hydrogen on Climate

In India, the method for hydrogen production significantly influences climate change. Utilizing renewable energy sources such as solar and wind to generate green hydrogen through the process of electrolysis can greatly reduce greenhouse gas emissions, thereby creating a cleaner environment. Instead, blue hydrogen is generated from natural gas through the use of carbon capture technology. It may serve as an interim solution but still relies on fossil fuels, posing potential environmental hazards. Grey hydrogen, derived from natural gas without carbon capture, emits considerable amounts of CO_2, exacerbating climate concerns.

The shift to green hydrogen generation possesses the ability to generate employment opportunities and enhance energy security, but it necessitates considerable investment in infrastructure and supportive policies. In summary, the move towards sustainable

Technology Area	Key Issues and Challenges
Refueling Site Compression Storage, and Dispensing	• Compressor cost, reliability, and efficiency • Storage cost and footprint • Dispenser cost and reliability • Cost of equipment for -40°C precooling • Meter accuracy and cost
Pipelines	• Installed capital cost • Management of pipeline integrity (e.g., potential for hydrogen embrittlement) • Pipeline compressor cost and reliability
Compressed Gas Tube Trailers	• High capital cost of composite tube trailers • DOT weight limit of 36.3 metric tons • Cost and footprint of tube trailer terminals (comprising buffer storage and high-volume compressors)
Liquid Tankers	• Capital cost of liquefaction • Energy intensity of liquefaction • Boil-off losses
Carriers	• Storage capacity and economic feasibility • Ability to undergo multiple hydrogen addition and removal cycles without degradation • Energy requirements for hydrogen loading and unloading processes • Designing robust catalysts and regeneration systems capable of withstanding numerous hydrogen addition and removal cycles

Figure 50.3 Key issues and challenges of green hydrogen

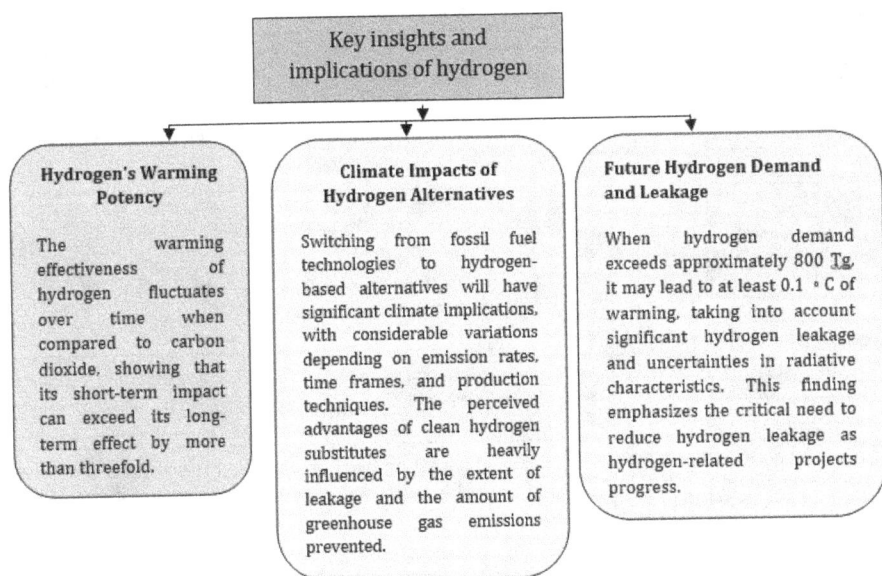

Key insights and implications of hydrogen

Hydrogen's Warming Potency	Climate Impacts of Hydrogen Alternatives	Future Hydrogen Demand and Leakage
The warming effectiveness of hydrogen fluctuates over time when compared to carbon dioxide, showing that its short-term impact can exceed its long-term effect by more than threefold.	Switching from fossil fuel technologies to hydrogen-based alternatives will have significant climate implications, with considerable variations depending on emission rates, time frames, and production techniques. The perceived advantages of clean hydrogen substitutes are heavily influenced by the extent of leakage and the amount of greenhouse gas emissions prevented.	When hydrogen demand exceeds approximately 800 Tg, it may lead to at least 0.1 °C of warming, taking into account significant hydrogen leakage and uncertainties in radiative characteristics. This finding emphasizes the critical need to reduce hydrogen leakage as hydrogen-related projects progress.

Figure 50.4 Key Insights and Implications of hydrogen

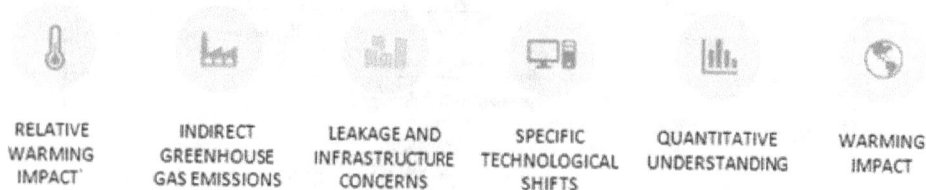

RELATIVE	INDIRECT	LEAKAGE AND	SPECIFIC	QUANTITATIVE	WARMING
WARMING	GREENHOUSE	INFRASTRUCTURE	TECHNOLOGICAL	UNDERSTANDING	IMPACT
IMPACT	GAS EMISSIONS	CONCERNS	SHIFTS		

Figure 50.5 Climate consequences of increasing deployment of hydrogen technologies

hydrogen production presents obstacles and prospects for India's climate objectives. The possibility of hydrogen leakage, particularly during its transit through existing natural gas systems, leaks significant amounts of methane. It poses challenges such as the implications for specific technological shifts, need to be carefully evaluated to understand their climate and environmental impacts, need for a better quantitative understanding of the climatic and environmental problems related to the implementation of hydrogen. emissions of nitrogen oxides, local water availability for the generation of green hydrogen, as well as the efficiency and permanence of CCUS for blue hydrogen. The key insights and implications of hydrogen are shown in Figure 50.4, and the climate consequences of increasing deployment of hydrogen technologies are shown in Figure 50.5.

There are several possible climate advantages to the effort to create and apply green hydrogen technology. These are presented in Figure 50.6.

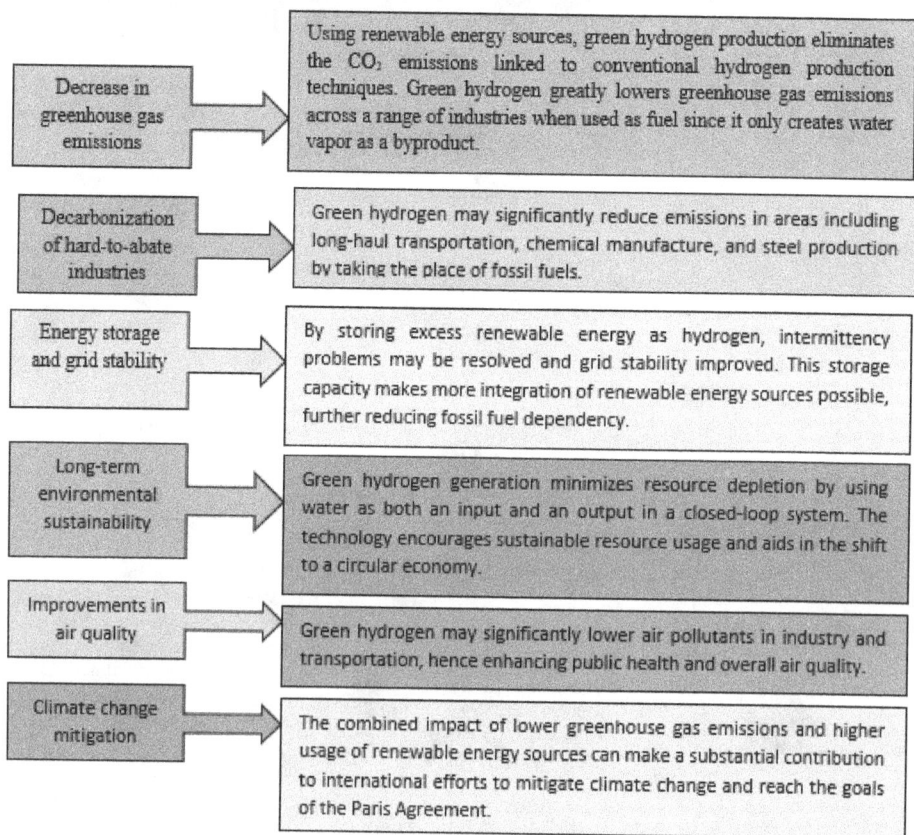

Decrease in greenhouse gas emissions	Using renewable energy sources, green hydrogen production eliminates the CO_2 emissions linked to conventional hydrogen production techniques. Green hydrogen greatly lowers greenhouse gas emissions across a range of industries when used as fuel since it only creates water vapor as a byproduct.
Decarbonization of hard-to-abate industries	Green hydrogen may significantly reduce emissions in areas including long-haul transportation, chemical manufacture, and steel production by taking the place of fossil fuels.
Energy storage and grid stability	By storing excess renewable energy as hydrogen, intermittency problems may be resolved and grid stability improved. This storage capacity makes more integration of renewable energy sources possible, further reducing fossil fuel dependency.
Long-term environmental sustainability	Green hydrogen generation minimizes resource depletion by using water as both an input and an output in a closed-loop system. The technology encourages sustainable resource usage and aids in the shift to a circular economy.
Improvements in air quality	Green hydrogen may significantly lower air pollutants in industry and transportation, hence enhancing public health and overall air quality.
Climate change mitigation	The combined impact of lower greenhouse gas emissions and higher usage of renewable energy sources can make a substantial contribution to international efforts to mitigate climate change and reach the goals of the Paris Agreement.

Figure 50.6 Potential climatic benefit

5. Conclusion and Future Scope

India's shift towards renewable energy and green hydrogen is crucial for meeting its climate goals. By 2030, India plans to boost its renewable energy capacity, targeting 50% of electricity from renewables. This transition aims to cut carbon emissions and enhance energy independence, especially in rural areas. Green hydrogen, produced via renewable energy electrolysis, provides a cleaner alternative to fossil-fuel-based hydrogen. Despite its potential to decarbonize various sectors, challenges like high production costs, infrastructure needs, and safety issues in storage and transportation persist. Significant financial investment is required to scale green hydrogen production, with global estimates suggesting $150 billion is needed over the next decade to achieve cost competitiveness with fossil-fuel-derived hydrogen. While transitioning to a sustainable hydrogen economy is challenging, green hydrogen development offers India a transformative opportunity to meet climate goals, boost energy security, and drive economic growth. Success relies on overcoming barriers, supportive policies, and innovation. These efforts could position India as a leader in the global transition to sustainable energy.

References

[1] Canada Hydrogen Strategy. (2023, August 2). Retrieved October 4, 2024, from https://www.iea.org/policies/17710-canada-hydrogen-strategy

[2] Chouhan, K., Sinha, S., Kumar, S., & Kumar, S. (2021). Simulation of steam reforming of biogas in an industrial reformer for hydrogen production. *International Journal of Hydrogen Energy*, 46(53), 26809–26824. https://doi.org/10.1016/j.ijhydene.2021.05.152

[3] Grostern, J., & Kyllmann, C. (2024, July 24). German govt adopts import strategy for green hydrogen. Retrieved October 4, 2024, from https://www.cleanenergywire.org/news/german-govt-adopts-import-strategy-green-hydrogen

[4] Hydrogen Industry Development Plan (2021–2035). (2023, January 13). Retrieved October 5, 2024, from https://www.iea.org/policies/16977-hydrogen-industry-development-plan-2021–2035.

[5] Hydrogen. Retrieved October 4, 2024, from https://energy.ec.europa.eu/topics/energy-systems-integration/hydrogen_en

[6] IRENA (2024), Green hydrogen strategy: A guide to design, International Renewable Energy Agency, Abu Dhabi. Retrieved October 28, 2024 from https://www.irena.org/Publications/2024/Jul/Green-hydrogen-strategy-A-guide-to-design

[7] Lebrouhi, B. E., Djoupo, J. J., Lamrani, B., Benabdelaziz, K., & Kousksou, T. (2022). Global hydrogen development-A technological and geopolitical overview. *International Journal of Hydrogen Energy*, 47(11), 7016–7048. https://doi.org/10.1016/j.ijhydene.2021.12.076

[8] Li, H., Ma, C., Zou, X., Li, A., Huang, Z., & Zhu, L. (2021). On-board methanol catalytic reforming for hydrogen Production-A review. *International Journal of Hydrogen Energy*, 46(43), 22303–22327. https://doi.org/10.1016/j.ijhydene.2021.04.062

[9] Midilli, A., Kucuk, H., Topal, M. E., Akbulut, U., & Dincer, I. (2021). A comprehensive review on hydrogen production from coal gasification: Challenges and Opportunities. *International Journal of Hydrogen Energy*, 46(50), 25385–25412. https://doi.org/10.1016/j.ijhydene.2021.05.088

[10] National Green Hydrogen Mission. (2023, March 6). Retrieved October 5, 2024, from https://www.india.gov.in/spotlight/national-green-hydrogen-mission

[11] National Hydrogen Strategy 2024. (2024). Retrieved October 5, 2024, from https://www.dcceew.gov.au/sites/default/files/documents/national-hydrogen-strategy-2024.pdf

[12] Net-zero emissions by 2050. (2024, September 3). Retrieved October 4, 2024, from https://www.canada.ca/en/services/environment/weather/climatechange/climate-plan/net-zero-emissions-2050.html

[13] US National Clean Hydrogen Strategy. (2024, July 24). Retrieved October 4, 2024, from https://dilo.com/blog/article/news/new-hydrogen-roadmap

51 SPWM Fed MLI (Multi-level Inverter) Configuration for Induction Motor Drive with Minimal Switch Count

Shaik Rahimpasha[1], V. Sivarama Krishna[2], Karamalla Haleema[1], and Farooq Sunar Mahammad[3]

[1]Vardhaman college of engineering Hyderabad, India
[2]Department of Electrical and Electronics Engineering, Chaitanya Bharathi Institute of Technology Hyderabad, India
[3]Santhiram Engineering College, Nandyal, Andhra Pradesh, India

Abstract

Induction motors are widely employed in numerous industrial applications due to a wide range of advantages, such as their robust construction, low maintenance requirements, affordability, etc. Although they have many advantages, Speed control is a significant concern. The article proposes a novel MLI (multi-level inverter) circuit for an induction motor driving system to address this problem. This state-of-the-art multi-level inverter is configured utilizing merely three standard two-level inverters. In a standard five-level inverter, 24 switching devices are required. However, this proposed five-level inverter requires only eighteen switching devices. Consequently, the inverter circuit's design is less expensive and more complex. In addition, switching losses are reduced since one inverter is always clamped while the other inverters are switched. This circuit uses four times less DC source voltage than a typical five-level NPC inverter circuit. This arrangement remains operational even if a two-level inverter switch's switching device fails. Comparing the suggested configuration to the conventional five-level inverter architecture, both its Total Harmonic Distortion (THD) is reduced and its reliability is increased. To simulate the proposed inverter developed with a five-horsepower induction motor driving mechanism, MATLAB/Simulink is utilized. The results demonstrate that this arrangement is valid.

Keywords: MLI, Space Vector PWM, induction motor

1. Introduction

Currently, the demand for inverter-operated AC drive systems is growing significantly. Variable speed Enabling AC drives to change motor speed without affecting air gap current distribution, variable AC voltage, and frequency are essential. It is typical to generate variable voltage and frequency using inverters. Although the inverter generates the intended fundamental component of output voltage, its excessive harmonic content amplifies the machine's losses. Subsequently, reducing

Email: rahimdmcl@gmail.com[1]; skrishna_nteee@cbit.ac.in[2]; haleema218@gmail.com[1]; farooq.cse@srecnandyal.edu.in[3]

DOI: 10.1201/9781003661917-51

harmonics is a significant concern in all AC drives fed by a power converter. These harmonics are minimized using PWM techniques. The task of the PWM method is to control the output voltage level and operating frequency of the inverter, as well as to minimize the harmonic content in the inverter output voltages [1–4]. Various PWN techniques can be used to minimize the percentage of harmonics in the inverter output voltage.

One of the most important methods for harmonic elimination is presented in [5], which minimizes harmonics by correctly determining the switching angles using the distribution factor. Nevertheless, as the number of layers increases, more offline computations are required, increasing the system complexity. An interesting PWM technique presented in [6] considers trapezoidal modulation signals and variable frequency sine wave carrier signals, where harmonic content is minimized by varying the frequency of the carrier signal with the slope of the trapezoidal modulation wave. A study has been conducted on eight different advanced PWM methods presented in [7], discussing the advantages and disadvantages of each method in terms of its complexity and its implementation. Rice. I. Stator winding of an induction motor, (a) General winding arrangement, (b) Multilevel inverters have been utilized to boost the system output voltages' harmonic characteristics and winding design for the suggested topology.

Many different multi-level inverter systems have been proposed in [8–9]. The fundamental design of multi-level inverters comprises switching elements connected in parallel or series with the DC sources to produce a step voltage waveform. This waveform is small, though, because the higher voltage stress on the switching elements is reduced and the waveform output is improved. Diode clamp multilevel inverter topology (or NPC) is one of the most popular and basic inverter topologies [10]. The NPC inverter configuration can provide more effective voltage waveforms, but it needs additional clamping diodes

and has limitations with capacitor voltage imbalance issues. Therefore, distinctive methods are used to obtain equal capacitor voltages. Another traditional topology is a multilevel inverter with a flying capacitor (or capacitor clamp) [11]. In this topology, no clamping diodes are required, but more capacitor banks are required. The voltages of these capacitor banks must be controlled within acceptable voltage ripple limits. An interesting topology without problems of voltage imbalance of capacitors and clamping diodes is the H-bridge cascade inverter [12]. However, in this topology, the need for isolated DC voltage sources is greater. A multi-level voltage can obtained by using two two-level inverters by supply power to the winding of an induction motor from both sides [13–14].

When compared to traditional multi-level inverters, this has the advantage of halving the DC source's voltage need. A new inverter circuit is proposed employing only three standard two-level inverters, which are described below.

2. Modulation Method and Proposed MLI

(Multi-level Inverter) Topology

The power circuit for a 4 pole induction motor driven by an MLI topology is shown in Figure 51.1. The 4 pole induction motor's stator winding comprised two such windings, which were connected to two-level inverters using the technique depicted in the figure, which involves separating and connecting similar voltage characteristics for the windings separately. As with the arrangement that uses two back-to-back inverters fed by an open-wound induction motor, inverters 1 and 2 share a single winding in this design [15]. Three inverters employ the second winding. Inverters 1, 2, and 3 are fed by a single DC source, consequently, zero-sequence currents cannot flow. Four times more voltage is obtained from the DC source. The action of the inverter is thus limited to the linear region of modulation. As a result, the circuit becomes simpler, and a novel PWM

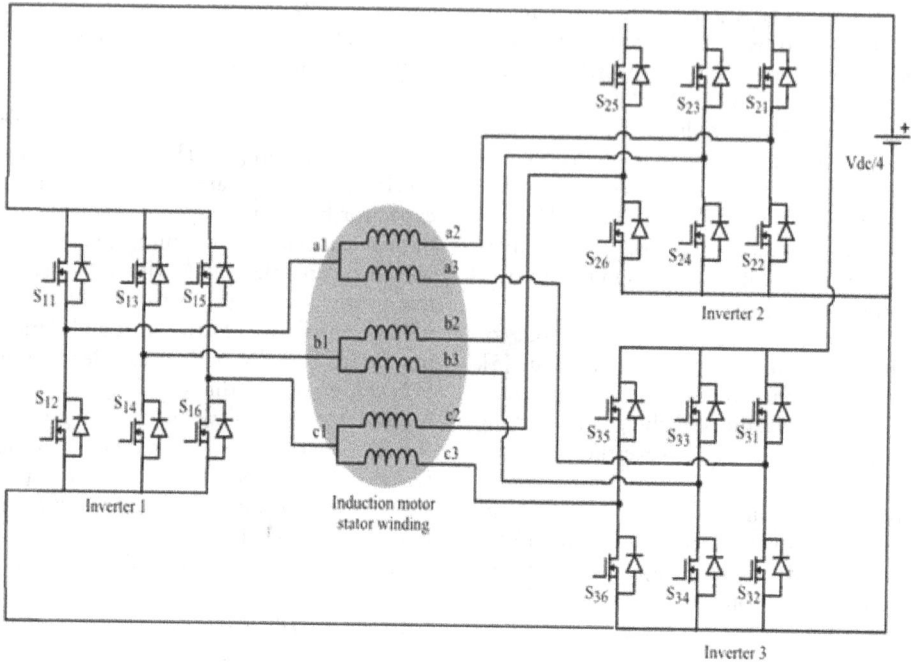

Figure 51.1 MLI circuit design

technique is put forward for drive control. Reference phase voltages are utilized to produce signals for the inverter, and the performance is demonstrated to be comparable to that of previously employed PWM approaches. The reference voltage is raised by a certain percentage of the total applied voltage in the linear modulation zone using this previously mentioned innovative modulation technique. Common-mode voltages are added to generate a reference performance that is equivalent to reference phase voltages, much to SVPWM.

Introducing carrier-based PWM pulse width modulation technology, where sinusoidal reference voltages are added to the actual offset voltages before comparison with the carrier.

The corresponding modulation signals for phase A have been shown in Figure 51.2, with the modulation signal for inverter 1 at the top, the general topology modulation signal in the middle, and the modulation signals for inverters 2 and 3 at the bottom. The general topology modulation signal (effective) is the difference between

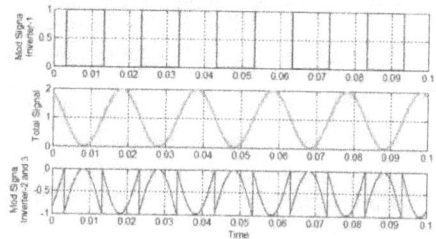

Figure 51.2 Modulating signals for the inverter-1 (top signal), modulating signals for inverter-2 and 3 (bottom signal), effective modulating signal of the inverter 1 and 2 (middle signal)

the modulation signals of inverters 1 and 2. While the carrier signal for inverter 3 is 1800 out of phase with the carrier signal for inverter 2, the modulation signal for inverters 2 and 3 is identical. Due to this phase shift between the two carrier waves, the harmonics in the first midband region are canceled [16–17]. The benefit of this pulse width modulation (PWM) technology is that it is simple to generate the modulation signals because they resemble sine waveforms. This significantly reduces circuit complexity and control overhead. The primary advantage of this configuration is that it significantly decreases switching operations by saving stator winding in induction motors that produce multi-level voltage over the entire winding. Furthermore, the PWM techniques currently in use make handling such a configuration simple.

Harmonics in the first mid-band may be suppressed by ensuring that the two carrier signals have an appropriate phase shift. Although one inverter is constantly locked in that topology, switching losses are significantly reduced. As discussed in greater detail in the next section, this architecture provides the added advantage of being capable to continue operating even in the case of an inverter switch failure.

3. Reliability Issues

This arrangement is more reliable than the traditional topologies. A multilayer voltage waveform is achieved by connecting many of the switches in series for traditional NPC arrangements. However, this configuration makes it vulnerable because the system should be totally shut down if any switch faults. Two inverters coupled to an open-winding induction motor drive system can continue to function if one of the inverters fails, as in the example of Figure 51.4(a), inverter 1 fails. As demonstrated in Figure 51.4(b), the system may be configured to function as a two-level inverter-driven induction motor system in the case that, for illustration, inverters 2 or 3 fail. PWM schemes need to be modified in these scenarios.

As Table 51.1 shows, there are certainly advantages to the above arrangement over conventional five-level inverter configurations. The traditional system calls for 24 switching devices, whereas this configuration only needs 18. Additionally, it isn't recommended to utilize either the capacitor-clamped inverter configuration or the diode-clamped inverter configuration, as the latter requires additional capacitor banks while the former requires

Table 51.1 A Comparison Between Suggested and Standard Topologies

	NPC	FC	H-bridge inverter	Suggested Topology
Basic switches	24	24	24	18
Diodes for Clamping	36	0	0	0
Isolated voltage sources	1 (VdC)	1 (VdC)	6 (VdC) 4	1 (VdC) 4
Capacitors Banks	4	18	0	0
Reliability During any Switching device failure	No	No	Yes	Yes

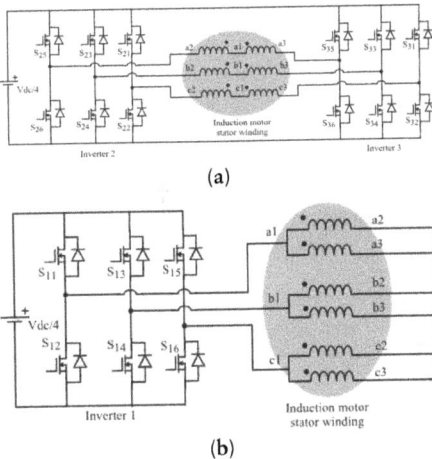

(a)

(b)

Figure 51.3 The circuit configuration in case of

a) Inverter one fault condition

b) Inverter two or three fault condition

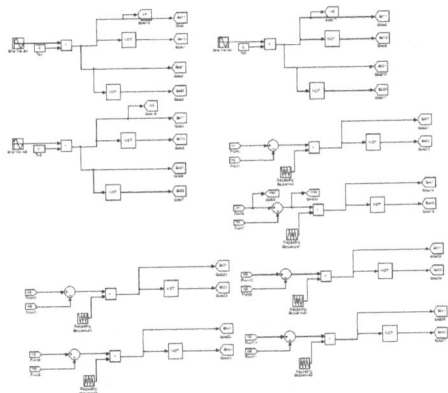

Figure 51.4 A simulation circuit typically produces the inverter circuit's total gating signals

more clamping diodes. Additionally, unlike diode clamp and capacitor clamped arrangements, this structure can operate even in case the switching device fails.

4. Results-Analysis

The developed simulation has been deployed to assess the proposed topology over a range of modulation indices from 0 to 0.866. Figure 51.4 illustrates a simulation system utilized for gate signal generation. In Figure 51.5, the simulation results for the case where the modulation index is 0.4 are illustrated. The total voltage across the induction motor is equal to the sum of the voltages across the motor winding, as stated in the previous section. Thus, along the stator phase winding of the induction motor, a five-step voltage plot with a modulation index of 0.4 is produced.

The outcomes for an index of modulation of 0.6 and 0.8, respectively, are shown in the Figures 51.6 and 51.7. The accumulation of flux in the air gap is primarily caused by a multilayer voltage waveform that is shown here (third from the top) along the motor's phase winding.

From Figure 51.8, inverter 1 switches at the reference frequency while inverters two and three switches at the carrier frequency. This results in low switching loss and the operation switching device can be as small as possible for the inverter-1.

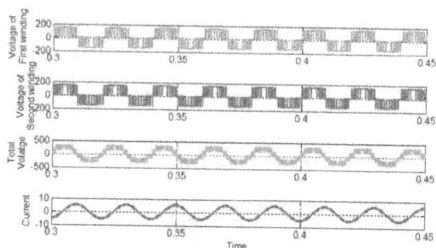

Figure 51.6 Top trace is voltage appearing between stator winding of points a1-a2, second trace is voltage appearing between stator winding of points a1-a3, third trace is the voltage appraing across the total phase winding and bottom trace is current flowing through the phase winding of the motor of the modulation index 0.6 [X-axis 10 ms/div, Y-axis 500 V/div.]

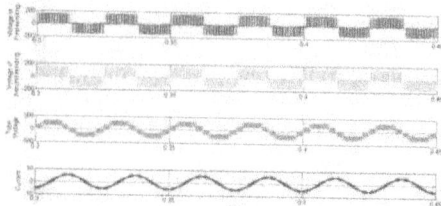

Figure 51.7 Top trace is the voltage appearing between, stator winding of points a1-ar, second trace is voltage appearing between stator winding of points a1-a3, third trace is voltage appearing across total phase winding and bottom trace is current flowing through the phase winding of motor of the modulation index 0.8 [X-axis 10 ms/div, Y-axis 500 V/div.]

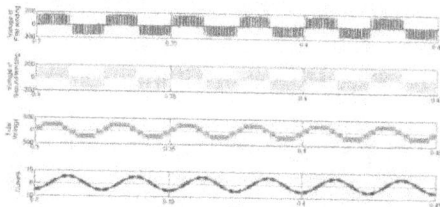

Figure 51.5 Top trace is volage appearing between the stator winding of points a1-a2, second trace is volage appearing between stator winding of points a1-a3, third trace is the voltage appearing across total phase winding and bottom trace is current flowing through the phase winding of motor of the modulation index 0.4 [X-axis 10 ms/div, Y-axis 200 V/div.]

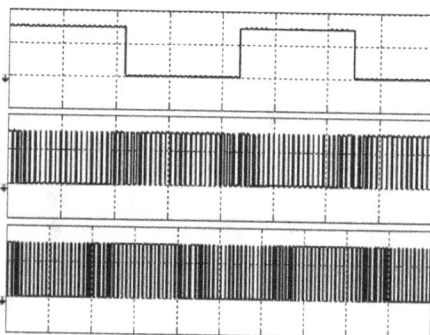

Figure 51.8 The top trace is inverter-1 pole voltage, middle trace is inverter-2 pole voltage and bottom trace is inverter-3 pole voltage of the modulation index 0.8 [X-axis 5 ms/div., Y-axis 50 V/div.]

The motor stator phase's common winding current's harmonic spectrum for modulation indices 0.4, 0.6, 0.8, and 1 is shown in Figure 51.9. As the modulation index increases, overall THD (harmonic distortion) reduces, as can be observed in the above image. This idea is explained extensively by examining harmonic spectra for four distinct modulation indices.

(a)

(b)

(c)

(d)

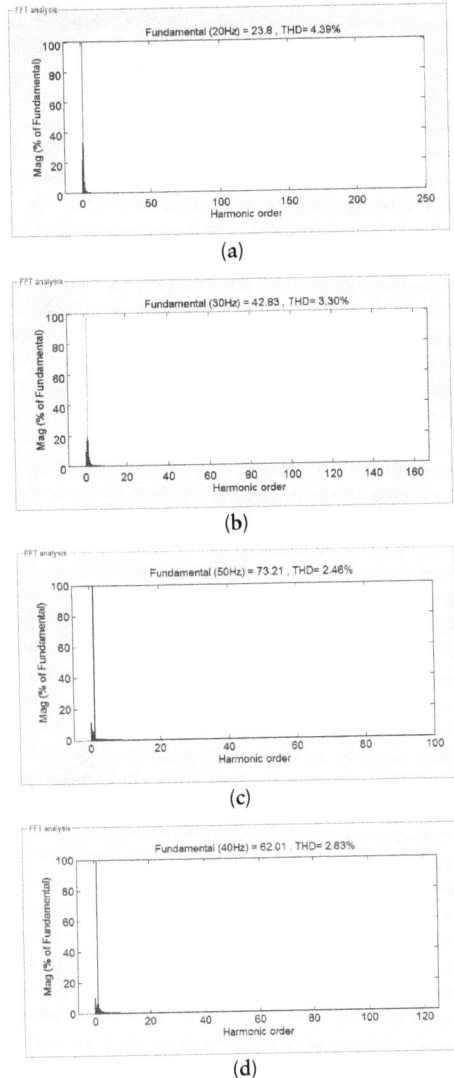

Figure 51.9 Harmonic spectrum for the current flowing through the stator winding of the motor of the

(a) Modulation index 1

(b) Modulation index 0.8

(c) Modulation index 0.6

(d) Modulation index 0.4

5. Conclusion

The three-standard two-level inverter is explained in this paper. Actually, the standard five-level inverter uses twenty-four switching devices, but eighteen switching devices are used here for the five-level proposed inverter. It reduces the cost and complexity of the circuit of an inverter. This topology significantly reduces switching loss because a typical two-level inverter operates at a fundamental frequency. This proposed topology reduces the DC source voltage level by a factor of four down from the DC source voltage requirement in a normal five-level inverter. Thus, it has the benefit of operating even if the switching device fails in one inverter, thus increasing reliability in the system. Here are the MATLAB/Simulink simulation results for the suggested four-pole induction motor drive system. Additionally, since only two-level inverters are taken into account in the specified configuration, the structure does not include any balancing concerns at the neutral point. MATLAB/Simulink is used to simulate a five horsepower, four-pole induction motor for the recommended topology design. All the harmonics that fall in the first mid-band region are suppressed.

References

[1] Rodriguez, J., Burnet, S., Wu, B., Pontt, J. O., & Kuro, S. (2007). Multi-level voltage source-converter topologies for industrial medium-voltage drives. *IEEE Trans. Ind. Electron.*, 54(6), 2930–2945.

[2] Evanchuk, J., & Salmon, J. (2013). Three-bar coupled inductor power for parallel multilevel three-phase voltage converters. *IEEE Trans. Ind. Electron.*, 60(5), 1979–1988.

[3] Boller, T., Holtz, J., & Rathore, A. K. (2014). Neutral Potential Balancing Using Synchronous Optimal Width – Pulse Modulation of Multi-Level Inverters in Medium Power AC Drives. *IEEE Trans. Ind. Appl.*, 50(1), 549–557.

[4] Holmes, D. G., & McGrath, B. P. (2001). Harmony suppression capabilities of carrier-based PWM for two-level and multi-level cascaded inverters. *IEEE Trans. Ind. Appl.*, 37(2), 574–582.

[5] Boller, T., Holtz, J., & Rathore, A. K. (2014). Neutral Potential Balancing Using Synchronous Optimal Pulse Width modulation of multi-level Inverters in Medium Power AC Drives. *IEEE Trans. Ind. Appl.*, 50(1), 549–557.

[6] Khazraei, M., Sepahvand, H., Corzine, K. A., & Ferdowsi, M. (2011). Active capacitor voltage balancing in single-phase flying-capacitor multilevel power converters. *IEEE Transactions on Industrial Electronics*, 59(2), 769–778.

[7] Babaei, E., Alilu, S., & Laali, S. (2013). A new general topology for cascaded multilevel inverters with reduced number of components based on developed H-bridge. *IEEE Transactions on industrial electronics*, 61(8), 3932–3939.

[8] Chatzakis, J., Vogiatzaki, M., Rigakis, H., und Manitis, M., & Antonidakis, E. (2006). A new high capacity pulse width modulation inverter. In *Proceedings of the 10th WSEAS International Conference on Circuits*. Vouliagmeni, Athens, Greece, July 10–12, pp. 280–285.

[9] Mwiniwiwa, B., Boon-Tek Ooi, & Wolanski, Z. (1998). UPFC using a multiconverter operating on the phase-shifted delta carrier SPWM strategy. *IEEE Trans. Ind. Appl.*, 34(3), 495–500.

[10] Somasekhar, V. T., Venugopal Reddy, B., & Sivakumar, K. (2014). Four-stage inversion scheme for 6-pole open-winding induction motors for improved DC link utilization. *IEEE Trans. Ind. Electron.*, 61(9), 4565–4572.

[11] Bodo, N., Jones, M., & Levy, E. (2014). Space Vector PWM with Common Mode Voltage Elimination for Five-Phase Open-Wind Drives with a Single DC Source. *IEEE Trans. Ind. Electron.*, 61(5), 2197–2207,

[12] Shiny, G., & Baiju, M. R. (2012). Low-Computation Fractal-based Space Vector Generation Scheme for Four-Level Inverter Using Open Winding Induction Motor. *IET Electric Power Appl.*, 6(9). 652–660.

[13] Kumar, K., & Sivakumar, K. (2014). Four-Level Inverter Configuration for Four-Pole Single DC Link Induction Motor. *IEEE Trans. Ind. Electron.* doi: 10.1109/TIE.2014.2327577.

[14] Somasekhar, V. T., Srinivas, S., Prakash Reddy, B., Nagarjuna Reddy, C., & Sivakumar, K. (2007). Pulse Wide Modulation Switching Strategy for Dynamic Residual Current Balancing for Open End Induction Motor Drive Winding Fed by Two Inverters. *IET Electric Power Appl.*, 1(4), 591–600.

[15] Kim, J.-S., & Sul, S.-K. (1995). A new method for space vector PWM voltage modulation. *Proceedings. IPEC*, 742–747.

[16] Mei, J., Xiao, B., Shen, K., Tolbert, L. M., & Zheng, J. Y. (2013). Modular multilevel inverter with new modulation method and its application to photovoltaic grid-connected generator. *IEEE Transactions on Power Electronics*, 28(11), 5063–5073.

[17] Li, C., Ji, S. M., & Tan, D. P. (2012). Multiple-loop digital control method for a 400-Hz inverter system based on phase feedback. *IEEE Transactions on Power Electronics*, 28(1), 408–417.

For Product Safety Concerns and Information please contact our EU
representative GPSR@taylorandfrancis.com
Taylor & Francis Verlag GmbH, Kaufingerstraße 24, 80331 München, Germany

www.ingramcontent.com/pod-product-compliance
Lightning Source LLC
Chambersburg PA
CBHW061616220326
41598CB00026BA/3780